The Optimal Homotopy Asymptotic Method

Vasile Marinca • Nicolae Herisanu

The Optimal Homotopy Asymptotic Method

Engineering Applications

Springer

Vasile Marinca
Politehnica University of Timisoara
Timisoara
Romania

Center for Advanced and Fundamental Technical Research
Romanian Academy, Timisoara Branch
Timisoara
Romania

Nicolae Herisanu
Politehnica University of Timisoara
Timisoara
Romania

Center for Advanced and Fundamental Technical Research
Romanian Academy, Timisoara Branch
Timisoara
Romania

ISBN 978-3-319-15373-5 ISBN 978-3-319-15374-2 (eBook)
DOI 10.1007/978-3-319-15374-2

Library of Congress Control Number: 2015936718

Springer Cham Heidelberg New York Dordrecht London

Printed on acid-free paper

Springer International Publishing AG Switzerland is part of Springer Science+Business Media (www.springer.com)

Preface

This is an engineering book written for researchers, engineers, natural scientists, and students studying engineering at undergraduate and postgraduate levels. Also, this can be used by physicians, mathematicians, and so on. Its aim is to allow the reader to learn and gain a comprehensive understanding of nonlinear problems by reading and then problem solving. This book was written alongside of several years, with the motto: "There is always more to discover about nonlinear dynamical systems". We assume that the reader has some prior knowledge on linear differential equations, integrals, trigonometric functions, polynomials, series expansions, and algebraic calculus. The book contains a relative great amount of models and it was necessary to make selections between these models to prepare our material.

Analytical solutions of nonlinear differential equations or linear differential equations with variable coefficients play an important role in the study of nonlinear dynamical systems, but sometimes it is difficult to find these solutions, especially when strong nonlinearities are involved. Also, computers skills would be essential because computer simulation is a powerful tool for examination, for confirmation, and sometimes for refutation of the obtained results. In general, most of the known analytical methods are restricted to limited cases depending on the parameters which appear in the governing equations and are valid only for nonlinear problems with weak nonlinearity.

Some references include recently published articles, but listening to lectures is not enough. All processes of learning are somehow connected to active participation and the learning and applying of nonlinear dynamical systems are no exceptions. Most models will demand some independent thinking and are selected in order to illustrate the main ideas of our procedure, which allowed the reader to understand the present material.

The whole book consists of concrete examples from various domains of engineering, physics, or applied mathematics. The authors believe that the problem of motion of different dynamical systems can be assimilated only by working with the differential equations applied to concrete examples. All the analytical developments within the sections of this book are accompanied by comparisons with numerical or exact results or with other known results in the literature. A better knowledge of different methods in the literature, especially the Lindstedt–Poincare method, the method of harmonic balance, the method of Krylov–Bogoliubov, or the method of multiple scales, led to a better choice of the so-called optimal auxiliary functions. These functions are absolutely necessary to obtain very efficient results and therefore are cornerstone of our procedure. Unlike all known analytic approaches, this optimal method provides us with a simple way to control and adjust the convergence region of solutions of nonlinear dynamical systems.

Actually, the capital strength of our procedure is its fast convergence, since after only one iteration and rarely after two iterations, the analytical solutions converge to the exact ones, which proves that this optimal method is very efficient in practice.

The book is divided into five chapters. The first chapter is introductory and the second one is devoted to a short history of the development of homotopy and gives the basic ideas of our procedure. The third chapter deals with the first alternative of the optimal homotopy asymptotic method with two iterations. The fourth chapter treats the optimal homotopy asymptotic method with a single iteration and solving the linear equation in the first approximation. The last chapter is devoted to the optimal homotopy asymptotic method with a single iteration but without solving the equation in the first approximation.

All models and results presented in this book were obtained and formulated by the authors.

We shall be satisfied if this book will serve as an access point to the basic concepts of nonlinear dynamical systems and thereby will stimulate interest of researchers in engineering as well as physicians or applied mathematicians with a basic mathematical background and a good deal of intellectual curiosity.

Last but not least, the authors are deeply indebted and grateful to members of their families. Without their support and inspiration, this book would never have been conceived and written.

Timisoara, Romania

Vasile Marinca
Nicolae Herisanu

Contents

1 **Introduction** . . . 1
References . . . 6

2 **Optimal Homotopy Asymptotic Method** . . . 9
2.1 A Short History of the Homotopy . . . 9
2.2 Basic Idea of OHAM . . . 13
2.3 Convergence of the Homotopy-Series 2.28 . . . 16
2.4 Convergence of the Approximate Solution of Order m Given by Eq. 2.29 . . . 17
References . . . 22

3 **The First Alternative of the Optimal Homotopy Asymptotic Method** . . . 23
3.1 Thin Film Flow of a Fourth-Grade Fluid Down a Vertical Cylinder . . . 24
3.1.1 Numerical Examples . . . 27
3.2 The Jeffery–Hamel Flow Problem . . . 28
3.2.1 Numerical Examples . . . 36
3.3 Oscillations of a Mass Attached to a Stretched Wire . . . 40
3.3.1 Numerical Examples . . . 45
3.4 The Motion of a Particle on a Rotating Parabola . . . 46
3.4.1 Numerical Examples . . . 54
3.5 Nonlinear Oscillator with Discontinuities and Fractional-Power Restoring Force . . . 57
References . . . 67

4 **The Second Alternative of the Optimal Homotopy Asymptotic Method** . . . 69
4.1 The Flow of a Walters-Type B' Viscoelastic Fluid in a Vertical Channel with Porous Wall . . . 70
4.1.1 Problem Statement and Governing Equation . . . 72

4.1.2 Solution of Walters-Type B' Viscoelastic Fluid in a Vertical Channel with OHAM 76
4.1.3 Governing Equation of the Temperature and Its Solution 82
4.1.4 Numerical Results and Discussions 84
4.2 Thin Film Flow of an Oldroyd 6-Constant Fluid over Moving Belt 115
4.2.1 Governing Equations 116
4.2.2 Application of OHAM to Thin Film Flow of an Oldroyd 6-Constant Fluid 118
4.2.3 Numerical Results and Discussions 126
4.3 Falkner-Skan Equation 146
4.3.1 The Governing Equation 147
4.3.2 Application of OHAM to Falkner-Skan Equation 148
4.3.3 Numerical Examples 153
4.4 Viscous Flow Due to a Stretching Surface with Partial Slip 173
4.4.1 Equation of Motion 174
4.4.2 Application of OHAM to Viscous Fluid Given by Eq. 4.220 175
4.4.3 Numerical Examples 178
4.5 The Flow and Heat Transfer in a Viscous Fluid Over an Unsteady Stretching Surface 199
4.5.1 Equations of Motion 200
4.5.2 Application of OHAM to Flow and Heat Transfer 202
4.5.3 Numerical Examples 210
4.6 Blasius' Problem 225
4.6.1 Solution of Blasius' Problem by Optimal Homotopy Asymptotic Method 228
4.7 Thermal Radiation on MHD Flow over a Stretching Porous Sheet 230
4.7.1 Solution of the Problem with Optimal Homotopy Asymptotic Method 231
4.7.2 Numerical Examples 234
4.8 Nonlinear Equations Arising in Heat Transfer 236
4.8.1 Cooling of a Lumped System with Variable Specific Heat 237
4.8.2 The Temperature Distribution Equation in a Thick Rectangular Fin Radiation to Free Space 241
4.8.3 A Heat Transfer Problem 243
4.9 The Nonlinear Age-Structured Population Models 246
4.9.1 Analytical Solution for Nonlinear Age-Structured Population Models Using OHAM 248
4.10 Volterra's Population Model 260
4.10.1 Numerical Examples 263

4.11 Lotka-Volterra Model with Three Species . . . 264
4.11.1 Numerical Examples . . . 268
4.12 Bratu's Problem . . . 271
4.12.1 The Exact Solution of Bratu's Problem 4.548 . . . 272
4.12.2 Solutions of the Bratu's Problem by Means of OHAM . . . 274
4.12.3 Numerical Examples . . . 275
4.13 Lane-Emden Equation . . . 281
4.13.1 Solution of Eq. 4.584 by Means of OHAM . . . 282
4.13.2 Numerical Examples . . . 284
4.14 Oscillations of a Uniform Cantilever Beam Carrying an Intermediate Lumped Mass and Rotary Inertia . . . 287
4.14.1 Numerical Examples . . . 291
4.15 Nonlinear Jerk Equations . . . 294
4.15.1 Numerical Examples . . . 297
4.16 Nonlinear Oscillator with Discontinuities . . . 299
4.16.1 Numerical Examples . . . 302
4.17 Truly Nonlinear Oscillators . . . 303
4.17.1 Solution of Eqs. 4.673 Using OHAM . . . 304
4.17.2 Numerical Examples . . . 306
4.18 The Nonlinear Oscillator $\ddot{x} + (1 + \dot{x}^2)x = 0$ . . . 310
4.18.1 Approximate Periodic Solution of Eq. 4.690 by Means of OHAM . . . 310
4.18.2 Numerical Examples . . . 312
4.19 Nonlinear Oscillators with Quadratic and Cubic Nonlinearities . . . 314
4.19.1 Solutions of Eq. 4.715 by Means of OHAM . . . 316
4.19.2 Numerical Examples . . . 318
4.20 Damped Oscillator with Fractional-Order Restoring Force . . . 319
4.20.1 Numerical Examples . . . 322
4.21 The Oscillator with Cubic and Harmonic Restoring Force . . . 325
4.21.1 Numerical Examples . . . 328
4.22 Duffing Harmonic Oscillator . . . 330
4.22.1 Solutions of Duffing-Harmonic Oscillator Using OHAM . . . 332
4.22.2 Numerical Examples . . . 333
4.23 The Oscillator with Linear and Cubic Elastic Restoring Force and Quadratic Damping . . . 335
4.23.1 Numerical Examples . . . 338
4.24 Generalized Duffing Equation . . . 340
4.24.1 Numerical Examples . . . 344
4.25 Generalized Van der Pol Equation . . . 348
4.25.1 Solutions of the Van der Pol Equation 4.835 by Means of OHAM . . . 350
4.25.2 Numerical Examples . . . 353

4.26 Oscillations of an Electrical Machine 357
4.26.1 Numerical Example 360
4.27 Dynamic Analysis of a Rotating Electric Machine 360
4.27.1 Numerical Examples 365
4.28 A Non-conservative Oscillatory System of a Rotating Electrical Machine 367
4.29 Nonlinear Dynamics of an Electrical Machine Rotor-Bearing System 371
4.29.1 Numerical Examples 378
References 380

5 The Third Alternative of the Optimal Homotopy Asymptotic Method 391
5.1 Overview 391
5.1.1 The Convergence of the Approximate Solution 5.6 394
5.2 Thomas-Fermi Equation 394
5.2.1 Approximate Solution of the Thomas-Fermi Equation by OHAM 396
5.2.2 Numerical Examples 398
5.3 Multiple Solutions for the Upper-Convected Maxwell Fluid Over a Porous Stretching Plate 402
5.3.1 Multiple Solutions with OHAM 404
5.3.2 Numerical Results 409
5.4 Dual Solutions of the Unsteady Viscous Flow Over a Shrinking Cylinder 416
5.4.1 Multiple-Dual Approximate Solutions of the Unsteady Viscous Flow by OHAM 421
5.4.2 Numerical Examples 423
5.5 Axisymmetric Flow of an Incompressible Fluid Between Two Parallel Plates 434
5.5.1 Solutions of the Problem 5.174 and 5.175 with OHAM 436
5.5.2 Numerical Examples 438
5.6 Heat Transfer with Variable Thermal Conductivity 442
5.6.1 Approximate Solutions of Eqs. 5.194 and 5.195 by OHAM 443
5.6.2 Numerical Examples 445
5.7 Nonlinear Dynamical Model of a Permanent Magnet Synchronous Generator 446
5.7.1 Dynamic Model 447
5.7.2 Approximate Solution with OHAM 449
5.7.3 Results and Discussions 452
References 463

Chapter 1
Introduction

A nonlinear dynamical system is not a tidy subject, but is vital in engineering practice. "Nonlinear systems are surely the rule, not the exception, not only in research, but also in the engineering world" [1]. The world around us is inherently nonlinear and the most nonlinear phenomena are models of our real-life problems. The most successful class of rules for describing practice phenomena are differential equations. All major theories of engineering are stated in terms of differential equations. Differential equations lie at the basis of scientific mathematical philosophy [2] of our scientific world. This scientific philosophy began with the discovery of the calculus by Newton and Leibnitz and continues to the present days. The first two hundred years of this scientific philosophy, from Newton and Euler, through to Hamilton and Maxwell, produced many stunning successes in formulating the "rules of the world". Nonlinear differential equations are widely used as models to describe complex physical phenomena in various fields of science as fluid dynamics, solid state physics, heat transfer, vibrations, electrical machines, chemical kinetics and so on. The evolution of some systems can be studied by means of the linear differential equations, but majority of the physics systems do not lead to linear differential equations but to nonlinear differential equations. A system of linear differential equations is one for which the dependent quantities or variables only appear to the first power. If terms are present which involve products of the dependent variables, or other powers or other mathematical forms, the system is said to be nonlinear. A linear dynamical system is one which the dynamic rule is linearly proportional to the system variables. Linear systems can be analyzed by breaking the problem into pieces and then adding these pieces together to build a complete solution. This property of linear differential equations is called the principle of superposition. It is the cornerstone from which all linear theory is built. But, unfortunately, the solutions of a nonlinear equation cannot usually be added together to build a larger solution. The principle of superposition fails to hold for nonlinear systems. A general feature of all nonlinear dynamical equations including nonlinear differential equations is the "breakdown" of linear additivity or superposition. Because of this breakdown, many mathematical techniques

V. Marinca, N. Herisanu, *The Optimal Homotopy Asymptotic Method*,
DOI 10.1007/978-3-319-15374-2_1

(among which Laplace transform, Fourier analysis etc) for solving linear differential equations no longer work or are useful for attempting to solve nonlinear differential equations [3].

Generally speaking, dynamics is a concise term referring to the study of time-evolving processes and the corresponding system of equations, which describe this evolution, is called a dynamical system. In this monograph, nonlinear system refers either to a dynamical process with a physical existence, or to an equation which may be a model of the process. A nonlinear dynamical system receives the name in accordance with the nonlinear coefficient from the examined differential equation. In this way, there are known the systems of nonlinear mass or nonlinear damping, or nonlinear restoring force, and so on. Most engineering problems are nonlinear and their linearization only has a value of prime approximation which corresponds to a more or less schematization satisfying the laws of nature. This fact is required by the insufficiency of analysis and experimentation techniques. The engineering sciences of modern era with their fine instrumentation and advances calculus methods have shown that the establishment of movement laws neglect nonlinear terms lead to inadequate or even false results. Considering only linear problems lead not only to more quantitative errors, but also deform very often the quality of the phenomena.

In many instances the linear analysis is insufficient to describe the behavior of engineering systems adequately. In fact, one of the most fascinating features of a study of nonlinear problems is the occurrence of new and unsuspected phenomena, i.e. new in the sense the phenomena are not predicted, or even hinted at the linear theory. On the other hand, certain phenomena observed physically are unexplainable, except by some known aspects due to nonlinearities present in the systems. In the field of nonlinear dynamical systems, rapid progress in the last two centuries has occurred due in large measure to the ability of investigators to respect physical laws in terms of rather simple equations. In many cases, the governing equations were not so simple, therefore certain assumptions, more or less consistent with the physical situation, were employed to reduce the equations to types more easily solvable. Thus, the process of linearization has become an intrinsic part of rational analysis of physical problems. An analysis based on linearized equations, then may be thought of as an analysis of the corresponding but idealized problem.

The mathematician Henri Poincare (1878–1900) was first to appreciate the true sources of the problem [4]: the difficulty lay not in the rules but rather in specifying the initial conditions. He wrote: "if we knew exactly the laws of nature and the situation of the universe at the initial moment, we could predict exactly the situation of the same universe at a succeeding moment". Poincaré discovery of sensitive dependence on initial conditions in what are now termed chaotic dynamical systems has only been fully appreciated by the layer scientific community during the past three decades.

Initially, the theory of nonlinear systems builds up on its own on the foundations of the results of Poincaré, Lyapunov (1983), Birkhoff (1908–1944). Very strong interactions between theoretical researches and practical implications in engineering, mathematical or physical systems were the reason for the success in well-known Soviet schools: Moscow, Gorki and Kiev. The most important component of theory of dynamical systems was the theory of nonlinear oscillations.

The end of the nineteenth century was dominated by Poincaré's works on periodic solutions of ordinary differential equations, which constitute the foundation of the most part of results obtained until now. The Poincaré's contribution of the analytical method of nonlinear dynamics must also be noted with the Poincaré's method of the small parameter, and the notion of generating solution. It permits to define the nontabulated transcendental functions, solutions of differential equations, by a convergent series expansion. This way was considerably developed and extended (asymptotic methods) later by the dynamic schools of the former Soviet Union.

Under the name of theory of oscillations, the study of nonlinear dynamics had a new phase of growth in the twentieth century, beginning due in particular to the results of Rayleigh [5] and Van der Pol [6]. It was associated with the development of electrical and radio engineering [7]. There are known two relatively independent branches of the nonlinear oscillation theory. So, the first branch corresponds to the qualitative methods. The second branch corresponds to the analytical methods. The complex transcendental functions are defined by convergent, or at least asymptotically convergent series expansions, or in "the mean". The method of Poincaré's small parameter, the asymptotic method of Krylov-Bogoliubov-Mitropolski are analytical. So are the averaging methods [8, 9] and the methods of harmonic linearization in the theory of nonlinear oscillations. These two independent branches of the nonlinear oscillations theory have the same aims: construction of mathematical tools for the solutions of concrete problems, development of a general theory of nonlinear systems.

Since 1960, the important development of computers has meant a large extension to the numerical approach of the nonlinear systems problems. Such an approach constitutes a powerful tool when it is associated with the qualitative or analytical methods.

The problem of the construction of mathematical tools fitted to the study of nonlinear oscillations was first formulated by Mandelstham in 1920. This was done in connection with the study of nonlinear systems belonging to radio-engineering. Mandelsthan and Papaleski formulated the fundamental problems solved later by his disciples. Such formulations constitute a decisive step in understanding concrete nonlinear systems.

In the beginning, the most popular approach to nonlinear problems was the fitting method. This method is based on the approximation of a nonlinear characteristic, with a piecewise linear characteristic. So the solution of a nonlinear problem is changed into the solution of a set of linear problems corresponding to different linear segments with conditions of continuity at the junction of the segments. His professor Andronov amplified his activity with a precise purpose: the elaboration of a theory of nonlinear oscillations [10] in order to dispose of mathematical tools, common to different scientific disciplines. To be physically significant, a model of dynamic systems must respect the following conditions: a solution should exist, this solution should be unique, the unique solution should be continuous with respect to the initial conditions or boundary conditions, and the dynamic system should be structurally stable. The last concept was introduced in 1937 by Andronov and Pontrjagin. At this point, we should emphasize an important remark: the number of initial and/or boundary conditions must be the same with the sum of the orders of derivations of the nonlinear differential equations.

For Andronov's school, the analytical methods have been always an auxiliary tool for qualitative method studies or for the understanding of specific problems given by engineering.

So, as early as 1932, Mandelstham and Papaleski used the Poincaré's small parameter method to study of nonlinear resonance, subharmonic resonances, synchronization phenomenon, etc.

In 1956, Malkin published his famous book [11] dealing with the method of small parameter. This book also gives a method of successive approximations in the Lyapunov sense via analytical methods. Using the theoretical results of Andronov's school, the practical applications are very large and very important for analysis and synthesis purpose. Now, they go beyond the limit of engineering or physics and concern also natural sciences, dynamics of population, economy, etc.

The Krylov-Bogoliubov school from Kiev has developed essentially analytical methods. The foundation of their results is the classical method of perturbations which has been generalized to nonconservative systems. In 1932, the Krylov-Bogoliubov method gave a close foundation to the Van der Pol studies about oscillators. Later, the asymptotic method due to Mitropolski constitutes an improvement, with the use of only asymptotically convergent series expansions. It is the same for the averaging method and method of accelerating the convergence [8, 9]. With respect to the Poincaré's small parameter method, these methods are such that the "full" determination of the first harmonic and of the following harmonics of a periodic solution does not depend on the determination of the upper harmonics.

The Hayashi's school on nonlinear oscillations developed many studies especially oriented toward electric circuits [12]. Analytical methods as well as qualitative and numerical ones have been intensively used by him and his two disciples Kawakami and Ueda.

In the last years efforts were also made for collecting the main results concerning nonlinear vibration in monographs. In this respect, we mention the books by Minorsky [13], Stoker [14], McLachlan [15], Kauderer [16], Sansone and Conti [17], Nayfeh and Mook [18], Hagedorn [19], etc.

The theory of random vibration has a comparatively late development. From the publication in 1905 of Einstein's studies on Brownian motion [20], considered by him as a particular type of random vibration, more than two decades elapsed until the appearance of the first work applying the theory of random process to the study of beam and string solutions. Crandal [21] proposed and applied a perturbation technique which is an extension to random vibrations of the perturbation method used for weakly nonlinear deterministic systems.

Nonlinearities are commonplace in engineering systems. They result from structural properties which give rise to nonlinear elastic forces, geometrical and kinematic configurations which create autoparametric interactions or combined bending and torsional motion, inertial effects including rotational loadings, deformations such as curvatures and buckled states, machine tool chatter vibrations due to tool and work-piece interactions, aerodynamic effects, the elastic foundations, strain displacement of plates, systems with elastic deformations, fluid-structure interaction, finite belt stretching, internal combustion and so on.

The study of nonlinear dynamical systems is eclectic, selecting what appears to be most useful among various and diverse theories and methods. This poses an additional challenge since the skills required for research in nonlinear dynamics can range from knowledge of some sophisticated mathematics to a detailed understanding of the nuts and bolts of computer hardware. The common thread throughout all nonlinear dynamical systems is the need and desire to solve nonlinear problems.

An exact solution for strongly nonlinear systems are frequently scarce, at least at the present state of knowledge in nonlinear systems, an approach of accurate semi analytical or approximate analytical solutions is most significant appealing. In this respect, new and innovative approaches capable to solve nonlinear dynamical systems beyond the restrictions of the classical methods of perturbations and harmonic balancing should be known [22, 23].

In the last years, some fruitful results have been obtained for solving various nonlinear problems. There exists some well-known analytical approaches applicable to nonlinear problems such as the averaging method [9], the Lindstedt-Poincare method [18, 19], a modified Lindstedt-Poincare method [24], the method of harmonic balance [18, 19], the weighted linearization method [25], Adomian decomposition method [26], variational iteration method [27], the boundary element method [28], the optimal variational method [29], the optimal iteration method [30], the optimal homotopy perturbation method [31], the optimal parametric iteration method [32], and so on.

All the above mentioned methods work very well for weakly nonlinear dynamical systems and some of them work even for strongly nonlinear problems. It is very important in case of strongly nonlinear dynamical systems to prove and to ensure the condition of convergence of the solutions. Every approach used in the study of nonlinear dynamical systems must be rigorously proved, even if the results obtained by the approach are in good agreement with other known results in the literature.

It is also worthy to note that the majority of scientists have not been led their discoveries by a process of deduction from general postulates, or general principles, but rather by a thorough examination of properly chosen particular cases [22, 33]. The generalizations have come later, because it is far easier to generalize an established result than to discover a new line of argument. Generalization is the temptation of a lot of researchers working now with nonlinear dynamical systems.

Nonlinear dynamical systems which appear in engineering practice are a vast subject, so we will only present enough so that the readers can appreciate and understand the various topics that will be presented in subsequent chapters as we explore various domain of our nonlinear world. Our intention is to make the reader familiar and to understand and analyze the different nonlinear models that will be presented in the following chapters. For this purpose we deal with the Optimal

Homotopy Asymptotic Method in a proper manner entirely different in comparison with any other approaches. To solve a nonlinear dynamical system, our procedure does not need restrictive hypotheses, is very rapid convergent usually after only one iteration and rarely after two iterations, and the convergence of the solutions is ensured in a rigorous way. The cornerstone of the validity and flexibility of our procedure is the choice of the linear operator and the optimal auxiliary functions which contribute to very accurate solutions. The parameters which are involved in the composition of the optimal auxiliary functions are optimally identified via various methods in a rigorous way from mathematical point of view. This technique is very effective, explicit and easy to apply. It was introduced for the first time in 2008 and developed in a series of papers [33–49], and was successfully applied by many authors [50–60].

It should also be mentioned that the frontiers of nonlinear dynamical systems are continuously being pushed out with new ideas and applications continuously appearing on a regular basis in various research publications [3]. At present, there is a somewhat "piecemeal" approach to tackling nonlinear dynamical equations, but undoubtedly, as the subject matures, new mathematical techniques and concepts will be discovered and further "unification" will occur.

In this book, the mathematical level of the text assures a good work of basic calculus (ordinary and partial derivatives, integrals, series expansion, etc.) and a reasonable familiarity with differential equations. This book intended to be open-ended at whetting the appetite of the reader to more fully explore our nonlinear world.

The important development of the theory of nonlinear dynamical systems during these centuries, has essentially its origin in the studies of the "natural effects" encountered in these systems, and rejection of non-essential generalizations, i.e. the study of concrete nonlinear systems have been possible due to the foundation of results from the theory of nonlinear dynamical system field.

References

1. R.M. May, Simple mathematical models with very complicated dynamics. Nature **261**, 459–467 (1976)
2. V.I. Arnold, *Mathematical Methods of Classical Mechanics* (Springer, Berlin, Heidelberg, 1997)
3. R.E. Enns, *It's a Nonlinear World* (Springer, New York, 2011)
4. H. Poincare, Les nouvelles methods de la mecanique celeste, Tomm 1, 2 et 3, Paris (1891, 1893, 1894)
5. J. Rayleigh, *The Theory of Sound* (Dover, New York, 1945)
6. B. Van der Pol, On relaxation oscillations. Philos. Mag. **7**, 978–992 (1926)
7. A. Lienard, Etude des oscillations extremes. Rev. Generale de l'Electricite **23**, 901–912 (1928)
8. N. Bogoliubov, Y. Mitropolski, A.M. Samoilenko, *Method of the Acceleration of Convergence in Nonlinear Mechanics* (Naukova Dumka, Kiev, 1969)
9. Y. Mitropolski, *The Average Method in Nonlinear Mechanics* (Naukova Dumka, Kiev, 1971)

10. A.A. Andronov, A.A. Witt, S.E. Schaikin, *Theory of Oscillations* (Pergamon Press, Oxford, 1966)
11. I.G. Malkin, *Some problems of the Theory of Nonlinear Oscillations* (Gostehizdat, Moscow, 1956)
12. C. Hayashi, *Nonlinear Oscillations in Physical Systems* (McGraw-Hill, New York, 1964)
13. N. Minorsky, *Introduction in Nonlinear Mechanics* (Edwards, Ann Arbor, MI, 1947)
14. J.J. Stoker, *Nonlinear Vibration in Mechanical and Electrical Systems* (Willey, New York, 1950)
15. N. McLachlan, *Theory of Vibration* (Dover, New York, 1991)
16. N. Kauderer, *Nichtlinear Mechanic* (Springer, Berlin Gotingen, Heidelberg, 1958)
17. G. Sansone, R. Conti, *Nonlinear Differential Equation* (Pergamon, Oxford, 1964)
18. A.H. Nayfeh, D.T. Mook, *Nonlinear Oscillations* (Willey, New York, 1974)
19. P. Hagedorn, *Nonlinear Oscillations* (Clarendon Press, Oxford, 1981)
20. A. Einstein, *The Theory of Brownian Movement* (Dover, New York, 1956)
21. S.N. Crandall, W.D. Mark, *Random Vibration in Mechanical System* (Academic, New York, London, 1963)
22. C. Mira, Some historical aspects of nonlinear dynamics possible trends for the future. J. Franklin Inst. **334**(5–6), 1075–1113 (1997)
23. C. Mira, *Systemes asservis non lineaires* (Hermes, Paris, 1990)
24. S.N. Chen, Y.K. Cheung, A modified Lindstedt-Poincare method for strongly non-linear two degree-of-freedom system. J. Sound Vibr. **193**, 751–762 (1996)
25. V.P. Agrawal, N.N. Denman, Weighted linearization technique for period approximation in large amplitude nonlinear oscillations. J. Sound Vibr. **99**, 463–473 (1985)
26. G. Adomian, A review of decomposition method in applied mathematics. J. Math. Anal. Appl. **135**, 501–544 (1998)
27. J.H. He, Variational iteration method, a kind of nonlinear analytical technique. Some examples. Int. J. Non-linear Mech. **34**, 699–708 (1999)
28. Y.Y. Wu, S.J. Liao, X.Z. Zhao, Some notes on the general boundary element method for highly nonlinear problems. Commun. Nonlinear. Sci. Numer. Simul. **10**, 725–735 (2005)
29. V. Marinca, N. Herisanu, Optimal variational method for truly nonlinear oscillators, J. Appl. Math. ID 620267 (2013)
30. V. Marinca, N. Herisanu, An optimal iteration method for strongly nonlinear oscillators, J. Appl. Math. ID 906341 (2012)
31. N. Herisanu, V. Marinca, Optimal homotopy perturbation method for a non-conservative dynamical system of a rotating electrical machine. Z. Naturforsh. **67a**, 509–516 (2012)
32. V. Marinca, N. Herisanu, Optimal parametric iteration method for solving multispecies Lotka-Volterra equation, Discr. Dynam. Nat. Soc. ID 842121 (2012)
33. V. Marinca, N. Herisanu, *Nonlinear Dynamical Systems in Engineering. Some Approximate Approaches* (Springer, Berlin, Heidelberg, 2011)
34. V. Marinca, N. Herisanu, Application of optimal homotopy asymptotic method for solving nonlinear equations arising in heat transfer. Int. Commun. Heat Mass Transf. **35**, 710–715 (2008)
35. V. Marinca, N. Herişanu, I. Nemeş, Optimal homotopy asymptotic method with application to thin film flow. Cent. Eur. J. Phys. **6**, 648–653 (2008)
36. N. Herisanu, V. Marinca, T. Dordea, G. Madescu, A new analytical approach to nonlinear vibration of an electrical machine. Proc. Rom. Acad. Ser. A **9**, 229–236 (2008)
37. V. Marinca, N. Herişanu, C. Bota, B. Marinca, An optimal homotopy asymptotic method applied to the steady flow of a fourth grade fluid past a porous plate. Appl. Math. Lett. **22**, 245–251 (2009)
38. N. Herisanu, V. Marinca, Accurate analytical solutions to oscillators with discontinuities and fractional-power restoring force by means of optimal homotopy asymptotic method. Comput. Math. Appl. **60**, 1607–1615 (2010)

39. V. Marinca, N. Herisanu, Determination of periodic solutions for the motion of a particle on a rotating parabola by means of the optimal homotopy asymptotic method. J Sound Vibr. **329**, 1450–1459 (2010)
40. N. Herisanu, V. Marinca, Explicit analytical approximation to large-amplitude nonlinear oscillations of a uniform cantilever beam carrying an intermediate lumped mass and rotary inertia. Meccanica **45**, 847–855 (2010)
41. V. Marinca, N. Herisanu, D. Bala, Some optimal approximate methods with application to thin film flow. WSEAS Trans. Syst. **7**(9), 744–753 (2010)
42. V. Marinca, N. Herisanu, An optimal homotopy asymptotic approach to nonlinear MHD Jeffery-Hamel flow, Math. Probl. Eng. Article ID 169056 (2011)
43. V. Marinca, N. Herisanu, An approximate solution for the nonlinear Lane-Emden type equation on a semi-infinite domain. AIP Conf. Proc. **1479**, 2387–2390 (2012)
44. R.D. Ene, V. Marinca, R. Negrea, B Caruntu, Optimal homotopy asymptotic method for solving a nonlinear problem in elasticity, in *14-th SYNASC-2012* (2012), pp. 98–102
45. V. Marinca, N. Herisanu, Optimal homotopy asymptotic approach to nonlinear oscillators with discontinuities. Scientific Res. Essays **8**(4), 161–167 (2013)
46. V. Marinca, N. Herisanu, T. Marinca, Approximate solutions to a cantilever beam using optimal homotopy asymptotic method. Appl. Mech. Mater. **430**, 22–26 (2013)
47. N. Herisanu, V. Marinca, Oprimal homotopy asymptotic approach to self-excited vibrations. Appl. Mech. Mater. **430**, 27–31 (2013)
48. R.D. Ene, V. Marinca, B. Marinca, Approximate analytical solutions of nonlinear vibrations of a thin elastic plate. Appl. Mech. Mater. **430**, 40–44 (2013)
49. V. Marinca, N. Herisanu, Optimal homotopy asymptotic method for solving Blasius equation. Appl. Math. Comput. **231**, 134–139 (2014)
50. R. Nawaz, H. Ullah, S. Islam, M. Idress, Application of optimal homotopy asymptotic method in Burgers equations, J. Appl. Math. ID 387478 (2011)
51. H. Ullah, S. Islam, M. Idress, M. Arif, Solution of boundary layer problems with heat transfer by optimal homotopy asymptotic method, Abstract Appl. Anal. ID 324869 (2013)
52. R. Nawaz, S. Islam, I.A. Shah, H. Ullah, Optimal homotopy asymptotic method to nonlinear damped generalized regularized long-wave equation, Math. Probl. Eng. ID 503137 (2013)
53. M. Idrees, S. Haq, S. Islam, Application of the optimal homotopy asymptotic method to squeezing flow. Comput. Math. Appl. **59**, 3858–3866 (2010)
54. M. Idrees, S. Haq, S. Islam, Application of optimal homotopy asymptotic method to special sixth-order boundary value problems. J. World Appl. Sci. **9**, 138–143 (2010)
55. A. Golbabai, M. Fardi, K. Sayevand, Application of optimal homotopy asymptotic method for solving strongly nonlinear oscillatory system. Math. Comput. Model. **58**, 1837–1843 (2013)
56. M. Esmaeilpour, D.D. Ganji, Solution of the Jeffery-Hamel flow problem by optimal homotopy asymptotic method. Comput. Math. Appl. **59**, 3405–3411 (2010)
57. S. Iqbal, M. Idrees, A.M. Siddiqui, A.R. Ansari, Some solutions of the linear and nonlinear Klein-Gordon equations using optimal homotopy asymptotic method. Appl. Math. Comput. **216**, 2898–2909 (2010)
58. N. Ratib Anakira, A. K. Alomari, I. Hashim, Optimal homotopy asymptotic method for solving delay differential equations, Math. Probl. Eng. ID 498902 (2013)
59. A. Zeb, S. Iqbal, A.M. Siddiqui, T. Haroon, Application of the optimal homotopy asymptotic method to flow with heat transfer of a pseudoplastic fluid inside a circular pipe. J. Chin. Inst. Eng. **36**(6), 797–805 (2013)
60. M. Ghoreishi, A.I.B.M. Ismail, A.K. Alomari, Comparison between homotopy analysis method and optimal homotopy asymptotic method for nth-order integro-differential equation. Math. Meth. Appl. Sci. **34**(15), 1833–1842 (2011)

Chapter 2
Optimal Homotopy Asymptotic Method

2.1 A Short History of the Homotopy

The notion of homotopy is an important part of topology and thus of differential geometry. The homotopy continuation method or shortly speaking homotopy was known as early as in the 1930s. Thus, in 1892, Lyapunov [1] introduced the so-called "artificial small parameters method" considering a linear differential equation with variable coefficients in the form

$$\frac{du}{dt} = M(t)u$$

with M(t) a time periodic matrix. He replaced this equation with the equation

$$\frac{du}{dt} = \varepsilon M(t)u$$

To get the solution of the last equation, Lyapunov developed the power series over ε for the variable u and then setting $\varepsilon = 1$.

Later, this method was used by kinematicians in the 1960s in the US for solving mechanism synthesis problems [2]. The latest development was done by Morgan at General Motors [3]. There are also two important literature studies by Garcia and Zangwill [4] and Allgower and Georg [5].

The continuation method gives a set of certain answers and some iteration processes to obtain the solution more exactly. Considering the following nonlinear algebraic equation

$$f(x) = 0 \tag{2.1}$$

for a given set of equations in n variables $x_1, x_2, \ldots, x_n$, we modifiy the Eq. 2.1 by omitting some of the terms and adding new ones until we have a new system of

V. Marinca, N. Herisanu, *The Optimal Homotopy Asymptotic Method*,
DOI 10.1007/978-3-319-15374-2_2

equations, the solutions to which may be easily known. We then deform the coefficients of the new system into the coefficients of the original system by a series of small increments to obtain our solutions. This is called homotopy continuation technique.

To find the solution of Eq. 2.1 we choose a new starting system or the auxiliary homotopy function

$$g(x) = 0 \tag{2.2}$$

The auxiliary homotopy function g(x) must be known or controllable and easy to solve. Then, we define the homotopy continuation function as

$$H(x,t) = tf(x) + (1-t)g(x) = 0 \tag{2.3}$$

where t is an arbitrary parameter which varies from 0 to 1, i.e. $t \in [0, 1]$. Therefore we have the following two boundary conditions

$$H(x,0) = g(x) \tag{2.4}$$

$$H(x,1) = f(x) \tag{2.5}$$

This is the famous homotopy continuation method. It is also called the Bootstrap method or parameter-perturbation method, but these names did not become popular.

Traditionally, the perturbation procedures require the existence of a small parameter which is not always the case. Thus, it is necessary to develop some analytic approximation methods, which are independent of any small or large physical parameters.

It should be emphasized here that there exist two remarkable moments in the history of the homotopy. First of all, the concept of homotopy is employed to gain analytic approximations of nonlinear differential equations by Shijun Liao [6–8] in 1992. For a given nonlinear differential equation

$$N(u(x)) = 0, \quad x \in D \tag{2.6}$$

where N is a nonlinear differential operator and u(x) is an unknown function, Liao constructed a one-parameter family of equations in the embedding parameter p ∈ [0,1] called the zeroth-order deformation equation

$$(1-p)L[\varphi(x,p) - u_0(x)] + pN[\varphi(x,p)] = 0, \quad x \in D, p \in [0,1] \tag{2.7}$$

where L is an auxiliary linear operator and $u_0(x)$ is an initial guess. The homotopy provides us with larger freedom than the traditional non-perturbation methods to choose, both auxiliary linear operator L and the initial guess. The embedding parameter p increases from 0 to 1, such that $\varphi(x,p)$ varies from the initial approximation (guess) $u_0(x) = \varphi(x,0)$ to the unknown exact solution $u(x) = \varphi(x,1)$ of the original equation. Since the embedding parameter $p \in [0,1]$ has no physical meaning, one can always constructs a kind of zeroth-order deformation equation, no matter whether or not there exist small or large physical parameters.

So, if the Taylor series

$$\varphi(x,p) = \sum_{k=0}^{\infty} u_k(x)p^k \tag{2.8}$$

converge at p = 1, then we obtain the so-called homotopy-series solution

$$u(x) = \sum_{k=0}^{\infty} u_k(x) \tag{2.9}$$

which must satisfy the original Eq. 2.6. Here $u_k(x)$ is governed by a linear differential equation related to the auxiliary linear operator L and therefore is easy to solve. One has an extremely large freedom to choose the auxiliary linear operator L which is unnecessary to be the whole linear part of a given nonlinear differential equation. But this early homotopy-analysis method cannot always guarantee the convergence approximation series, especially for nonlinear differential equations with strong nonlinearity.

Secondly, exceptionally moment in the history of homotopy is due to the freedom on constructing the zeroth-order deformation equation that the so-called convergence-control parameter $\hbar$ (later c_0) was introduced by Liao [9] in 1997 to construct a more generalised zeroth-order deformation equation

$$(1-p)L[\varphi(x,p) - u_0(x)] = c_0 p N[\varphi(x,p)], \quad x \in D, \quad p \in [0,1] \tag{2.10}$$

In this way, the homotopy-series solution Eq. 2.9 is dependent upon x and the auxiliary parameter c_0, which has no physical meaning at all. The use of the convergence-control parameter c_0 is indeed a great progress, which provides us one more "artificial" degree of freedom in essence. It seems that more "artificial" degrees of freedom imply larger possibility to get better approximations by means of this Homotopy Analysis Method (HAM). In 1999 Liao introduced the zeroth-order deformation equation in a more general form [10]

$$[1-B(p)]L[\varphi(x,p) - u_0(x)] = c_0 A(p) N[\varphi(x,p)], \quad x \in D, \quad p \in [0,1] \tag{2.11}$$

where A(p) and B(p) are called deformation functions satisfying

$$A(0) = B(0) = 0, \quad A(1) = B(1) = 1 \tag{2.12}$$

In addition Liao introduced in 2004 [11] a non-zero auxiliary function H(x) to further generalise the zeroth-order deformation equation, i.e.

$$[1 - B(p)]L[\varphi(x,p) - u_0(x)] = c_0 A(p) H(x) N[\varphi(x,p)], \quad x \in D, \quad p \in [0,1] \tag{2.13}$$

The so-called rule of solution expression, rule of coefficient ergodicity and rule of solution existence play important roles and greatly simplify the application of the homotopy analysis method. The convergence of the homotopy-series solution 2.9 depends upon the auxiliary parameter c_0, the auxiliary function H(x), the initial guess $u_0(x)$ and the auxiliary linear operator L. The combination of the convergence theorem and the freedom of the choice of auxiliary parameter c_0, the auxiliary function H(x), the initial guess $u_0(x)$ and the auxiliary operator L establish the cornerstone of the validity and flexibility of the homotopy analysis method. Liao suggested investigating the convergence of some special quantities which often have important meanings. For example are used so-called "c_0 curves" (originally $\hbar$-curves) by plotting curves $du/dx(0)$ or $d^2u/dx^2(0)$ and so on, versus c_0. Obviously, some rigorous mathematical theorems are needed to find the best convergence-control parameter c_0 for Eq. 2.10 [12]. Unfortunately, we cannot give a mathematical proof in general for "c_0-curves" [11]. However, such kind of c_0-curves cannot tell us the best convergence-control parameter c_0, which corresponds to the fastest convergence series [13].

Frankly speaking, the HAM is a method for the time of computers [12]: "without high-performance computers and symbolic computation software such as Mathematica, Maple and so on, it is impossible to solve high-order deformation equations quickly so as to get approximations at high enough order. Without computers and symbolic computation software, it is also impossible to choose a proper value of the convergence-control parameter c_0 by means of analyzing the high-order approximations. It is true that expressions given by the HAM are often lengthy and thus can be hardly expressed on only one page. However, by means of computer and symbolic computation software it often needs only a few seconds to calculate these lengthy results. Note that one needs much more time to calculate a traditional "analytic" expression in a length of half page by means of a traditional computational tool such as slide rule". This is an original concept of "analytic solution" unlike the traditional concept of analytic solution, which has been formed 100 years ago.

The essence of the homotopy analysis method was emphasized by Liu [15]:" The generalised Taylor expansion including a secret auxiliary parameter c_0 (originally $\hbar$) which can control and adjust the convergence region of the series is the foundation of the homotopy analysis method proposed by Liao. The secret of c_0 cannot be understood in the frame of the homotopy analysis method. This is a serious shortcoming of Liao's method" [14].

In 2010, Liao developed an optimal homotopy analysis method with only three convergence-control parameters [16]. "Like Marinca and Herisanu's approach"

[13], these three parameters are determined by minimizing the (averaged) squared residual. Liao suggested in [13] that the basic optimal homotopy-analysis method with only one convergence-control parameter c_0 should be used first in practice.

In 1998 He proposed the Homotopy Perturbation Method (HPM) which is valid in general for nonlinear differential equations.

In a whole different manner, we proposed another technique called the Optimal Homotopy Asymptotic Method (OHAM). Instead of an infinite series, we need only a few terms, mostly two terms. Our procedure is applied successfully to obtain the analytical (in classical sense) approximate solutions of currently important problems in practice and we have also shown its effectiveness, generalization and reliability.

The OHAM is not only useful for nonlinear differential equations but also it is useful for linear differential equation with variable or constant coefficients or nonlinear partial differential equations. OHAM is a powerful method for solving nonlinear problems without depending on small or large parameters, which shows its validity and potential for the solution of nonlinear problems in science and engineering applications.

2.2 Basic Idea of OHAM

The OHAM was proposed for the first time in 2008 by Marinca and Herisanu [17–20] and is based on the concept of homotopy.

To apply OHAM we consider the following differential equation

$$L[u(x)] + g(x) + N[u(x)] = 0, \quad x \in D \tag{2.14}$$

subject to the initial/boundary conditions

$$B\left(u(x), \frac{du(x)}{dx}\right) = 0 \tag{2.15}$$

where L is a linear operator, x denotes independent variable, $u(x)$ is an unknown function, $g(x)$ is a known function, $N[u(x)]$ is a nonlinear operator, D is the domain of interest and B is a boundary operator.

By means of OHAM one first constructs a family of equations:

$$\begin{aligned}&(1-p)[L(\varphi(x,p,C_i)) + g(x)] = \\&= H(x,p,C_i)[L(\varphi(x,p,C_i)) + g(x) + N(\varphi(x,p,C_i))]\end{aligned} \tag{2.16}$$

where $p \in [0,1]$ is an embedding parameter, $H(x,p,C_i)$ is a nonzero auxiliary function for $p \neq 0$ and $H(x,0,C_i) = 0$, $\varphi(x;p,C_i)$ is an unknown function, C_i, $i = 1,2,\ldots,s$ are unknown parameters which will be determined later.

Obviously, when $p = 0$ and $p = 1$ it holds

$$\varphi(x,0,C_i) = u_0(x) \tag{2.17}$$

$$\varphi(x;1,C_i) = u(x,C_i), \quad i = 1,2,\ldots,s \tag{2.18}$$

respectively. Therefore, as p increases from 0 to 1, the solution $\varphi(x;p,C_i)$ varies from $u_0(x)$ to the solution $u(x,C_i)$, where the initial approximation $u_0(x)$ is obtained from Eq. 2.16 for $p=0$:

$$L[u_0(x)] + g(x) = 0 \tag{2.19}$$

and

$$B\left(u_0, \frac{du_0}{dx}\right) = 0 \tag{2.20}$$

We choose the auxiliary function $H(x;p,C_i)$ in the form

$$H(x;p,C_i) = pH_1(x,C_i) + p^2H_2(x,C_j) + \ldots \tag{2.21}$$

where $H_i(x,C_j)$, $i=1,2,\ldots$ are the auxiliary functions depending on x and on the parameters C_j, $j=1,2,\ldots,s$.

Let us consider the solution of Eq. 2.16 in the form

$$\varphi(x;p,C_i) = u_0(x) + \sum_{i\geq 1} u_i(x,C_j)p^i \quad j = 1,2,\ldots,s \tag{2.22}$$

Now, substituting Eq. 2.22 into Eq. 2.16 and equating the coefficients of like powers of p, we obtain the governing equation of the initial approximation $u_0(x)$ given by Eq. 2.19 with initial/boundary condition 2.20 and then the governing equation of the approximation of order k, $u_k(x,C_i)$, k = 1,2,...., i.e.

$$L(u_1(x,C_i)) = H_1(x,C_j)N_0(u_0(x)) \tag{2.23}$$

with

$$B\left(u_1, \frac{du_1}{dx}\right) = 0 \tag{2.24}$$

$$\begin{aligned} &L(u_k(x,C_j) - u_{k-1}(x,C_j)) = H_k(x,C_j)N_0(x) + \\ &+\sum_{i=1}^{k-1} H_i(x,C_j)\left[L(u_{k-i}(x,C_j)) + \right. \\ &\left. +N_{k-i}(u_0(x), u_1(x,C_j), \ldots, u_{k-i}(x,C_j))\right], \quad k = 2,3,\ldots \end{aligned} \tag{2.25}$$

and

$$B\left(u_k, \frac{du_k}{dt}\right) = 0 \tag{2.26}$$

where $N_m(u_0(x), u_1(x, C_j), \ldots, u_m(x, C_j))$ is the coefficient of p^m, obtained expanding $N(\varphi(x; p, C_i))$ in series with respect to the embedding parameter p:

$$\begin{aligned} N(\varphi(x;p,C_j)) &= N_0(u_0(x)) + N_1(u_0(x), u_1(x,C_j))p+ \\ &+N_2(u_0(x), u_1(x,C_j), u_2(x,C_j))p^2 + \ldots \end{aligned} \tag{2.27}$$

where $\varphi(x; p, C_j)$ is given by Eq. 2.22.

It should be emphasized that $u_k(x,C_i)$ for $k \geq 0$ are governed by the linear Eqs. 2.19, 2.23, 2.25 and with the initial/boundary conditions given by Eqs. 2.20, 2.24 and 2.26 respectively. These conditions come from the original problem, which in general, can be easily solved.

The convergence of the series 2.22 depends upon the auxiliary functions $H_i(x, C_j)$, $i = 1,2,\ldots$ There are many possibilities to choose the functions $H_i(x, C_j)$. The convergence of the solutions $u_k(x,C_j)$ and consequently the convergence of the solution $u(x,C_j)$ depend on the auxiliary function $H_i(x, C_j)$. Basically, the shape of $H_i(x, C_j)$ must follow the terms appearing in Eq. 2.25. Therefore, we try to choose $H_i(x, C_j)$ so that in Eq. 2.25, the product

$$H_i(x, C_j)\left[L(u_{k-i}(x, C_j)) + N_{k-i}(u_0(x), u_1(x, C_j), \ldots, u_{k-i}(x, C_j))\right]$$

be of the same shape with the terms which appear into $N_{k-i}(u_0, u_1, \ldots, u_{k-i})$. For example $H_i(x, C_j)$ could be chosen as exponential functions, trigonometric functions, polynomial functions and so on, depending on the shape of terms already present in the iteration given by Eq. 2.25.

If the series 2.22 is convergent at $p = 1$, one has

$$u(x, C_j) = u_0(x) + \sum_{k \geq 1} u_k(x, C_j), \quad j = 1, 2, \ldots, s \tag{2.28}$$

Generally speaking, the solution of Eq. 2.14 can be determined approximately in the form

$$\overline{u}(x, C_j) = u_0(x) + \sum_{k=1}^{m} u_k(x, C_j), \quad j = 1, 2, \ldots s \tag{2.29}$$

The equation 2.29 is named approximate solution of order m. In this book we consider only the cases $m \leq 2$.

2.3 Convergence of the Homotopy-Series 2.28

In what follows, we suppose that the series 2.28 is convergent. From the convergence of series 2.28 one can get

$$\lim_{k\to\infty} u_k(x, C_i) = 0, \quad i = 1, 2, \ldots, s \tag{2.30}$$

From the identity

$$u_k(x, C_i) = u_0(x) + [u_1(x, C_i) - u_0(x, C_i)] + [u_2(x, C_i) - u_1(x, C_i)] + \ldots + \\ +[u_k(x, C_i) - u_{k-1}(x, C_i)] = u_0(x) + \sum_{j=1}^{k}[u_j(x, C_i) - u_{j-1}(x, C_i)], i = 1, 2, \ldots, s \tag{2.31}$$

and taking into account Eq. 2.30 it holds that

$$0 = \lim_{k\to\infty} u_k(x, C_i) = u_0(x) + \sum_{j=1}^{\infty}[u_j(x, C_i) - u_{j-1}(x, C_i)] \tag{2.32}$$

Using the linear operator L and in consequence the property L(0) = 0 and thereafter Eqs. 2.23 and 2.25, from Eq. 2.32 it follows that

$$\begin{aligned}
&0 = L(u_0(x)) + g(x) + L(u_1(x, C_i)) - [L(u_0(x)) + g(x)] + \\
&+\sum_{j=2}^{\infty}[L(u_j(x, C_i)) - L(u_{j-1}(x, C_i))] = 0 + H_1(x, C_i)N_0(u_0(x)) + \\
&+\sum_{j=2}^{\infty}[H_j(x, C_i)N_0(u_0(x)) + \sum_{q=1}^{j-1}H_q(x, C_i)[L(u_{j-q}(x, C_i)) + \\
&+N_{j-q}(u_0(x), u_1(x, C_i), \ldots. u_{j-q}(x, C_i))] = \left[\sum_{j=1}^{\infty}H_j(x, C_i)\right]N_0(u_0(x)) + \\
&+\sum_{j=2}^{\infty}\sum_{q=1}^{j-1}H_q(x, C_i)[L(u_{j-q}(x, C_i)) + N_{j-q}(u_0(x), u_1(x, C_i), \ldots. u_{j-q}(x, C_i))] = \\
&= H(x, C_i)N_0(u_0(x)) + \sum_{q=1}^{\infty}\sum_{j=q+1}^{\infty}H_q(x, C_i)[L(u_{j-q}(x, C_i)) + \\
&+N_{j-q}(u_0(x), u_1(x, C_i), \ldots, u_{j-q}(x, C_i))] = H(x, C_i)N_0(u_0(x)) +
\end{aligned}$$

$$
\begin{aligned}
&+\sum_{j=1}^{\infty}H_j(x,C_j)\sum_{m=1}^{\infty}[L(u_m(x,C_i))+N_m(u_0(x),u_1(x,C_i),\ldots,u_m(x,C_i))]=\\
&=H(x,C_i)N_0(u_0(x))+H(x,C_i)[L(\sum_{m=1}^{\infty}u_m(x,C_i)+\\
&+\sum_{m=1}^{\infty}N_m(u_0(x),u_1(x,C_i),\ldots,u_m(x,C_i))]=H(x,C_i)N_0(u_0(x))+\\
&+H(x,C_i)[L(u(x,C_i)-L(u_0(x))+N(u(x,C_i))-N_0(u_0(x))]=\\
&=H(x,C_i)N_0(u_0(x))+H(x,C_i)[L(u(x,C_i))-(L(u_0(x))+g(x))+g(x)+\\
&+N(u(x,C_i))-N_0(u_0(x))]=H(x,C_i)[L(u(x,C_i))+g(x)+N(u(x,C_i))],\\
&i=1,2,\ldots,s
\end{aligned}
\tag{2.33}
$$

But $H(x,C_i)\neq 0$, such that

$$L(u(x,C_i))+g(x)+N(u(x,C_i))=0,\quad i=1,2,\ldots,s \tag{2.34}$$

The equation 2.34 confirms that $u(x,C_i)$ is the exact solution of Eq. 2.14.

In conclusion, we obtained an important result: if the solution series 2.28 converge, where $u_k(x,C_i)$ is governed by Eqs. 2.19, 2.23 and 2.25, then Eq. 2.28 is a solution of Eq. 2.14.

2.4 Convergence of the Approximate Solution of Order m Given by Eq. 2.29

Inserting Eq. 2.29 into Eq. 2.14 we get the following residual

$$R(x,C_i)=L(\overline{u}(x,C_i))+g(x)+N(\overline{u}(x,C_i)),\quad i=1,2,\ldots s,x\in D \tag{2.35}$$

If $R(x,C_i)=0$ then $\bar{u}(x,C_i)$ happens to be the exact solution. Generally such case will not arise for nonlinear problems. At this moment, the mth-order approximation given by Eq. 2.29 depends on the parameters C_i, $i=1,2,\ldots,s$ and these parameters can be optimally identified via various methods, such as the least square method, the Galerkin method, the collocation method, the Ritz method, the Kantorovich method and so on. The first option always should be minimizing the square residual error:

$$J(C_1,C_2,\ldots,C_s)=\int_a^b R^2(x,C_1,C_2,\ldots,C_s)dx \tag{2.36}$$

where a and b are two values depending on the given problem, and R is given by Eq. 2.35. The unknown parameters $C_1, C_2, \ldots, C_s$ can be identified from the conditions

$$\frac{\partial J}{\partial C_1} = \frac{\partial J}{\partial C_2} = \ldots = \frac{\partial J}{\partial C_s} = 0 \tag{2.37}$$

With these parameters known (namely convergence-control parameters or optimal convergence-control parameters), the approximate solution of order m given by Eq. 2.29 is well-determined.

Another convenient possibility to determine the optimal values of the parameters $C_1, C_2, \ldots, C_s$ is to solve the following system:

$$R(x_1, C_i) = R(x_2, C_i) = \ldots = R(x_s, C_i) = 0, \quad x_i \in D, i = 1, 2, \ldots, s \tag{2.38}$$

where the residual R is given by Eq. 2.35.

It should be emphasized that our procedure contains the optimal auxiliary functions $H_i(x, C_j)$, which provides us with a simple way to adjust and control convergence of the solution. It is very important to properly choose the functions $H_i(x, C_j)$ which appear in the mth-order approximation Eq. 2.29.

In particular, we consider a nonlinear differential equation of the form

$$\ddot{u}(t) + \omega^2 u(t) = f(u(t), \dot{u}(t), \ddot{u}(t)) \tag{2.39}$$

where the dot denotes derivative with respect to time, ω is a constant and f is in general a nonlinear function. The initial conditions for Eq. 2.39 are

$$u(0) = A, \quad \dot{u}(0) = 0 \tag{2.40}$$

where A is the amplitude of the oscillations. Note that it is unnecessary to assume the existence of any small or large parameter in Eqs. 2.14 or 2.39.

The equation 2.39 describes a system oscillating with an unknown period T. We switch to a scalar time $\tau = \frac{2\pi t}{T} = \Omega t$. Under the transformations

$$\tau = \Omega t, \quad u(t) = Ax(\tau) \tag{2.41}$$

the original Eq. 2.39 becomes

$$\Omega^2 x''(\tau) + \omega^2 x(\tau) = \frac{1}{A} f\left(Ax(\tau), A\Omega x'(\tau), A\Omega^2 x''(\tau)\right) \tag{2.42}$$

and the initial conditions can be written as

$$x(0) = 1, \quad x'(0) = 0 \tag{2.43}$$

where the prime denotes the derivative with respect to τ and Ω is the frequency of the considered system.

The family of Eqs. 2.16 can be written in this case in the form

$$(1-p)L(\varphi(\tau,p,C_i)) = H(\tau,p,C_i)[N(\varphi(\tau,p,C_i),\Omega(\lambda,p,C_i))] \tag{2.44}$$

where L is a linear operator

$$L(\varphi(\tau,p,C_i)) = \Omega_0^2\left[\frac{\partial^2\varphi(\tau,p,C_i)}{\partial\tau^2} + \varphi(\tau,p,C_i)\right] \tag{2.45}$$

while N is a nonlinear operator

$$\begin{aligned} &N(\varphi(\tau,p,C_i),\Omega(\lambda,p,C_i)) = \Omega^2(\lambda,p,C_i)\frac{\partial^2\varphi(\tau,p,C_i)}{\partial\tau^2} + \\ &+(\omega^2+\lambda)\varphi(\tau,p,C_i) - \frac{1}{A}f(A\varphi(\tau,p,C_i),A\Omega(\lambda,p,C_i)\frac{\partial\varphi(\tau,p,C_i)}{\partial\tau}, \\ &A\Omega^2(\lambda,p,C_i)\frac{\partial^2\varphi(\tau,p,C_i)}{\partial\tau^2}) - p\lambda\varphi(\tau,p,C_i) \end{aligned} \tag{2.46}$$

where λ is an arbitrary unknown parameter and Ω_0 will be given later by Eq. 2.50. From Eqs. 2.40 and 2.41 we obtain the initial conditions

$$\varphi(0,p) = 1, \quad \left.\frac{\partial\varphi(\tau,p)}{\partial\tau}\right|_{\tau=0} = 0 \tag{2.47}$$

In this case, when $p=0$ and $p=1$ it holds

$$\begin{aligned} &\varphi(\tau,0) = x_0(\tau), \quad \varphi(\tau,1,C_i) = x(\tau,C_i), \\ &\Omega(\lambda,0,C_i) = \Omega_0(\lambda,C_i), \Omega(\lambda,1,C_i) = \Omega(\lambda,C_i) \end{aligned} \tag{2.48}$$

where $x_0(\tau)$ is an initial approximation of $x(\tau)$. Therefore, as the embedding parameter p increases from 0 to 1, $\varphi(\tau,p,C_i)$ varies from the initial approximation $x_0(\tau)$ to the solution $x(\tau,C_i)$, so does $\Omega(\lambda,pC_i)$ from the initial approximation $\Omega_0(\lambda)$ to the exact frequency $\Omega(\lambda,C_i)$.

Expanding $\varphi(\tau,p,C_i)$ and $\Omega(\lambda,p,C_i)$ in series with respect to the parameter p, one has respectively

$$\varphi(\tau,p,C_i) = x_0(\tau) + px_1(\tau,C_i) + p^2x_2(\tau,C_i) + \ldots., \quad i=1,2,\ldots,s \tag{2.49}$$

$$\Omega(\lambda,p,C_i) = \Omega_0(\lambda,C_i) + p\Omega_1(\lambda,C_i) + p^2\Omega_2(\lambda,C_i) + \ldots, \quad i=1,2,\ldots,s \tag{2.50}$$

If the initial approximation $x_0(\tau)$ and the auxiliary function $H(\tau,p,C_i)$ are properly chosen so that the above series converges at $p=1$, one has

$$x(\tau, C_i) = x_0(\tau) + x_1(\tau, C_i) + x_2(\tau, C_i) + \ldots, \quad i = 1, 2, \ldots, s \tag{2.51}$$

$$\Omega(\lambda, C_i) = \Omega_0(\lambda, C_i) + \Omega_1(\lambda, C_i) + \Omega_2(\lambda, C_i) + \ldots., \quad i = 1, 2, \ldots, s \tag{2.52}$$

The series 2.51 and 2.52 contain the auxiliary function $H(\tau, p, C_i)$. The results of the mth-order approximations are given by

$$\bar{x}(\tau, C_i) \approx x_0(\tau) + x_1(\tau, C_i) + \ldots + x_m(\tau, C_i), \quad i = 1, 2, \ldots., s \tag{2.53}$$

$$\overline{\Omega}(\lambda, C_i) \approx \Omega_0(\lambda, C_i) + \Omega_1(\lambda, C_i) + \ldots + \Omega_{m-1}(\lambda, C_i), \quad i = 1, 2, \ldots, s \tag{2.54}$$

If we substitute Eqs. 2.53 and 2.54 into Eq. 2.44 and then equating to zero the coefficients of various powers of p we obtain the following $\mathrm{m}+1$ linear equations

$$L(x_0(\tau)) = 0, \quad x_0(0) = 1, \quad x_0'(0) = 0 \tag{2.55}$$

$$\begin{aligned} &L(x_i(\tau, C_j)) - L(x_{i-1}(\tau, C_j)) = \sum_{l=1}^{i} H_l N_{i-l}(x_0(\tau), x_1(\tau, C_j), \ldots, \\ &x_{i-l}(\tau, C_j), \Omega_0(\lambda, C_j), \Omega_1(\lambda, C_j), \ldots \Omega_{i-l}(\lambda, C_j)); \\ &x_i(0) = x_i'(0) = 0, \quad i = 1, 2, \ldots m-1 \quad j = 1, 2, \ldots s \end{aligned} \tag{2.56}$$

$$\begin{aligned} &L(x_m(\tau, C_j)) - L(x_{m-1}(\tau, C_j)) = \\ &= \sum_{l=1}^{m-1} H_l(\tau, C_j) N_{m-1-l}(x_0(\tau), x_1(\tau, C_j), \ldots, x_{m-1-l}(\tau, C_j) \\ &, \Omega_0(\lambda, C_j), \Omega_1(\lambda, C_j) \ldots. \Omega_{m-1-l}(\lambda, C_j)) + H_m(\tau, C_j) N_0(u_0(x)), \\ &x_m(0) = x_m'(0) = 0, j = 1, 2, \ldots, s \end{aligned} \tag{2.57}$$

where N_{k} are obtained from the equation

$$\begin{aligned} &N(\varphi(\tau, p, C_j)) = N_0(x_0, \Omega_0, A, \lambda) + \\ &+ pN_1(x_0(\tau), x_1(\tau, C_j), \Omega_0(\lambda, C_j), \Omega_1(\lambda, C_j), A, \lambda) + \\ &+ p^2 N_2(x_0(\tau), x_1(\tau, C_j), x_2(\tau, C_j), \Omega_0(\lambda, C_j), \Omega_1(\lambda, C_j), \Omega_2(\lambda, C_j), A, \lambda) + \ldots \end{aligned} \tag{2.58}$$

Note that $\Omega_k(\lambda, C_j)$ can be determined avoiding the presence of secular terms in Eqs. 2.56 and 2.57.

If the frequency $\Omega(\lambda, C_j)$ depends upon the arbitrary parameter λ then we apply the so-called "principle of minimal sensitivity" [21] in order to fix the value of λ. We do this imposing that

$$\frac{d\Omega(\lambda, C_j)}{d\lambda} = 0 \tag{2.59}$$

This principle of the minimal sensitivity appears for the first time in the quantum field theory, $\lambda\varphi^4$ theory or quantum chromodynamics [21]. In its original formulation a Lagrangian density $\mathcal{L}$ which is not exactly solvable is interpolated with a solvable Lagrangian $\mathcal{L}_0(\lambda)$ depending upon one or more parameters $\lambda : \mathcal{L}\delta = \mathcal{L}_0(\lambda) + \delta(\mathcal{L} - \mathcal{L}_0(\lambda))$, δ being a parameter. We notice that the interpolation of the full Lagrangian with the solvable one, $\mathcal{L}_0(\lambda)$, brings an artificial dependence upon the arbitrary parameter λ. Such dependence, which would vanish if all perturbative orders were calculated, can be made weaker to a finite perturbative order, by requiring some physical observable P to be locally insensitive to λ, i.e. $\partial P/\partial\lambda = 0$.

In the above application this physical observable P is in fact the frequency Ω. This condition is known as the principle of minimal sensitivity and is normally seen to improve the convergence to the exact solution.

Unlike homotopy analysis method, in the proposed procedure (OHAM) the construction of the homotopy is quite different. In the frame of OHAM the linear operator L is well defined by Eqs. 2.19 or 2.45 and the initial approximation is rigorously determined from Eqs. 2.55, while in the homotopy analysis method these ones are arbitrarily chosen. Instead of an infinite series (as is the case of the homotopy analysis method) the OHAM searches for only a few terms (mostly two terms). The way to ensure the convergence in OHAM is quite different and more rigorous. Unlike other homotopy procedures, OHAM ensure a very rapid convergence since it needs at the most two approximations for achieving a very accurate solution. This is in fact the true power of the method. OHAM does not need a recurrence formula as other homotopy procedures (such as homotopy analysis method) do. OHAM is an iterative procedure which often converges to the exact solution after only one or two approximations. Iterations are performed in a very simple manner by identifying some coefficients. OHAM does not need high-order approximations. OHAM does not use the rules established in the frame of homotopy analysis method, it is a self-sustained method which has no "open questions" as is the case of other homotopy procedures. OHAM does not need the restrictive conditions $A(1) = 1$, $B(1) = 1$ as the homotopy analysis does. The homotopy analysis method and the homotopy perturbation methods are special cases of OHAM when $H(\tau,p) = pc_0$ and $H(\tau,p) = -p$, respectively, where the parameter c_0 is chosen from the so-called "c_0-curves" or minimizing the discrete residual. An important feature of the OHAM is that using Eqs. 2.37, 2.38 or 2.59 a minimization of errors is obtained. Finally, OHAM provides an analytic solution for complicated nonlinear problems expressed on only few rows, unlike other homotopy procedures which need few pages to express an analytic solution.

In the following, several applications are given considering the cases with two approximations (m = 2 into Eq. 2.29) or with a single approximation (m = 1 into Eq. 2.29) to demonstrate the general validity and the great potential of the OHAM. These will further demonstrate that OHAM is more general and absolute different in structure comparing to other methods, including other homotopy approaches.

References

1. A.M. Lyapunov, *General Problem on Stability of Motion* (Taylor and Francis, London, 1992)
2. T.M. Wu, A new formula of solving nonlinear equations by Adomian decomposition method and homotopy methods. Appl. Math. Comput. **172**, 903–907 (2006)
3. A.P. Morgan, *Solving Polynomial Systems using Continuation for Engineering and Scientific Problems* (Prentice Hall, Englewood, New York, 1987)
4. C.B. Garcia, W.S. Zangwill, *Pathways to Solutions. Fixed Points and Equilibrum* (Prentice-Hall, Book Company, Englewood Clifts, NY, 1981)
5. E.L. Allgower, K. Georg, *Numerical Continuation Methods, an Introduction* (Springer, New York, NY, 1990)
6. S.J. Liao, The proposed homotopy analysis technique for the solution of nonlinear problems, PhD thesis, Shanghai Jiao Tong University (1992)
7. S.J. Liao, *A Kind of Linearity Invariance Under Homotopy and Some Simple Applications of it in Mechanics, Bericht, no. 520* (Institut fur Schiffbau der Universitaet Hamburg, Germany, 1992)
8. S.J. Liao, Application of process analysis method to the solution of 2D non-linear progressive gravity waves. J. Ship Res. **36**, 30–37 (1992)
9. S.J. Liao, A kind of approximate solution technique which does not depend upon small parameters (II)—An application in fluid mechanics. Int. J. Nonlinear Mech. **32**, 815–822 (1997)
10. S.J. Liao, An explicit, totally analytic approximate solution for Blasius viscous flow problems. Int. J. Nonlinear Mech. **34**, 759–778 (1999)
11. S.J. Liao, *Beyond Perturbation-Introduction to the Homotopy Analysis Method* (Chapman and Hall, CRC, Boca Raton, FL, 2004)
12. S.J. Liao, Notes on the homotopy analysis method. Some definitions and theorems. Commun. Nonlinear Sci. Numer. Simul. **14**, 983–997 (2009)
13. S.J. Liao, *Homotopy Analysis Method in Nonlinear Differential Equations* (Higher Education Press, Springer, Beijing, Berlin, 2012)
14. C.-S. Liu, The essence of the homotopy analysis method. Appl. Math. Comput. **216**, 1299–1303 (2010)
15. C.-S. Liu, The essence of the generalized Taylor theorem as the foundation of the homotopy analysis method. Commun. Nonlinear Sci. Numer. Simul. **16**(1254–1262) (2011)
16. S.J. Liao, An optimal homotopy-analysis approach for strongly nonlinear differential equations. Commun. Nonlinear Sci. Numer. Simul. **15**, 2003–2016 (2010)
17. V. Marinca, N. Herişanu, I. Nemeş, Optional homotopy asymptotic method with application to thin film flow. Cent. Eur. J. Phys. **6**, 648–653 (2008)
18. V. Marinca, N. Herişanu, Application of optimal homotopy asymptotic method for solving nonlinear equations arising in heat transfer. Int. Commun. Heat Mass Transf. **35**, 710–715 (2008)
19. N. Herişanu, V. Marinca, T. Dordea, Gh. Madescu, A new analytical approach to nonlinear vibration of an electrical machine. Proc. Rom. Acad. Ser. A. **9**, 229–236 (2008)
20. V. Marinca, N. Herisanu, *Nonlinear Dynamical Systems in Engineering. Some Approximate Approaches* (Springer, Berlin-Heidelberg, 2011)
21. P. Amore, A. Aranda, Improved Lindstedt-Poincaré method for the solution of nonlinear problems. J. Sound Vib. **283**, 1115–1136 (2009)

Chapter 3
The First Alternative of the Optimal Homotopy Asymptotic Method

In this chapter, we consider m = 2 into Eq. 2.29 such that we obtain the second-order approximate solution in the form

$$\overline{u}(x, C_i) = u_0(x) + u_1(x, C_i) + u_2(x, C_i) \tag{3.1}$$

where the terms u_0, u_1 and u_2 are given by the linear differential equations 2.19, 2.20, 2.23, 2.25 and 2.26 respectively or, more precisely, by the following equations:

$$L(u_0(x)) + g(x) = 0, \quad B\left(u_0, \frac{du_0}{dx}\right) = 0 \tag{3.2}$$

$$L(u_1(x, C_i)) = H_1(x, C_i)N_0(u_0(x)), \quad B\left(u_1, \frac{du_1}{dx}\right) = 0 \tag{3.3}$$

$$\begin{gathered} L(u_2(x, C_i)) - L(u_1(x, C_i)) = \\ = H_1(x, C_i)[L(u_1(x, C_i)) + N_1(u_0(x), u_1(x, C_i))] + H_2^*(x, C_i)N_0(u_0(x)), \\ B\left(u_2, \frac{du_2}{dx}\right) = 0 \end{gathered}$$

Taking into account Eq. 3.3, the last equation can be written in the equivalent form

$$\begin{gathered} L(u_2(x, C_i)) - L(u_1(x, C_i)) = \\ = H_1(x, C_i)N_1(u_0(x), u_1(x, C_i)) + H_2(x, C_i)N_0(u_0(x)), \\ B\left(u_2, \frac{du_2}{dx}\right) = 0 \end{gathered} \tag{3.4}$$

where $H_2(x, C_i) = H_1^2(x, C_i) + H_2^*(x, C_i)$.

In what follows we illustrate the efficiency of the presented procedure.

V. Marinca, N. Herisanu, *The Optimal Homotopy Asymptotic Method*,
DOI 10.1007/978-3-319-15374-2_3

3.1 Thin Film Flow of a Fourth-Grade Fluid Down a Vertical Cylinder

The study of non-Newtonian fluids is important for various branches of engineering and technology. Some examples of such fluids are blood, drilling muds, clay coatings and other suspensions, certain oils and greases, polymer melts, elastomers and many emulsions. For such fluids it is very difficult to suggest a single model which exhibits all properties of non-Newtonian fluids. Several fluid models have been proposed to predict the non-Newtonian behavior of various types of materials. Among these, the most popular model for non-Newtonian fluids is the "second grade" fluid. Although the "second grade fluid" describes the normal stress effects but for a steady flow it does not exhibit the property of shear thinning or thickening. Due to this fact some experiments may be well described by the fluids of grade three or four [1–5]. Here, we consider a fourth grade fluid falling on the outside of an infinitely long vertical cylinder of a radius R. The flow is considered in thin, uniform, axisymmetric film with thickness δ, in contact with stationary air. In cylindrical coordinates, we have

$$\frac{\partial p}{\partial r} = (2\alpha_1 + \alpha_2)\frac{1}{r}\frac{d}{dr}\left[r\left(\frac{du}{dr}\right)^2\right] + \frac{4}{r}\left(\gamma_3 + \gamma_4 + \gamma_5 + \frac{\gamma_6}{2}\right)\frac{d}{dr}\left[r\left(\frac{du}{dr}\right)^4\right] \tag{3.5}$$

$$\frac{\partial p}{\partial z} = \frac{\mu}{r}\frac{d}{dr}\left(r\frac{du}{dr}\right) + \frac{2}{r}(\beta_2 + \beta_3)\frac{d}{dr}\left[r\left(\frac{du}{dr}\right)^3\right] + \rho g \tag{3.6}$$

where $p \neq p(z)$ is pressure, $\alpha_1, \alpha_2, \beta_2, \beta_3, \gamma_3, \gamma_4, \gamma_5$ and γ_6 are known parameters and Eq. 3.6 further gives

$$r\frac{d^2u}{dr^2} + \frac{du}{dr} + \frac{2(\beta_2 + \beta_3)}{\mu}\left[3r\left(\frac{du}{dr}\right)^2\frac{d^2u}{dr^2} + \left(\frac{du}{dr}\right)^3\right] + \frac{\rho g}{\mu}r = 0 \tag{3.7}$$

The boundary conditions are:

$$u(R) = 0, \quad \frac{du}{dr} = 0 \quad at\ r = R + \delta \tag{3.8}$$

Defining

$$\eta = \frac{r}{R}, \ f = \frac{R}{\nu}u, \ k = \frac{\rho g R^3}{\mu^2}, \ b = \frac{\mu(\beta_2 + \beta_3)}{R^4}, \ d = 1 + \frac{\delta}{R}, \tag{3.9}$$

Equations 3.7 and 3.8 reduce to

$$\eta\frac{d^2f}{d\eta^2}+\frac{df}{d\eta}+k\eta+2b\left[\left(\frac{df}{d\eta}\right)^3+3\eta\left(\frac{df}{d\eta}\right)^2\frac{d^2f}{d\eta^2}\right]=0 \tag{3.10}$$

$$f(1)=0, f'(d)=0 \tag{3.11}$$

In accordance with Eqs. 3.10 and 3.11, the linear operator is chosen as:

$$\mathrm{L}(f(\eta))=\eta\frac{\partial^2 f(\eta)}{\partial\eta^2}+\frac{\partial f(\eta)}{\partial\eta} \tag{3.12}$$

and we define a nonlinear operator as

$$N(f(\eta)=2b\left[\left(\frac{\partial f(\eta)}{\partial\eta}\right)^3+3\eta\left(\frac{\partial f(\eta)}{\partial\eta}\right)^2\frac{\partial^2 f(\eta)}{\partial\eta^2}\right] \tag{3.13}$$

and the function $g(\eta)$ as

$$g(\eta)=k\eta \tag{3.14}$$

The initial conditions are

$$f(1)=0, f'(d)=0 \tag{3.15}$$

Equation 3.2 can be written as

$$\eta f_0''+f_0'+k\eta=0 \tag{3.16}$$

$$f_0'(d)=0 \tag{3.17}$$

It is obtained

$$f_0'(\eta)=\frac{k}{2}\left(\frac{d^2}{\eta}-\eta\right) \tag{3.18}$$

Equation 3.3 becomes

$$\eta f_1''+f_1'=2H_1(\eta,C_i)b\left(f_0'^3+3\eta f_0'^2 f_0''\right) \tag{3.19}$$

$$f_1'(d)=0 \tag{3.20}$$

If we choose $H_1(\eta,C_i)=C_1$, where C_1 is a unknown parameter, then the derivative of solution of Eq. 3.19 subject to condition 3.20 is given by

$$f_1'(\eta) = \frac{1}{4}C_1 bk^3\left(\frac{d^2}{\eta} - \eta\right)^3 \tag{3.21}$$

Equation 3.4 reduces to

$$\begin{aligned}&\eta f_2'' + f_2' = \eta f_1'' + f_1' + C_1\{\eta f_1'' + f_1' + 6b[f_0'^2 f_1' + \\ &+\eta(f_0'^2 f_1'' + 2f_0' f_1' f_0'')]\} + 2bH_2(\eta, C_i)(f_0'^3 + 3\eta f_0'^2 f_0'')\end{aligned} \tag{3.22}$$

$$f_2'(d) = 0 \tag{3.23}$$

Choosing

$$\begin{aligned}H_2(\eta, C_i) = C_2 + bf_0'^2(\eta)\frac{\frac{2d^2}{\eta} + 3\eta}{\frac{d^2}{\eta} + 2\eta}C_3 + b^2k^2 f_0'^4(\eta)\frac{\frac{3d^2}{\eta} + 4\eta}{\frac{d^2}{\eta} + 2\eta}C_4 + \\ +b^3k^4 f_0'^6(\eta)\frac{\frac{3d^2}{\eta} + 5\eta}{\frac{d^2}{\eta} + 2\eta}C_5,\end{aligned}$$

with C_2, C_3, C_4 and C_5 unknown parameters and substituting Eqs. 3.18 and 3.21 into 3.22 and then integrating Eq. 3.22 with condition 3.23, we obtain the derivative of second solution:

$$\begin{aligned}f_2'(\eta) = \frac{1}{4}(C_1^2 + C_1 + C_2)bk^3\left(\frac{d^2}{\eta} - \eta\right)^3 + \frac{b^2k^7}{16}(6C_1^2 + C_3)\left(\frac{d^2}{\eta} - \eta\right)^5 + \\ + \frac{b^3k^9}{64}6C_4)\left(\frac{d^2}{\eta} - \eta\right)^7 + \frac{b^4k^{11}}{256}C_5\left(\frac{d^2}{\eta} - \eta\right)^9\end{aligned} \tag{3.24}$$

Thus the derivative of solution up to second order is given by

$$\begin{aligned}\overline{f'}(\eta) = f_o' + f_1' + f_2' = \alpha\left(\frac{d^2}{\eta} - \eta\right) + \beta\left(\tfrac{d^2}{\eta} - \eta\right)^3 + \gamma\left(\tfrac{d^2}{\eta} - \eta\right)^5 + \\ +\delta\left(\tfrac{d^2}{\eta} - \eta\right)^7 + \phi\left(\tfrac{d^2}{\eta} - \eta\right)^9\end{aligned} \tag{3.25}$$

where

$$\alpha = \frac{k}{2}, \beta = \frac{1}{4}(C_1^2 + 2C_1 + C_2)bk^3,$$
$$\gamma = \frac{b^2k^7}{16}(6C_1^2 + C_3), \delta = \frac{b^3k^9}{64}C_4, \phi = \frac{b^4k^{11}}{256}C_5 \tag{3.26}$$

Substituting the derivative of the solution up to second order 3.25 into Eq. 2.35, results in the residual:

$$R(\eta, C_1, C_2, C_3, C_4, C_5) = \eta \overline{f}'' + \overline{f}' + k\eta + 2b\left[\overline{f}'^3 + 3\eta \overline{f}'^2 \overline{f}''\right] \tag{3.27}$$

The functional 2.36 can be written as:

$$J(C_1, C_2, C_3, C_4, C_5) = \int_1^d R^2(\eta, C_1, \ldots, C_5)d\eta \tag{3.28}$$

The values of the parameters C_1, C_2, C_3, C_4, and C_5 result from the conditions:

$$\frac{\partial J(C_1, \ldots, C_5)}{\partial C_i} = 2\int_1^d R(\eta, C_1, \ldots, C_5)\frac{\partial R(\eta, C_1, \ldots, C_5)}{\partial C_i}d\eta = 0,$$
$$i = 1, 2, \ldots 5 \tag{3.29}$$

3.1.1 Numerical Examples

In order to show the validity of our procedure, we consider two particular cases. In a first particular case when $k = b = 1$, $d = 1.1$ (in [2], $b \geq 0.3$ is considered a parameter corresponding to strong nonlinearity), from Eqs. 3.29 are obtained:

$C_1 = -0.468279131$, $C_2 = -0.282737082$, $C_3 = 4.684287933$, $C_4 = -48$, $C_5 = 399.2$

From Eq. 3.25 the derivative of the approximate solution is obtained up to second order in the form:

$$\overline{f}'(\eta) = 0.5\left(\frac{1.21}{\eta} - \eta\right) - 0.25\left(\frac{1.21}{\eta} - \eta\right)^3 +$$
$$+0.375\left(\frac{1.21}{\eta} - \eta\right)^5 - 0.75\left(\frac{1.21}{\eta} - \eta\right)^7 + 1.559375\left(\frac{1.21}{\eta} - \eta\right)^9 \tag{3.30}$$

For the second case we consider $b = 1$, $k = 1.5$. From Eqs. 3.29 it holds that

$$C_1 = -0.61349821,\ C_2 = -0.149383633,\ C_3 = 0.408386344,$$
$$C_4 = -21.33333334,\ C_5 = 157.6724053271$$

The derivative of the second-order approximate solution may be written as

$$\overline{f}'(\eta) = 0.75\left(\frac{1.21}{\eta} - \eta\right) - 0.84375\left(\tfrac{1.21}{\eta} - \eta\right)^3 + 2.84765625\left(\tfrac{1.21}{\eta} - \eta\right)^5 - $$
$$-12.81445313\left(\tfrac{1.21}{\eta} - \eta\right)^7 + 53.2745239\left(\tfrac{1.21}{\eta} - \eta\right)^9 \tag{3.31}$$

Now, integrating Eqs. 3.30 and 3.31 one can get the second-order approximate solution of Eqs. 3.10 and 3.11. Figures 3.1, 3.2, 3.3 and 3.4 show the second-order approximate solutions $\overline{f}(\eta)$ and their derivatives $\overline{f}'(\eta)$, which are compared with numerical results obtained from Eqs. 3.10 and 3.11 for b = 1, k = 1 and k = 1.5, respectively.

A comparison of our results with numerical solutions obtained through a fourth-order Runge–Kutta method is presented in Tables 3.1 and 3.2.

From the Tables 3.1 to 3.2 it can be seen that the analytical solutions of the thin film flow of a fourth grade fluid down a vertical cylinder obtained by OHAM are very accurate. It should be emphasized that the traditional perturbation method assumes the parameter k as a small parameter while OHAM do not make use of this assumption.

3.2 The Jeffery–Hamel Flow Problem

From mathematical point of view, the study of an incompressible, viscous fluid between non-parallel walls (or fluid flow through convergent-divergent channels) was pioneered by Jeffery [6] and Hamel [7]. They studied an exact similarity solution of the Navier–Stokes equations. Jeffery–Hamel flow is well-known in fluid mechanics, civil, environmental, mechanical and bio-medical engineering and has been studied extensively by several authors [8–12].

We consider a system of cylindrical polar coordinates (r, θ, z) with a steady two-dimensional flow of an incompressible conducting viscous fluid from a source or sink at channel walls lying in planes, with angle 2α, as shown in Fig.3.5.

Assuming that the velocity is only along the radial direction and depends on r and θ, $V(u(r, \theta),0)$ using the continuity and the Navier–Stokes equations in polar coordinates, the governing equations are [11, 12]:

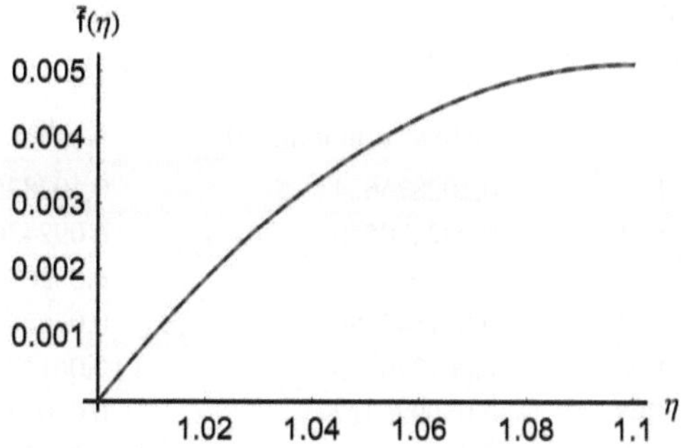

Fig. 3.1 Comparison between the approximate solution obtained by integrating Eq. 3.30 and numerical solution of Eq. 3.10 for $k = b = 1$, $d = 1.1$: (*red line*) numerical solution and (*blue dashed line*) approximate solution (colour figure online)

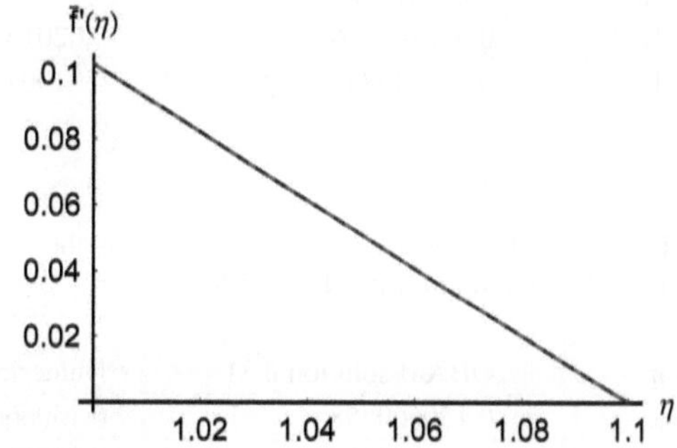

Fig. 3.2 Comparison between the derivative of the approximate solution 3.30 and numerical derivative of the solution of Eq. 3.10 for $k = b = 1$, $d = 1.1$: (*red line*) numerical results and (*blue dashed line*) approximate solution (colour figure online)

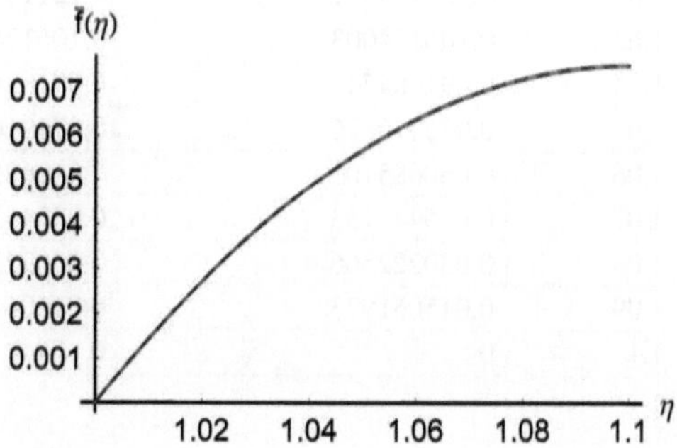

Fig. 3.3 Comparison between the approximate solution obtained by integrating Eq. 3.31 and numerical solution of Eq. 3.10 for $k = 1.5$, $b = 1$, $d = 1.1$: (*red line*) numerical solution and (*blue dashed line*) approximate solution (colour figure online)

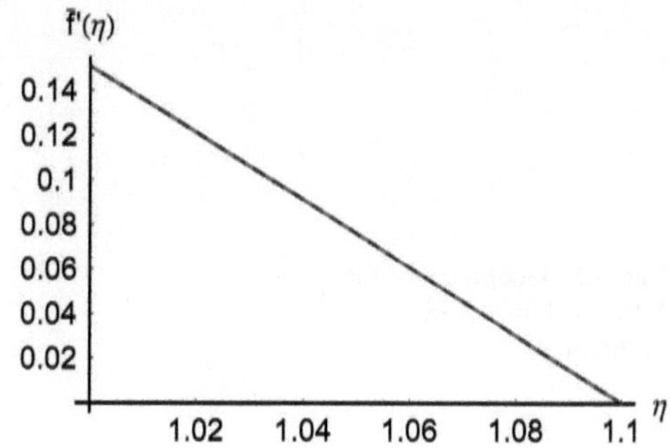

Fig. 3.4 Comparison between the derivative of the approximate solution 3.31 and numerical derivative of the solution of Eq. 3.10 for $k = 1.5$, $b = 1$, $d = 1.1$: (*red line*) numerical results and (*blue dashed line*) approximate solution (colour figure online)

Table 3.1 The results of the derivative of the second-order approximate solution 3.30 and numerical solution for k = b = 1

η	OHAM solution 3.30	Numerical solution	$\varepsilon_1 = \left|\tilde{f}'(\eta) - f'_{num}(\eta)\right|$
1	0.102825634	0.102825639	5.66×10^{-9}
1.01	0.092430550	0.092430575	2.48×10^{-8}
1.02	0.082033174	0.082033194	2.00×10^{-8}
1.03	0.071643186	0.071643217	3.15×10^{-8}
1.04	0.061270735	0.061270658	7.70×10^{-8}
1.05	0.050926322	0.050926333	1.10×10^{-8}
1.06	0.040620665	0.040620669	4.14×10^{-9}
1.07	0.030364568	0.030364565	2.32×10^{-9}
1.08	0.020168776	0.020168772	4.30×10^{-9}
1.09	0.010043845	0.010043844	5.11×10^{-10}
1.1	0	0	0

Table 3.2 The results of the derivative of the second-order approximate solution 3.31 and numerical solution for b = 1, k = 1.5

η	OHAM solution 3.31	Numerical solution	$\varepsilon_1 = \left|\tilde{f}'(\eta) - f'_{num}(\eta)\right|$
1	0.150660558	0.150660459	9.86×10^{-8}
1.01	0.135984959	0.135985624	6.65×10^{-7}
1.02	0.121149211	0.121149622	4.10×10^{-7}
1.03	0.106174003	0.106174131	1.28×10^{-7}
1.04	0.091084743	0.091084815	7.11×10^{-8}
1.05	0.075910836	0.075910873	3.66×10^{-8}
1.06	0.060685105	0.060685104	1.10×10^{-9}
1.07	0.045443153	0.045443114	3.86×10^{-8}
1.08	0.030222566	0.030222564	2.64×10^{-9}
1.09	0.015061973	0.015061973	6.58×10^{-10}
1.1	0	0	0

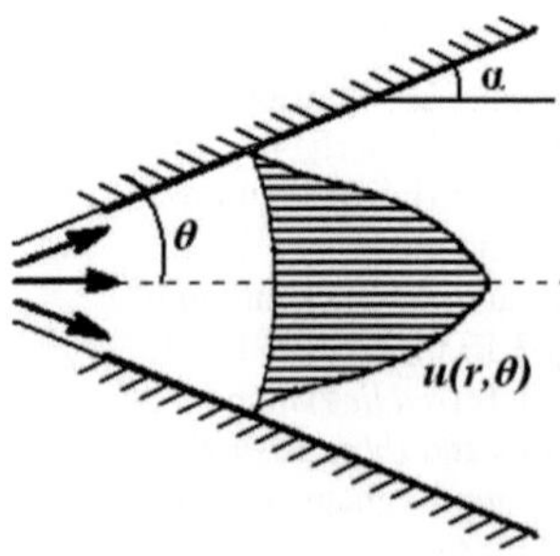

Fig. 3.5 Geometry of the Jeffery–Hamel flow problem

$$\frac{\rho}{r}\frac{\partial}{\partial r}(ru(r,\theta)) = 0 \tag{3.32}$$

$$\begin{aligned} u(r,\theta)\frac{\partial u(r,\theta)}{\partial r} &= -\frac{1}{\rho}\frac{\partial p}{\partial r} + \\ +\nu&\left[\frac{\partial^2 u(r,\theta)}{\partial r^2} + \frac{1}{r}\frac{\partial u(r,\theta)}{\partial r} + \frac{1}{r^2}\frac{\partial^2 u(r,\theta)}{\partial \theta^2} - \frac{u(r,\theta)}{r^2}\right] \end{aligned} \tag{3.33}$$

$$-\frac{1}{\rho r}\frac{\partial p}{\partial \rho} + \frac{2\nu}{r^2}\frac{\partial u(r,\theta)}{\partial \theta} = 0 \tag{3.34}$$

where ρ is the fluid density, p is the pressure and ν is the kinematic viscosity. From Eq. 3.32 and using dimensionless parameters we get:

$$f(\theta) = ru(r,\theta) \tag{3.35}$$

$$F(x) = \frac{f(\theta)}{f_{\max}}, \quad x = \frac{\theta}{\alpha} \tag{3.36}$$

Substituting Eq. 3.36 into Eqs. 3.33 and 3.34 and eliminating the pressure, we obtain an ordinary differential equation for the normalized function profile $F(x)$:

$$F'''(x) + 2\alpha \mathrm{Re} F(x)F'(x) + 4\alpha^2 F'(x) = 0 \tag{3.37}$$

where prime denotes derivative with respect to x and the Reynolds number is

$$\mathrm{Re} = \frac{\alpha f_{\max}}{\nu} = \frac{u_{\max} r\alpha}{\nu}\begin{pmatrix} \textit{divergent channel}: & \alpha > 0, & u_{\max} > 0 \\ \textit{convergent channel}: & \alpha < 0, & u_{\max} < 0 \end{pmatrix} \tag{3.38}$$

and $u_{\max}$ is the maximum velocity at the centre of the channel.

The boundary conditions for Eq. 3.37 are

$$F(0) = 1, \quad F'(0) = 0, \quad F(1) = 0 \tag{3.39}$$

We choose $g(x) = 0$ into Eq. 3.2 and the linear operator

$$L(F(x)) = \frac{\partial^3 F(x)}{\partial x^3} \tag{3.40}$$

The nonlinear operator is

$$N(F(x)) = 2\alpha \mathrm{Re} F(x)F'(x) + 4\alpha^2 F'(x) \tag{3.41}$$

and the boundary conditions are

$$F(0) = 1, \quad F'(0) = 0, \quad F(1) = 0 \tag{3.42}$$

Equation 3.2 becomes

$$\begin{aligned} &F_0'''(x) = 0 \\ &F_0(0) = 1, \quad F_0'(0) = 0, \quad F_0(1) = 0 \end{aligned} \tag{3.43}$$

It is obtained that

$$F_0(x) = 1 - x^2 \tag{3.44}$$

From Eqs. 3.41 and 2.58, we obtain the following expression

$$N_0(F_0(x)) = 2\alpha \mathrm{Re} F_0(x) F_0'(x) + 4\alpha^2 F_0'(x) \tag{3.45}$$

Note that substituting Eq. 3.44 into Eq. 3.45 it follows that

$$N_0(F_0(x)) = 4\alpha \mathrm{Re} x^3 - 4(\alpha \mathrm{Re} + 2\alpha^2)x \tag{3.46}$$

There are many possibilities to choose the functions $H_j(x,C_i)$, $j = 1,2,\ldots$ The convergence of the solutions F_j, $j = 1,2$ and consequently the convergence of the approximate solution $\overline{F}(x)$ given by Eq. 3.1 depend on the optimal auxiliary function $H_j(x,C_i)$. Basically, the shape of $H(x,C_i)$ should follow the terms appearing in Eq. 3.46 which are polynomial functions. We consider the following cases in Eq. 3.3:

Case 3.2.a
If H_1 is of the form

$$H_1(x, C_i) = C_1 \tag{3.47}$$

where C_1 is an unknown parameter at this moment, then Eq. 3.3 becomes:

$$F_1'''(x) - H_1(x, C_i) N_0(F_0(x)) = 0 \tag{3.48}$$

Substituting Eqs. 3.46 and 3.47 into Eq. 3.48, we obtain the equation in F_1:

$$F_1'''(x) - 4C_1\alpha \mathrm{Re} x^3 - 4C_1(\alpha \mathrm{Re} + 2\alpha^2)x = 0, \; F_1(0) = F_1'(0) = F_1(1) = 0 \tag{3.49}$$

The solution of Eq. 3.49 is given by

$$F_1(x) = \frac{C_1\alpha\mathrm{Re}}{30}x^6 - \frac{C_1(\alpha\mathrm{Re}+2\alpha^2)}{6}x^4 + \frac{2\alpha\mathrm{Re}+5\alpha^2}{15}C_1x^2 \tag{3.50}$$

Equation 3.4 can be written in the form

$$\begin{aligned} &F_2'''(x) - F_1'''(x) - C_1N_1(F_0,F_1) - H_2(x,C_i)N_0(F_0) = 0,\\ &F_2(0) = F_2'(0) = F_2(1) = 0 \end{aligned} \tag{3.51}$$

where N_1 is obtained from Eq. 2.58:

$$N_1(F_0,F_1) = 2\alpha\mathrm{Re}\left(F_0F_1' + F_0'F_1\right) + 4\alpha^2F_1' \tag{3.52}$$

If we consider

$$H_2(x,C_i) = C_2 \tag{3.53}$$

where C_2 is an unknown parameter, then from Eqs. 3.44, 3.50, 3.51, 3.52 and 3.53 we obtain the following equation in F_2:

$$\begin{aligned} F_2''' = &\left[-4(2\alpha^2+\alpha\mathrm{Re})(C_1+C_2) + \frac{40\alpha^4+36\alpha^3\mathrm{Re}+8\alpha^2\mathrm{Re}^2}{15}C_1^2\right]x+\\ &+\left[4\alpha\mathrm{Re}(C_1+C_2) - \frac{80\alpha^4+120\alpha^3\mathrm{Re}+36\alpha^2\mathrm{Re}^2}{15}C_1^2\right]x^3+\\ &+\frac{24\alpha^3\mathrm{Re}+12\alpha^2\mathrm{Re}^2}{5}C_1^2x^5 - \frac{8\alpha^2\mathrm{Re}^2}{15}C_1^2x^7 \end{aligned} \tag{3.54}$$

so that the solution of Eq. 3.54 is given by

$$\begin{aligned} F_2(x) = &-\frac{\alpha^2\mathrm{Re}^2C_1^2}{1350}x^{10} + \frac{2\alpha^3\mathrm{Re}+\alpha^2\mathrm{Re}^2}{140}C_1^2x^8+\\ &+\left[\frac{\alpha\mathrm{Re}(C_1+C_2)}{30} - \frac{20\alpha^4+30\alpha^3\mathrm{Re}+9\alpha^2\mathrm{Re}^2}{450}C_1^2\right]x^6+\\ &+\left[-\frac{(\alpha\mathrm{Re}+2\alpha^2)(C_1+C_2)}{6} + \frac{2\alpha^2\mathrm{Re}^2+9\alpha^3\mathrm{Re}+10\alpha^4}{90}C_1^2\right]x^4+\\ &+\left[\frac{2\alpha\mathrm{Re}+5\alpha^2}{15}(C_1+C_2) - \frac{1260\alpha^4+900\alpha^3\mathrm{Re}+163\alpha^2\mathrm{Re}^2}{18900}C_1^2\right]x^2 \end{aligned} \tag{3.55}$$

The second-order approximate solution is obtained from the Eq. 3.1

$$\overline{F}(x) = F_0(x) + F_1(x) + F_2(x) \tag{3.56}$$

where F_0, F_1 and F_2 are given by Eqs. 3.44, 3.50 and 3.55 respectively.

Case 3.2.b
In this case we consider

$$H_1(x, C_i) = C_1 \tag{3.57}$$

$$H_2(x, C_i) = C_2 + C_3 x \tag{3.58}$$

where C_1, C_2 and C_3 are unknown parameters. It is clear that the function F_1 is given by Eq. 3.50. Equation 3.4 becomes

$$\begin{aligned} F_2''' = & \left[-4(2\alpha^2 + \alpha\mathrm{Re})(C_1 + C_2) + \frac{40\alpha^4 + 36\alpha^3\mathrm{Re} + 8\alpha^2\mathrm{Re}^2}{15} C_1^2\right] x + \\ & + 4(2\alpha^2 + \alpha\mathrm{Re}) C_3 x^2 + \left[4\alpha\mathrm{Re}(C_1 + C_2) - \right. \\ & \left. - \frac{80\alpha^4 + 120\alpha^3\mathrm{Re} + 36\alpha^2\mathrm{Re}^2}{15} C_1^2\right] x^3 + 4\alpha\mathrm{Re} C_3 x^4 + \\ & + \frac{24\alpha^3\mathrm{Re} + 12\alpha^2\mathrm{Re}^2}{5} C_1^2 x^5 - \frac{8\alpha^2\mathrm{Re}^2}{15} C_1^2 x^7 \end{aligned} \tag{3.59}$$

and has the solution

$$\begin{aligned} F_2(x) = & -\frac{\alpha^2\mathrm{Re}^2 C_1^2}{1350} x^{10} + \frac{\alpha^2\mathrm{Re}^2 + 2\alpha^3\mathrm{Re}}{140} C_1^2 x^8 + \frac{2\alpha\mathrm{Re}}{105} C_3 x^7 + \\ & + \left[\frac{\alpha\mathrm{Re}(C_1 + C_2)}{30} - \frac{9\alpha^2\mathrm{Re}^2 + 30\alpha^3\mathrm{Re} + 20\alpha^4}{450} C_1^2\right] x^6 - \\ & - \frac{\alpha\mathrm{Re} + 2\alpha^2}{15} C_3 x^5 + \left[-\frac{(\alpha\mathrm{Re} + 2\alpha^2)(C_1 + C_2)}{6} + \right. \\ & \left. + \frac{2\alpha^2\mathrm{Re}^2 + 9\alpha^3\mathrm{Re} + 10\alpha^4}{90} C_1^2\right] x^4 + \\ & + \left[\frac{2\alpha\mathrm{Re} + 5\alpha^2}{15}(C_1 + C_2) + \frac{5\alpha\mathrm{Re} + 14\alpha^2}{105} C_3 - \right. \\ & \left. - \frac{1{,}260\alpha^4 + 900\alpha^3\mathrm{Re} + 163\alpha^2\mathrm{Re}^2}{18{,}900} C_1^2\right] x^2 \end{aligned} \tag{3.60}$$

The second-order approximate solution becomes

$$\overline{F}(x) = F_0(x) + F_1(x) + F_2(x) \tag{3.61}$$

where F_0, F_1 and F_2 are given by Eqs. 3.44, 3.50 and 3.60 respectively

Case 3.2.c
In the third case we consider

$$\begin{aligned} H_1(x, C_i) &= C_1 + C_2 x \\ H_2(x, C_i) &= C_3 + C_4 x + C_5 x^2 \end{aligned} \tag{3.62}$$

Equation 3.3 can be written as

$$F_1''' - 4\alpha \text{Re} C_2 x^4 - 4\alpha \text{Re} C_1 x^3 + 4(\alpha \text{Re} + 2\alpha^2) C_2 x^2 + \\ +4(\alpha \text{Re} + 2\alpha^2) C_1 x = 0, \quad F_1(0) = F_1'(0) = F_1(1) = 0 \tag{3.63}$$

From Eq. 3.63 we have

$$F_1(x) = \frac{2\alpha \text{Re} C_2}{105} x^7 + \frac{\alpha \text{Re} C_1}{30} x^6 - \frac{(\alpha \text{Re} + 2\alpha^2) C_2}{15} x^5 - \\ - \frac{(\alpha \text{Re} + 2\alpha^2) C_1}{6} x^4 + \left[\frac{(\alpha \text{Re} + 2\alpha^2)(5C_1 + 2C_2)}{30} - \frac{\alpha \text{Re}(7C_1 + 4C_2)}{210} \right] x^2 \tag{3.64}$$

Equation 3.4 becomes

$$F_2'''(x) = [B_1 C_1 - 4(2\alpha^2 + \alpha \text{Re})(C_1 + C_3)]x + \left[B_1 C_2 - 4\left(2\alpha^2 + \right.\right. \\ \left.\left. +\alpha \text{Re}\right)\left(C_2 + C_4\right]x^2 + [B_3 C_1 + 4\alpha \text{Re}(C_1 + C_3) - 4(2\alpha^2 + \alpha \text{Re}) C_5]x^3 + \\ +[B_4 C_1 + B_3 C_2 + 4\alpha \text{Re}(C_2 + C_4)]x^4 + (B_5 C_1 + B_4 C_2 + 4\alpha \text{Re} C_5)x^5 + \\ +(B_6 C_1 + B_5 C_2)x^6 + (B_7 C_1 + B_6 C_2)x^7 + (B_8 C_1 + B_7 C_2)x^8 + B_8 C_2 x^9, \\ F_2(0) = F_2'(0) = F_2(1) = 0 \tag{3.65}$$

where

$$\begin{aligned} B_1 &= \frac{40\alpha^4 + 36\alpha^3 \text{Re} + 8\alpha^2 \text{Re}^2}{15} C_1 + \frac{112\alpha^4 + 96\alpha^3 \text{Re} + 20\alpha^2 \text{Re}^2}{115} C_2 \\ B_3 &= -\frac{80\alpha^4 + 120\alpha^3 \text{Re} + 36\alpha^2 \text{Re}^2}{15} C_1 - \frac{112\alpha^3 \text{Re} + 40\alpha^2 \text{Re}^2}{105} C_2 \\ B_4 &= -\frac{2(2\alpha^2 + \alpha \text{Re})^2}{3} C_2; B_5 = \frac{12(2\alpha^3 \text{Re} + \alpha^2 \text{Re}^2)}{5} C_1; \\ B_6 &= \frac{6(2\alpha^3 \text{Re} + \alpha^2 \text{Re}^2)}{5} C_2; B_7 = -\frac{8\alpha^2 \text{Re}^2}{15} C_1; B_8 = -\frac{12\alpha^2 \text{Re}^2}{35} C_2 \end{aligned} \tag{3.66}$$

The solution of Eq. 3.65 is given by

$$F_2(x) = \left[\frac{5\alpha^2 + 2\alpha \text{Re}}{15}(C_1 + C_3) + \frac{14\alpha^2 + 5\alpha \text{Re}}{105}(C_2 + C_4) + \right. \\ + \frac{28\alpha^2 + 9\alpha \text{Re}}{420} C_5 - C_1 \left(\frac{B_1}{24} + \frac{B_3}{120} + \frac{B_4}{210} + \frac{B_5}{336} + \frac{B_6}{504} + \frac{B_7}{720} + \frac{B_8}{990} \right) - \\ \left. - C_2 \left(\frac{B_1}{60} + \frac{B_3}{210} + \frac{B_4}{336} + \frac{B_5}{504} + \frac{B_6}{720} + \frac{B_7}{990} + \frac{B_8}{1320} \right) \right] x^2 +$$

$$+\frac{B_1C_1-4(2\alpha^2+\alpha\text{Re})(C_1+C_3)}{24}x^4+\frac{B_1C_2-4(2\alpha^2+\alpha\text{Re})(C_2+C_4)}{60}x^5+$$
$$+\frac{B_3C_1+4\alpha\text{Re}(C_1+C_3)-4(2\alpha^2+\alpha\text{Re})C_5}{120}x^6+$$
$$+\frac{B_4C_1+B_3C_2+4\alpha\text{Re}(C_2+C_4)}{210}x^7+\frac{B_5C_1+B_4C_2+4\alpha\text{Re}C_5}{336}x^8+$$
$$+\frac{B_6C_1+B_5C_2}{504}x^9+\frac{B_7C_1+B_6C_2}{720}x^{10}+\frac{B_8C_1+B_7C_2}{990}x^{11}+\frac{B_8C_2}{1320}x^{12} \tag{3.67}$$

The second-order approximate solution is

$$\overline{F}(x)=F_0(x)+F_1(x)+F_2(x) \tag{3.68}$$

where F_0, F_1, and F_2 are given by Eqs. 3.44, 3.64 and 3.67.

3.2.1 *Numerical Examples*

In what follows we will show that the error of the solution decreases when the number of terms in the optimal auxiliary functions H_1 and H_2 increases.

Example 3.2.1.a For Re = 50 and $\alpha=\pi/36$ in the Case 3.2.a it is obtained the optimal convergence-control parameters:

$$C_1=1.1164009886051645 \quad C_2=-3.1878118970124516$$

such that the second-order approximate solution 3.56 becomes:

$$\overline{F}(x)=1-1.7646483845121845x^2+1.2283825066455862x^4-$$
$$-0.6162400883273966x^6+0.1700828308582972x^8-$$
$$-0.017576864664302104x^{10} \tag{3.69}$$

Example 3.2.1.b For Re = 50 and $\alpha=\pi/36$ in the Case 3.2.b, we obtain

$$C_1=0.32644909312102754, \quad C_2=-2.5573184155107833 5,$$
$$C_3=1.70917159426260534$$

The second-order approximate solution 3.61 in this case, can be written as

$$\begin{aligned}\overline{F}(x) = 1 - 1.773585746050146x^2 + 1.43520979332775x^4 - \\ -0.4989133414941081x^5 - 0.31780151640373655x^6 + \\ +0.1420508181021099x^7 + 0.014542898859664956x^8 - \\ -0.0015029063415338511x^{10}\end{aligned} \quad (3.70)$$

Example 3.2.1.c For Re = 50 and $\alpha = \pi/36$ in the Case 3.2.c, we have

$$\begin{aligned}\overline{F}(x) = 1 - 1.770374066052662x^2 + 1.337949235348047x^4 - \\ -0.203423907087277x^5 - 0.569143539334867x^6 + \\ +0.061972513688921x^7 + 0.177364873860666x^8 - \\ -0.002394488028487x^9 - 0.032243761721675x^{10} + \\ +0.000295658728838x^{11} - 2.519401504 \cdot 10^{-6}x^{12}\end{aligned} \quad (3.71)$$

In order to obtain better approximate solutions, we can decompose the initial domain [0,1] into two sub-domains: [0,1/2] and [1/2,1]. On each of these sub-domains we get a set of optimal values of the parameters C_i, as well as the corresponding approximate solutions. In the following two examples we consider the Case 3.2.c for different values of the parameter Re and α.

Example 3.2.1.d For Re = 50 and $\alpha = \pi/36$ we obtain the second-order approximate solution for $x \in [0,1/2]$ in the form

$$\begin{aligned}\overline{F}(x) = 1 - 1.769702520049971x^2 + 1.289551605410544x^4 + \\ +0.022239471285127x^5 - 0.936849855007278x^6 + \\ +0.225352934921464x^7 + 0.320409083569552x^8 - \\ -0.140811040565325x^9 - 0.025630461284894x^{10} + \\ +0.01738660322047x^{11} - 1.945821112 \cdot 10^{-3}x^{12}\end{aligned} \quad (3.72)$$

For $x \in (1/2,1]$, the second-order approximate solution becomes

$$\begin{aligned}\overline{F}(x) = 1 - 1.7694306056438x^2 + 1.272306346176572x^4 + \\ +0.115419048785007x^5 - 1.138935233525897x^6 + \\ +0.394401822236987x^7 + 0.356109563720727x^8 - \\ -0.267744040640376x^9 + 0.01229535862981x^{10} + \\ +0.0330596193356307x^{11} - 7.481879075 \cdot 10^{-3}x^{12}\end{aligned} \quad (3.73)$$

Example 3.2.1.e For Re = 80 and $\alpha = -\pi/36$ the second-order approximate solution may be written as

$$\begin{aligned}\overline{F}(x) &= 1 - 0.399280142902042x^2 - 0.464804424026152x^4 + \\ &+0.013728467428812x^5 - 0.238509551263266x^6 + \\ &+0.11473626504029x^7 - 0.047048090371222x^8 - \\ &-0.119295799878836x^9 + 0.149339685921657x^{10} + \\ &+0.014813751033227x^{11} - 0.02368016108x^{12}\end{aligned} \tag{3.74}$$

For x ∈ (1/2,1], we obtain the second-order approximate solution in the form

$$\begin{aligned}\overline{F}(x) &= 1 - 0.399289894277253x^2 - 0.475273491097486x^4 + \\ &+0.089792187892071x^5 - 0.437258631641853x^6 + \\ &+0.312069931032005x^7 - 0.028138248428475x^8 - \\ &-0.27033367710964x^9 + 0.208740959524978x^{10} + \\ &+0.033569126430928x^{11} - 0.033878262325275x^{12}\end{aligned} \tag{3.75}$$

It can be seen from Tables 3.3, 3.4, 3.5 and 3.6 that the analytical solutions of Jeffery–Hamel flows obtained by OHAM are very accurate.

The examples presented in this section lead to the very important conclusion that the accuracy of the obtained results is growing along with increasing the number of parameters in the optimal auxiliary functions. However, an excellent accuracy can be obtained if the optimal convergence-control parameters and the approximate solutions are determined on a few sub-domains of the initial domain (in our example we used only two sub-domains).

Some other methods such as DTM, HPM or HAM give a good accuracy, but OHAM is by far the best method delivering faster convergence and better accuracy. In this procedure, iterations are performed in a very simple manner by identifying some coefficients and therefore very good approximations are obtained in few terms.

Table 3.3 The results of the second-order approximate solutions 3.69, 3.70, 3.71, 3.72, 3.73, and numerical solution of $F(x)$ for Re = 50, $\alpha = \pi/36$

x	$\overline{F}(x)$, Eq. 3.69	$\overline{F}(x)$, Eq. 3.70	$\overline{F}(x)$, Eq. 3.71	$\overline{F}(x)$, Eqs. 3.72 and 3.73	Numerical solution
0	1	1	1	1	1
0.1	0.9824757398	0.9824023709	0.9824274588	0.9824312411	0.98243124
0.2	0.9313404708	0.9311747695	0.9312254780	0.9312259703	0.93122597
0.3	0.8506933600	0.8505904572	0.8506194504	0.8506106331	0.85061063
0.4	0.7466883536	0.7467991730	0.7467911543	0.7467908108	0.74679081
0.5	0.6266302803	0.6269125974	0.6269194165	0.6269481807	0.62694818
0.6	0.4979241147	0.4981011627	0.4981872062	0.4982344626	0.49823446
0.7	0.3670653350	0.3667898713	0.3669365083	0.3669663523	0.36696635
0.8	0.2388747293	0.2380421053	0.2381242542	0.2381237516	0.23812375
0.9	0.1161677302	0.1152192724	0.1151522198	0.1151519316	0.11515193
1	1.73472×10^{-16}	1.4962×10^{-16}	1.4772×10^{-16}	3.3946×10^{-13}	0

Table 3.4 Comparison between the OHAM and numerical solutions for Re = 50 and $\alpha = \pi/36$ (error = $|F_{num} - \overline{F}_{app}|$)

x	Error of the solution 3.69	Error of the solution 3.70	Error of the solution 3.71	Error of the solution 3.72 and 3.73
0	0	0	0	0
0.1	4.44×10^{-5}	2.88×10^{-5}	3.78×10^{-6}	1.10×10^{-9}
0.2	1.14×10^{-4}	5.12×10^{-5}	4.91×10^{-7}	3.99×10^{-10}
0.3	8.27×10^{-5}	2.01×10^{-5}	8.82×10^{-7}	3.13×10^{-9}
0.4	1.02×10^{-4}	8.36×10^{-6}	3.44×10^{-7}	8.99×10^{-10}
0.5	3.17×10^{-4}	3.55×10^{-5}	2.87×10^{-5}	7.93×10^{-10}
0.6	3.10×10^{-4}	1.33×10^{-4}	4.72×10^{-5}	2.65×10^{-9}
0.7	9.89×10^{-5}	1.76×10^{-4}	2.98×10^{-5}	2.38×10^{-9}
0.8	7.50×10^{-4}	8.16×10^{-5}	5.04×10^{-7}	1.68×10^{-9}
0.9	1.01×10^{-3}	6.73×10^{-5}	2.89×10^{-7}	1.63×10^{-9}
1	1.73×10^{-16}	1.49×10^{-16}	1.47×10^{-16}	3.39×10^{-13}

Table 3.5 Comparison between Differential Transformation Method (DTM) [11], Homotopy Perturbation Method (HPM) [11], Homotopy Analysis Method [11] and OHAM—Eqs. 3.74 and 3.75 for Re = 80, $\alpha = -\pi/36$

x	$\overline{F}(x)$ (DTM)	$\overline{F}(x)$ (HPM)	$\overline{F}(x)$ (HAM)	$\overline{F}(x)$ (OHAM)	Numerical solution
0	1	1	1	1	1
0.1	0.9959603887	0.9960671874	0.9995960242	0.9959606278	0.9959606278
0.2	0.9832745481	0.9836959424	0.9832755258	0.9832755383	0.9832755383
0.3	0.9601775551	0.9610758773	0.9601798911	0.9601799114	0.96017991139
0.4	0.9235170706	0.9249245156	0.9235215737	0.9235215894	0.9235215894
0.5	0.8684511349	0.8701997697	0.8684588997	0.8684588780	0.86845887772
0.6	0.7880785402	0.7898325937	0.7880910186	0.7880909200	0.78809092032
0.7	0.673248448	0.6745334968	0.6731437690	0.6731436345	0.6731436346
0.8	0.5119644061	0.5128373095	0.5119909939	0.5119911038	0.5119910891
0.9	0.2915280122	0.2918936991	0.2915580178	0.2915587524	0.29155874261
1	0	0	−0.000001149	0	0

Table 3.6 Comparison between OHAM results and numerical solutions for Re = 80, $\alpha = -\pi/36$

X	$\overline{F}(x)$, Eq. 3.74 and 3.75	Numerical solution	Error of the solutions
0	1	1	0
0.1	0.995960627802	0.9959606278	2.60×10^{-12}
0.2	0.983275538304	0.9832755383	4.17×10^{-12}
0.3	0.960179911407	0.96017991139	1.73×10^{-11}
0.4	0.923521589494	0.9235215894	9.46×10^{-11}
0.5	0.868458878024	0.86845887772	3.04×10^{-10}
0.6	0.788090920067	0.78809092032	2.52×10^{-10}
0.7	0.673143634591	0.6731436346	6.49×10^{-12}
0.8	0.511991103837	0.5119910891	1.47×10^{-8}
0.9	0.291558752370	0.29155874261	9.81×10^{-9}
1	0	0	0

The Optimal Homotopy Asymptotic Method is employed to propose new analytic approximate solutions for some nonlinear dynamical systems. This procedure is valid even if the nonlinear differential equations contain any small or large parameters.

3.3 Oscillations of a Mass Attached to a Stretched Wire

In what follows we consider an example of a conservative nonlinear oscillatory system with an irrational elastic term. A mass attached to the centre of a stretching elastic wire, in dimensionless form, has the equation of motion [13–15]:

$$\ddot{u} + u - \frac{\alpha u}{\sqrt{1+u^2}} = 0, \quad 0 < \alpha \le 1 \tag{3.76}$$

subject to initial conditions

$$u(0) = A, \quad \dot{u}(0) = 0 \tag{3.77}$$

The motions corresponding to Eq. 3.76 are periodic [14], the system oscillating between symmetric bounds [−A, A] and the frequency is dependent on the amplitude A. For small values of u, Eq. 3.76 becomes [14]

$$\ddot{u} + (1-\alpha)u \approx 0 \tag{3.78}$$

such that $\omega \approx \sqrt{1-\alpha}$, and for large values of u, it follows that

$$\ddot{u} + u \approx 0 \tag{3.79}$$

and therefore $\omega \approx 1$.

Mickens, using the lowest order harmonic balance method [13, 16], obtained the periodic solution as follows

$$u(t) = A\cos\sqrt{1-\alpha/\sqrt{1+A^2/2}} \tag{3.80}$$

but using this procedure it is very difficult to construct higher-order analytical approximations.

In [13], the approximate analytical solution of Eqs. 3.76 and 3.77 is obtained combining Newton's method with the harmonic balance method, and in [14] is applied the parameter-expansion method.

Under the transformations

$$\begin{aligned} \tau &= \Omega t \\ u(t) &= Ax(\tau) \end{aligned} \tag{3.81}$$

equations 3.76 and 3.77 become

$$\Omega^2 x'' + x - \frac{\alpha x}{\sqrt{1 + A^2 x^2}} = 0 \tag{3.82}$$

respectively

$$x(0) = 1, \quad x'(0) = 0 \tag{3.83}$$

where dot denotes differentiation with respect to t and prime denotes differentiation with respect to τ.

In this case, the linear and nonlinear operators for Eqs. 3.76 and 3.77 are respectively ($g(\tau) = 0$):

$$L(x(\tau)) = \Omega_0^2 \left[x''(\tau) + x(\tau) \right] \tag{3.84}$$

$$\begin{aligned} N[x(\tau), \Omega(\lambda, p)] &= \Omega^2(p) x''(\tau) + (1 + \lambda) x(\tau) - \frac{\alpha x(\tau)}{\sqrt{1 + A^2 x^2(\tau)}} - \\ &- p\lambda x(\tau) - \Omega_0^2 \left[x''(\tau) + x(\tau) \right] \end{aligned} \tag{3.85}$$

where x and Ω are given by Eqs. 2.49, 2.50 respectively and λ is an unknown parameter. We obtain the following three equations from Eqs. 3.2, 3.3, and 3.4:

$$\Omega_0^2 \left(x_0'' + x_0 \right) = 0, \quad x_0(0) = 1, \quad x'(0) = 0 \tag{3.86}$$

$$\begin{aligned} &\Omega_0^2 (x_1'' + x_1) - H_1(\tau, C_i) [\Omega_0^2 x_0'' + (1 + \lambda) x_0 - \\ &- \frac{\alpha x_0}{\sqrt{1 + A^2 x_0^2}}] = 0, \qquad x_1(0) = 0, \quad x_1'(0) = 0 \end{aligned} \tag{3.87}$$

$$\begin{aligned}&\Omega_0^2(x_2''+x_2)-\Omega_0^2(x_1''+x_1)-H_1(\tau,C_i)\big[2\Omega_0\Omega_1x_0''+\Omega_0^2x_1''+(1+\lambda)x_1-\\&-\frac{\alpha x_1}{\left(1+A^2x_0^2\right)^{3/2}}-\lambda x_0\big]-H_2(\tau,C_i)\big[\Omega_0^2x_0''+(1+\lambda)x_0-\\&-\frac{\alpha x_0}{\sqrt{1+A^2x_0^2}}\big]=0,\qquad x_2(0)=0,x_2'(0)=0\end{aligned}\tag{3.88}$$

Equation 3.86 has the following solution:

$$x_0(\tau)=\cos\tau\tag{3.89}$$

If this result is substituted into Eq. 3.87, and if we choose $H_1(\tau,C_i)=C_1$ we obtain the following equation:

$$\begin{aligned}&\Omega_0^2(x_1''+x_1)-C_1\big[\left(1-\Omega_0^2+\lambda-a_1\alpha\right)\cos\tau-\alpha a_3\cos 3\tau-\\&-\alpha a_5\cos 5\tau-\alpha a_7\cos 7\tau-\alpha a_9\cos 9\tau\ldots\big]=0\end{aligned}\tag{3.90}$$

where the term $\cos\tau\left(1+A^2\cos^2\tau\right)^{-1/2}$ was expanded into a Fourier series:

$$\frac{\cos\tau}{\sqrt{1+A^2\cos^2\tau}}=\sum_{k=0}^{\infty}a_{2k+1}\cos(2k+1)\tau\tag{3.91}$$

and

$$a_{2k+1}=\frac{4}{\pi}\int_0^{\pi/2}\frac{\cos x\cos(2k+1)x}{\sqrt{1+A^2\cos^2x}}dx\tag{3.92}$$

and C_1 is an unknown parameter in this moment. Avoiding the presence of a secular term needs:

$$\Omega_0^2=1+\lambda-a_1\alpha\tag{3.93}$$

With this requirement, the solution of Eq. 3.90 is

$$\begin{aligned}x_1(\tau,C_1)=&\frac{C_1\alpha a_3}{8\Omega_0^2}(\cos 3\tau-\cos\tau)+\frac{C_1\alpha a_5}{24\Omega_0^2}(\cos 5\tau-\cos\tau)+\\&+\frac{C_1\alpha a_7}{48\Omega_0^2}(\cos 7\tau-\cos\tau)+\frac{C_1\alpha a_9}{80\Omega_0^2}(\cos 9\tau-\cos\tau)\end{aligned}\tag{3.94}$$

If we substitute Eqs. 3.91, 3.93 and 3.94 into Eq. 3.88 and if we expand the term $\frac{x_1}{\left(1+A^2x_0^2\right)^{3/2}}$ into a Fourier series:

$$\frac{x_1(\tau)}{\sqrt{1+A^2x_0^2}^{\,3}} = \frac{C_1\alpha}{8\Omega_0^2}\left[\sum_{k=0}^{\infty} b_{2k+1}\cos(2k+1)\tau\right] \tag{3.95}$$

where

$$\begin{aligned} b_{2k+1} &= \frac{2}{\pi}\int_0^{\frac{\pi}{2}} \frac{f(x)\cos(2k+1)x}{\left(1+A^2\cos^2 x\right)^{3/2}}dx, \\ f(x) &= a_3(\cos 3x - \cos x) + \frac{a_5}{3}(\cos 5x - \cos x) + \\ &+ \frac{a_7}{6}(\cos 7x - \cos x) + \frac{a_9}{10}(\cos 9x - \cos x) \end{aligned} \tag{3.96}$$

then we obtain the following equation in x_2, in the case that $H_2(\tau, C_i) = C_2 + 2C_3\cos 2\tau$:

$$\begin{aligned} \Omega_0^2(x_2'' + x_2) = &-\left[2\Omega_0\Omega_1 + \frac{C_1 a_1\alpha^2}{8\Omega_0^2}\left(a_3 + \frac{a_5}{3} + \frac{a_7}{6} + \frac{a_9}{10}\right) + \alpha a_3 C_3 + \lambda + \right. \\ &\left. + b_1\frac{C_1\alpha^2}{8\Omega_0^2}\right]\cos\tau + H.O.T \end{aligned} \tag{3.97}$$

where H.O.T. stands for higher order terms.

No secular terms in $x_2(\tau)$ requires that

$$\Omega_1 = -\frac{\lambda}{2\Omega_0} - \frac{C_1\alpha^2 K}{16\Omega_0^2} - \frac{\alpha a_3 C_3}{2C_1\Omega_0} \tag{3.98}$$

where

$$K = b_1 + a_1\left(a_3 + \frac{a_5}{3} + \frac{a_7}{6} + \frac{a_9}{10}\right). \tag{3.99}$$

From Eqs. 2.52, 3.93 and 3.98, we obtain the frequency in the form

$$\Omega = \Omega_0 - \frac{\lambda}{2\Omega_0} - \frac{C_1\alpha^2 K}{16\Omega_0^3} - \frac{\alpha a_3 C_3}{2C_1\Omega_0} \tag{3.100}$$

where Ω_0 is given by Eq. 3.93.

The parameter λ can be determined applying the "principle of minimal sensitivity". From Eq. 2.59 it holds that

$$\lambda = -\frac{3C_1\alpha^2K}{8\Omega_0^2} - \frac{2\alpha a_3C_3}{C_1} \tag{3.101}$$

This result is substituted into Eqs. 3.100 and 3.93 and we find the following:

$$\Omega = \Omega_0 + \frac{C_1\alpha^2K}{8\Omega_0^3} + \frac{\alpha a_3C_3}{2C_1\Omega_0}$$
$$\Omega_0^2 = \frac{1-a_1\alpha}{2} - \frac{\alpha a_3C_3}{C_1} + \sqrt{\left(\frac{1-a_1\alpha}{2} - \frac{\alpha a_3C_3}{C_1}\right)^2 - \frac{3}{8}C_1\alpha^2K} \tag{3.102}$$

Substituting Eqs. 3.89 and 3.94 into 3.88 one retrieves:

$$\begin{aligned}
&\Omega_0^2(x_2'' + x_2) = \big[-\alpha(C_1+C_2)a_3 - \alpha C_3a_5 - \frac{\alpha^2C_1^2}{8\Omega_0^2}(a_1a_3+b_3)+\\
&+a_3C_1^2\big]\cos 3\tau + \big[-\alpha a_5(C_1+C_2) - \alpha(a_3+a_7)C_3 - \frac{\alpha^2C_1^2}{24\Omega_0^2}(a_1a_5+\\
&+b_5) + a_5C_1^2\big]\cos 5\tau + \big[-\alpha a_7(C_1+C_2) - \alpha(a_5+a_9)C_3-\\
&-\frac{\alpha^2C_1^2}{48\Omega_0^2}(a_1a_7+b_7) + a_7C_1^2\big]\cos 7\tau + \big[-\alpha a_1(C_1+C_2) - \alpha a_7C_3-\\
&-\frac{\alpha^2C_12}{80\Omega_0^2}(a_1a_9+b_9) + a_9C_1^2\big]\cos 9\tau - \alpha a_9C_3\cos 11\tau
\end{aligned} \tag{3.103}$$

The above equation has the solution

$$\begin{aligned}
&x_2(\tau, C_i) = \left[\frac{\alpha a_3(C_1+C_2) - a_3C_1^2 + \alpha a_5C_3}{8\Omega_0^2}+\right.\\
&\left.+\frac{\alpha^2(a_1a_3+b_3)C_1^2}{64\Omega_0^4}\right](\cos 3\tau - \cos\tau) + \left[\frac{\alpha a_5(C_1+C_2) - a_5C_1^2}{24\Omega_0^2}+\right.\\
&\left.+\frac{\alpha(a_5+a_7)C_3}{24\Omega_0^2} + \frac{\alpha(a_1a_5+b_5)C_1^2}{576\Omega_0^4}\right](\cos 5\tau - \cos\tau)+\\
&+\left[\frac{\alpha a_7(C_1+C_2) - a_7C_1^2 + \alpha(a_5+a_9)C_3}{48\Omega_0^2} + \frac{\alpha^2(a_1a_7+b_7)C_1^2}{2304\Omega_0^4}\right](\cos 7\tau-\\
&-\cos\tau) + \left[\frac{\alpha a_9(C_1+C_2) - a_9C_1^2 + \alpha a_7C_3}{80\Omega_0^2}+\right.\\
&\left.+\frac{\alpha^2(a_1a_9+b_9)C_1^2}{6400\Omega_0^4}\right](\cos 9\tau - \cos\tau) + \frac{\alpha a_9C_3}{120\Omega_0^2}(\cos 11\tau - \cos\tau)
\end{aligned} \tag{3.104}$$

The second-order approximate solution is

$$\overline{x}(\tau, C_i) = x_0(\tau) + x_1(\tau, C_1) + x_2(\tau, C_1, C_2, C_3) \tag{3.105}$$

where x_0, x_1 and x_2 are given by Eqs. 3.89, 3.94 and 3.104 respectively. Using the transformations 3.81, the second-order approximate solution of Eqs. 3.76 and 3.77 is given by

$$\overline{u}(\tau, C_1, C_2, C_3) = A\cos\Omega t + \sum_{k=1}^{5} D_{2k+1}[\cos(2k+1)\Omega t - \cos\Omega t] \tag{3.106}$$

where Ω is given by Eq. 3.102 and

$$\begin{aligned}
D_3 &= \frac{\alpha a_3(2C_1 + C_2) - a_3 C_1^2 + \alpha a_5 C_3}{8\Omega_0^2} + \frac{\alpha^2(a_1 a_3 + b_3)C_1^2}{64\Omega_0^4} \\
D_5 &= \frac{\alpha a_5(2C_1 + C_2) - a_5 C_1^2 + \alpha(a_3 + a_7)C_3}{24\Omega_0^2} + \frac{\alpha^2(a_1 a_5 + b_5)C_1^2}{576\Omega_0^4} \\
D_7 &= \frac{\alpha a_7(2C_1 + C_2) - a_7 C_1^2 + \alpha(a_5 + a_9)C_3}{48\Omega_0^2} + \frac{\alpha^2(a_1 a_7 + b_7)C_1^2}{2304\Omega_0^4} \\
D_9 &= \frac{\alpha a_9(2C_1 + C_2) - a_9 C_1^2 + \alpha a_7 C_3}{80\Omega_0^2} + \frac{\alpha^2(a_1 a_9 + b_9)C_1^2}{6400\Omega_0^4}, D_{11} = \frac{\alpha a_9 C_3}{120\Omega_0^2}
\end{aligned} \tag{3.107}$$

Substitution of Eq. 3.106 into Eq. 3.76 results in a residual, which reads

$$R(t, C_1, C_2, C_3) = \ddot{\overline{u}} + \overline{u} - \frac{\alpha\overline{u}}{\sqrt{1 + u^2}} \tag{3.108}$$

3.3.1 Numerical Examples

As numerical examples we consider the following four cases:

Case 3.3.1.a) In the first case, for A = 1, α = 1/2, we obtain

$$C_1 = -0.874427354209, C_2 = 2.195545432429, C_3 = -0.017262473399,$$
$$\Omega = 0.786117117465$$

The second-order approximate solution given by Eq. 3.106 can be written 3.109

$$\begin{aligned}
\overline{u}(t) &= 0.993832153\cos\Omega t + 0.006394109\cos 3\Omega t - 0.000242361\cos 5\Omega t + \\
&+ 0.000017735\cos 7\Omega t - 1.61\times 10^{-6}\cos 9\Omega t - 1.96\times 10^{-8}\cos 11\Omega t
\end{aligned} \tag{3.109}$$

Case 3.3.1.b) If we consider A = 1, α = 3/4, it is obtained

$$C_1 = 0.488692092553, C_2 = -1.781222855475, C_3 = -0.043204882202, \\ \Omega = 0.652737820378$$

$$\overline{u}(t) = 0.986472035 \cos \Omega t + 0.013933104 \cos 3\Omega t - 0.000434722 \cos 5\Omega t + \\ +0.0000327201 \cos 7\Omega t - 3.03 \times 10^{-6} \cos 9\Omega t - 1.06 \times 10^{-7} \cos 11\Omega t \tag{3.110}$$

Case 3.3.1.c) For A = 10, α = 1/2, yields

$$C_1 = -1.189248262547, C_2 = -8.354987200868, C_3 = 0.003959841496, \\ \Omega = 0.968053752342$$

$$\overline{u}(t) = 9.970036132 \cos \Omega t + 0.035129406 \cos 3\Omega t - 0.006356285 \cos 5\Omega t + \\ +0.002015834 \cos 7\Omega t - 0.000825248 \cos 9\Omega t + 1.6 \times 10^{-7} \cos 11\Omega t \tag{3.111}$$

Case 3.3.1.d) In the last case, for A = 10, α = 3/4, we obtain

$$C_1 = -1.043677214759, C_2 = -10.120518622724, C_3 = 0.035477178464, \\ \Omega = 0.951164010235$$

$$\overline{u}(t) = 9.953108979 \cos \Omega t + 0.055028624 \cos 3\Omega t - 0.0100177009 \cos 5\Omega t + \\ +0.0031755102 \cos 7\Omega t - 0.001297652 \cos 9\Omega t + 2.23 \times 10^{-6} \cos 11\Omega t \tag{3.112}$$

Figures 3.6, 3.7, 3.8 and 3.9 show the comparison between the present solution and the numerical integration results obtained using a fourth-order Runge–Kutta method. From these figures it can be seen that the solution obtained through OHAM is nearly identical with that given by the numerical method. Also, the principle of minimal sensitivity improves the convergence of the frequency and approximate solution.

3.4 The Motion of a Particle on a Rotating Parabola

In this section we consider the motion of a ring of mass m sliding freely on the wire described by the parabola $z = qx^2$ which rotates with a constant angular velocity ω about the z-axis as shown in Fig. 3.10 [16–18]. Using the Euler-Lagrange equations, the motion of this particle is given by the equation:

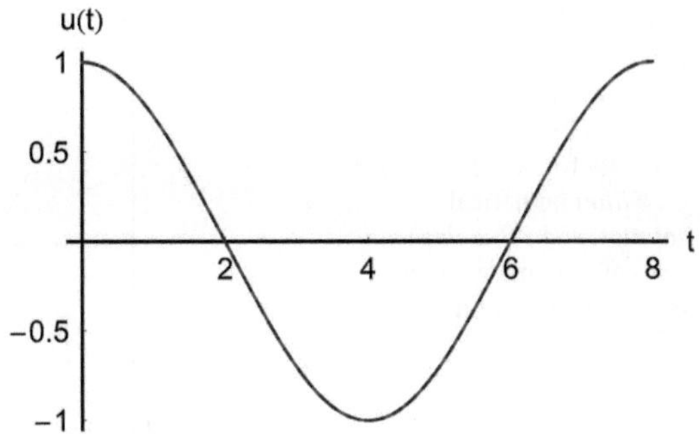

Fig. 3.6 Comparison between the approximate solution 3.109 and numerical solution of Eq. 3.94 for A = 1, α = 1/2: (*red line*) numerical solution and (*blue dashed line*) approximate solution (colour figure online)

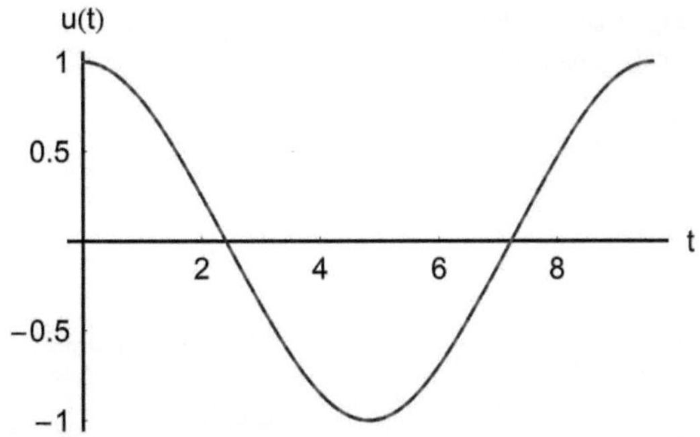

Fig. 3.7 Comparison between the approximate solution 3.110 and numerical solution of Eq. 3.94 for A = 1, α = 3/4: (*red line*) numerical solution and (*blue dashed line*) approximate solution (colour figure online)

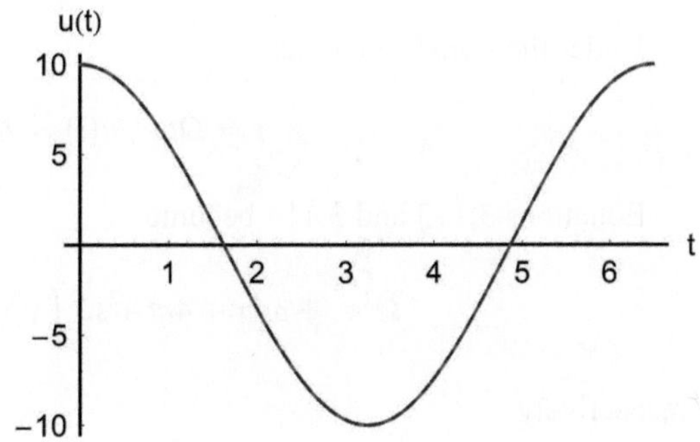

Fig. 3.8 Comparison between the approximate solution 3.111 and numerical solution of Eq. 3.94 for A = 10, α = 1/2: (*red line*) numerical solution and (*blue dashed line*) approximate solution (colour figure online)

$$\left(1+4q^2u^2\right)\frac{d^2u}{dt^2}+\Lambda u+4q^2\left(\frac{du}{dt}\right)^2u=0 \tag{3.113}$$

with the boundary conditions:

$$u(0)=A,\quad \frac{du}{dt}(0)=0 \tag{3.114}$$

where q, Λ and A are known constants and need not to be small.

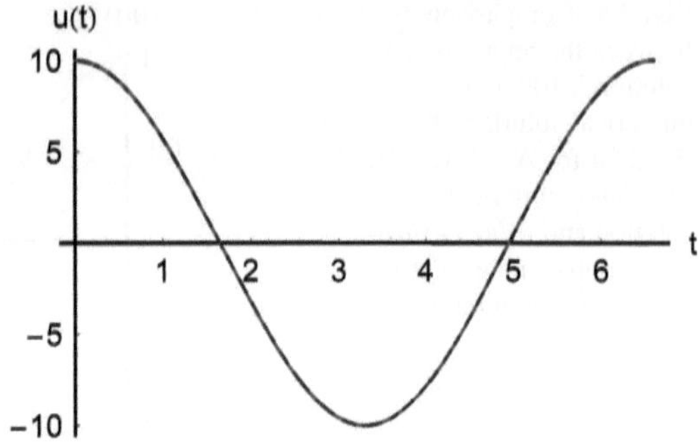

Fig. 3.9 Comparison between the approximate solution 3.112 and numerical solution of Eq. 3.94 for A = 1, α = 3/4: (*red line*) numerical solution and (*blue dashed line*) approximate solution (colour figure online)

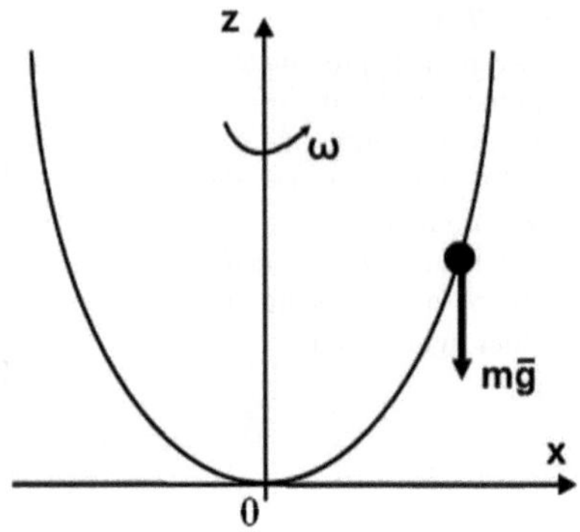

Fig. 3.10 Particle on a rotating parabola

Under the transformations

$$\tau = \Omega t, \quad u(t) = Ax(\tau) \tag{3.115}$$

Equations 3.113 and 3.114 become

$$\Omega^2 x'' + \omega_0^2 x + 4q^2 A^2 \Omega^2 \left(x^2 x'' + xx'^2 \right) = 0 \tag{3.116}$$

respectively

$$x(0) = 1, \quad x'(0) = 0 \tag{3.117}$$

where $\Lambda = \omega_0^2$ and $' = \frac{d}{d\tau}$.

The operators 2.45 and 2.46 are respectively:

$$L(x(\tau)) = \Omega_0^2 \left[x''(\tau) + x(\tau) \right] \tag{3.118}$$

$$N[x(\tau),\Omega(\lambda,p)] = \Omega^2(p)x''(\tau) + (\omega_0^2+\lambda)x(\tau)+ \\ +4q^2A^2\Omega^2(p)\left[x^2(\tau)x''(\tau)+x(\tau)x'^2(\tau)\right] - p\lambda x(\tau) - \Omega_0^2\left[x''(\tau)+x(\tau)\right] \tag{3.119}$$

where x and Ω are given by Eq. 2.49 and 2.50, respectively, and λ is an unknown parameter. From Eqs. 3.2, 3.3 and 3.4, we obtain the following three equations:

$$\Omega_0^2\left(x_0''+x_0\right)=0,\quad x_0(0)=1,\quad x'(0)=0 \tag{3.120}$$

$$\Omega_0^2(x_1''+x_1) - H_1(\tau,C_i)\left[\Omega_0^2x_0''+(\omega_0^2+\lambda)x_0+ \\ +4q^2A^2\Omega_0^2(x_0x_0''+x'{}_0^2x_0)\right]=0,\quad x_1(0)=x_1'(0)=0 \tag{3.121}$$

$$\Omega_0^2(x_2''+x_2)-\Omega_0^2(x_1''+x_1)-H_1(\tau,C_i)\{2\Omega_0\Omega_1x_0''+\Omega_0^2x_1''+(\omega_0^2+\lambda)x_1+ \\ +4q^2A^2\left[\Omega_0^2\left(2x_0x_0''x_1+x_0^2x_1''+2x_0x_0'x_1'+2x_0x_0'x_1'+x_0'^2x_1\right)+ \\ +2\Omega_0\Omega_1\left(x_0^2x_0''+x_0'^2x_0\right)\right]-\lambda x_0\}-H_2(\tau,C_j)\left[\Omega_0^2x_0''+(\omega_0^2+\lambda)x_0+ \\ +4q^2A^2\Omega_0^2\left(x_0^2x_0''+x_0'^2x_0\right)\right]=0,\quad x_2(0)=x_2'(0)=0 \tag{3.122}$$

Equation 3.120 has the following solution:

$$x_0(\tau) = \cos\tau \tag{3.123}$$

If this result is substituted into Eq. 3.121 and assuming that $H_I = C_I =$ constant, we obtain the following equation:

$$\Omega_0^2(x_1''+x_1) - C_1\left[(\omega_0^2+\lambda-\Omega_0^2-2q^2A^2\Omega_0^2)\cos\tau- \\ -2q^2A^2\Omega_0^2\cos 3\tau\right]=0,\quad x_1(0)=x_1'(0)=0 \tag{3.124}$$

where C_I is an unknown parameter at this moment. Avoiding the presence of a secular term needs:

$$\Omega_0^2 = \frac{\omega_0^2+\lambda}{1+2q^2A^2} \tag{3.125}$$

With this requirement, the solution of Eq. 3.124 is:

$$x_1(\tau) = \frac{1}{4}C_1q^2A^2(\cos 3\tau - \cos\tau) \tag{3.126}$$

If we substitute Eqs. 3.123, 3.125 and 3.126 into Eq. 3.122, we obtain the equation in x_2:

$$\begin{aligned}
&\Omega_0^2\left(x_2''+x_2\right)+\frac{2C_1q^2A^2\left(\omega_0^2+\lambda\right)}{1+2q^2A^2}\cos 3\tau+\\
&+C_1\left\{\left[\frac{C_1q^4A^4\left(\omega_0^2+\lambda\right)}{2\left(1+2q^2A^2\right)}+2\Omega_0\Omega_1\left(1+2q^2A^2\right)+\lambda\right]\cos\tau+\right.\\
&+\left[\frac{\left(\omega_0^2+\lambda\right)C_1q^2A^2(3q^2a^2+16)}{2\left(1+2q^2A^2\right)}+2\Omega_0\Omega_1q^2A^2\right]\cos 3\tau+\\
&\left.+\frac{9C_1q^4A^4\left(\omega_0^2+\lambda\right)}{2\left(1+2q^2A^2\right)}\cos 5\tau\right\}+H_2\left(\tau,C_j\right)\left[\frac{2q^2A^2\left(\omega_0^2+\lambda\right)}{1+2q^2A^2}\cos 3\tau\right]=0,\\
&x_2(0)=x_2'(0)=0
\end{aligned}\tag{3.127}$$

No secular term in $x_2(\tau)$ requires that

$$2\Omega_0\Omega_1=-\frac{\lambda}{1+2q^2a^2}-\frac{C_1q^4A^4\left(\omega_0^2+\lambda\right)}{2\left(1+2q^2A^2\right)^2}\tag{3.128}$$

From Eqs. 3.128 and 2.52, we obtain the frequency in the form:

$$\Omega=\Omega_0-\frac{\lambda}{\Omega_0\left(1+2q^2A^2\right)}-\frac{C_1q^4A^4\Omega_0}{4\left(1+2q^2A^2\right)}\tag{3.129}$$

where Ω_0 is given by Eq. 3.125.

The parameter λ can be determined applying the "principle of minimal sensitivity" and thus we obtain

$$\lambda=\frac{C_1\omega_0q^4A^4}{2+4q^2A^2-C_1q^4A^4}\tag{3.130}$$

This result is substituted into Eq. 3.129 and we have:

$$\Omega=\frac{\omega_0}{1+2q^2A^2}\sqrt{1+2q^2A^2-\frac{1}{2}C_1q^4A^4}\tag{3.131}$$

Substituting Eqs. 3.129, 3.130 and 3.131 into Eq. 3.127, we obtain:

$$\begin{aligned}
&x_2''+x_2+2C_1q^2A^2\cos 3\tau+\frac{C_1^2q^2A^2\left(5q^4A^4+7q^2A^2+2\right)}{1+2q^2A^2}\cos 3\tau+\\
&+\frac{9}{2}C_1^2q^4A^4\cos 5\tau+2H_2\left(\tau,C_j\right)q^2A^2\cos 3\tau=0,\qquad x_2(0)=x_2'(0)=0
\end{aligned}\tag{3.132}$$

As we shown, there are many possibilities to choose the function $H_2(\tau,C_j)$. The convergence of the solution $x_2(\tau)$ and consequently the convergence of the approximate solution $\bar{x}(\tau)$ depend on the optimal auxiliary function $H_2(\tau,C_j)$. Basically, the shape of $H_2(\tau,C_j)$ must follow the terms appearing in Eq. 3.127, which are $\cos\tau$, $\cos 3\tau$, $\cos 5\tau$ (odd-order harmonics). Therefore we try to choose $H_2(\tau,C_j)$ so that in Eq. 3.127 the product

$$H_2\left[\frac{2q^2A^2(\omega_0^2+\lambda)}{1+2q^2A^2}\cos 3\tau\right]$$

be of the same shape with the other terms (a combination of functions $\cos\tau$, $\cos 3\tau$, $\cos 5\tau \ldots$).

All three cases presented in this section demonstrate the importance of the function $H_2(\tau,C_j)$ on the accuracy of the solution. In the same time, a larger number of constants in $H_2(\tau,C_j)$ lead to a better accuracy of the results. If the error obtained using a certain $H_2(\tau,C_j)$ is unsatisfactory, one can choose other shapes for this function.

We will consider three cases:

Case 3.4.a
We consider the function H_2 of the form:

$$H_2(\tau,C_j)=C_2' \tag{3.133}$$

where C_2' is an unknown parameter.

Substituting Eq. 3.133 into Eq. 3.132, we obtain the equation in x_2:

$$\begin{gathered}x_2''+x_2+\left[2(C_1+C_2')q^2A^2+\frac{C_1^2q^2A^2(5q^4A^4+7q^2A^2+2)}{1+2q^2A^2}\right]\cos 3\tau+\\+\frac{9}{2}C_1^2q^4A^4\cos 5\tau=0,\quad x_2(0)=x_2'(0)=0\end{gathered} \tag{3.134}$$

The solution of Eq. 3.134 becomes:

$$\begin{gathered}x_2(\tau)=\left[\frac{C_1+C_2'}{4}+\frac{C_1^2q^2a^2(5q^4A^4+7q^2A^2+2)}{8(1+2q^2A^2)}\right](\cos 3\tau-\\-\cos\tau)+\frac{3}{16}C_1^2q^4A^4(\cos 5\tau-\cos\tau)\end{gathered} \tag{3.135}$$

The second-order approximate solution is

$$\bar{x}(\tau)=x_0(\tau)+x_1(\tau)+x_2(\tau)$$

where x_0, x_1 and x_2 are given by Eqs. 3.123, 3.126 and 3.135. Using the transformations 3.115, the second-order approximate solution of Eq. 3.113 becomes

$$\overline{u}(t) = M\cos\Omega t + N\cos 3\Omega t + P\cos 5\Omega t \tag{3.136}$$

where Ω is given by Eq. 3.131 and

$$\begin{aligned} M &= A - \frac{2C_1 + C_2'}{4}q^2A^3 - \frac{C_1^2q^2A^3(16q^4A^4 + 17q^2A^2 + 4)}{16(1 + 2q^2A^2)} \\ N &= \frac{2C_1 + C_2'}{4}q^2A^3 + \frac{C_1^2q^2A^3(5q^4A^4 + 7q^2A^2 + 2)}{8(1 + 2q^2A^2)} \\ P &= \frac{3}{16}C_1^2q^4A^5 \end{aligned} \tag{3.137}$$

Case 3.4.b
We consider the function $H_2(\tau,C_j)$ if the form

$$H_2(\tau, C_j) = C_2 + C_3\cos 2\tau + C_4\cos 4\tau \tag{3.138}$$

where C_2, C_3 and C_4 are unknown parameters.

Substituting Eq. 3.138 into Eq. 3.132 and avoiding the presence of a secular term, we obtain:

$$C_4 = -C_3 \tag{3.139}$$

respectively:

$$\begin{aligned} &x_2'' + x_2 + \left[2(C_1 + C_2)q^2A^2 + \frac{C_1^2q^2A^2(5q^4A^4 + 7q^2A^2 + 2)}{1 + 2q^2A^2}\right]\cos 3\tau + \\ &+\left[\frac{9}{2}C_1^2q^4A^4 + C_3q^2A^2\right]\cos 5\tau - C_3q^2A^2\cos 7\tau = 0 \\ &x_2(0) = x_2'(0) = 0 \end{aligned} \tag{3.140}$$

With these requirements, the solution of Eq. 3.140 becomes:

$$\begin{aligned} x_2(\tau) = &\left[\frac{C_1 + C_2}{4}q^2A^2 + \frac{C_1^2q^2A^2(5q^4A^4 + 7q^2A^2 + 2)}{8(1 + 2q^2A^2)}\right](\cos 3\tau - \\ &-\cos\tau) + \left[\frac{3}{16}C_1^2q^4A^4 + \frac{1}{24}C_3q^2A^2\right](\cos 5\tau - \cos\tau) - \\ &-\frac{1}{48}C_3q^2A^2(\cos 7\tau - \cos\tau) \end{aligned} \tag{3.141}$$

The second-order approximate solution in this case is:

$$\overline{u} = \overline{M}\cos\Omega t + \overline{N}\cos 3\Omega t + \overline{P}\cos 5\Omega t + \overline{Q}\cos 7\Omega t \tag{3.142}$$

where Ω is given by Eq. 3.131 and the coefficients are:

$$\overline{M} = A - \frac{2C_1 + C_2}{4} q^2 A^3 - \\ - \frac{C_1^2 q^2 A^3 (16q^4 A^4 + 17q^2 A^2 + 4)}{16(1 + 2q^2 A^2)} - \frac{1}{48} C_3 q^2 A^3 \\ \overline{N} = \frac{2C_1 + C_2}{4} q^2 A^3 + \frac{C_1^2 q^2 A^3 (5q^4 A^4 + 7q^2 A^2 + 2)}{8(1 + 2q^2 A^2)} \\ \overline{P} = \frac{3}{16} C_1^2 q^4 A^5 + \frac{1}{24} C_3 q^2 A^3 \\ \overline{Q} = -\frac{1}{48} C_3 q^2 A^3 \tag{3.143}$$

Case 3.4.c

We consider the function $H_2(\tau, C_j)$ of the form:

$$H_2(\tau, C_j) = C_2^* + C_3^* \cos 2\tau + C_4^* \cos 4\tau + C_5^* \cos 6\tau + C_6^* \cos 8\tau \\ + C_7^* \cos 10\tau \tag{3.144}$$

where C_2^*, C_3^*, C_4^*, C_5^*, C_6^* and C_7^* are unknown parameters.

Substituting Eq. 3.143 into Eq. 3.132, we obtain:

$$C_4^* = -C_3^* \tag{3.145}$$

and then:

$$x_2(\tau) = \left[\frac{2C_1 + 2C_2^* + C_5^*}{8} q^2 A^2 + \right. \\ \left. + \frac{C_1^2 q^2 A^2 (5q^4 A^4 + 7q^2 A^2 + 2)}{8(1 + 2q^2 A^2)} \right] (\cos 3\tau - \cos\tau) + \left[\frac{3}{16} C_1^2 q^4 A^5 + \right. \\ \left. + \frac{1}{24} (C_3^* + C_6^*) q^2 A^2 \right] (\cos 5\tau - \cos\tau) + \frac{1}{48} (C_3^* - C_7^*) q^2 A^2 (\cos\tau - \\ - \cos 7\tau) + \frac{1}{80} C_5^* q^2 A^2 (\cos 9\tau - \cos\tau) + \frac{C_6^*}{120} q^2 A^2 (\cos 11\tau - \cos\tau) + \\ + \frac{C_7^* q^2 A^2}{168} (\cos 13\tau = \cos\tau) \tag{3.146}$$

The second-order approximate solution of Eq. 3.113 becomes:

$$\overline{u}(t) = M^* \cos \Omega t + N^* \cos 3\Omega t + P^* \cos 5\Omega t + \\ + Q^* \cos 7\Omega t + R^* \cos 9\Omega t + S^* \cos 11\Omega t + T^* \cos 13\Omega t \tag{3.147}$$

where

$$\begin{aligned}
M^* &= A - \frac{q^2A^3}{240}\left(120C_1 + 60C_2^* + 5C_3^* + 3C_4^* + 32C_5^* + 10C_6^* + 45/7C_7^*\right) - \\
&- \frac{C_1^2q^2A^3\left(16q^4A^4 + 17q^2A^2 + 4\right)}{16\left(1 + 2q^2A^2\right)} \\
N^* &= \frac{4C_1 + 2C_2^* + C_5^*}{8}q^2A^3 + \frac{C_1^2q^2A^3\left(5q^4A^4 + 7q^2A^2 + 2\right)}{8\left(1 + 2q^2A^2\right)} \\
P^* &= \frac{C_3^* + C_6^*}{24}q^2A^3 + \frac{3}{16}C_1^2q^4A^5 \\
Q^* &= -\frac{1}{48}(C_3^* - C_7^*)A^3q^2 \\
R^* &= \frac{1}{80}C_4^*q^2A^3,\ S^* = \frac{1}{120}C_5^*q^2A^3,\ T* = \frac{1}{168}C_7^*q^2A^3
\end{aligned} \tag{3.148}$$

3.4.1 Numerical Examples

We will show through six numerical examples that the error of the solutions decreases when the number of terms in the auxiliary function $H(\tau,p)$ increases. In Eqs. 3.113 and 3.114, we consider $\Lambda = \omega_0 = 1, A = 1$ and two cases for q in every of the Cases 3.4.a, 3.4.b and 3.4.c. The optimal values of the optimal convergence-control parameters C_i are obtained using the collocation method.

3.4.1.a) For $q = 1$ in the Case 3.4.a, it is obtained:

$$C_1 = -0.401483291,\ C_2' = -0.065781508$$

The second-order approximate solution 3.136 becomes in this case:

$$\begin{aligned}
\overline{u}(t) = {} & 1.092937297\cos\Omega t - 0.123160203\cos 3\Omega t \\
& + 0.030222906\cos 5\Omega t
\end{aligned} \tag{3.149}$$

where Ω is obtained from Eq. 3.131: $\Omega = 0.596353888$.
3.4.1.b) For $q = 1$ in the Case 3.4.b, it is obtained

$$C_1 = -0.398431527;\ C_2 = -0.052485317;\ C_3 = 0.0341786762$$

$$\begin{aligned}
\overline{u}(t) = {} & 1.089257032\cos\Omega t - 0.119734278\cos 3\Omega t + \\
& + 0.031189301\cos 5\Omega t - 0.00712055\cos 7\Omega t
\end{aligned} \tag{3.150}$$

where $\Omega = 0.596211722$.

3.4.1.c) For $q = 1$ in the Case 3.4.c, we obtain the following results

$$C_1 = -0.395739605284;\ C_2^* = 0.058122467034;\ C_3^* = 0.323248880814;$$
$$C_4^* = 0.157197995091;\ C_5^* = -0.056136143239;\ C_6^* = -0.457745502123;$$
$$C_7^* = 0.011563495249$$

$$\begin{aligned}\bar{u}(t) &= 1.080167589\cos\Omega t - 0.099000466\cos 3\Omega t + 0.023760318\cos 5\Omega t - \\ &-0.006493445\cos 7\Omega t + 0.001964974\cos 9\Omega t - 0.000467801\cos 11\Omega t + \\ &+0.0000688303\cos 13\Omega t\end{aligned} \tag{3.151}$$

where $\Omega = 0.596086291$.

3.4.1.d) For $q = 2$ in the Case 3.4.a we obtain

$$C_1 = -0.167434521,\ C_2' = -0.002382096$$

$$\begin{aligned}\bar{u}(t) = &1.081827344\cos\Omega t - 0.165930301\cos 3\Omega t \\ &+ 0.084102956\cos 5\Omega t\end{aligned} \tag{3.152}$$

where $\Omega = 0.357278398$.

3.4.1.e) For $q = 2$ in the Case 3.4.b it is obtained

$$C_1 = -0.164357411;\ C_2 = 0.017447955;\ C_3 = -0.073610524$$

$$\begin{aligned}\bar{u}(t) &= 1.071279366\cos\Omega t - 0.146185231\cos 3\Omega t + \\ &+0.068771654\cos 5\Omega t + 0.00613421\cos 7\Omega t\end{aligned} \tag{3.153}$$

where $\Omega = 0.356852829$.

3.4.1.f) For $q = 2$ in the Case 3.4.c it is obtained

$$C_1 = -0.161236247379;\ C_2^* = 0.089827491529;\ C_3^* = 0.31938160121;$$
$$C_4^* = 0.215959283345;\ C_5^* = -0.137595505379;\ C_6^* = -0.483816774592;$$
$$C_7^* = 0.050797321278$$

$$\begin{aligned}\bar{u}(t) &= 1.106947017\cos\Omega t - 0.142571421\cos 3\Omega t + 0.0505855201\cos 5\Omega t - \\ &-0.022382023\cos 7\Omega t + 0.010797964\cos 9\Omega t - 0.004586516\cos 11\Omega t + \\ &+0.00120946003\cos 13\Omega t\end{aligned} \tag{3.154}$$

where $\Omega = 0.356420648$.

It is easy to verify the accuracy of the obtained solutions if we graphically compare these analytical solutions with the numerical ones. Figures 3.11, 3.12, 3.13, 3.14, 3.15 and 3.16 show the comparison between the present solutions and the numerical integration results obtained by a fourth-order Runge–Kutta method.

It can be seen from Figs. 3.11, 3.12, 3.13, 3.14, 3.15 and 3.16 that the solutions obtained by OHAM are very accurate being nearly identical with the solutions

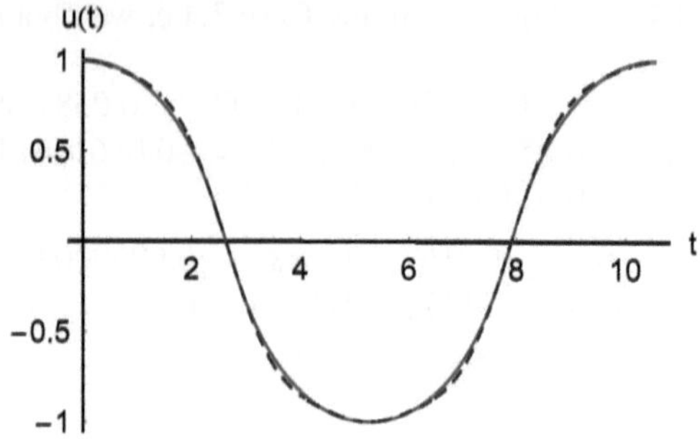

Fig. 3.11 Comparison between the approximate and numerical results of Eq. 3.113 in Case 3.4.1.a), for $A = \omega_0 = a = q = 1$: (*red line*) numerical solution and (*blue dashed line*) approximate solution 3.149 (colour figure online)

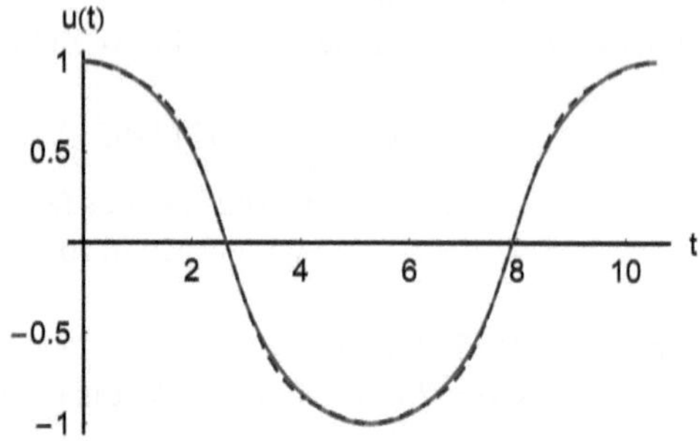

Fig. 3.12 Comparison between the approximate and numerical results of Eq. 3.113 in Case 3.4.1.b), for $A = \omega_0 = a = q = 1$: (*red line*) numerical solution and (*blue dashed line*) approximate solution 3.150 (colour figure online)

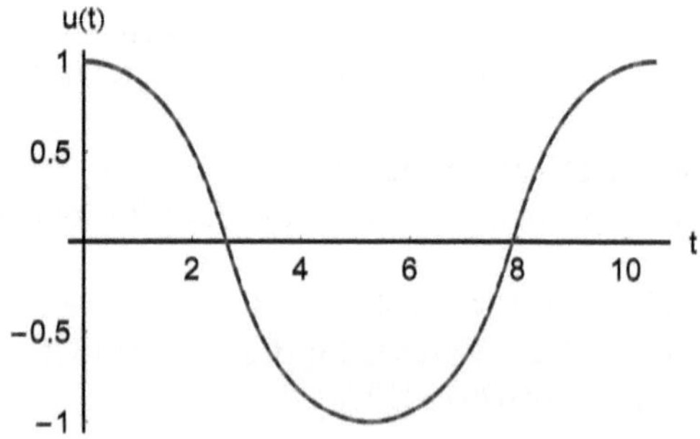

Fig. 3.13 Comparison between the approximate and numerical results of Eq. 3.113 in Case 3.4.1.c), for $A = \omega_0 = a = q = 1$: (*red line*) numerical solution and (*blue dashed line*) approximate solution 3.151 (colour figure online)

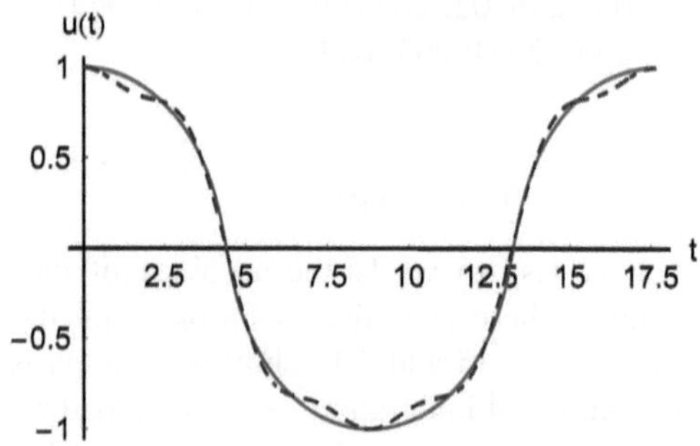

Fig. 3.14 Comparison between the approximate and numerical results of Eq. 3.113 in Case 3.4.1.d), for $A = \omega_0 = a = 1$, $q = 2$: (*red line*) numerical solution and (*blue dashed line*) approximate solution 3.152 (colour figure online)

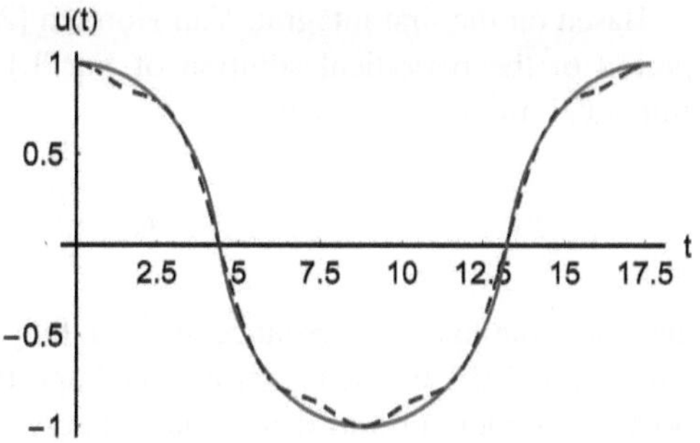

Fig. 3.15 Comparison between the approximate and numerical results of Eq. 3.113 in Case 3.4.1.e), for $A = \omega = a = 1$, $q = 2$: (*red line*) numerical solution and (*blue dashed line*) approximate solution 3.153 (colour figure online)

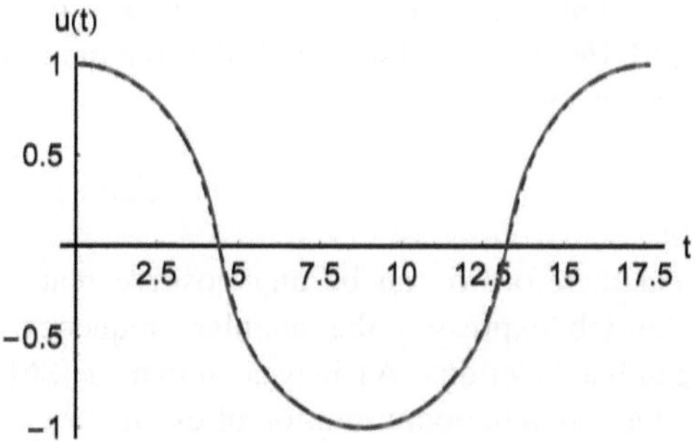

Fig. 3.16 Comparison between the approximate and numerical results of Eq. 3.113 in Case 3.4.1.f), for $A = \omega = a = 1$, $q = 2$: (*red line*) numerical solution and (*blue dashed line*) approximate solution 3.154 (colour figure online)

obtained by a fourth-order Runge–Kutta method. Moreover, the analytical solutions obtained by our procedure prove to be more accurate along with an increased number of terms in the optimal auxiliary function $H(\tau,p)$.

3.5 Nonlinear Oscillator with Discontinuities and Fractional-Power Restoring Force

In a series of papers, Mickens and his co-workers [19–21] considered nonlinear oscillator equations with elastic force as a fractional order function

$$\ddot{x} + x^{\alpha} = 0 \tag{3.155}$$

where $\alpha = 1/(2n+1)$, with n a positive integer number. For $n = 1$, he applied the harmonic balance method and later he used a generalised harmonic balance to determine an approximate periodic solution and angular frequency. Awrejcewicz and Andrianov [22] examined periodic solutions of the differential equation

$$\ddot{x} + sign(x) = 0 \tag{3.156}$$

Based on the first integral, Van Horssen [23] found the exact expression for the period of the periodical solution of Eq. 3.155. Hu and Xing [24] extended the fractional model 3.155 with

$$\alpha = \frac{2m + 1}{2n + 1} \tag{3.157}$$

and using the harmonic balance method, they determined the approximate periodic solution and the frequency. It is very important to emphasize that the oddness of both the numerator and denominator is crucial. If one of the parts of the ratio is even, the system 3.157 is not an oscillator. The case when in Eq. 3.155 α is an odd ratio but also if α has even nominator or denominator was studied by Gottlieb [25]. He introduced so called "the sign function" of the force according to deflection sign change

$$\ddot{x} + sign(x) \cdot |x|^{\alpha} = 0 \tag{3.158}$$

and therefore α can be any positive real value. By harmonic balance method, Gottlieb expressed the angular frequency of the periodic solutions in terms gamma functions. As it was shown in [26] and [27], for arbitrary positive odd integer α, a periodic solution of oscillator given by Eq. 3.155 can be represented in the form

$$x = X(\tau), \quad \tau = \tau(t/a) \tag{3.159}$$

where $X(-\tau) = -X(\tau)$ and

$$\tau(\phi) = \frac{2}{\pi}\arcsin\left(\sin\frac{\pi\phi}{2}\right), \quad \tau(\phi) = \tau(\phi + 4) \tag{3.160}$$

is the periodic saw-tooth function of the period $T = 4a$. Both the function $X(\tau)$ and the parameter a are determined by successive approximations. In [27], Pilipchuck generalised solution 3.159 as

$$x = sign(\tau)X(\tau) \tag{3.161}$$

and then he considered a more general class of oscillators given by Eq. 3.158. In the case of Eq. 3.157, the oscillator given by Eq. 3.161 is absolutely identical to the original one, Eq. 3.155, as well as the solution given by Eq. 3.161 due to its oddness with respect to τ. Also, in [27] is showed that Eq. 3.158 is an oscillator with α odd, even rational or irrational. For example, $\ddot{x} + sign(x) \cdot |x|^{3/4} = 0$ is an oscillator, whereas $\ddot{x} + |x|^{3/4} = 0$ is not an oscillator at all. Using the modified Lindstedt-Poincare method, Liu [28] determined first-order approximate solutions and the period with 1.76 % accuracy. An improved harmonic balance method has been presented by Lim and Wu [29] to construct more accurate approximations to frequencies. Unlike the classical harmonic balance method, linearization is carried out prior to harmonic balancing thus resulting in simple linear algebraic equations.

In the theory of oscillations it is known that in a conservative system the sum of a particle's kinetic and potential energy will determine the physical limits of the motion. These limits are the turning points since the velocity will be zero. A particle oscillates if it moves repeatedly backward and forward between two points, and the equilibrium point is situated midway between these. The motion will be symmetrical and the displacement of the particle from the midpoint, a periodic function of time which is not necessarily sinusoidal. In what follows, we consider such restoring forces on a particle and offer a comparison between approximate solutions obtained through OHAM and numerical solution of a non-linear equation of motion. Attention here is restricted to rational powers less than unity.

This kind of nonlinear oscillator was studied up to now using different methodologies [30–35]. We consider an ordinary differential equation a with single-term positive-power nonlinear oscillator with fractional-power restoring force:

$$\ddot{u} + sign(u)|u|^{\alpha} = 0, \quad \alpha > 0 \tag{3.162}$$

with initial conditions

$$u(0) = A, \ \ \dot{\mathrm{u}}(0) = 0 \tag{3.163}$$

With a new independent variable $\tau = \Omega t$, Eq. 3.162 can be written as

$$\Omega^2 u'' + sign(u)|u|^{\alpha} = 0 \tag{3.164}$$

with initial conditions

$$u(0) = A, \ \ \mathrm{u}'(0) = 0 \tag{3.165}$$

For Eq. 3.164, the linear operator 2.45 is defined by

$$L(u(\tau)) = \Omega_0^2\left[\frac{\partial^2 u(\tau)}{\partial\tau^2} + u(\tau)\right] \tag{3.166}$$

and the nonlinear operator 2.46 is given by the equation:

$$\begin{aligned} N[u(\tau), \Omega(\lambda, p)] = \Omega^2(\lambda, p)u''(\tau) + \lambda u(\tau) + sign(u(\tau))|u(\tau)|^{\alpha} - \\ -p\lambda u(\tau) - \Omega_0^2\left[u''(\tau) + u(\tau)\right] \end{aligned} \tag{3.167}$$

Eq. 2.55 can be written as

$$\Omega_0^2\left(u_0'' + u_0\right) = 0, \ \ \mathrm{u}_0(0) = A, \ \ \mathrm{u}'(0) = 0 \tag{3.168}$$

and has the solution

$$u_0(\tau) = A\cos\tau \tag{3.169}$$

If $f(u(\tau)) = sign(u(\tau))|u(\tau)|^{\alpha}$, where $u = u_0 + pu_1 + p^2u_2 + \ldots$, then in the following we have taken into account the identity:

$$\begin{aligned} f(u) &= f(u_0 + pu_1 + p^2u_2 + \ldots.) = \\ &= f(u_0) + pu_1 f'(u_0) + p^2\left[u_2 f'(u_0) + \frac{1}{2}u_1^2 f''(u_0)\right] + 0(p^3) \end{aligned} \tag{3.170}$$

where for example

$$f'(u_0) = \alpha|u_0|^{\alpha-1} \tag{3.171}$$

Taking into account Eq. 3.170, the first term in Eq. 3.167 is given by:

$$N_0(u_0, \Omega_0, \lambda) = \Omega_0^2 u_0'' + \lambda u_0 + sign(u_0)|u_0|^{\alpha} \tag{3.172}$$

Equation 3.3 can be written in the form:

$$\begin{aligned} &\Omega_0^2(u_1'' + u_1) - H_1(\tau, C_i)\left[\Omega_0^2 u_0'' + \lambda u_0 + sign(u_0)|u_0|^{\alpha}\right] = 0, \\ &u_1(0) = u_1'(0) = 0 \end{aligned} \tag{3.173}$$

Using Eq. 3.169, we obtain the following Fourier series expansions:

$$sign(u_0)|u_0|^{\alpha} = A^{\alpha}(a_{1\alpha}\cos\tau + a_{3\alpha}\cos 3\tau + \ldots.) \tag{3.174}$$

where

$$a_{2k+1\alpha} = \frac{4}{\pi}\int_0^{\frac{\pi}{2}} (\cos\tau)^{\alpha}\cos(2k+1)\tau d\tau, \quad \text{k} = 0, 1, 2, \ldots. \tag{3.175}$$

Considering $H_1(\tau, C_i) = C_1$ (unknown parameter) and substituting Eqs. 3.169 and 3.174 into Eq. 3.173, we have:

$$\begin{aligned} &\Omega_0^2(u_1'' + u_1) - C_1(-A\Omega_0^2 + \lambda A + a_{1\alpha}A^{\alpha})\cos\tau - C_1A^{\alpha}a_{3\alpha}\cos 3\tau - \\ &-C_1A^{\alpha}a_{5\alpha}\cos 5\tau - C_1A^{\alpha}a_{7\alpha}\cos 7\tau - \ldots. = 0 \end{aligned} \tag{3.176}$$

No secular terms in u_1 requires eliminating contributions proportional to $\cos\tau$ in Eq. 3.176:

$$\Omega_0^2 = \lambda + a_{1\alpha}A^{\alpha-1} \quad , \quad \lambda \geq 0 \tag{3.177}$$

The solution of Eq. 3.176 can be written as

$$u_1(\tau) = \frac{C_1 a_{3\alpha} A^\alpha}{8\Omega_0^2}(\cos\tau - \cos 3\tau) + \frac{C_1 a_{5\alpha} A^\alpha}{24\Omega_0^2}(\cos\tau - \cos 5\tau) + \\ + \frac{C_1 a_{7\alpha} A^\alpha}{48\Omega_0^2}(\cos\tau - \cos 7\tau) + \ldots.. \tag{3.178}$$

If we consider $H_2(\tau, C_i) = C_2 + 2C_3 \cos 2\tau$ with C_2 and C_3 unknown parameters, then Eq. 3.4 has the form:

$$\Omega_0^2(u_2'' + u_2) - \Omega_0^2(u_1'' + u_1) - C_1\left[\Omega_0^2 u_1'' + 2\Omega_0\Omega_1 u_0'' + \lambda(u_1 - u_0) + \right. \\ \left. + \alpha|u_0|^{\alpha-1}u_1\right] - (C_2 + 2C_3\cos 2\tau)\left[\Omega_0^2 u_0'' + \lambda u_0 + sign(u_0)|u_0|^\alpha\right] = 0, \\ u_2(0) = u_2'(0) = 0 \tag{3.179}$$

Having in view the identities:

$$|\alpha| = \alpha sign(\alpha) \tag{3.180}$$

$$\frac{\cos(2k+1)\tau}{\cos\tau} = \\ = 2\cos 2k\tau - 2\cos 2(k-1)\tau + 2\cos 2(k-2)\tau + \ldots. + (-1)^k\cos\tau \tag{3.181}$$

and Eq. 3.169, we can write Eq. 3.182

$$|u_0|^{a-1} = A^{a-1}\left[a_{1a} - a_{3a} + a_{5a} - a_{7a} + \ldots + \right. \\ \left. + 2(a_{3a} - a_{5a} + a_{7a} + \ldots)\cos 2\tau + 2(a_{5a} - a_{7a} + \ldots)\cos 4\tau + \ldots\right] \tag{3.182}$$

Substitution of Eqs. 3.169, 3.174, 3.178 and 3.182 into Eq. 3.179 yields:

$$\Omega_0^2(u_2'' + u_2) = \left[2AK_1\frac{\lambda - \Omega_0^2}{\Omega_0^2} - 2A\Omega_0\Omega_1 - \lambda A + \frac{C_1^2\alpha A^{2\alpha-1}}{48\Omega_0^2}\beta_1 + C_3A^\alpha a_{3\alpha} + \right. \\ \left. + \frac{2AK_2}{\Omega_0^2}\right]\cos\tau + \left[\frac{C_1^2A^\alpha a_{3\alpha}}{8}\left(9 - \frac{\lambda}{\Omega_0^2}\right) + \frac{\alpha C_1^2A^{2\alpha-1}\beta_3}{48\Omega_0^2} + C_3A^\alpha a_{5\alpha} + \right. \\ \left. + (C_1 + C_2)A^\alpha a_{3\alpha}\right]\cos 3\tau + \left[\frac{C_1^2A^\alpha a_{5k}}{24}\left(25 - \frac{\lambda}{\Omega_0^2}\right) + A^\alpha(a_{3\alpha} + a_{7\alpha})C_3 + \right. \\ \left. + \frac{\alpha C_1^2A^{2\alpha-1}\beta_5}{48\Omega_0^2} + (C_1 + C_2)A^\alpha a_{5\alpha}\right]\cos 5\tau + \left[\frac{C_1^2A^\alpha a_{7\alpha}}{48}\left(49 - \frac{\lambda}{\Omega_0^2}\right) + \right. \\ \left. + \frac{\alpha C_1^2A^{2\alpha-1}\beta_7}{48\Omega_0^2} + (C_1 + C_2)A^\alpha a_{7\alpha} + A^\alpha(a_{5\alpha} + a_{9\alpha})C_3\right]\cos 7\tau + \ldots. \tag{3.183}$$

where

$$\begin{aligned}
K_1 &= \frac{C_1 A^{\alpha-1}(6a_{3\alpha} + 2a_{5\alpha} + a_{7\alpha})}{96} \\
K_2 &= \frac{C_1 A^{2\alpha-2}\left(6a_{1\alpha}a_{3\alpha} - 6a_{3\alpha}^2 + 2a_{1\alpha}a_{5\alpha} - 2a_{5\alpha}^2 + a_{1\alpha}a_{7\alpha} - a_{7\alpha}^2\right)}{96} \\
\beta_1 &= 6a_{1\alpha}a_{3\alpha} + 2a_{1\alpha}a_{5\alpha} + a_{1\alpha}a_{7\alpha} - 6a_{3\alpha}^2 - 6a_{5\alpha}^2 + 6a_{5\alpha}a_{7\alpha} \\
\beta_3 &= -6a_{1\alpha}a_{3\alpha} + 12a_{3\alpha}^2 - 6a_{3\alpha}a_{5\alpha} + 2a_{5\alpha}^2 + 7a_{3\alpha}a_{7\alpha} - 3a_{5\alpha}a_{7\alpha} \\
&\quad + a_{7\alpha}^2 \\
\beta_5 &= -6a_{3\alpha}^2 + 14a_{3\alpha}a_{5\alpha} - 2a_{1\alpha}a_{5\alpha} - 6a_{3\alpha}a_{7\alpha} + 2a_{5\alpha}a_{7\alpha} - 2a_{7\alpha}^2 \\
\beta_7 &= -8a_{3\alpha}a_{5\alpha} + 2a_{5\alpha}^2 + 5a_{3\alpha}a_{7\alpha} - a_{5\alpha}a_{7\alpha} - 3a_{7\alpha}^2
\end{aligned} \tag{3.184}$$

The secular term in the solution of u_2 can be eliminated from Eq. 3.183 if

$$\Omega_1 = \frac{K_1\left(\lambda - \Omega_0^2\right)}{\Omega_0^3} - \frac{\lambda}{2\Omega_0} + \frac{K_2}{\Omega_0^3} + \frac{C_1^2 \alpha A^{2\alpha-2}}{96\Omega_0^3}\beta_1 + \frac{A^{\alpha-1}a_{3\alpha}C_3}{2\Omega_0} \tag{3.185}$$

From Eqs. 2.52 and 3.185 we obtain the frequency in the form:

$$\Omega = \Omega_0 + \frac{K_1\left(\lambda - \Omega_0^2\right)}{\Omega_0^3} - \frac{\lambda}{2\Omega_0} + \frac{K_2}{\Omega_0^3} + \frac{C_1^2 \alpha A^{2\alpha-2}}{96\Omega_0^3}\beta_1 + \frac{A^{\alpha-1}a_{3\alpha}C_3}{2\Omega_0} \tag{3.186}$$

where Ω_0 is given by Eq. 3.177.

The parameter λ can be determined applying the "principle of minimal sensitivity". From Eq. 2.59 we obtain:

$$\begin{aligned}
\lambda = &-\frac{1}{2}(a_{1\alpha} - a_{3\alpha}C_3)A^{\alpha-1} + \\
&+\sqrt{\frac{1}{4}(a_{1\alpha} - a_{3\alpha}C_3)^2 A^{2\alpha-2} + 6K_2 - 6K_1 a_{1\alpha}A^{\alpha-1} + \frac{C_1^2 \alpha A^{2\alpha-2}}{16}\beta_1 + a_{1\alpha}a_{3\alpha}A^{2\alpha-2}C_3}
\end{aligned} \tag{3.187}$$

From Eqs. 3.287 and 3.177 it follows that

$$\begin{aligned}
\Omega_0^2 = &\frac{1}{2}(a_{1\alpha} + a_{3\alpha}C_3)A^{\alpha-1} + \\
&+\sqrt{\frac{1}{4}(a_{1\alpha} - a_{3\alpha}C_3)^2 A^{2\alpha-2} + 6K_2 - 6K_1 a_{1\alpha}A^{\alpha-1} + \frac{C_1^2 \alpha A^{2\alpha-2}}{16}\beta_1 + a_{1\alpha}a_{3\alpha}A^{2\alpha-2}C_3}
\end{aligned} \tag{3.188}$$

where K_1, K_2 and β_i are given by Eq. 3.184.

Now, we can write the solution of Eq. 3.183 in the form

$$\begin{aligned}u_2(\tau) = {} & \left[\frac{C_1^2A^\alpha a_{3\alpha}}{64\Omega_0^2}\left(9-\frac{\lambda}{\Omega_0^2}\right)+\frac{\alpha C_1^2A^{2\alpha-1}\beta_3}{384\Omega_0^4}+\frac{C_3A^\alpha a_{5\alpha}}{8\Omega_0^2}+\right.\\ & \left.+\frac{(C_1+C_2)A^\alpha a_{3\alpha}}{8\Omega_0^2}\right](\cos\tau-\cos 3\tau)+\left[\frac{C_1^2A^\alpha a_{5\alpha}}{576\Omega_0^2}\left(25-\frac{\lambda}{\Omega_0^2}\right)+\right.\\ & \left.+\frac{\alpha C_1^2A^{2\alpha-1}\beta_5}{1152\Omega_0^4}+\frac{(C_1+C_2)A^\alpha a_{5\alpha}}{24\Omega_0^2}+\frac{C_3A^\alpha(a_{3\alpha}+a_{7\alpha})}{24\Omega_0^2}\right](\cos\tau-\\ & -\cos 5\tau)+\left[\frac{C_1^2A^\alpha a_{7\alpha}}{2304\Omega_0^2}\left(49-\frac{\lambda}{\Omega_0^2}\right)+\frac{\alpha C_1^2A^{2\alpha-1}\beta_7}{2304\Omega_0^4}+\right.\\ & \left.+\frac{(C_1+C_2)A^\alpha a_{7\alpha}}{48\Omega_0^2}+\frac{C_3A^\alpha(a_{5\alpha}+a_{9\alpha})}{48\Omega_0^2}\right](\cos\tau-\cos 7\tau)\end{aligned} \tag{3.189}$$

In order to determine the second-order approximate solution it is necessary to substitute Eqs. 3.169, 3.178 and 3.189 into the equation:

$$\bar{u}(\tau) = u_0(\tau)+u_1(\tau)+u_2(\tau) \tag{3.190}$$

By means of the transformation $\tau = \Omega t$, the second-order approximate solution of Eq. 3.162 is:

$$\begin{aligned}\bar{u}(t) = {} & A\cos\Omega t+\left[\frac{C_1^2A^\alpha a_{3\alpha}}{64\Omega_0^2}\left(9-\frac{\lambda}{\Omega_0^2}\right)+\frac{\alpha C_1^2A^{2\alpha-1}\beta_3}{384\Omega_0^4}+\frac{C_3A^\alpha a_{5\alpha}}{8\Omega_0^2}\right.\\ & \left.+\frac{(2C_1+C_2)A^\alpha a_{3\alpha}}{8\Omega_0^2}\right](\cos\Omega t-\cos 3\Omega t)+\left[\frac{C_1^2A^\alpha a_{5\alpha}}{576\Omega_0^2}\left(25-\frac{\lambda}{\Omega_0^2}\right)+\right.\\ & \left.+\frac{\alpha C_1^2A^{2\alpha-1}\beta_5}{1152\Omega_0^4}+\frac{(2C_1+C_2)A^\alpha a_{5\alpha}}{24\Omega_0^2}+\frac{C_3A^\alpha(a_{3\alpha}+a_{7\alpha})}{24\Omega_0^2}\right](\cos\Omega t-\\ & -\cos 5\Omega t)+\left[\frac{C_1^2A^\alpha a_{7\alpha}}{2304\Omega_0^2}\left(49-\frac{\lambda}{\Omega_0^2}\right)+\frac{\alpha C_1^2A^{2\alpha-1}\beta_7}{2304\Omega_0^4}+\right.\\ & \left.+\frac{(2C_1+C_2)A^\alpha a_{7\alpha}}{48\Omega_0^2}+\frac{C_3A^\alpha(a_{5\alpha}+a_{9\alpha})}{48\Omega_0^2}\right](\cos\Omega t-\cos 7\Omega t)\end{aligned} \tag{3.191}$$

where $\beta_1, \beta_3, \beta_5, \beta_7$ are given by Eq. 3.184 and Ω, λ, Ω_0 depends on the parameters C_1, C_2 and C_3, which will be optimally determined.

We illustrate the accuracy of the OHAM by comparing previously obtained approximate solutions with the numerical integration results obtained by means of a fourth-order Runge–Kutta method.

In the first case we consider $\alpha = 3/5$ and from Eq. 3.175 we obtain:

$$a_{1\frac{3}{5}} = 1.0872931957; \quad a_{3\frac{3}{5}} = -\frac{1}{9}a_{1\frac{3}{5}}; \quad a_{5\frac{3}{5}} = \frac{1}{21}a_{1\frac{3}{5}}; \quad a_{7\frac{3}{5}} = -\frac{11}{399}a_{1\frac{3}{5}}$$
$$a_{9\frac{3}{5}} = \frac{22}{1197}a_{1\frac{3}{5}}$$

3.5.1.a) For $A = 1$, applying the collocation method we obtain

$$C_1 = -0.343257069557; \; C_2 = -0.26406442041, \; C_3 = 0.030222115337;$$
$$\Omega = 1.040742431$$

The frequency in the case $A = 1$ is Ω_{num} =1.040748 and therefore the relative error between the approximate and the numerical frequency is 0.0005 %.

The second-order approximate solution 3.191 becomes:

$$\begin{aligned}\bar{u}(t) &= 1.010438818\cos\Omega t - 0.011758271\cos 3\Omega t + \\ &+ 0.001842245\cos 5\Omega t - 0.000522791\cos 7\Omega t\end{aligned} \tag{3.192}$$

3.5.1.b) For $A = 5$, we obtain the following expressions:

$$C_1 = -0.155385844709; \; C_2 = -0.509684110981; \; C_3 = 0.03418440749,$$
$$\Omega = 0.754333389$$

The frequency in this case is $\Omega_{num} = 0.754313$ and therefore the relative error between the approximate and the numerical frequency is 0.0027 %.

The second-order approximate solution 3.191 becomes in this case:

$$\begin{aligned}\bar{u}(t) &= 5.050083579\cos\Omega t - 0.056478969\cos 3\Omega t + \\ &+ 0.008926955\cos 5\Omega t - 0.002531564\cos 7\Omega t\end{aligned} \tag{3.193}$$

In the second case we consider $\alpha = 1/3$. From Eq. 3.175 it is obtained:

$$a_{1\frac{1}{3}} = 1.1595952669; \quad a_{3\frac{1}{3}} = -\frac{1}{5}a_{1\frac{1}{3}}; \quad a_{5\frac{1}{3}} = \frac{1}{10}a_{1\frac{1}{3}}; \quad a_{7\frac{1}{3}} = -\frac{7}{110}a_{1\frac{1}{3}}$$
$$a_{9\frac{1}{3}} = \frac{1}{22}a_{1\frac{1}{3}}$$

3.5.1.c) For $A = 1$, we find

$$C_1 = -0.303870496857; \; C_2 = 0.093089844013; \; C_3 = 0.063059497344,$$
$$\Omega = 1.070456263$$

The frequency for $A = 1$ is $\Omega_{num} = 1.0704507$ and therefore the relative error between the approximate and the numerical frequency is 0.0005 %.

The second-order approximate solution 3.191 becomes:

$$\bar{u}(t) = 1.009603051 \cos \Omega t - 0.011327353 \cos 3\Omega t + \\ +0.002483710 \cos 5\Omega t - 0.000759407 \cos 7\Omega t \tag{3.194}$$

3.5.1.d) For $A = 5$, we obtain the values:

$$C_1 = -0.962540396758;\ C_2 = 0.48119576393;\ C_3 = 0.040538802187, \\ \Omega = 0.625987366$$

The frequency is $\Omega_{num} = 0.626003359$ and therefore the relative error between the approximate and the numerical frequency is 0.0025 %.

The second-order approximate solution 3.191 becomes in this case:

$$\bar{u}(t) = 5.052159619 \cos \Omega t - 0.061569752 \cos 3\Omega t + \\ +0.0135672601 \cos 5\Omega t - 0.004157127 \cos 7\Omega t \tag{3.195}$$

Figures 3.17, 3.18, 3.19 and 3.20 show a comparison between the present analytical solutions and the numerical integration results obtained using a fourth-order Runge–Kutta method.

One can observe that the second-order approximate analytical results obtained through OHAM are almost identical with the numerical simulation results in all considered cases for various values of the parameters α and A. Moreover, the

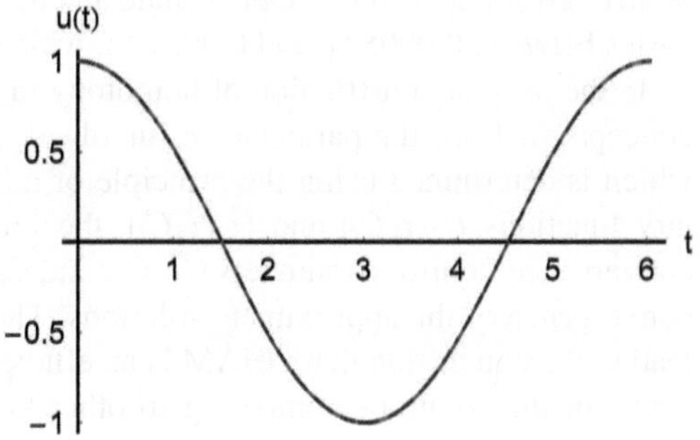

Fig. 3.17 Comparison between the approximate solution 3.192 and numerical results of Eq. 3.162 in the case $\alpha = 3/5$, $A = 1$: (*red line*) numerical solution and (*blue dashed line*) approximate solution (colour figure online)

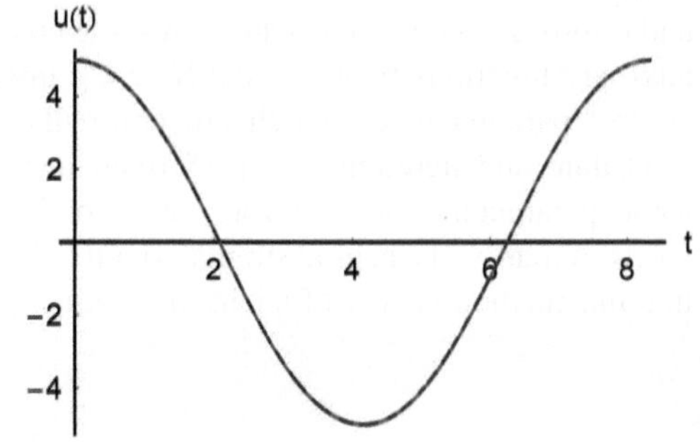

Fig. 3.18 Comparison between the approximate solution 3.193 and numerical results for Eq. 3.162 in the case $\alpha = 3/5$, $A = 5$: (*red line*) numerical solution and (*blue dashed line*) approximate solution (colour figure online)

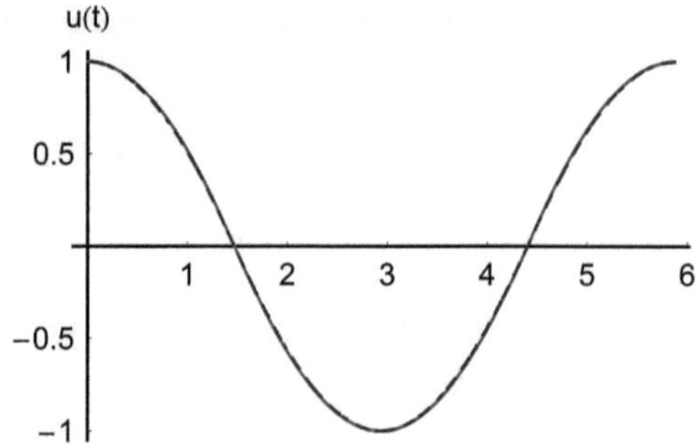

Fig. 3.19 Comparison between the approximate solution 3.194 and numerical results for Eq. 3.162 in the case $\alpha = 1/3$, $A = 1$: (*red line*) numerical solution and (*blue dashed line*) approximate solution (colour figure online)

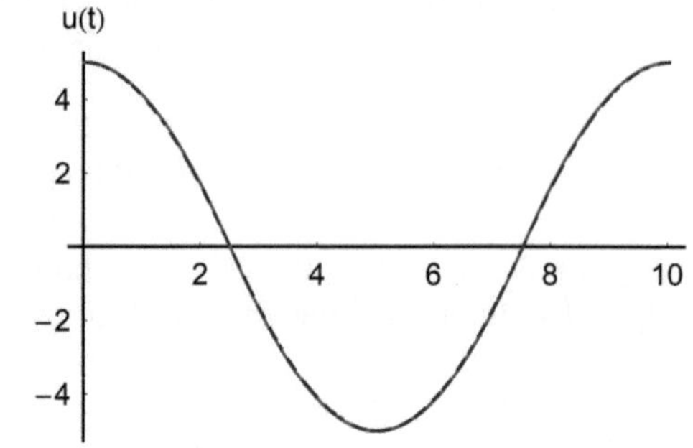

Fig. 3.20 Comparison between the approximate solution 3.195 and numerical results for Eq. 3.162 in the case $\alpha = 1/3$, $A = 5$: (*red line*) numerical solution and (*blue dashed line*) approximate solution (colour figure online)

relative error between the approximate and the numerical frequency presented here varies between 0.0005 % and 0.0027 %, which proves the accuracy of the method.

In the present construction of homotopy in this chapter appear some distinctive concepts such as: the parameter λ (involved in the case of nonlinear oscillations) which is determined using the principle of minimal sensitivity, the optimal auxiliary functions $H_1(\tau, C_i)$ and $H_2(\tau, C_i)$, the linear operator L and several optimal convergence-control parameters C_1, C_2, ..., C_s, whose optimal values ensure a fast convergence of the approximate solutions. The examples presented in this chapter lead to the conclusion that OHAM is an efficient method, ensuring a faster convergence of the solutions comparing to other known methods. The accuracy of the obtained results is growing along with increasing the number of convergence-control parameters in the auxiliary functions. The OHAM provides us with a simple and rigorous way to control and adjust the convergence of the solutions through the auxiliary functions $H_1(\tau, C_i)$ and $H_2(\tau, C_i)$ involving several optimal convergence-control parameters C_i which are optimally determined. OHAM is an iterative procedure and iterations are performed in a very simple manner by identifying some parameters and therefore very good approximations are obtained in few terms. Actually, the capital strength of OHAM is its fast convergence, which proves that this method is very efficient in practice.

References

1. V. Marinca, N. Herişanu, I. Nemeş, Optimal homotopy asymptotic method with application to thin film flow. Cent. Eur. J. Phys. **6**, 648–653 (2008)
2. T. Hayat, M. Sajid, On analytic solution of thin film flow of fourth grade fluid down a vertical cylinder. Phys. Lett. A **361**, 316–322 (2007)
3. M. Sajid, T. Hayat, S. Asghar, On the analytic solution of the steady flow of a fourth grade fluid. Phys. Lett. A **355**, 18–26 (2000)
4. V. Marinca, N. Herisanu, An optimal homotopy perturbation approach to thin film flow of a fourth grade fluid. AIP Conf. Proc. **1479**, 2383–2386 (2012)
5. A.M. Siddiqui, R. Mahmood, Q.K. Ghori, Homotopy perturbation method for thin film flow of a fourth grade fluid down a vertical cylinder. Phys. Lett. A **352**, 404–410 (2006)
6. G.B. Jeffery, The two-dimensional steady motion of a viscous fluid. Philos. Mag. **6**, 455–465 (1915)
7. G. Hamel, Spiralformige Bewegungen Zaher Flussigkeiten. Jahresber. Deutsch. Math. Verein. **25**, 34–60 (1916)
8. L. Rosenhead, The steady two-dimensional radial flow of viscous fluid between two inclined plane walls. Proc. R. Soc. A **175**, 436–467 (1940)
9. H. Schlichting, *Boundary-Layer Theory* (McGraw-Hill, New York, 2000)
10. V. Marinca, N. Herisanu, An optimal homotopy asymptotic approach to nonlinear MHD Jeffery-Hamel flow. Math. Probl. Eng. Article ID 169056 (2011)
11. A.A. Joneidi, G. Domairry, M. Babaelahi, Three analytical method applied to Jeffery-Hamel flow. Commun. Nonlinear Sci. Numer. Simul. **15**, 3423–3434 (2010)
12. S.M. Moghimi, D.D. Ganji, H. Bararnia, M. Hosseini, M. Jalaal, Homotopy perturbation method for nonlinear Jeffery-Hamel problem. Comput. Math. Appl. **61**, 2213–2216 (2011)
13. R.E. Mickens, *Oscillations in Planar Dynamic Systems* (World Scientific Publishing, Singapore, 1996)
14. W.P. Sun, B.S. Wu, C.W. Lim, Approximate analytical solutions for oscillations of a mass attached to a stretched elastic wire. J. Sound Vibr. **300**, 1042–1047 (2007)
15. L. Xu, Application of He's parameter-expansion method to an oscillation of a mass attached to a stretched elastic wire. Phys. Lett. A **368**, 259–261 (2007)
16. A.H. Nayfeh, D.T. Mook, *Nonlinear Oscillations* (Wiley, New York, 1979)
17. V. Marinca, N. Herisanu, Determination of periordic solutions for the motion of a particle on a rotating parabola by means of the optimal homotopy asymptotic method. J Sound Vibr. **329**, 1450–1459 (2010)
18. V. Marinca, N. Herisanu, *Nonlinear Dynamical Systems in Engineering. Some Approximate Approaches* (Springer, Berlin, Heidelberg, 2011)
19. R.E. Mickens, D. Semwogerere, Fourier analysis of a rational harmonic balance approximation for periodic solutions. J Sound Vibr. **195**, 528–530 (1996)
20. R.E. Mickens, Oscillations in an $x^{4/3}$ potential. J Sound Vibr. **246**, 375–378 (2001)
21. K. Cooper, R.E. Mickens, Generalised harmonic balance numerical method for determining analytical approximation in the periodic solutions of the $x^{4/3}$ potential. J. Sound Vibr. **250**, 951–954 (2002)
22. J. Awrejcewicz, I.V. Andrianov, Oscillations of non-linear system with restoring force close to sign(x). J. Sound Vibr. **252**, 962–966 (2002)
23. W.T. van Horssen, On the periods of the periodic solutions of the non-linear oscillator equation $\ddot{x}+x^{1/(2n+1)}=0$. J. Sound Vibr. **269**, 961–964 (2003)
24. H. Hu, Z.-G. Xiang, Oscillations in an $x^{(2n+2)/(2n+1)}$ potential. J. Sound Vibr. **192**, 43–64 (1996)
25. H.P.W. Gotlieb, Frequencies of oscillations with fractional power linearities. J. Sound Vibr. **26**, 557–566 (2003)
26. V.N. Pilipciuk, Analytical study of vibrating systems with strong nonlinearities by employing saw-tooth time transformations. J Sound Vibr. **192**, 43–64 (1996)

27. V.N. Pilipciuk, Oscillations with a generalised power-form elastic term. J Sound Vibr **270**, 270–272 (2004)
28. H.-M. Liu, Approximate period of nonlinear oscillation with discontinuities by modified Lindstedt-Poincare method, Chaos. Solitons Fractals **23**, 577–579 (2005)
29. C.W. Lim, B.S. Wu, Accurate higher-order approximations to frequencies of nonlinear oscillators with fractional powers. J. Sound Vibr. **281**, 1157–1162 (2005)
30. F.O. Zengin, M.O. Kaya, S.A. Demirbag, Applications of parameter-expansion method to nonlinear oscillators with discontinuities. Int. J. Nonlin. Sci. Numer. Simul. **9**, 267–270 (2008)
31. J.E. Bolwell, Solutions of free nonlinear oscillators with power law in one dimension. J Sound Vibr. **252**, 955–961 (2002)
32. J.I. Ramos, Piecewise-linearized method for oscillators with fractional power nonlinearities. J Sound Vibr. **300**(502–521) (2007)
33. L. Cveticanin, Oscillator with fraction order restoring force. J. Sound Vibr. **320**, 1064–1077 (2008)
34. I. Kovacic, On the motion of nonlinear oscillators with fractional-order restoring force and time variable parameters. Phys. Lett. A **373**, 1839–1840 (2009)
35. A. Belendez, C. Pascual, M. Ortuno, T. Belendez, S. Gallego, Application of a modified homotopy perturbation method to obtain higher-order approximations to a nonlinear oscillator with discontinuities. Nonlin Anal. Real World Appl. **10**, 601–610 (2009)

Chapter 4
The Second Alternative of the Optimal Homotopy Asymptotic Method

The purpose of this chapter is to obtain an approximate solution by means of OHAM for the nonlinear differential equation

$$L[u(x)] + g(x) + N[u(x)] = 0, \tag{4.1}$$

subject to the initial conditions

$$B\left[u(x), \frac{du(x)}{dx}\right] = 0 \tag{4.2}$$

after only one iteration, solving a simple linear differential equation. If $\phi(x,p;C_i)$ is a unknown function such as

$$\phi(x, p; C_i) = u_0(x) + pu_1(x, C_i) \tag{4.3}$$

where it is known that $p \in [0, 1]$, and $u_0(x)$ is an initial approximation of $u(x)$ which has the properties

$$L[u_0(x)] + g(x) = 0, \quad B\left[u_0(x), \frac{du_0(x)}{dx}\right] = 0 \tag{4.4}$$

then, the first-order approximate solution of Eq. 4.1 is given by

$$\overline{u}(x, C_i) = \phi(x, 1, C_i) = u_0(x) + u_1(x, C_i) \tag{4.5}$$

where C_i are unknown parameters at this moment, and the first approximation $u_1(x, C_i)$ will be determined as described below.

By means of the linear operator L and nonlinear operator N from Eq. 4.1, we can construct a family of equations

V. Marinca, N. Herisanu, *The Optimal Homotopy Asymptotic Method*,
DOI 10.1007/978-3-319-15374-2_4

$$\begin{aligned}&\mathrm{H}\left[L(\phi(x,p,C_i))+g(x),H(x,C_i),N(\phi(x,p,C_i)\right]=\\&=L[u_0(x)]+g(x)+p[L(u_1(x,C_i))-H(x,C_i)N(u_0(x))]\end{aligned} \tag{4.6}$$

which satisfies the properties:

$$\begin{aligned}&\mathrm{H}\left[L(\phi(x,0,C_i))+g(x),H(x,C_i),N(\phi(x,0,C_i)\right]=\\&=L[u_0(x)]+g(x)=0\end{aligned} \tag{4.7}$$

$$\begin{aligned}&\mathrm{H}\left[L(\phi(x,1,C_i))+g(x),H(x,C_i),N(\phi(x,1,C_i)\right]=\\&=H(x,C_i)\left(L(\overline{u}(x,C_i))+g(x)-N(\overline{u}(x,C_i))\right]=0\end{aligned} \tag{4.8}$$

with $H(x, C_i)$ an arbitrary optimal auxiliary function.

Identifying the coefficients of p^i, $i=0,1$ into Eq. 4.6 we can obtain the governing equation of $u_0(x)$ from Eq. 4.4 and the governing equation of $u_1(x, C_i)$:

$$L(u_1(x,C_i))=H(x,C_i)N(u_0(x))] \tag{4.9}$$

with the initial/boundary conditions

$$B\left[u_1(x,C_i),\frac{du_1(x,C_i)}{dx}\right]=0 \tag{4.10}$$

In the same manner as well as in Sect. 2.2, it is natural to choose the unknown optimal auxiliary function $H(x, C_i)$ so that the product $H(x, C_i)N(u_0(x))$ and $N(u_0(x))$ be of the same shape. The parameters C_1, C_2,... which appear in the first-order approximate solution given by Eq. 4.5 can be determined in many ways (see Sect. 2.4). It is clear that in this way, the approximate solution 4.5 is well-determined.

We will illustrate our procedure through some cases of nonlinear dynamical systems arising in engineering, applied mathematics, biology and so on.

4.1 The Flow of a Walters-Type B' Viscoelastic Fluid in a Vertical Channel with Porous Wall

The flow of Newtonian and non-Newtonian fluids in a porous surface channel is important for engineers and applied mathematicians. In the last few decades, these fluids have attracted considerable attention from researchers in many branches of nonlinear dynamical systems in science and technology. Some studies were made having in view its numerous industrial applications. Examples of such applications are gaseous diffusion through porous wall by a pressure gradient, boundary layer control (the fluid particles are removed through slits in the wall into the interior of the body), chemical processes (removal of the reactants), or also transpiration cooling, prevent corrosion and reduce drag and so on.

Many mathematical models have been developed to study the behavior of Newtonian and non-Newtonian fluids. There are few models for these fluids that have elastic and viscous properties. The flow of Walters' viscoelastic fluid has been used in several studies and for modelling many phenomena. For example, Chen et al. [1] studied the flow and heat transfer in the boundary layer of a viscoelastic fluid with short memory (Walters' liquid B' model) over a stretching surface subject to either constant temperature or uniform heat flux. Steady three-dimensional flow of a Walters' B' fluid in a vertical channel is investigated by Bariş in [2]. The basic equation governing flow and heat transfer were solved with a numerical technique taking the elastic number as the perturbation parameter.

The steady laminar flow of a viscoelastic fluid modeled by Walters' B' fluid near a rotating disk is considered in [3]. For this study, the solution takes into account and eliminates the error introduced by replacing numerical infinity with a finite value. The steady two-dimensional stagnation point flow of an incompressible viscoelastic fluid (obeying Walters' B model) over a flat deformable surface when the surface is stretched in its own plane is investigated in [4]. The influence of Soret and Dufour effects is studied by Postelnicu in [5] for the heat and mass transfer characteristics over a plate. A transient model for the free convective, nonlinear, steady, laminar and mass transfer in a viscoelastic fluid from a vertical porous plate is presented in [6]. The Walters' B model is employed which introduces supplementary terms into the momentum conservation equation. The transformed conservation equations are solved using the finite difference method. Peristaltic flow of Walters' B fluid in an asymmetric channel is investigated in [7]. The governing equations are first modeled and then the perturbation solutions are obtained by taking wave number as small parameter. Viscoelastic Walters' B fluid flows for three problems: stagnation-point flow, Blasius flow and Sakiadis flow have been investigated in [8]. In each of these problems, Cauchy equations are changed to nondimensional differential equations using stream functions and with assumption of boundary layer flow. Amatousse et al. [9] proposed a simplified two-equations model for a Walters-type B" viscoelastic liquid flowing down an inclined plane. The analysis was restricted to the case of traveling waves. The weighted-residual integral method is employed to investigate the flow of a thin layer of viscoelastic fluid. In the [10] is investigated the stability of a thin incompressible viscoelastic fluid designated as Walters' B" during spin coating. The long-wave perturbation method is proposed to derive a generalized kinematic model of thin film flow. The double dispersion effects on MHD viscoelastic fluid flow over a vertical cone and a flat plate saturated with non-Darcy porous medium in the presence of Soret and Dufour effects are analyzed in [11]. The Crank-Nicholson method is used to solve this problem.

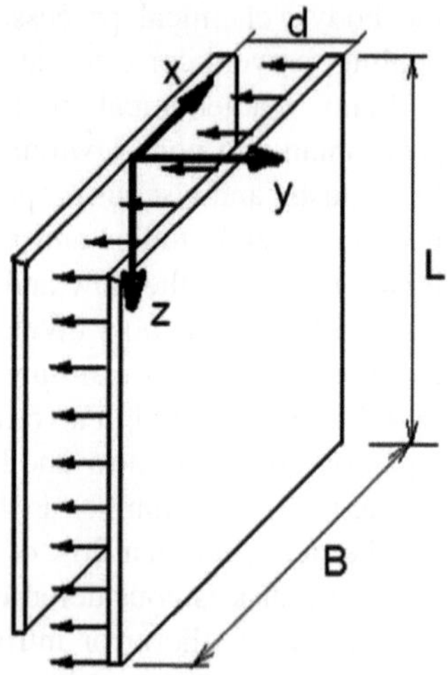

Fig. 4.1 Flow geometry and coordinate system

4.1.1 Problem Statement and Governing Equation

In what follows, we assume that the steady flow of Walters' B' fluid in a vertical channel with porous wall and the flow is steady and laminar; the fluid is incompressible, the body force per unit mass is equal to the gravitational acceleration. The viscosity, the specific heat and thermal conductivity of the fluid remain invariable throughout the fluid, the heat flux vector can be represented by Fourier's law and the effect of radial heating and viscous dissipation are negligible [2, 12–14].

Figure 4.1 shows the flow geometry and coordinate system. In this case, a fluid is injected through a vertical porous plate at y = d with uniform velocity and we assume that the distance between the walls, d, is small compared to the dimensions B and L: L ≫ B ≫ d. It follows that the edge effects can be ignored and the isobars are parallel to the z-axis. The fluid strikes another vertical impermeable plate at y = 0. Due to the action of gravity along the z-axis, the fluid flows out through the opening of plates.

The basic equations are the following:

The continuity equation:

$$\nabla V = 0 \tag{4.11}$$

with the equation of motion as

$$\rho(V\nabla V) = \nabla T + \rho g \tag{4.12}$$

and the energy equation:

$$\rho c_p(V\nabla T) = k\Delta T \tag{4.13}$$

where ∇ is the gradient operator, V is the velocity vector, ρ is the density, T is temperature, g is the gravitational acceleration vector, c_p the specific heat at constant pressure, k the thermal conductivity and Δ the Laplacian operator.

After some manipulations, we get:

$$\rho(V\nabla V) = -\nabla p + \rho g + \eta_0 \nabla^2 V - 2k_0 V\nabla\nabla^2 V + k_0 \nabla^2(V\nabla V) \tag{4.14}$$

where η_0 and k_0 are the limiting viscosity at small rate of shear and the short memory coefficient, respectively:

$$\eta_0 = \int_0^\infty N(\tau)d\tau, k_0 = \int_0^\infty \tau N(\tau)d\tau \tag{4.15}$$

N(τ) being the distribution function with relaxation time τ. The model of Walters' B' fluid (or with short memory) is valid if the terms

$$\int_0^\infty \tau^n N(\tau)d\tau, n \geq 2 \tag{4.16}$$

have been neglected.

The velocity components corresponding to the x, y and z directions are denotes by u, v and w respectively. If $\eta = y/d$, then we look for a solution of the form

$$u = \frac{U}{d} x F'(\eta), v = -UF(\eta), w = \frac{d^2 g\rho}{\eta_0} h(\eta) \tag{4.17}$$

where the prime denotes the differentiation with respect to the new variable η and U is the uniform injection velocity.

The boundary conditions corresponding to the velocity field are:

$$F(0) = 0, F'(0) = 0, F(1) = 1, F'(1) = 0, h(0) = 0, h(1) = 0 \tag{4.18}$$

From Eqs. 4.14 and 4.17 it follows that

$$\frac{\partial p}{\partial x} = \frac{Ux}{d^2}\left(-U\rho F'^2 + U\rho FF'' + \frac{Uk_0}{d^2}F''^2 + \frac{\eta_0}{d}F''' - \frac{2Uk_0}{d^2}F'F''' + \frac{Uk_0}{d^2}FF^{IV}\right) \tag{4.19}$$

$$\frac{\partial p}{\partial \eta} = -U^2 \rho F F' - \frac{U\eta_0}{d} F'' + \frac{3U^2 k_0}{d^2} F' F'' - \frac{U^2 k_0}{d^2} F F''' \tag{4.20}$$

$$h'' + \mathrm{Re}\, F h' + \mathrm{Re}\, S\left(F h''' - F'' h' - 2F' h''\right) + 1 = 0 \tag{4.21}$$

where Re is the cross-flow Reynolds number and S is the elastic number given by

$$\mathrm{Re} = \frac{U d \rho_0}{\eta_0}, S = \frac{k_0}{\rho d^2} \tag{4.22}$$

From Eq. 4.20 we obtain

$$p(x,\eta) = -\frac{1}{2} U^2 \rho F^2 - \frac{U\eta_0}{d} F' + \frac{2U^2 k_0}{d^2} F'^2 - \frac{U^2 k_0}{d^2} F F'' + \psi(x) \tag{4.23}$$

where $\psi(x)$ is an arbitrary function of x, obtained through integration of Eq. 4.20.
In Eq. 4.23 F is independent of x and therefore it follows that

$$\frac{\partial p}{\partial x} = \frac{d\psi}{dx} \tag{4.24}$$

From Eqs. 4.24 and 4.19 we obtain

$$\frac{d\psi}{dx} = \frac{U x \eta_0}{d^3}\left[F''' + \mathrm{Re}\left(F F'' - F'^2\right) + \mathrm{Re}\, S\left(F F^{IV} + F''^2 - 2F' F'''\right)\right] \tag{4.25}$$

We remark that into Eq. 4.25, the function ψ depends only on x and F depends only on η, such that the Eq. 4.25 is true only in the case when

$$\begin{aligned} &F'''(\eta) + \mathrm{Re}\left(F(\eta)F''(\eta) - F'^2(\eta)\right) + \\ &+ \mathrm{Re}\, S\left(F(\eta)F^{IV}(\eta) + F''^2(\eta) - 2F'(\eta)F'''(\eta)\right) = C \end{aligned} \tag{4.26}$$

where C is a constant. For η = 0 into Eq. 4.26, we obtain the value

$$C = F'''(0) + \mathrm{Re}\, S F''^2(0) \tag{4.27}$$

Inserting Eq. 4.26 into Eq. 4.25 and then integrating, we obtain

$$\psi(x) = \frac{U\eta_0 C}{2d^3} x^2 + \psi_0 \tag{4.28}$$

with ψ_0 constant of integration. Substituting Eq. 4.28 into Eq. 4.23, we have:

$$p(x,\eta) = -\frac{1}{2}\rho U^2 F^2(\eta) - \frac{U\eta_0}{d}F'(\eta) + \frac{2U^2 k_0}{d^2}F'^2(\eta) - \\ -\frac{U^2 k_0}{d^2}F(\eta)F''(\eta) + \frac{U\eta_0 C}{2d^3}x^2 + \psi_0 \tag{4.29}$$

Equation 4.29 gives the pressure at any point if one can solve Eq. 4.26 with the initial conditions 4.18. On the other hand, the pressure variations in the x and y directions, in non-dimensional form can be written as:

$$p_x = \frac{p(0,\eta) - p(x,\eta)}{\rho U^2} = -\frac{C}{2\mathrm{Re}}\left(\frac{x}{d}\right)^2 \tag{4.30}$$

$$p_y = \frac{p(x,0) - p(x,\eta)}{\rho U^2} = \frac{1}{2}F^2(\eta) + \frac{1}{\mathrm{Re}}F'(\eta) + S\left(F(\eta)F''(\eta) - 2F'^2(\eta)\right) \tag{4.31}$$

The shear stress on the walls of the vertical channel has the components

$$\tau_{xy}(0) = -\frac{T_{xy}(0)d^2}{\eta_0 Ux} = F''(0), \tau_{xy}(1) = \frac{T_{xy}(1)d^2}{\eta_0 Ux} = F''(1) + \mathrm{Re}\ Sf'''(1) \tag{4.32}$$

$$\tau_{zy}(0) = \frac{T_{zy}(0)}{\rho g d} = h'(0), \tau_{zy}(1) = \frac{T_{zy}(1)}{\rho g d} = h'(1) + \mathrm{Re}\ Sh''(1) \tag{4.33}$$

We remark that into Eq. 4.26 for the elastic number S = 0, the fluid is Newtonian and the order of this equation is three, while for S ≠ 0 the fluid is non-Newtonian and the order of equation is four.

In general, Eq. 4.26 is very difficult to solve, because of the appearance of the constant C which is unknown *apriori* and this is a nonlinear differential equation. To eliminate the constant C some researchers differentiate Eq. 4.25 with respect to η, but in this case the new equation thus obtained is of order five and an additional boundary condition must be employed in order to obtain an accurate solution, which those researchers are omitting. Therefore, from mathematical point of view such a procedure is wrong. Our procedure (OHAM) allows us for the first time to straightly solve the Eq. 4.26 with initial conditions 4.18.

4.1.2 Solution of Walters-Type B' Viscoelastic Fluid in a Vertical Channel with OHAM

In what follows we apply our procedure to obtain an approximate solution of Eqs. 4.26 and 4.18, and later for Eqs. 4.21 and 4.18. For this purpose, we choose the linear operator L of the form ($g(\eta) = 0$):

$$L^F(\phi(\eta, \mathrm{p})) = \frac{\partial^4 \phi(\eta, p)}{\partial \eta^4} \tag{4.34}$$

and the nonlinear operator becomes

$$\begin{aligned} N^F(\phi(\eta, \mathrm{p})) = & -\frac{\partial^4 \phi(\eta, p)}{\partial \eta^4} + \frac{\partial^3 \phi(\eta, p)}{\partial \eta^3} + \mathrm{Re}\left[\phi(\eta, p)\frac{\partial^2 \phi(\eta, p)}{\partial \eta^2} - \right. \\ & \left. -\left(\frac{\partial \phi(\eta, p)}{\partial \eta}\right)^2\right] + \mathrm{Re}\, S\left[\phi(\eta, p)\frac{\partial^4 \phi(\eta, p)}{\partial \eta^4} + \left(\frac{\partial^2 \phi(\eta, p)}{\partial \eta^2}\right)^2 - \right. \\ & \left. -2\frac{\partial \phi(\eta, p)}{\partial \eta} - \frac{\partial^3 \phi(\eta, p)}{\partial \eta^3}\right] - C \end{aligned} \tag{4.35}$$

The initial conditions 4.18 become

$$\phi(0, p) = 0, \frac{\partial \phi(0, p)}{\partial \tau} = 0, \phi(1, p) = 1, \frac{\partial \phi(1, p)}{\partial \tau} = 0 \tag{4.36}$$

Considering only the approximation of the first-order, Eqs. 4.3 and 4.5 become, respectively

$$\phi(\eta, p) = F_0(\eta) + pF_1(\eta) \tag{4.37}$$

$$\overline{F}(\eta) = F_0(\eta) + F_1(\eta) \tag{4.38}$$

where $F_0(\eta)$ is obtained from Eq. 4.4:

$$\frac{d^4 F_0(\eta)}{d\eta^4} = 0, F_0(0) = 0, F_0'(0) = 0, F_0(1) = 1, F_0'(1) = 0 \tag{4.39}$$

Its solution is

$$F_0(\eta) = 3\eta^2 - 2\eta^3 \tag{4.40}$$

The second function, $F_1(\eta)$ is obtained from Eq. 4.9 and this is written in the form

$$\frac{d^4F_1(\eta)}{d\eta^4} = H^F(\eta, C_i)N^F(F_0(\eta)),$$
$$F_1(0) = 0, F_1'(0) = 0, F_1(1) = 0, F_1'(1) = 0 \tag{4.41}$$

where

$$N^F(F_0(\eta)) = \frac{\partial^3 F_0(\eta)}{\partial\eta^3} - \frac{\partial^4 F_0(\eta)}{\partial\eta^4} + \text{Re}\left[F_0(\eta)\frac{\partial^2 F_0(\eta)}{\partial\eta^2} - \left(\frac{\partial F_0(\eta)}{\partial\eta}\right)^2\right] +$$
$$+\text{Re}\, S\left[F_0(\eta)\frac{\partial^4 F_0(\eta)}{\partial\eta^4} + \left(\frac{\partial^2 F_0(\eta)}{\partial\eta^2}\right)^2 - 2\frac{\partial F_0(\eta)}{\partial\eta}\frac{\partial^3 F_0(\eta)}{\partial\eta^3}\right] - C \tag{4.42}$$

Substituting Eq. 4.40 into Eq. 4.42, the nonlinear operator N^F becomes

$$N^F(F_0(\eta)) = 6\left[\text{Re}(4\eta^4 - 8\eta^3 + 3\eta^2) + 6\text{Re}\, S - 2 - C\right] \tag{4.43}$$

The expression 4.43 is a polynomial one and the optimal auxiliary function H^F from Eq. 4.41 is chosen such that the product $H^F N^F$ and N^F be of the same form. Therefore we choose the optimal auxiliary function H^F of a polynomial form:

$$H^F(\eta) = \frac{1}{6}(C_1 + C_2\eta + C_3\eta^2 + \ldots + C_q\eta^{q-1}) \tag{4.44}$$

where q is an arbitrary positive integer number and C_i, $i = 1,2,\ldots,q$ are unknown parameters.

Inserting Eqs. 4.40, 4.43 and 4.44 into Eq. 4.41 we obtain the equation:

$$\frac{d^4F_1(\eta)}{d\eta^4} = \text{Re}\left[3C_1\eta^2 + (3C_2 - 8C_1)\eta^3 + (3C_3 - 8C_2 + 4C_1)\eta^4 +\right.$$
$$+(3C_4 - 8C_3 + 4C_2)\eta^5 + \ldots\ldots + (3C_q - 8C_{q-1} + 4C_{q-2})\eta^{q+1} + \tag{4.45}$$
$$\left. +(-8C_q + 4C_{q-1})\eta^{q+2} + 4C_q\eta^{q+3}\right] + (6\text{Re}\, S - 2 - C)(C_1 + C_2 q +$$
$$+C_3 q^2 + \ldots + C_q\eta^{q-1}), F_1(0) = F_1'(0) = F_1(1) = F_1'(1) = 0$$

By solving Eq. 4.45 we obtain for q = 8

$$
\begin{aligned}
F_1(\eta) = \mathrm{Re}\Bigg[\Bigg(&-\frac{C_1}{840}-\frac{C_2}{630}-\frac{C_3}{720}-\frac{31C_4}{27720}-\frac{7C_5}{7920}-\frac{C_6}{1430}-\frac{67C_7}{120120}- \\
&-\frac{9C_8}{20020}\Bigg)\eta^2+\left(\frac{11C_2}{7560}+\frac{11C_3}{7560}+\frac{67C_4}{53440}+\frac{2C_5}{2079}+\frac{47C_6}{61776}+\frac{109C_7}{180180}+\right. \\
&\left.+\frac{5C_8}{10296}\right)\eta^3+\frac{C_1}{120}\eta^6+\frac{3C_2-8C_1}{840}\eta^7+\frac{4C_1-8C_2+3C_3}{1680}\eta^8+ \\
&+\frac{3C_4-8C_3+4C_2}{3024}\eta^9+\frac{3C_5-8C_4+4C_3}{5040}\eta^{10}+\left(\frac{C_4}{1980}-\frac{C_5}{990}+\frac{C_6}{2640}\right)\eta^{11}+ \\
&+\left(\frac{C_5}{2970}-\frac{C_6}{1485}+\frac{C_7}{3960}\right)\eta^{12}+\left(\frac{C_6}{4290}-\frac{C_7}{2145}+\frac{C_8}{5720}\right)\eta^{13}+\left(\frac{C_7}{6006}-\frac{C_8}{3003}\right)\eta^{14}+ \\
&+\frac{C_8}{8190}\eta^{15}\Bigg]+(6\mathrm{Re}\,S-2-C)\Bigg[\left(\frac{C_1}{24}+\frac{C_2}{60}+\frac{C_3}{120}+\frac{C_4}{210}+\frac{C_5}{336}+\frac{C_6}{504}+\frac{C_7}{720}+\right. \\
&\left.+\frac{C_8}{990}\right)\eta^2-\left(\frac{C_1}{12}+\frac{C_2}{40}+\frac{C_3}{90}+\frac{C_4}{168}+\frac{C_5}{280}+\frac{C_6}{432}+\frac{C_7}{630}+\frac{C_8}{880}\right)\eta^3+\frac{C_1}{24}\eta^4+ \\
&+\frac{C_2}{120}\eta^5+\frac{C_3}{360}\eta^6+\frac{C_4}{840}\eta^7+\frac{C_5}{1680}\eta^8+\frac{C_6}{3024}\eta^9+\frac{C_7}{5040}\eta^{10}+\frac{C_8}{7920}\eta^{11}\Bigg]
\end{aligned}
\tag{4.46}
$$

The first-order approximate solution of Eqs. 4.26 and 4.18 is

$$\overline{F}(\eta) = F_0(\eta) + F_1(\eta) \tag{4.47}$$

where $F_0(\eta)$ and $F_1(\eta)$ are given by Eqs. 4.40 and 4.46 respectively.
The residual given by Eq. 2.35 becomes

$$
\begin{aligned}
R^F(\eta, C_1, C_2, \ldots, C_8, C) = \overline{F}'''(\eta) + \mathrm{Re}\left[\overline{F}(\eta)\overline{F}''(\eta) - \overline{F}'^2(\eta)\right]+ \\
+\mathrm{Re}\,S\left[\overline{F}(\eta)\overline{F}^{IV}(\eta) + \left(\overline{F}''(\eta)\right)^2 - 2\overline{F}'(\eta)\overline{F}'''(\eta)\right] - C
\end{aligned}
\tag{4.48}
$$

with $\overline{F}$ obtained from Eq. 4.47.

For every values of Re and S, the parameters C_i, i = 1,2,...,8 and C can be determined from the system 2.37. In this way the first-order approximate solution 4.47 is determined in a closed form.

Now, to solve Eqs. 4.21 and 4.18, we choose the following linear operator L^h in the form

$$L^h(\varphi(\eta, p)) = \frac{\partial^3 \varphi(\eta, p)}{\partial \eta^3} \tag{4.49}$$

and the nonlinear operator as

$$N^h(\varphi(\eta,p)) = \frac{\partial^2\varphi(\eta,p)}{\partial\eta^2} - \frac{\partial^3\varphi(\eta,p)}{\partial\eta^3} + \mathrm{Re}\,\phi(\eta,p)\frac{\partial\varphi(\eta,p)}{\partial\eta} +$$
$$+\mathrm{Re}\,S\left[\phi(\eta,p)\frac{\partial^3\varphi(\eta,p)}{\partial\eta^3} - \frac{\partial^2\phi(\eta,p)}{\partial\eta^2}\frac{\partial\varphi(\eta,p)}{\partial\eta} - 2\frac{\partial\phi(\eta,p)}{\partial\eta}\frac{\partial^2\varphi(\eta,p)}{\partial\eta^2}\right] + 1 \tag{4.50}$$

The initial conditions 4.18 can be written as

$$\varphi(0,p) = 0, \varphi(1,p) = 0, \frac{\partial^2\varphi(0,p)}{\partial\eta^2} - \mathrm{Re}\,S\frac{\partial^2\phi(0,p)}{\partial\eta^2}\frac{\partial\varphi(0,p)}{\partial\eta} + 1 = 0 \tag{4.51}$$

The last condition from Eq. 4.51 is obtained by using Eq. 4.21 with the substitution $\eta = 0$. In this case Eqs. 4.3 and 4.5 are, respectively

$$\varphi(\eta,p) = h_0(\eta) + ph_1(\eta) \tag{4.52}$$
$$\overline{h}(\eta) = h_0(\eta) + h_1(\eta) \tag{4.53}$$

The linear equation in $h_0(\eta)$ is obtained from Eq. 4.4:

$$\frac{d^3h_0(\eta)}{d\eta^3} = 0; h_0(0) = 0; h_0(1) = 0; h_0''(0) - \mathrm{Re}\,SF_0''(0)h_0'(0) + 1 = 0 \tag{4.54}$$

and has the solution

$$h_0(\eta) = \alpha(\eta^2 - \eta) \tag{4.55}$$

where α is obtained from Eq. 4.54 in the form:

$$\alpha = -[2 + \mathrm{Re}SF''(0)]^{-1} \tag{4.56}$$

From Eq. 4.9 we can obtain the equation which give the function $h_1(\eta)$:

$$\frac{d^3h_1(\eta)}{d\eta^3} = H^h(\eta,D_i)N_0^h(h_0(\eta)); h_1(0) = 0, h_1(1) = 0 \tag{4.57}$$

where

$$\begin{aligned} N^h(h_0(\eta)) &= \frac{d^2h_0(\eta)}{d\eta^2} + \text{Re}\, F_0(\eta)\frac{dh_0(\eta)}{d\eta} + \\ &+\text{Re}\, S\left[F_0(\eta)\frac{d^3h_0(\eta)}{d\eta^3} - \frac{d^2F_0(\eta)}{d\eta^2}\frac{dh_0(\eta)}{d\eta} - 2\frac{\partial F_0(\eta)}{\partial \eta}\frac{d^2h_0(\eta)}{d\eta^2}\right] + 1 \end{aligned} \tag{4.58}$$

It is interesting to remark that the third initial condition for h_1 can be obtained from Eqs. 4.51 and 4.53 for p = 1 and this becomes:

$$h_1''(0) = \text{Re}\, SF_0''(0)h_1'(0) \tag{4.59}$$

Inserting Eqs. 4.55 and 4.40 into Eq. 4.58, the nonlinear operator N^h has the form

$$N^h(\eta) = \alpha_0 + \alpha_1\eta + \alpha_2\eta^2 + \alpha_3\eta^3 + \alpha_4\eta^4 \tag{4.60}$$

where the coefficients which appear in this equation are given by

$$\begin{aligned} &\alpha_0 = 2\alpha + 1 + 6\alpha\text{Re}\, S;\, \alpha_1 = -48\alpha\text{Re}\, S;\, \alpha_2 = 48\alpha\text{Re}\, S - 3\alpha\text{Re}; \\ &\alpha_3 = 8\alpha\text{Re};\, \alpha_4 = -4\alpha\text{Re} \end{aligned} \tag{4.61}$$

In the same manner like Eq. 4.44, we choose the optimal auxiliary function $H^h(\eta)$ of polynomial form:

$$H^h(\eta, D_i) = D_1 + D_2\eta + D_3\eta^2 + D_4\eta^3 + D_5\eta^4 + D_6\eta^5 + D_7\eta^6 + D_8\eta^7 \tag{4.62}$$

where D_i, i = 1,2,...,8 are unknown parameters.

Using Eqs. 4.60, 4.62 and 4.57, then solving the last equation, it follows that

$$\begin{aligned} h_1(\eta) &= \frac{1}{6}\alpha_0D_1\eta^3 + \frac{1}{24}(\alpha_0D_2 + \alpha_1D_1)\eta^4 + \frac{1}{60}(\alpha_0D_3 + \alpha_1D_2 + \\ &+\alpha_2D_1)\eta^5 + \frac{1}{120}(\alpha_0D_4 + \alpha_1D_3 + \alpha_2D_2 + \alpha_3D_1)\eta^6 + \frac{1}{210}(\alpha_0D_5 + \\ &+\alpha_1D_4 + \alpha_2D_3 + \alpha_3D_2 + \alpha_4D_1)\eta^7 + \frac{1}{336}(\alpha_0D_6 + \alpha_1D_5 + \alpha_2D_4 + \alpha_3D_3 + \\ &+\alpha_4D_2)\eta^8 + \frac{1}{504}(\alpha_0D_7 + \alpha_1D_6 + \alpha_2D_5 + \alpha_3D_4 + \alpha_4D_3)\eta^9 + \\ &+\frac{1}{720}(\alpha_0D_8 + \alpha_1D_7 + \alpha_2D_5 + \alpha_3D_5 + \alpha_4D_4)\eta^{10} + \frac{1}{990}(\alpha_1D_8 + \alpha_2D_7 + \\ &+\alpha_3D_6 + \alpha_4D_5)\eta^{11} + \frac{1}{1320}(\alpha_2D_8 + \alpha_3D_7 + \alpha_4D_6)\eta^{12} + \\ &+\frac{1}{1716}(\alpha_3D_8 + \alpha_4D_7)\eta^{13} + \frac{1}{2184}\alpha_4D_8\eta^{14} + A\eta^2 + B\eta \end{aligned} \tag{4.63}$$

where

$$A = -(2\alpha + 1)M; B = 2\alpha M \tag{4.64}$$

$$\begin{aligned}
M = &\left(\frac{\alpha_0}{6} + \frac{\alpha_1}{24} + \frac{\alpha_2}{60} + \frac{\alpha_3}{120} + \frac{\alpha_4}{210}\right)D_1 + \left(\frac{\alpha_0}{24} + \frac{\alpha_1}{60} + \frac{\alpha_2}{120} + \frac{\alpha_3}{210} +\right.\\
&\left.+\frac{\alpha_4}{336}\right)D_2 + \left(\frac{\alpha_0}{60} + \frac{\alpha_1}{120} + \frac{\alpha_2}{240} + \frac{\alpha_3}{336} + \frac{\alpha_4}{504}\right)D_3 + \left(\frac{\alpha_0}{120} + \frac{\alpha_1}{210} + \frac{\alpha_2}{336} +\right.\\
&\left.+\frac{\alpha_3}{504} + \frac{\alpha_4}{720}\right)D_4 + \left(\frac{\alpha_0}{210} + \frac{\alpha_1}{336} + \frac{\alpha_2}{504} + \frac{\alpha_3}{720} + \frac{\alpha_4}{990}\right)D_3 + \left(\frac{\alpha_0}{336} + \frac{\alpha_1}{504} +\right.\\
&\left.+\frac{\alpha_2}{720} + \frac{\alpha_3}{990} + \frac{\alpha_4}{1320}\right)D_6 + \left(\frac{\alpha_0}{504} + \frac{\alpha_1}{720} + \frac{\alpha_2}{990} + \frac{\alpha_3}{1320} + \frac{\alpha_4}{1716}\right)D_7 +\\
&+\left(\frac{\alpha_0}{720} + \frac{\alpha_1}{990} + \frac{\alpha_2}{1320} + \frac{\alpha_3}{1716} + \frac{\alpha_4}{2184}\right)D_8
\end{aligned} \tag{4.65}$$

The first-order approximate solution of Eqs. 4.21 and 4.18 is obtained from Eqs. 4.53, 4.55 and 4.63:

$$\begin{aligned}
\overline{h}(\eta) = &h_0(\eta) + h_1(\eta) = (B - \alpha)\eta + (A + \alpha)\eta^2 + \frac{1}{6}\alpha_0 D_1 \eta^3 + \frac{1}{24}(\alpha_0 D_2 +\\
&+\alpha_1 D_1)\eta^4 + \frac{1}{60}(\alpha_0 D_3 + \alpha_1 D_2 + \alpha_2 D_3)\eta^5 + \frac{1}{120}(\alpha_0 D_4 + \alpha_1 D_3 +\\
&+\alpha_2 D_2 + \alpha_3 D_1)\eta^6 + \frac{1}{210}(\alpha_0 D_5 + \alpha_1 D_4 + \alpha_2 D_3 + \alpha_3 D_2 + \alpha_4 D_1)\eta^7 +\\
&+\frac{1}{336}(\alpha_0 D_6 + \alpha_1 D_5 + \alpha_2 D_4 + \alpha_3 D_3 + \alpha_4 D_2)\eta^8 + \frac{1}{504}(\alpha_0 D_7 + \alpha_1 D_6 +\\
&+\alpha_2 D_5 + \alpha_3 D_4 + \alpha_4 D_3)\eta^9 + \frac{1}{720}(\alpha_0 D_8 + \alpha_1 D_7 + \alpha_2 D_6 + \alpha_3 D_5 +\\
&+\alpha_4 D_4)\eta^{10} + \frac{1}{990}(\alpha_1 D_8 + \alpha_2 D_7 + \alpha_3 D_6 + \alpha_4 D_5)\eta^{11} + \frac{1}{1320}(\alpha_2 D_8 +\\
&+\alpha_3 D_7 + \alpha_4 D_6)\eta^{12} + \frac{1}{1716}(\alpha_3 D_8 + \alpha_4 D_7)\eta^{13} + \frac{1}{2184}\alpha_4 D_8 \eta^{14}
\end{aligned} \tag{4.66}$$

where A, B and M are given by Eqs. 4.64 and 4.65

The residual given by Eq. 2.35 for the function $\overline{h}(\eta)$ is

$$\begin{aligned}
R^{\overline{h}}(\tau, D_1, D_2, \ldots, D_8) = &\overline{h}''(\eta) + \mathrm{Re}\,\overline{F}(\eta)\overline{h}'(\eta) +\\
&+\mathrm{Re}\,S\left[\overline{F}(\eta)\overline{h}'''(\eta) - \overline{F}''(\eta)\overline{h}''(\eta) - 2\overline{F}'(\eta)\overline{h}''(\eta)\right] + 1
\end{aligned} \tag{4.67}$$

where $\overline{F}(\eta)$ is given by Eq. 4.47.

The parameters $D_1, D_2, \ldots, D_8$ will be determined from the Eqs. 2.37.

4.1.3 Governing Equation of the Temperature and Its Solution

In this section we introduce a temperature field. If T_0 and T_1 are the temperatures of walls of the impermeable and porous plate respectively, which are constant values, then the temperature field can be written as

$$T = (T_1 - T_0)\theta(\eta) + T_0 \tag{4.68}$$

where $\theta(\eta)$ is the dimensionless temperature. Inserting Eqs. 4.68 and 4.17 into Eq. 4.13 we obtain the ordinary differential equation

$$\theta''(\eta) + PeF(\eta)\theta'(\eta) = 0 \tag{4.69}$$

where $Pe = \frac{\rho U d c_p}{k}$ is the Peclet number and c_P is the specific heat at constant pressure. The initial conditions for Eq. 4.69 are:

$$\theta(0) = 0, \theta(1) = 1 \tag{4.70}$$

In order to solve Eqs. 4.69 and 4.70 we choose the linear operator L^{θ} in the form:

$$L^{\theta}(\Theta(\eta, p)) = \frac{\partial^2 \Theta(\eta, p)}{\partial \eta^2} \tag{4.71}$$

and the nonlinear operator as

$$N^{\theta}(\Theta(\eta, p) = Pe\Phi(\eta, p)\frac{\partial \Theta(\eta, p)}{\partial \eta} \tag{4.72}$$

The initial conditions (4.70) become

$$\Theta(0, p) = 0; \Theta(1, p) = 1 \tag{4.73}$$

The equations 4.3 and 4.5 are written respectively as

$$\Theta(\eta, p) = \theta_0(\eta) + p\theta_1(\eta) \tag{4.74}$$

$$\overline{\theta}(\eta, p) = \theta_0(\eta) + \theta_1(\eta) \tag{4.75}$$

For the function $\theta_0(\eta)$, the linear equation (4.4) is

$$\frac{d^2\theta_0(\eta)}{d\eta^2} = 0, \theta_0(0) = 0, \theta_0(1) = 1 \quad (4.76)$$

This equation has the solution

$$\theta_0(\eta) = \eta \quad (4.77)$$

The equation in $\theta_1(\eta)$ is obtained from Eq. 4.9:

$$\frac{d^2\theta_1(\eta)}{d\eta^2} = H^\theta(\eta, K_i)N^\theta(\theta_0(\eta)), \theta_1(0) = 0, \theta_1(1) = 0 \quad (4.78)$$

where

$$N^\theta(\theta_0(\eta)) = PeF_0(\eta)\frac{d\theta_0(\eta)}{d\eta} \quad (4.79)$$

Substituting Eqs. 4.77 and 4.40 into Eq. 4.79, we obtain the nonlinear operator N^θ in the form

$$N^\theta(\theta_0(\eta)) = Pe\left(3\eta^2 - 2\eta^3\right) \quad (4.80)$$

The optimal auxiliary function H^θ is chosen in the form

$$H^\theta(\eta, K_i) = \frac{1}{Pe}\left(K_1 + K_2\eta + K_3\eta^2 + K_4\eta^3 + K_5\eta^4 + K_6\eta^5 + K_7\eta^6 + K_8\eta^7\right) \quad (4.81)$$

with K_i, $i = 1,2,\ldots,8$ unknown parameters.

Using Eqs. 4.77, 4.80 and 4.81 into Eq. 4.78, the equation in $\theta_1(\eta)$ becomes

$$\begin{aligned}\frac{d^2\theta(\eta)}{d\eta^2} &= 3K_1\eta^2 + (3K_2 - 2K_1)\eta^3 + (3K_3 - 2K_2)\eta^4 + (3K_4 - 2K_3)\eta^5 + \\ &+ (3K_5 - 2K_4)\eta^6 + (3K_6 - 2K_5)\eta^7 + (3K_7 - 2K_6)\eta^8 + (3K_8 - 2K_7)\eta^9 - \\ &- 2K_8\eta^{10}, \theta_1(0) = 0, \theta_1(1) = 0\end{aligned} \quad (4.82)$$

From Eq. 4.82 is easy to determine the first-approximate function $\theta_1(\eta)$ such that the first-order approximate solution 4.75 of Eqs. 4.64 and 4.70 can be written in the form:

$$\begin{aligned}\overline{\theta}(\eta) &= \left(1-\frac{3K_1}{20}-\frac{K_2}{12}-\frac{11K_3}{210}-\frac{K_4}{28}-\frac{13K_5}{504}-\frac{7K_6}{360}-\frac{K_7}{66}-\frac{2K_8}{165}\right)\eta+\\&+\frac{K_1}{4}\eta^4+\frac{3K_2-2K_1}{20}\eta^5+\frac{3K_3-2K_2}{30}\eta^6+\frac{3K_4-2K_3}{42}\eta^7+\\&+\frac{3K_5-2K_4}{56}\eta^8+\frac{3K_6-2K_5}{72}\eta^9+\frac{3K_7-2K_6}{90}\eta^{10}+\\&+\frac{3K_8-2K_7}{110}\eta^{11}-\frac{K_8}{60}\eta^{12}\end{aligned} \tag{4.83}$$

From Eq. 2.35 one can obtain the residual for the function $\overline{\theta}(\eta)$:

$$R^{\overline{\theta}}(\eta,K_1,K_2,\ldots,K_8)=\overline{\theta}''(\eta)+Pe\overline{F}(\eta)\overline{\theta}'(\eta) \tag{4.84}$$

where $\overline{F}(\eta)$ is known from Eq. 4.47 and the parameters K_1, K_2, ..., K_8 can be determined from Eqs. 2.36.

4.1.4 *Numerical Results and Discussions*

We illustrate the accuracy of our procedure for different values of the coefficients Re and Pe. Also we represent graphically the behavior of the residual R given by Eq. 2.35 for every variable F, h and θ.

4.1.4.1 First, we will show through numerical examples that the error of the solutions $\overline{F}$ decreases when the number of terms in the optimal auxiliary function $H^{\overline{F}}$ given by Eq. 4.44 increases.

4.1.4.1.A In the case S = 0 and Re = 1, we consider the following three sub-cases (γ = 2):

a) If in the optimal auxiliary function $H^{\overline{F}}$ appear only three parameters C_1, C_2 and C_3, then the first-order approximate solution 4.47 becomes:

$$\begin{aligned}\overline{F}(\eta)=&\left[\left(-\frac{1}{840}-\frac{C+\gamma}{24}\right)C_1+\left(-\frac{1}{630}-\frac{C+\gamma}{60}\right)C_2+\right.\\&\left.+\left(-\frac{1}{720}-\frac{C+\gamma}{120}\right)C_3+3\right]\eta^2+\left[\frac{C+\gamma}{12}C_1+\left(\frac{11}{7560}+\frac{C+\gamma}{60}\right)C_2+\right.\\&\left.+\left(\frac{11}{7560}+\frac{C+\gamma}{90}\right)C_3-2\right]\eta^3-\frac{C+\gamma}{24}C_1\eta^4-\frac{C+\gamma}{120}C_2\eta^5+\\&+\left(\frac{C_1}{120}+\frac{C+\gamma}{360}C_3\right)\eta^6+\left(\frac{C_2}{280}-\frac{C_3}{105}\right)\eta^7+\left(\frac{C_1}{420}-\frac{C_2}{210}+\frac{C_3}{560}\right)\eta^8+\\&+\left(\frac{C_2}{756}-\frac{C_3}{378}\right)\eta^9+\frac{C_3}{1260}\eta^{10}\end{aligned}\tag{4.85}$$

From the system 2.37 the parameters C_i can be identified as

$$C_1=0.19309400178270444,\ C_2=0.9494362926999727,$$
$$C_3=-0.6839056036633757,\ C=-14.394364490334949$$

Substituting these values in expression 4.85 we obtain

$$\begin{aligned}\overline{F}(\eta)=&3.224419321\eta^2-2.399060748\eta^3+0.099720136\eta^4+\\&+0.09806412\eta^5-0.021937017\eta^6+0.00155485\eta^7-0.00528264\eta^8+\\&+0.00306534\eta^9-0.000542782\eta^{10}\end{aligned}\tag{4.86}$$

In Fig. 4.2 we present graphically the residual 4.48 for the solution 4.86. It can be observed that the residual error is less than 0.15.

Figures 4.3 and 4.4 show the approximate solutions 4.86 obtained by OHAM for $\overline{F}(\eta)$ and $\overline{F}'(\eta)$ respectively, with fixed values Re = 1 and S = 0. In Table 4.1 are presented some values of the functions F(η) and F'(η) for Re = 1 and S = 0 with three parameters in the solution 4.47.

b) In the second case we consider five parameters in the expression of $H^{\overline{F}}$ (q = 5), namely $C_1, C_2, \ldots, C_5$. In this case, the first-order approximate solution 4.47 can be written as

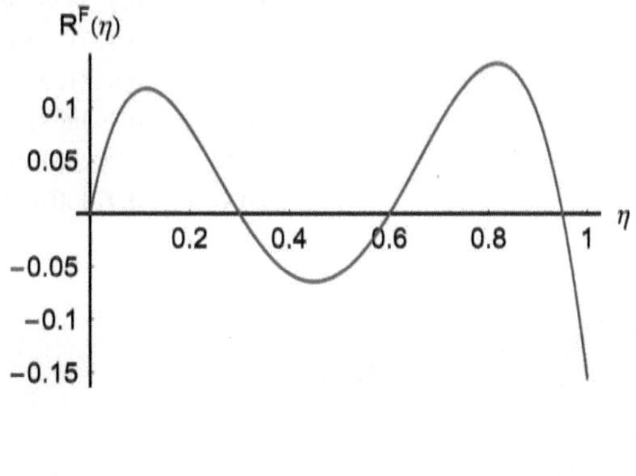

Fig. 4.2 The residual 4.48 of the solution 4.86 for Re = 1, S = 0

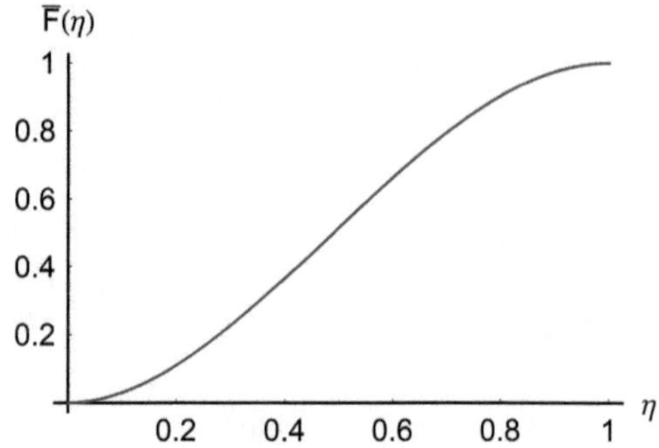

Fig. 4.3 The first-order approximate solution 4.86 for Re = 1, S = 0

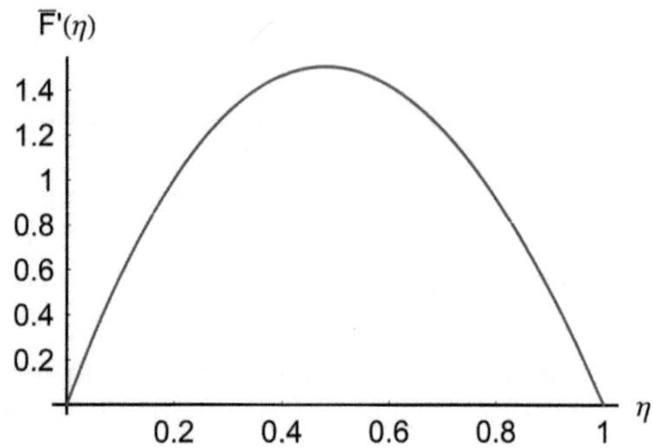

Fig. 4.4 Tangential velocity profile $\overline{F}'(\eta)$ obtained from 4.86 by OHAM for Re = 1, S = 0

Table 4.1 Values of the normal velocity $\overline{F}(\eta)$ and tangential velocity $\overline{F}'(\eta)$ for Re = 1, S = 0 and three parameters in Eq. 50

η	$\overline{F}(\eta)$	$\overline{F}'(\eta)$
0	0	0
0.1	0.029856093814847568	0.5733592550662049
0.2	0.10997394522480126	1.005815311289541
0.3	0.22645345783268747	1.3013288967336292
0.4	0.36583466870307335	1.4647133102335654
0.5	0.5151741691953773	1.5014552614906496
0.6	0.66210290768241	1.4175221263143234
0.7	0.7948642300066463	1.2191606085990163
0.8	0.9023316913087013	0.9126950448190263
0.9	0.9740070765821124	0.5043348613719163
1	1	0

$$\begin{aligned}\overline{F}(\eta) = \Bigg[&\left(-\frac{1}{840}-\frac{C+\gamma}{24}\right)C_1+\left(-\frac{1}{630}-\frac{C+\gamma}{60}\right)C_2+\left(-\frac{1}{720}-\frac{C+\gamma}{120}\right)C_3+\\
&+\left(-\frac{31}{27720}-\frac{C+\gamma}{210}\right)C_4+\left(-\frac{7}{7920}-\frac{C+\gamma}{336}\right)C_5+3\Bigg]\eta^2+\Bigg[\frac{C+\gamma}{12}C_1+\\
&+\left(\frac{11}{7560}+\frac{C+\gamma}{40}\right)C_2+\left(\frac{11}{7560}+\frac{C+\gamma}{90}\right)C_3+\left(\frac{67}{53440}+\frac{C+\gamma}{168}\right)C_4+\\
&+\left(\frac{2}{2079}+\frac{C+\gamma}{280}\right)C_5-2\Bigg]\eta^3-\frac{C+\gamma}{24}C_1\eta^4-\frac{C+\gamma}{120}C_2\eta^5+\left(\frac{C_1}{120}-\right.\\
&\left.-\frac{C_3(C+\gamma)}{360}\right)\eta^6+\left(\frac{C_2}{280}-\frac{C_3}{105}-\frac{C+\gamma}{840}C_4\right)\eta^7+\left(\frac{C_1}{420}-\frac{C_2}{210}+\frac{C_3}{560}-\right.\\
&\left.-\frac{C_5(C+\gamma)}{1680}\right)\eta^8+\left(\frac{C_2}{756}-\frac{C_3}{378}+\frac{C_4}{1008}\right)\eta^5+\left(\frac{C_3}{1260}-\frac{C_4}{630}+\frac{C_5}{1680}\right)\eta^{10}+\\
&+\left(\frac{C_4}{1980}-\frac{C_5}{990}\right)\eta^{11}+\frac{C_5}{2970}\eta^{12}\end{aligned}\tag{4.87}$$

The optimal convergence-control parameters C_i are:

$$\begin{aligned}&C_1 = -0.0011417381118812321,\ C_2 = 3.4132543261879045,\\ &C_3 = -7.9463528678664295,\ C_4 = 6.709946659576214,\\ &C_5 = -1.3947922992799233,\ C = -14.365979968537445\end{aligned}$$

With these values, the solution 4.87 becomes

$$\begin{aligned}\overline{F}(\eta) &= 3.227148663\eta^2 - 2.394329995\eta^3 - 0.0005882793\eta^4+\\ &+0.3517354619\eta^5 - 0.272966263\eta^6 + 0.110980919\eta^7 - 0.040712857\eta^8+\\ &+0.0321937\eta^9 - 0.0177876\eta^{10} + 0.00479774\eta^{11} - 0.000469617\eta^{12}\end{aligned}\tag{4.88}$$

In Fig. 4.5 is shown the residual 4.48 for the solution 4.88. In this case, the residual error is less than 0.0005.

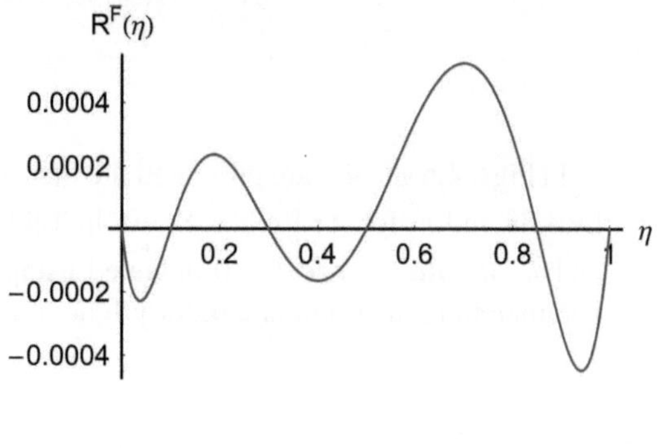

Fig. 4.5 The residual 4.48 of the solution 4.88 for Re = 1, S = 0

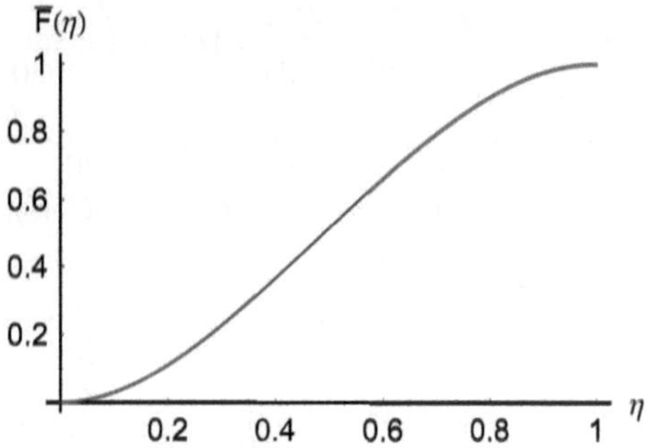

Fig. 4.6 The first-order approximate solution 4.88 for Re = 1, S = 0

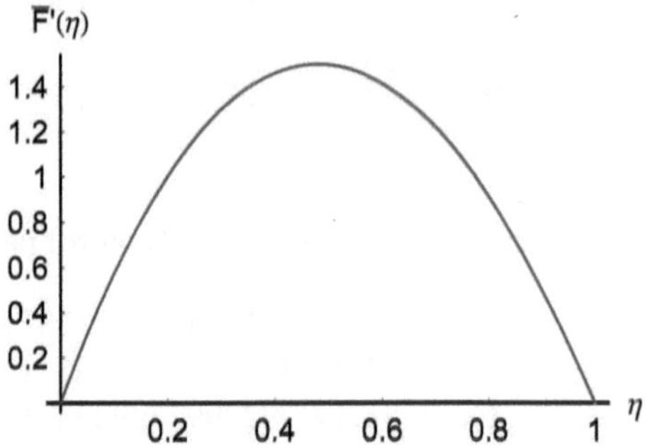

Fig. 4.7 Tangential velocity profile $\overline{F}'(\eta)$ obtained from 4.88 for Re = 1, S = 0

Table 4.2 Values of the normal velocity F and tangential velocity F′ for Re = 1, S = 0 and five parameters in Eq. 4.47

η	$\overline{F}(\eta)$	$\overline{F}'(\eta)$
0	0	0
0.1	0.02988033655185677	0.5737573890575263
0.2	0.11002671639907345	1.0058563797761386
0.3	0.22646941903561768	1.3005319287306907
0.4	0.3657369611585639	1.4633366593970931
0.5	0.5149411085391331	1.500268652447796
0.6	0.6617930327405382	1.417261613747082
0.7	0.7945853290173125	1.2200061971586125
0.8	0.902171106480404	0.9140624914052522
0.9	0.9739668049513972	0.5052129245853372
1	1	0

In Figs. 4.6 and 4.7 are presented the first-order approximate solutions $\overline{F}$ given by 4.88 and $\overline{F}'$ for for Re = 1, S = 0. In Table 4.2 are presented the values of $\overline{F}$ and $\overline{F}'$ for Re = 1 and S = 0 obtained using five optimal convergence-control parameters in the optimal auxiliary function 4.44.

c) In the third case, we consider eight optimal convergence-control parameters in Eq. 4.44 (q = 8). In this case, the first-order approximate solution 4.47 becomes

$$\begin{aligned}\overline{F}(\eta) &= \left[\left(-\frac{1}{840}-\frac{C+\gamma}{24}\right)C_1+\left(-\frac{1}{630}-\frac{C+\gamma}{60}\right)C_2+\left(-\frac{1}{720}-\frac{C+\gamma}{120}\right)C_3+\right.\\ &+\left(-\frac{31}{27720}-\frac{C+\gamma}{210}\right)C_4+\left(-\frac{7}{7920}-\frac{C+\gamma}{336}\right)C_5+\left(-\frac{1}{1430}-\frac{C+\gamma}{504}\right)C_6+\\ &\left.+\left(-\frac{67}{120120}-\frac{C+\gamma}{720}\right)C_7+\left(-\frac{9}{20020}-\frac{C+\gamma}{990}\right)C_8+3\right]\eta^2+\left[\frac{C+\gamma}{12}C_1+\right.\\ &+\left(\frac{11}{7560}-\frac{C+\gamma}{40}\right)C_2+\left(\frac{11}{7560}+\frac{C+\gamma}{90}\right)C_3+\left(\frac{67}{53440}+\frac{C+\gamma}{168}\right)C_4+\left(\frac{2}{2079}+\right.\\ &\left.+\frac{C+\gamma}{280}\right)C_5+\left(\frac{47}{61776}+\frac{C+\gamma}{432}\right)C_6+\\ &\left.+\left(\frac{109}{180180}+\frac{C+\gamma}{630}\right)C_7+\left(\frac{5}{10296}+\frac{C+\gamma}{880}\right)C_8-2\right]\eta^3-\frac{C+\gamma}{24}C_1\eta^4-\\ &-\frac{C+\gamma}{120}C_2\eta^5+\left(\frac{C_1}{120}-\frac{C+\gamma}{360}C_3\right)\eta^6+\left(\frac{C_2}{280}-\frac{C_3}{105}-\frac{C+\gamma}{840}C_4\right)\eta^7+\\ &+\left(\frac{C_1}{420}-\frac{C_2}{210}+\frac{C_3}{560}-\frac{C+\gamma}{1680}C_5\right)\eta^8+\left(\frac{C_2}{756}-\frac{C_3}{378}+\frac{C_4}{1008}-\right.\\ &\left.-\frac{C+\gamma}{3024}C_6\right)\eta^9+\left(\frac{C_3}{1260}-\frac{C_4}{630}+\frac{C_5}{1680}-\frac{C+\gamma}{5040}C_7\right)\eta^{10}+\left(\frac{C_4}{1980}-\frac{C_5}{990}+\right.\\ &\left.+\frac{C_6}{2640}-\frac{C+\gamma}{7920}C_8\right)\eta^{11}+\left(\frac{C_5}{2970}-\frac{C_6}{1485}+\frac{C_7}{3960}\right)\eta^{12}+\left(\frac{C_6}{4290}-\frac{C_7}{2145}+\right.\\ &\left.+\frac{C_8}{5720}\right)\eta^{13}+\left(\frac{C_7}{6006}-\frac{C_8}{3003}\right)\eta^{14}+\frac{C_8}{890}\eta^{15}\end{aligned} \tag{4.89}$$

From Eq. 2.37, we obtain the following set of parameters:

$C_1 = -0.0000134061531729063 76$, $C_2 = 3.3697440165764467$,
$C_3 = -7.51611441025494$, $C_4 = 4.879315944102504$,
$C_5 = 2.626681520311593$, $C_6 = -4.852105180959477$,
$C_7 = 3.0966346762608636$, $C_8 = -0.8248995116644152$,
$C = -14.365856936378094$

The solution 4.89 with the parameters given above becomes:

$$\begin{aligned}\overline{F}(\eta) &= 3.227133536\eta^2-2.394309489\eta^3-6.9074440261\cdot 10^{-6}\eta^4+\\ &+0.347248101\eta^5+0.34724799\eta^6+0.08386476\eta^7-0.010134082\eta^8+\\ &+0.009343621\eta^9-0.004548869\eta^{10}-0.003314783\eta^{11}+0.004933791\eta^{12}-\\ &-0.00271889\eta^{13}+0.000790282\eta^{14}-0.00010072\eta^{15}\end{aligned} \tag{4.90}$$

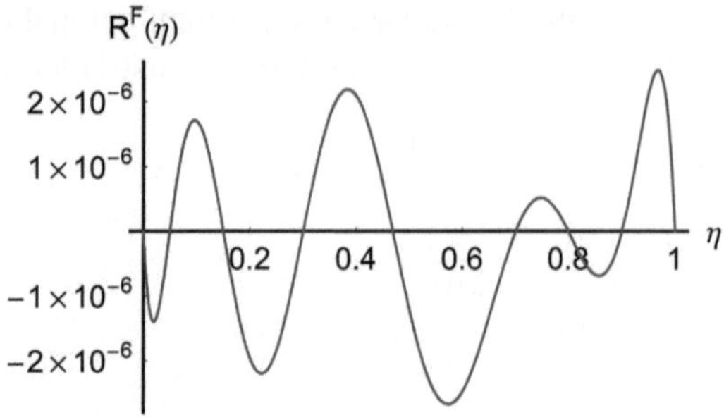

Fig. 4.8 The residual 4.48 obtained using the solution 4.90

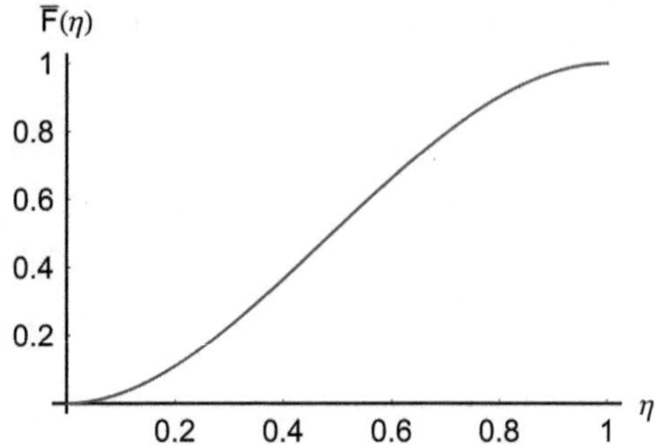

Fig. 4.9 The first-order approximate solution 4.90 for Re = 1, S = 0

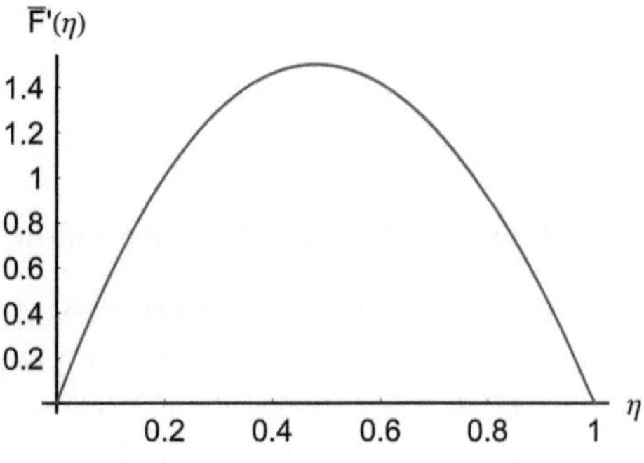

Fig. 4.10 Tangential velocity profile $\overline{F}'(\eta)$ obtained from 4.90 for Re = 1, S = 0

In Fig. 4.8 is shown the residual 4.48 obtained using the solution 4.90. In this third sub-case, the residual error is less than 3×10^{-6}.

In Figs. 4.9 and 4.10 are plotted the solutions $\overline{F}$ given by Eq. 4.90 and $\overline{F}'$, respectively. Table 4.3 presents values of $\overline{F}$ and $\overline{F}'$ for Re = 1 and S = 0 obtained using eight convergence-control parameters in the Eq. 4.44.

4.1.4.1.B In the case Re = 10 and S = 0 ($\gamma = 2$), from Eqs. 4.43, 4.45 and 2.37 employing eight optimal convergence-control parameters in the auxiliary function 4.44, we obtain

Table 4.3 Values of $\overline{F}$ and $\overline{F}'$ for Re = 1, S = 0 obtained using eight convergence-control parameters

η	$\overline{F}(\eta)$	$\overline{F}'(\eta)$
0	0	0
0.1	0.029880234285309434	0.5737558382596908
0.2	0.11002644843893622	1.005854525318963
0.3	0.2264689171028822	1.300529074044184
0.4	0.3657361601757617	1.4633339124097418
0.5	0.5149401434545289	1.5002684843275447
0.6	0.6617922297117911	1.4172649174340122
0.7	0.7945849514188462	1.2200108338240183
0.8	0.902171108858197	0.9140649462072055
0.9	0.9739668819518303	0.5052122245842189
1	1	0

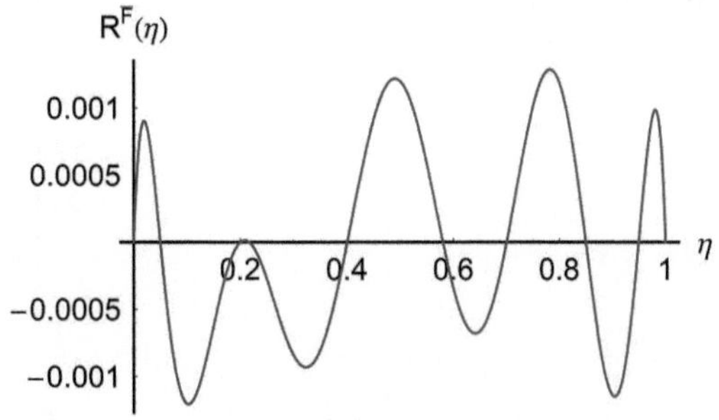

Fig. 4.11 The residual 4.48 obtained using the solution 4.91 for Re = 10, S = 0

$$C_1 = 0.0029973478951901326,\ C_2 = 27.677764444737328,$$
$$C_3 = -101.63465599093996,\ C_4 = 75.38940522138608,$$
$$C_5 = 187.4618153135904,\ C_6 = -411.5239871765859,$$
$$C_7 = 304.0821030268297,\ C_8 = -80.97791376399562$$
$$C = -38.1024429206438$$

The solution 4.47 in this case becomes

$$\overline{F}(\eta) = 5.018436362\eta^2 - 6.350407153\eta^3 + 0.004508798\eta^4 + 8.326956253\eta^5 - $$
$$-0.282068221\eta^6 + 1.077955301\eta^7 - 3.021236021\eta^8 - 1.110267047\eta^9 + $$
$$+1.290760571\eta^{10} - 3.440729428\eta^{11} + 4.170275\eta^{12} - 2.5184754131\eta^{13} + $$
$$+0.775954014\eta^{14} - 0.098874134\eta^{15} \tag{4.91}$$

In Fig. 4.11 is plotted the residual 4.48 obtained using the solution 4.91. One can be observed that the residual error is less than 0.0015. In Figs. 4.12 and 4.13 we show the solution $\overline{F}$ given by Eq. 4.91 and $\overline{F}'$, respectively, for Re = 10, S = 0.

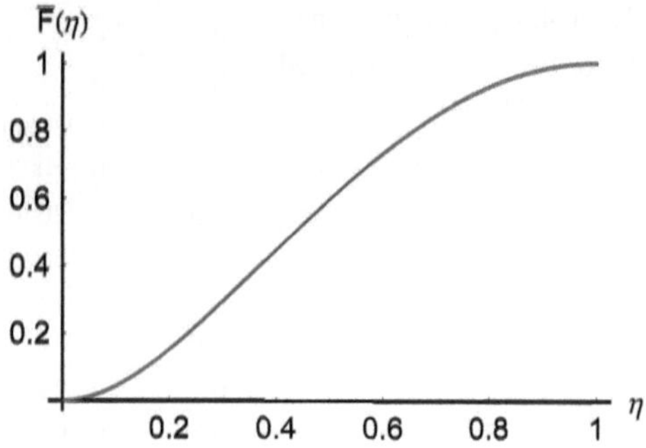

Fig. 4.12 The first-order approximate solution 4.91 for Re = 10, S = 0

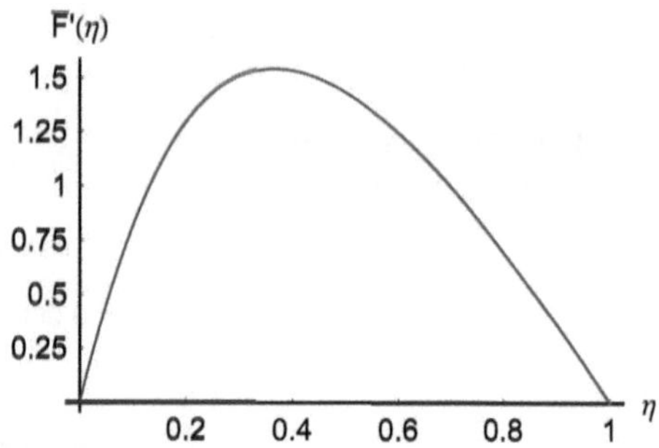

Fig. 4.13 Tangential velocity profile $\overline{F}'(\eta)$ obtained from Eq. 4.91 for Re = 10, S = 0

Table 4.4 Values of $\overline{F}$ and $\overline{F}'$ for Re = 10, S = 0 obtained using eight convergence-control parameters

η	$\overline{F}(\eta)$	$\overline{F}'(\eta)$
0	0	0
0.1	0.043907644671834084	0.8167698504812213
0.2	0.15200858185611146	1.2944699677637446
0.3	0.2940018342557843	1.5081553059358808
0.4	0.44742566547659623	1.534792248741867
0.5	0.5967483191342562	1.4343571195677371
0.6	0.731397435840506	1.246409938530867
0.7	0.8439245518795052	0.9950107001902657
0.8	0.9288019890482625	0.6955792406083332
0.9	0.9818391182925115	0.3600971953794443
1	1	0

In Table 4.4 are given values of $\overline{F}$ and $\overline{F}'$ for Re = 10 and S = 0 obtained using eight convergence-control parameters in the auxiliary function 4.44.

For the cases 4.1.4.1.A and subcases a), b) and c) presented above corresponding to S = 0 (Newtonian fluid) it is clear that the error of the normal velocity $\overline{F}$ decreases if the number of optimal convergence-control parameters involved in the optimal auxiliary function H^F increases. In the cases 4.1.4.1.A and subcase c) and 4.1.4.1.B

Table 4.5 Comparison between approximate and numerical solutions of Eq. 4.26 for Re = 1, S = 0

η	$\overline{F}(\eta)$ from Eq. 4.86	$\overline{F}(\eta)$ from Eq. 4.88	$\overline{F}(\eta)$ from Eq. 4.90	$F(\eta)$_*numerical*
0	0	0	0	0
0.1	0.0298560938148475	0.02988033655185677	0.0298802342853094	0.0298802476532127
0.2	0.1099739452248012	0.11002671639907345	0.1100264484389362	0.1100264530205633
0.3	0.2264534578326874	0.22646941903561768	0.2264689171028822	0.2264689161543199
0.4	0.3658346687030733	0.3657369611585639	0.3657361601757617	0.3657361660538983
0.5	0.5151741691953773	0.5149411085391331	0.5149401434545289	0.5149401503386936
0.6	0.66210290768241	0.6617930327405382	0.6617922297117911	0.661792237113393
0.7	0.7948642300066463	0.7945853290173125	0.7945849514188462	0.7945849522939098
0.8	0.9023316913087013	0.902171106480404	0.902171108858197	0.9021711071056759
0.9	0.9740070765821124	0.9739668049513972	0.9739668819518303	0.9739668798961164
1	1	1	1	1

Table 4.6 Analysis of the relative errors of approximate solutions for Re = 1, S = 0

η	$\varepsilon_1 = \lvert F_{num} - F_{4.86} \rvert$	$\varepsilon_2 = \lvert F_{num} - F_{4.88} \rvert$	$\varepsilon_3 = \lvert F_{num} - F_{4.90} \rvert$
0	0	0	0
0.1	0.000024153838365	0.0000000888986440	0.0000000133679033
0.2	0.000052507795762	0.0000002633785101	0.0000000045816271
0.3	0.000015458321632	0.0000005028812977	0.0000000009485623
0.4	0.000098502649175	0.0000007951046656	0.0000000058781366
0.5	0.000234018856683	0.0000009582004395	0.0000000068841647
0.6	0.000310670569017	0.0000007956271452	0.0000000074016019
0.7	0.000279277712736	0.0000003767234027	0.0000000008750636
0.8	0.000160584203025	0.0000000006252719	0.0000000017525211
0.9	0.000040196685996	0.0000000749447192	0.0000000020557139
1	0	0	0

it results that for S = 0 the error of the solution illustrated by the residual 4.48 decreases if the cross-flow Reynolds number increases in the condition that the number of optimal convergence-control parameters involved in the optimal auxiliary function $H^{\overline{F}}$ are the same. From the cases 4.1.4.1.A and subcases a), b) and c) we remark that the lower value of residual error is obtained when eight optimal convergence-control parameters are involved. Thus, it is natural to consider the correct value of the constant C as C = −14.365856936378094 in order to solve numerically the Eq. 4.26. For the best of our knowledge, for the first time, this method allows a rigorous determination of the constant C and thereafter solving Eq. 4.26 in a rigorous way from mathematical point of view.

Table 4.5 presents a comparison between approximate solutions given by Eqs. 4.86, 4.88 and 4.90 respectively, and the numerical solution of Eq. 4.26 obtained using a shooting method combined with a RK-4 algorithm, for Re = 1 and S = 0.

An analysis of the errors of approximate solutions considered above is presented in Table 4.6, where the relative error is computed as $\varepsilon_i = \lvert F_{num} - F_{OHAM} \rvert$.

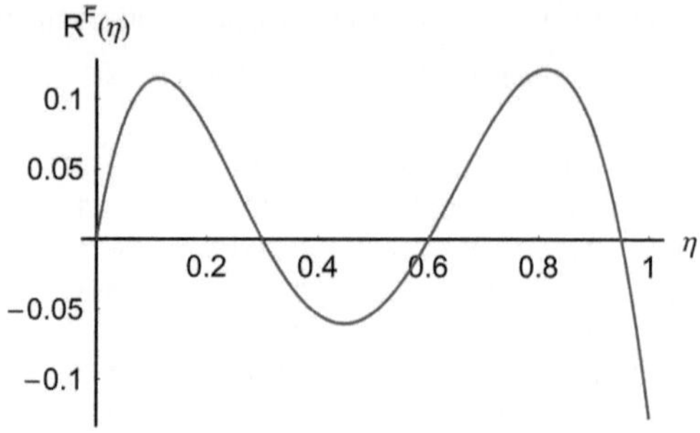

Fig. 4.14 The residual 4.48 obtained using the solution 4.92 for Re = 1, S = 0.2

One can remark that the corresponding residual errors obtained by replacing the approximate solutions into the initial differential equations are much higher than the relative errors presented in Table 4.6 for different cases (see Fig. 4.2 and the error ε_1 from Table 4.6: the residual error is less than 0.5 and ε_1 is less than 3.1×10^{-4}. This proves that OHAM leads to very good errors in all presented cases, even if the residual error shows moderate values.

4.1.4.1.C In the case Re = 1 and S = 0.2, we consider three sub-cases:

a) If in the optimal auxiliary function H^F are involved three optimal convergence-control parameters C_1, C_2 and C_3, then the first-order approximate solution for the normal velocity is given by Eq. 4.85 with $\gamma = 0.8$.

From the system 2.37 we obtain the values of the optimal convergence-control parameters

$$C_1 = 0.4207639577948163,\ C_2 = 3.126447815688742,$$
$$C_3 = -2.0911129135696975,\ C = -6.340416289581745$$

With these parameters known, the solution 4.47 becomes

$$\overline{F}(\eta) = 3.286724166\eta^2 - 2.497074058\eta^3 + 0.097133535\eta^4 + \\ +0.14434844\eta^5 - 0.02867599\eta^6 + 0.00715861\eta^7 - 0.0176202\eta^8 + \\ +0.00966796\eta^9 - 0.00165961\eta^{10} \qquad (4.92)$$

Figure 4.14 shows the residual 4.48 obtained by replacing the solution 4.92. In this case, the residual error is less than 0.13.

In Figs. 4.15 and 4.16 are graphically shown the normal velocity profile $\overline{F}$ given by Eq. 4.92 and the tangential velocity profile $\overline{F}'$ obtained from Eq. 4.92, respectively, for Re = 1 and S = 0.2.

In Table 4.7 are given some values of $\overline{F}$ and $\overline{F}'$ obtained for Re = 1 and S = 0.2 using three convergence-control parameters in the auxiliary function.

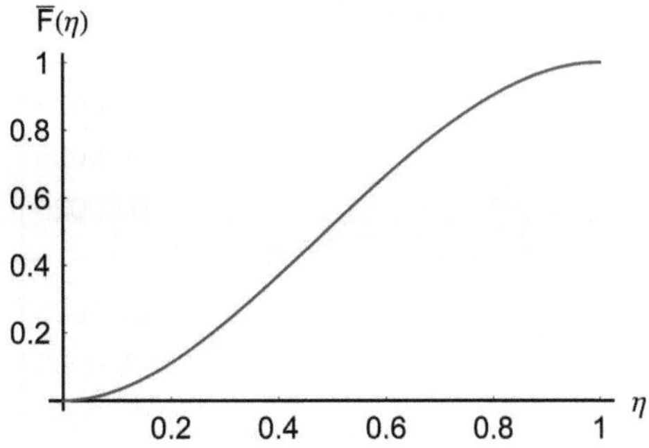

Fig. 4.15 The first-order approximate solution—normal velocity profile $\overline{F}$ given by Eq. 4.92 for Re = 1, S = 0.2

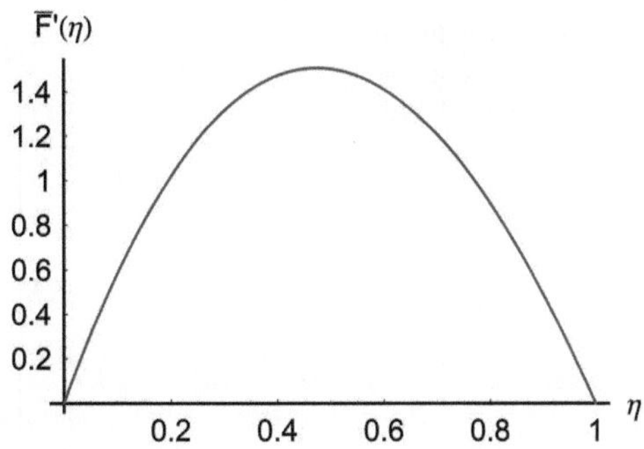

Fig. 4.16 Tangential velocity profile $\overline{F}'(\eta)$ obtained from Eq. 4.92 for Re = 1, S = 0.2

Table 4.7 Values of $\overline{F}$ and $\overline{F}'$ obtained from Eq. 4.92 for Re = 1, S = 0.2

η	$\overline{F}(\eta)$	$\overline{F}'(\eta)$
0	0	0
0.1	0.030381296768481388	0.5828916148469219
0.2	0.1116921850961173	1.0192501760677957
0.3	0.22950135889651146	1.3137534246087685
0.4	0.36991282173052326	1.4723905397478443
0.5	0.5196845303764669	1.502204842701805
0.6	0.666319159612472	1.410994466324505
0.7	0.7981233230003537	1.2069840883700644
0.8	0.9042334939950457	0.8984918178071792
0.9	0.9746095793193544	0.49362029327833556
1	1	0

b) If we consider five optimal convergence-control parameters in the auxiliary function H^F, C_1, C_2,. . .,C_5, then the first-order approximate solution is given by Eq. 4.47 with $\gamma = 0.8$.

The optimal convergence-control parameters C_i determined from Eqs. 2.37 are

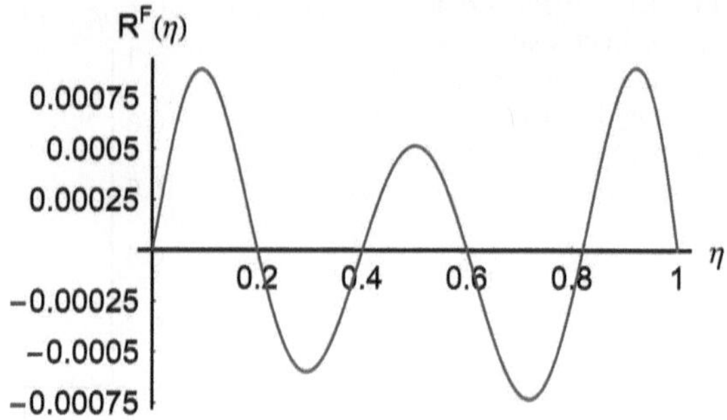

Fig. 4.17 The residual 4.48 obtained using the solution 4.93 for Re = 1, S = 0.2

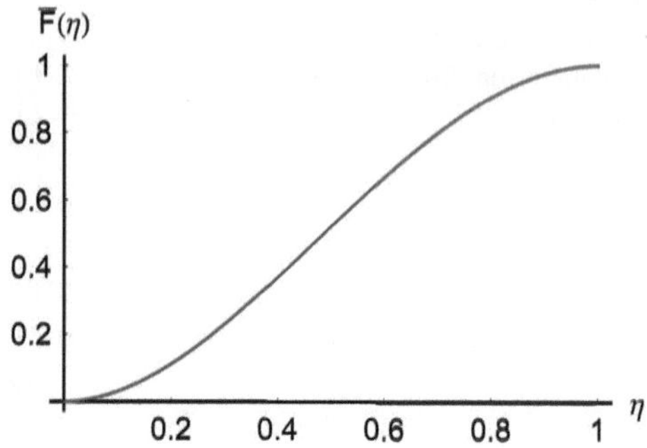

Fig. 4.18 The first-order approximate solution $\overline{F}$ given by Eq. 4.93 for Re = 1, S = 0.2

$$C_1 = 0.0022233658186828695,\ C_2 = 7.918027144402161,$$
$$C_3 = -13.97679009950881,\ C_4 = 9.124354518406268,$$
$$C_5 = -1.4385351756416804,\ C = -6.333124306683892$$

In this case, the solution 4.47 becomes

$$\overline{F}(\eta) = 3.300273455\eta^2 - 2.500285494\eta^3 + 0.0005125897424\eta^4 + \\ +0.365095464\eta^5 - 0.214801703\eta^6 + 0.088359956\eta^7 - 0.06739605348\eta^8 + \\ +0.0565012\eta^9 - 0.0264321\eta^{10} + 0.00606133\eta^{11} - 0.000484355\eta^{12} \tag{4.93}$$

In Fig. 4.17 is presented the residual 4.48 for the normal velocity $\overline{F}$ given by 4.93. We can observe that the residual error is less than 0.0009.

In Figs. 4.18 and 4.19 are graphically presented the normal velocity profile $\overline{F}$ given by Eq. 4.93 and the tangential velocity profile $\overline{F}'$ obtained from Eq. 4.93, respectively, for Re = 1 and S = 0.2.

In Table 4.8 we present some values of $\overline{F}$ and $\overline{F}'$ for Re = 1, S = 0.2 obtained using five optimal convergence-control parameters in the optimal auxiliary function.

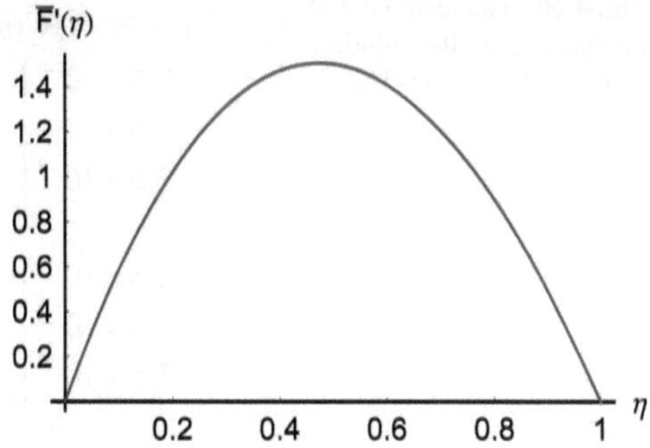

Fig. 4.19 Tangential velocity profile $\overline{F}'(\eta)$ obtained from Eq. 4.93 for Re = 1, S = 0.2

Table 4.8 Values of $\overline{F}$ and $\overline{F}'$ obtained from Eq. 4.93 for Re = 1, S = 0.2

η	$\overline{F}(\eta)$	$\overline{F}'(\eta)$
0	0	0
0.1	0.03050392683226389	0.5850980222638077
0.2	0.11207346048227051	1.0219128013499723
0.3	0.23010530665995502	1.3153568040832484
0.4	0.3705938255290741	1.4723060432804351
0.5	0.5202813836556429	1.5007135908772542
0.6	0.6667301430080141	1.4089256118346984
0.7	0.7983359569882735	1.2052091699131469
0.8	0.9043058795770955	0.8974962042035474
0.9	0.9746204599115392	0.49333182262879893
1	1	0

c) Considering eight optimal convergence-control parameters $C_1, C_2 \ldots, C_8$ in the optimal auxiliary function H^F, the first-order approximate solution is given by Eq. 4.89 with $\gamma = 0.8$.

The optimal values of the convergence-control parameters are obtained as:

$$C_1 = -0.00016890514876258,\ C_2 = 7.887480164858656,$$
$$C_3 = -12.918278938989813,\ C_4 = 2.4485335923720677,$$
$$C_5 = 16.4954090073686,\ C_6 = -23.91702648509455,$$
$$C_7 = 15.621837886355378,\ C_8 = -3.9883048598198605,$$
$$C = -6.333224002288458$$

The solution 4.47 in this case becomes:

$$\begin{aligned}\overline{F}(\eta) &= 3.300941439\eta^2 - 2.508368852\eta^3 - 3.894122589\cdot 10^{-5}\eta^4 + \\ &+ 0.36369328\eta^5 - 0.198556171\eta^6 + 0.044300108\eta^7 - 0.006299133489\eta^8 + \\ &+ .00327494404\eta^9 + 0.012830145\eta^{10} - 0.027270792\eta^{11} + 0.0256047\eta^{12} - \\ &- 0.0135552\eta^{13} + 0.00392915\eta^{14} - 0.000486973\eta^{15}\end{aligned} \tag{4.94}$$

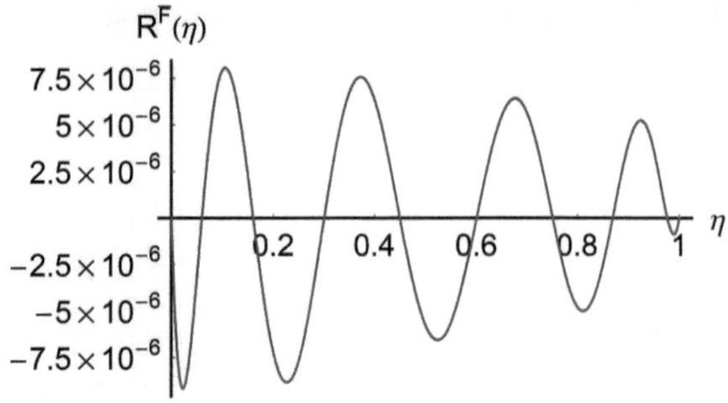

Fig. 4.20 The residual 4.48 obtained using the solution 4.94 for Re = 1, S = 0.2

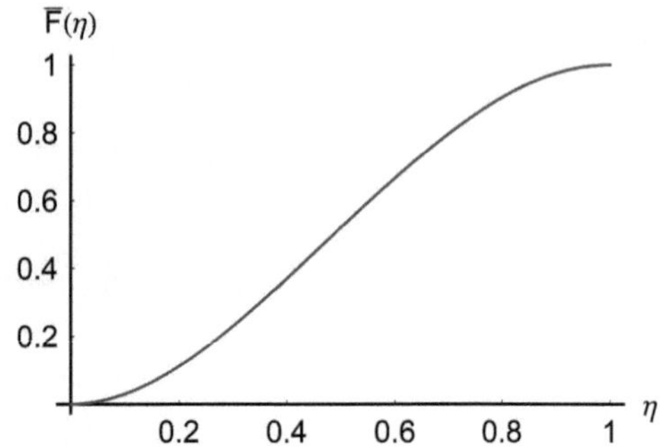

Fig. 4.21 Normal velocity profile $\overline{F}$ given by Eq. 4.94 for Re = 1, S = 0.2

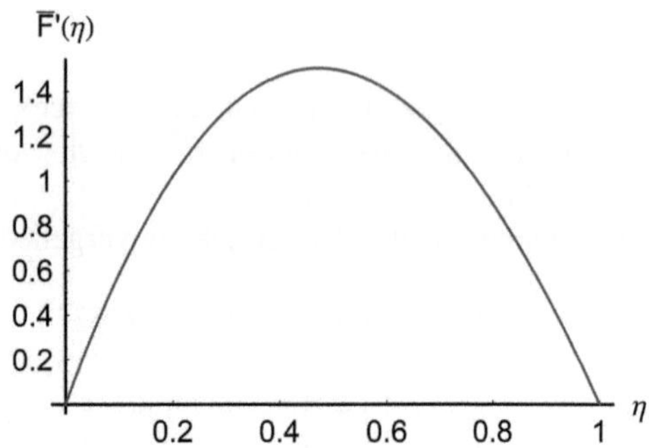

Fig. 4.22 Tangential velocity profile $\overline{F}'(\eta)$ obtained from Eq. 4.94 for Re = 1, S = 0.2

In Fig. 4.20 is presented the residual 4.48 obtained for the approximate solution 4.94 for Re = 1 and S = 0.2. It results that the residual error is less than 8×10^{-6}.

In Figs. 4.21 and 4.22 are shown the first-order approximate solution $\overline{F}$ given by Eq. 4.94 and the corresponding $\overline{F}'$, respectively, for Re = 1 and S = 0.2.

In Table 4.9 are presented some values of $\overline{F}$ and $\overline{F}'$ for Re = 1, S = 0.2, obtained using eight convergence-control parameters in the auxiliary function.

From these three sub-cases presented here for the non-Newtonian fluids ($S \neq 0$) we remark that the error of the first-order approximate solution $\overline{F}(\eta)$ decreases if

Table 4.9 Values of $\overline{F}$ and $\overline{F}'$ obtained from Eq. 4.94 for Re = 1, S = 0.2

η	$\overline{F}(\eta)$	$\overline{F}'(\eta)$
0	0	0
0.1	0.03050448299981235	0.5851073108755205
0.2	0.11207487997126836	1.021918915483842
0.3	0.23010690876762668	1.3153545057621339
0.4	0.37059496012089505	1.4723001922280714
0.5	0.5202820137748512	1.5007099022110941
0.6	0.666730546767414	1.408924408428650
0.7	0.798336254016446	1.2052078865556086
0.8	0.9043060139333776	0.8974944505757069
0.9	0.9746204676554932	0.4933312522296885
1	1	0

the number of optimal convergence-control parameters involved in the optimal auxiliary function H^F increases.

4.1.4.1.D In what follows we consider Re = 1 and S = 0.3 and we employ an optimal auxiliary function H^F containing eight optimal convergence-control parameters. In this case, the parameters C_i obtained from Eqs. 2.36 are

$$C_1 = 0.020669771298635353,\ C_2 = 23.834347709927023,$$
$$C_3 = -25.183995732113353,\ C_4 = -61.902691880355285,$$
$$C_5 = 237.44759712603766,\ C_6 = -327.2190471411699,$$
$$C_7 = 220.3319337511033,\ C_8 = -57.45562095656968,$$
$$C = -2.026485871127701$$

The first-order approximate solution in this case is

$$\begin{aligned}\overline{F}(\eta) &= 3.351501081\eta^2 - 2.584254544\eta^3 + 0.001573043552\eta^4 + \\ &+ 0.362775827\eta^5 - 0.127600564\eta^6 - 0.049634654\eta^7 + 0.099732494\eta^8 - \\ &- 0.160899366\eta^9 + 0.299456648\eta^{10} - 0.408306889\eta^{11} + 0.355937589\eta^{12} - \\ &- 0.189038375\eta^{13} + 0.055818044\eta^{14} - 0.007013338\eta^{15}\end{aligned} \tag{4.95}$$

In Fig. 4.23 we present the residual 4.48 obtained for the first-order approximate solution 4.95 for Re = 1 and S = 0.3, which is less than 6×10^{-4}.

In Figs. 4.24 and 4.25 we present graphically the first-order approximate solution $\overline{F}$ given by Eq. 4.95 and the corresponding $\overline{F}'$, respectively, for Re = 1 and S = 0.3.

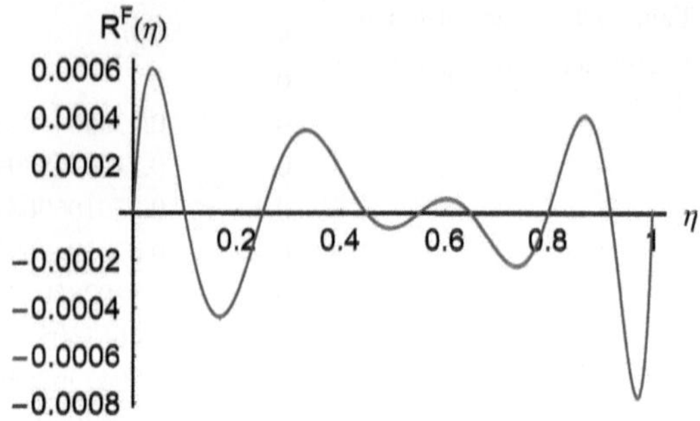

Fig. 4.23 The residual 4.48 obtained using the solution 4.95 for Re = 1, S = 0.3

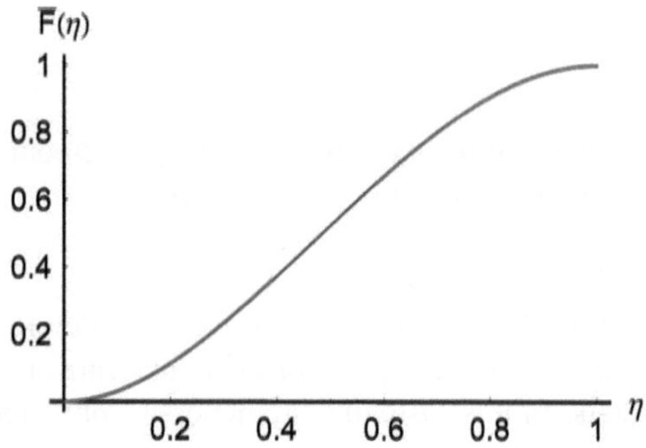

Fig. 4.24 Normal velocity profile $\overline{F}$ given by Eq. 4.95 for Re = 1, S = 0.3

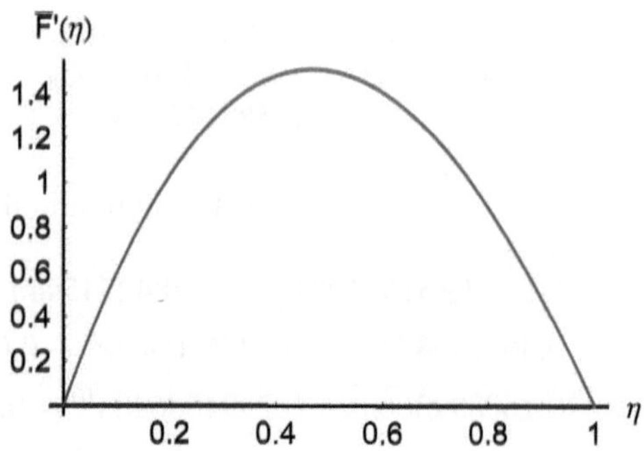

Fig. 4.25 Tangential velocity profile $\overline{F}'(\eta)$ obtained from Eq. 4.95 for Re = 1, S = 0.3

In Table 4.10 are presented some values of $\overline{F}$ given by Eq. 4.95 and corresponding $\overline{F}'$ for Re = 1, S = 0.3, obtained using eight convergence-control parameters in the auxiliary function.

From the last two cases 4.1.4.1.C and 4.1.4.1.D presented for non-Newtonian fluids results that if the cross-flow Reynolds numbers are the same and the elastic number S increases, then the error of the solutions decreases if the number of convergence-control parameters involved in the optimal auxiliary function H^F are the same.

Table 4.10 Values of $\overline{F}$ and $\overline{F}'$ obtained from Eq. 4.95 for Re = 1, S = 0.3

η	$\overline{F}(\eta)$	$\overline{F}'(\eta)$
0	0	0
0.1	0.030934404622845687	0.5929521740910892
0.2	0.11349596629683262	1.0331822405834932
0.3	0.23265507486372466	1.32601725874496
0.4	0.37404091953949	1.4791453628064843
0.5	0.5241314536581481	1.5017160019315088
0.6	0.670361084912139	1.403600121895620
0.7	0.8011631104625843	1.1948245258427717
0.8	0.9059640855903252	0.8851755336854674
0.9	0.9751469952268078	0.4839716070693227
1	1	0

Figures 4.3, 4.4, 4.6, 4.7, 4.9, 4.10, 4.12, 4.13, 4.15, 4.16, 4.18, 4.19, 4.21, 4.22, 4.24 and 4.25 correspond to the normal and tangential velocity components. All these are plotted versus η for different values of the cross-flow Reynolds number and elastic number. From Figs. 4.9, 4.21 and 4.24, and Tables 4.3, 4.9 and 4.10 it follows that if the elastic element of the fluid increases, then the normal velocity increases. The tangential velocity profiles are given in Figs. 4.10, 4.22 and 4.25 for the same values of cross-flow Reynolds number and elastic number. We must note that the tangential velocity increases along with increasing the elastic number S up to approximately η = 0.52 and thereafter decreases with increasing S. On the other hand, from Figs. 4.9 and 4.12 or Tables 4.3 and 4.4, we observe that increasing the value of the cross-flow Reynolds number, the normal velocity increases whereas the tangential velocity increases with an increase of the cross-flow Reynolds number up to approximately η = 0.45 and thereafter decreases with increasing Re. Again, from these figures it is clear that increasing the value of the cross-flow Reynolds number, the point at which the maximum of the tangential velocity occurs moves away from the porous plate and the elastic elements of the fluid move this point closer to the impermeable wall. By comparing Figs. 4.9, 4.10, 4.12, 4.13, 4.21, 4.22, 4.24 and 4.25 we deduce that the flow behavior remain the same, but the difference in magnitude of Newtonian and viscoelastic velocity profiles goes on increasing.

Table 4.11 presents a comparison between approximate solutions given by Eqs. 4.92, 4.93 and 4.94 respectively, and the numerical solution of Eq. 4.26 obtained using a shooting method combined with a RK-4 algorithm, for Re = 1 and S = 0.2 using the value of the constant C = −6.333224002288458.

An analysis of the errors of approximate solutions considered above is presented in Table 4.12, where the relative error is computed as $\varepsilon_i = |F_{num} - F_{OHAM}|$.

4.1.4.2 For the second variable h(η) we consider the approximate solutions 4.66 with eight optimal convergence-control parameters D_1, D_2,...,D_8 in the optimal auxiliary function H^h given by Eq. 4.62.

Table 4.11 Comparison between approximate and numerical solutions of Eq. 4.26 for Re = 1, S = 0.2

η	$\overline{F}(\eta)$ from Eq. 4.92	$\overline{F}(\eta)$ from Eq. 4.93	$\overline{F}(\eta)$ from Eq. 4.94	$F(\eta)\_num$
0	0	0	0	0
0.1	0.030381296768481	0.030503926832263	0.030504482999812	0.030504484486702
0.2	0.111692185096117	0.112073460482270	0.112074879971268	0.112074881450770
0.3	0.229501358896511	0.230105306659955	0.230106908767626	0.230106893712847
0.4	0.369912821730523	0.370593825529074	0.370594960120895	0.370594923158488
0.5	0.519684530376466	0.520281383655642	0.520282013774851	0.520281961708890
0.6	0.666319159612472	0.666730143008014	0.666730546767414	0.666730491409448
0.7	0.798123323000353	0.798335956988273	0.798336254016446	0.798336199442226
0.8	0.904233493995045	0.904305879577095	0.904306013933377	0.904305968482938
0.9	0.974609579319354	0.974620459911539	0.974620467655493	0.974620438568545
1	1	1	1	1

Table 4.12 Analysis of the relative errors of approximate solutions for Re = 1, S = 0.2

η	$\varepsilon_1 = \|F_{num} - F_{4.92}\|$	$\varepsilon_2 = \|F_{num} - F_{4.93}\|$	$\varepsilon_3 = \|F_{num} - F_{4.94}\|$
0	0	0	0
0.1	0.0001231877182209	0.00000055765443831	0.0000000014868899
0.2	0.0003826963546529	0.00000142096849969	0.0000000014795019
0.3	0.0006055348163362	0.00000158705289258	0.000000015054779
0.4	0.000682101427965	0.0000010976294141	0.0000000369624068
0.5	0.0005974313324238	0.0000005780532478	0.0000000520659605
0.6	0.0004113317969763	0.0000003484014342	0.0000000553579657
0.7	0.0002128764418725	0.0000002424539527	0.0000000545742198
0.8	0.0000724744878925	0.0000000889058427	0.0000000454504394
0.9	0.0000108592491907	−0.0000000213429941	0.0000000290869481
1	0	0	0

4.1.4.2.A In the case when Re = 1 and S = 0, from Eqs. 4.67 and 2.37 we obtain the values of the optimal convergence-control parameters:

$$\begin{aligned}
&D_1 = -528.8493915739472,\ D_2 = 14072.033802933543,\\
&D_3 = -131786.88971502372,\ D_4 = 581904.5306278821,\\
&D_5 = -1.3523628747177152 \times 10^6,\ D_6 = 1.7030296841955807 \times 10^6,\\
&D_7 = -1.0999348678727797 \times 10^6,\ D_8 = 285629.4840454936
\end{aligned}$$

The first-order approximate solution 4.66 can be written in the form

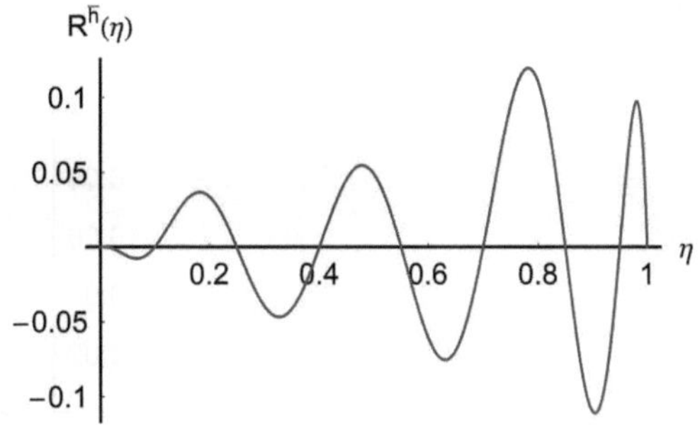

Fig. 4.26 The residual 4.67 obtained using the solution 4.87 for Re = 1, S = 0

Table 4.13 Comparison between the first-order approximate solution 4.96 and numerical solution

η	$\bar{h}(\eta)$	$h(\eta)\_num$	$\varepsilon = \lvert h_{OHAM} - h_{num} \rvert$
0	0	0	0
0.1	0.043923680317276025	0.0440790285583383	1.55×10^{-4}
0.2	0.07778843865641966	0.0780346155358546	1.46×10^{-4}
0.3	0.10156068530697011	0.10168158182393203	1.21×10^{-4}
0.4	0.11465844938354294	0.11492966159283365	2.71×10^{-4}
0.5	0.1175055630365538	0.11790609273398633	4×10^{-4}
0.6	0.11080165764973238	0.11101026956498468	2.08×10^{-4}
0.7	0.0944438561644887	0.09490025270293834	4.56×10^{-4}
0.8	0.06981814470407244	0.07042145942601283	6.03×10^{-4}
0.9	0.03843478790827248	0.03849672585420014	6.19×10^{-5}
1	0	0	0

$$\begin{aligned}\bar{h}(\eta) &= 0.48949305687\eta - 0.5\eta^2 - 13.22123478835\eta^5 + \\ &+ 193.52873558913\eta^6 - 1214.41032651151\eta^7 + \\ &+ 4250.441733356502\eta^8 - 9166.1428178204\eta^9 + \\ &+ 12677.50706447321\eta^{10} - 11279.542210128204\eta^{11} + \\ &+ 6238.06292875039\eta^{12} - 1947.778363594134\eta^{13} + 261.565461580122\eta^{14}\end{aligned} \tag{4.96}$$

In Fig. 4.26 is plotted the residual 4.67 corresponding to the solution 4.96 for Re = 1, S = 0. Analyzing Fig. 4.26, one can estimate that the first-order approximate solution $\bar{h}$ shows a good error, and this is definitely confirmed by Table 4.13, where the comparison performed in this case reveals a very good agreement between the numerical and the approximate analytical solution. Figure 4.27 depicts the axial velocity component $\bar{h}$ for Re = 1, S = 0.

4.1.4.2.B In this case, we consider Re = 1 and S = 0.2. In general, if S ≤ 0.025, then it is considered that the Eqs. 4.21 and 4.26 contains small parameters [2, 13]. The obtained optimal convergence-control parameters D_i are:

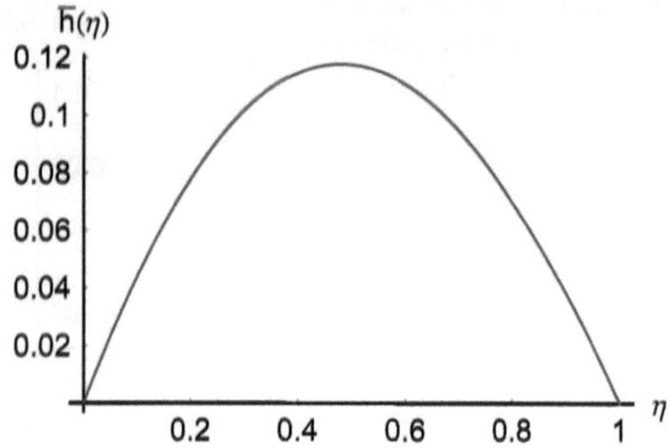

Fig. 4.27 Axial velocity $\bar{h}$ for Re = 1, S = 0

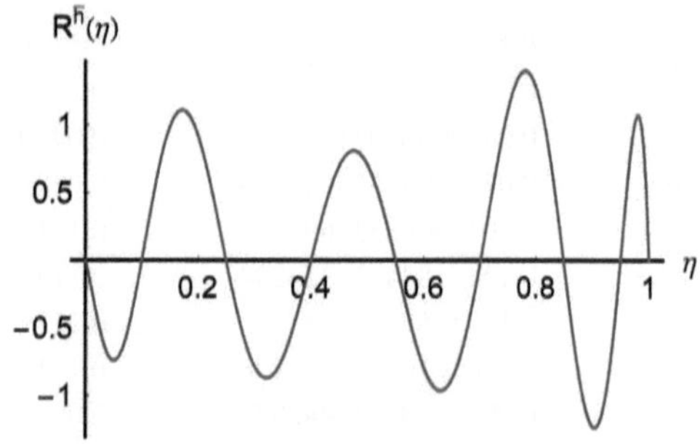

Fig. 4.28 The residual 4.67 obtained using the solution 4.97 for Re = 1, S = 0.2

$$
\begin{aligned}
&D_1 = -104.79334869657755,\ D_2 = 1610.2894466124394,\\
&D_3 = -8580.995083169182,\ D_4 = 17555.714505842498,\\
&D_5 = 738.0550796621604,\ D_6 = -55838.62297698337,\\
&D_7 = 77577.09721499111,\ D_8 = -33465.414996590436
\end{aligned}
$$

The first-order approximate solution of Eq. 4.21 becomes:

$$
\begin{aligned}
\bar{h}(\eta) &= 0.46818440801\eta - 0.19091506899\eta^2 - 0.633193497\eta^3 - \\
&-10.19180795998\eta^4 + 75.8822887501\eta^5 - 2226.01308840011\eta^6 + \\
&+303.97698260011\eta^7 - 36.22556186998\eta^8 - 422.08902290101\eta^9 + \\
&+490.89356820012\eta^{10} - 116.70078379989\eta^{11} - 142.16614150011\eta^{12} + \\
&+101.44883469899\eta^{13} - 18.45934368001\eta^{14}
\end{aligned}
\tag{4.97}
$$

In Fig. 4.28 we present the residual 4.67 corresponding to the solution 4.97 for Re = 1, S = 0.2. The axial component of velocity is due to the action of gravity along the z-axis. This component is shown in Fig. 4.29.

In Table 4.14 we present a comparison between the numerical and approximate solution obtained through OHAM for Eqs. 4.21 and 4.18. An excellent agreement between these two solutions is demonstrated.

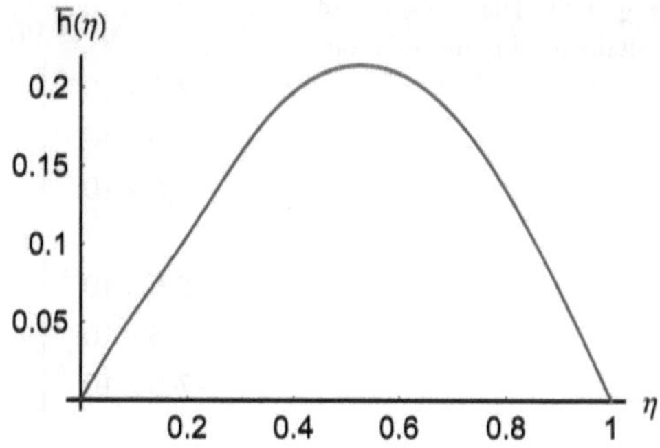

Fig. 4.29 The first-order approximate solution for axial velocity 4.97 for Re = 1, S = 0.2

Table 4.14 Values of numerical solution h(η) and approximate solution $\bar{h}(\eta)$ given by Eq. 4.97 and the relative error $\varepsilon = |h_{OHAM} - h_{num}|$

η	$\bar{h}(\eta)$ given by Eq. 4.97	h(η) numerical	$\varepsilon = \lvert h_{OHAM} - h_{num}\rvert$
0	0	0	0
0.1	0.043819386	0.044085602	2.66×10^{-4}
0.2	0.078029605	0.078045355	1.57×10^{-5}
0.3	0.101687023	0.101690745	1.05×10^{-5}
0.4	0.115049954	0.114930971	1.48×10^{-4}
0.5	0.117968851	0.117896193	7.26×10^{-5}
0.6	0.110995747	0.110990786	4.96×10^{-6}
0.7	0.094872471	0.094877139	4.66×10^{-6}
0.8	0.070408704	0.070402059	6.64×10^{-6}
0.9	0.038959230	0.0384863	4.72×10^{-4}
1	0	0	0

From Table 4.13 and 4.14 it is obvious that the axial velocity for viscoelastic fluid is higher than for a Newtonian fluid on the domain η ∈ [0, 0.452]. For Re = 1, the curves of the axial component of velocity for the viscoelastic and Newtonian fluid intersect at a common point around η = 0.452. Up to this point, the axial velocity increases and an opposite effect is observed beyond this point, when the axial velocity slightly decreases with an increase of the elasticity of the fluid.

On the other hand, the influence of elastic number S on the errors of the solutions is not meaningfully.

4.1.4.3 In the case of the last variable $\theta(\eta)$, we have in view the Eq. 4.83 and the residual 4.84. For the temperature $\theta(\eta)$ we used eight optimal convergence-control parameters K_1, K_2,. . .,K_8.

In what follows, we consider Re = 1 and different values of S and Pe.

4.1.4.3.A In this first sub-case we consider S = 0 and Pe = 2. From Eqs. 4.84 and 2.37 we obtain

Fig. 4.30 The residual 4.84 obtained using the solution 4.98 for Re = 1, S = 0, Pe = 2

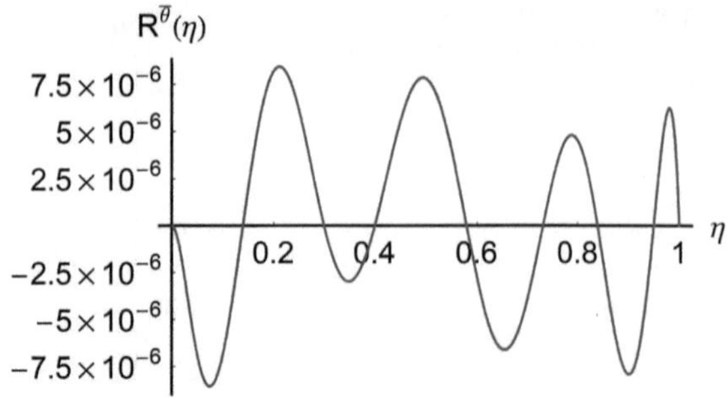

Fig. 4.31 The first-order approximate solution $\overline{\theta}(\eta)$ given by 4.98 for Re = 1, S = 0, Pe = 2

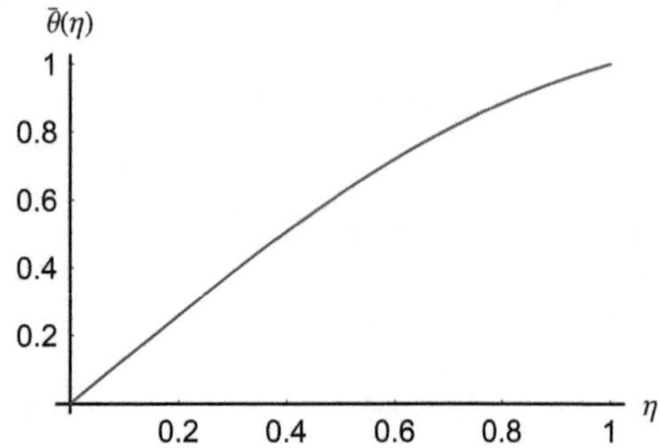

$$K_1 = -2.8032584037291977,\ K_2 = 0.2381739199709458,$$
$$K_3 = 0.08145622927694587,\ K_4 = 5.1999385170733845,$$
$$K_5 = 0.5597545358853374,\ K_6 = -11.65037497375808,$$
$$K_7 = 10.123435398935225,\ K_8 = -2.6919606409578236$$

The first-order approximate solution 4.83 for the temperature is:

$$\begin{aligned}\overline{\theta}(\eta) &= 1.302003486156817\eta - 0.7008146009322994\eta^4 + \\ &+0.31605192836856166\eta^5 - 0.0077326384037088\eta^6 + \\ &+0.36754531173\eta^7 - 0.1557252397587635\eta^8 - \\ &-0.5009810276811784\eta^9 + 0.5963450682702427\eta^{10} - \\ &-0.254795701885811\eta^{11} + 0.0407872821387549\eta^{12}\end{aligned} \tag{4.98}$$

In Fig. 4.30 is plotted the residual $R^{\overline{\theta}}$ given by Eq. 4.84 corresponding to the solution 4.98. The first-order approximate solutions $\overline{\theta}(\eta)$ and $\overline{\theta}'(\eta)$ are depicted in Figs. 4.31 and 4.32, respectively. In Table 4.15 are given some values of $\overline{\theta}$ and $\overline{\theta}'$ for the solution 4.98.

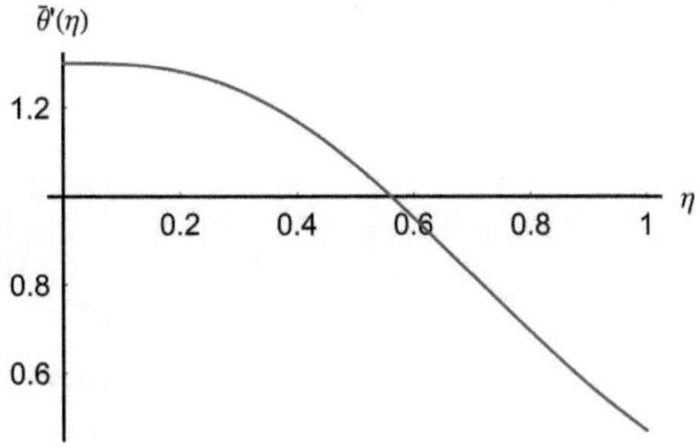

Fig. 4.32 The first-order approximate solution $\overline{\theta}'(\eta)$ obtained from 4.98 for Re = 1, S = 0, Pe = 2

Table 4.15 Values of $\overline{\theta}(\eta)$ and $\overline{\theta}'(\eta)$ for Re = 1, S = 0, Pe = 2 corresponding to 8 convergence-control parameters

η	$\overline{\theta}(\eta)$	$\overline{\theta}'(\eta)$
0	0	1.302003486156817
0.1	0.13013345469563203	1.2993601985927836
0.2	0.25938414097880014	1.2822309324060672
0.3	0.38575020780585334	1.2404116943852714
0.4	0.5064864566473091	1.1694035589640748
0.5	0.6187079018839721	1.0708878291136024
0.6	0.719977066603501	0.951873299456335
0.7	0.8087398684889628	0.8225955896427193
0.8	0.8845165210944285	0.6938662844163348
0.9	0.9478379830038848	0.5747778131561427
1	1	0.47141770865163

4.1.4.3.B In the second sub-case, we choose S = 0 and Pe = 5. The optimal convergence-control parameters are:

$$K_1 = -9.301761460785743,\ K_2 = 3.9541480214484825,$$
$$K_3 = -25.60892642587981,\ K_4 = 145.9519845167805,$$
$$K_5 == -200.79252192330227,\ K_6 = 74.79620951600414,$$
$$K_7 = 30.397886695059906,\ K_8 = -20.068326580497793$$

and the corresponding approximate solution becomes:

$$\begin{aligned}\overline{\theta}(\eta) &= 1.702080818482\eta - 2.3254403651964\eta^4 + \\ &+ 1.523298349295842\eta^5 - 2.8245025106845\eta^6 + 11.6446144381452\eta^7 - \\ &- 15.96931312149\eta^8 + 8.6940787832585\eta^9 - 0.6488750994092\eta^{10} - \\ &- 1.100006844631029\eta^{11} + 0.30406555424996656\eta^{12}\end{aligned} \tag{4.99}$$

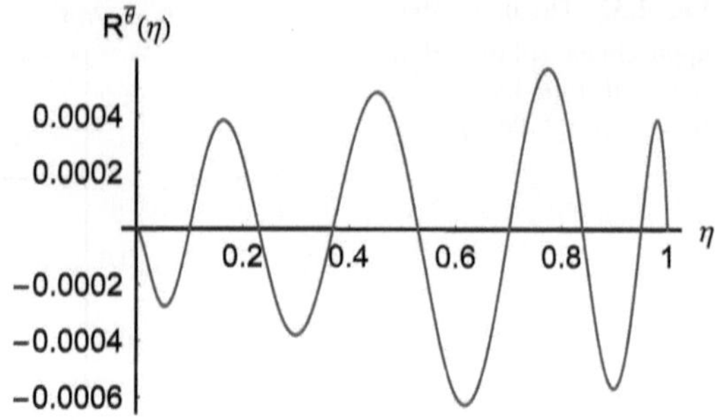

Fig. 4.33 The residual 4.84 obtained using the solution 4.99 for Re = 1, S = 0, Pe = 5

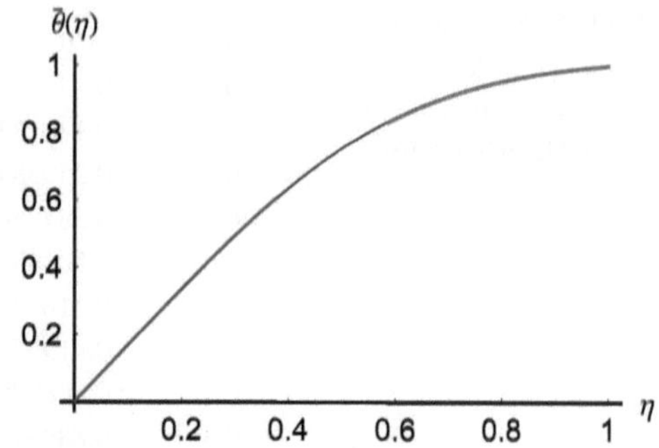

Fig. 4.34 The first-order approximate solution $\overline{\theta}(\eta)$ given by 4.99 for Re = 1, S = 0, Pe = 5

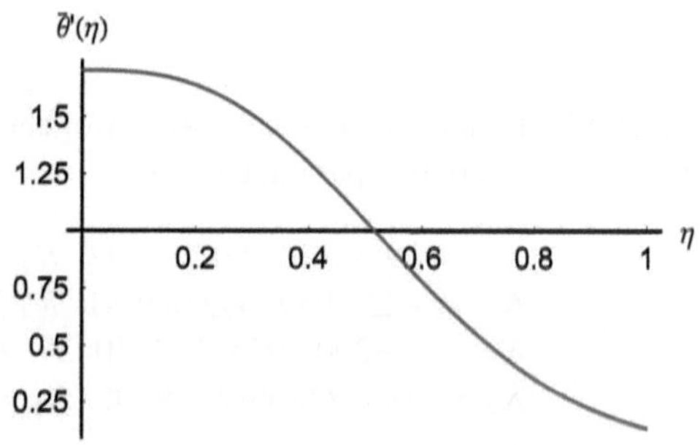

Fig. 4.35 The first-order approximate solution $\overline{\theta}'(\eta)$ obtained from 4.99 for Re = 1, S = 0, Pe = 5

In Figs. 4.33, 4.34 and 4.35 are plotted the residual 4.84, the first-order approximate solution $\overline{\theta}(\eta)$ given by 4.99 and the first-order approximate solution $\overline{\theta}'(\eta)$ obtained from 4.99, respectively. In Table 4.16 are given some values of $\overline{\theta}$ and $\overline{\theta}'$ for the solution 4.99.

4.1.4.3.C In the third sub-case we consider S = 0 and Pe = 8. The optimal convergence-control parameters are:

Table 4.16 Values of $\overline{\theta}(\eta)$ and $\overline{\theta}'(\eta)$ for Re = 1, S = 0, Pe = 5

η	$\overline{\theta}(\eta)$	$\overline{\theta}'(\eta)$
0	0	1.702080818482067
0.1	0.16998895967947722	1.6934407477006779
0.2	0.33711467968724634	1.6382074233883777
0.3	0.49509516876264015	1.5078692152983295
0.4	0.6361133789502281	1.3012372761189586
0.5	0.7536481065901425	1.0442878031119531
0.6	0.8446585261730079	0.7778529370392102
0.7	0.910190399272687	0.5399931771129365
0.8	0.9543712130571269	0.35290183064078207
0.9	0.9826034243880359	0.22038910716528992
1	1.0000000000000009	0.13426152823450366

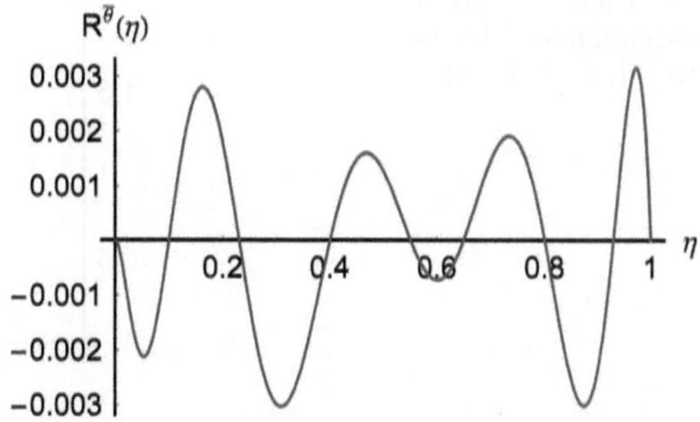

Fig. 4.36 The residual 4.84 obtained using the solution 4.100 for Re = 1, S = 0, Pe = 8

$$K_1 = -18.59089659266167,\ K_2 = 27.86557717109708,$$
$$K_3 = -221.68008480448853,\ K_4 = 1056.2031160775666,$$
$$K_5 = -2003.9885772576697,\ K_6 = 1831.1356608518008,$$
$$K_7 = -807.6839481032731,\ K_8 = 136.46039212661012$$

The first-order approximate solution for Eqs. 4.69 and 4.70 becomes:

$$\begin{aligned}
\overline{\theta}(\eta) &= 2.0251133107981024\eta - 4.647724148165418\eta^4 + \\
&+6.03842623493073\eta^5 - 24.025713623188658\eta^6 + \\
&+85.99967413432564\eta^7 - 145.07807844311\eta^8 + \\
&+131.96366857042696\eta^9 - 67.61470184459357\eta^{10} + \\
&+18.40680975078524\eta^{11} - 2.067581698888032\eta^{12}
\end{aligned} \quad (4.100)$$

The residual 4.84, the first-order approximate solution $\overline{\theta}(\eta)$ given by Eq. 4.100 and the first-order approximate solution $\overline{\theta}'(\eta)$ obtained from 4.100 are presented in Figs. 4.36, 4.37 and 4.38, respectively. In Table 4.17 are given some values of $\overline{\theta}$ and $\overline{\theta}'$ corresponding to the solution 4.100.

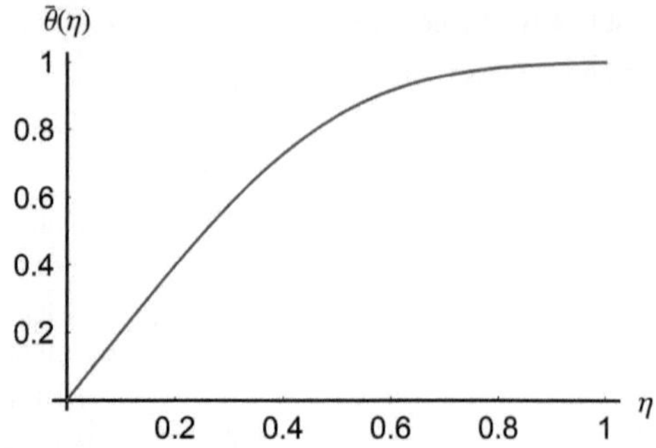

Fig. 4.37 The first-order approximate solution $\overline{\theta}(\eta)$ given by 4.100 for Re = 1, S = 0, Pe = 8

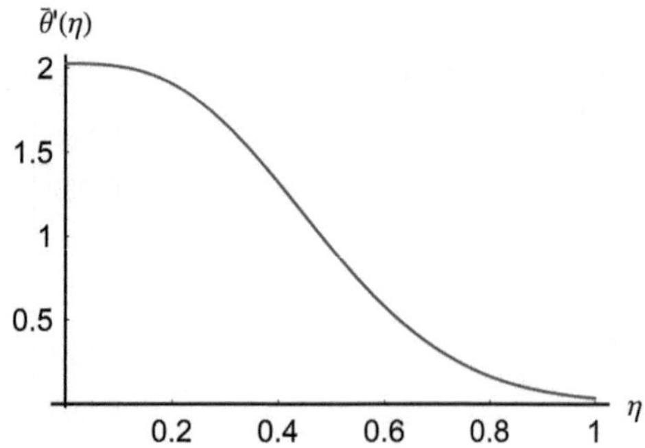

Fig. 4.38 The first-order approximate solution $\overline{\theta}^{'}(\eta)$ obtained from 4.100 for Re = 1, S = 0, Pe = 8

Table 4.17 Values of $\overline{\theta}(\eta)$ and $\overline{\theta}^{'}(\eta)$ for Re = 1, S = 0, Pe = 8

η	$\overline{\theta}(\eta)$	$\overline{\theta}^{'}(\eta)$
0		
0.1	0.20209019674464232	2.008597487551044
0.2	0.39877151525022103	1.904954339136693
0.3	0.5785664333574535	1.6682326814916273
0.4	0.7285566320658745	1.3176384326661992
0.5	0.840741158832685	0.9267483186405935
0.6	0.9153691170917215	0.5784987619379927
0.7	0.95957667975114	0.322668918878505
0.8	0.9831554363147219	0.16344479813363222
0.9	0.9946863896035358	0.07680318566130495
1		

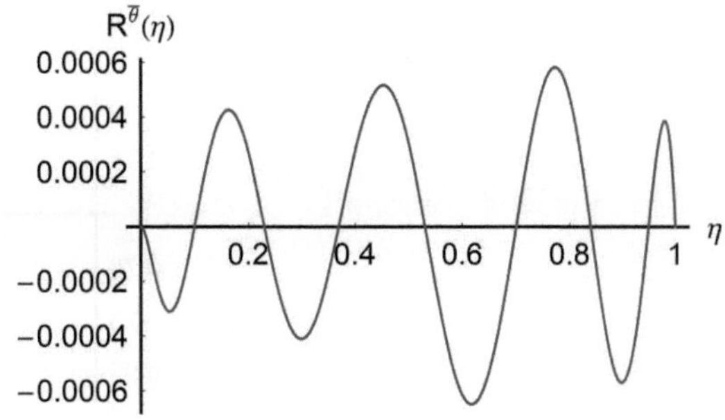

Fig. 4.39 The residual 4.84 obtained using the solution 4.101 for Re = 1, S = 0.2, Pe = 5

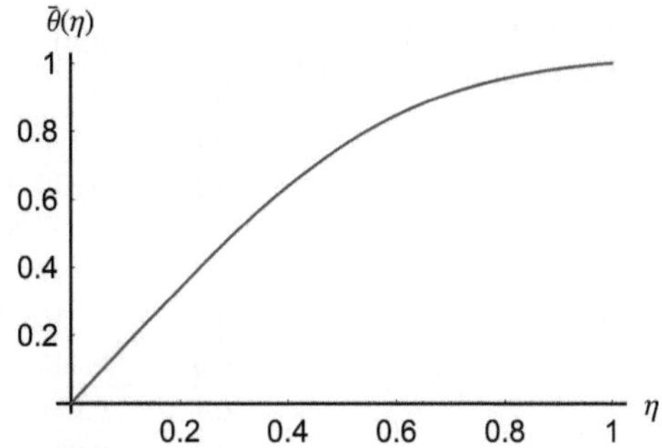

Fig. 4.40 The first-order approximate solution $\overline{\theta}(\eta)$ given by 4.101 for Re = 1, S = 0.2, Pe = 5

4.1.4.3.D In this last case, we consider S = 0.2 and Pe = 5. The optimal convergence-control parameters in this case are:

$$\begin{aligned}
&K_1 = -9.567865633247216,\ K_2 = 4.593134581768264,\\
&K_3 = -29.44674632890017,\ K_4 = 165.76527033704969,\\
&K_5 = -244.5240493091533,\ K_6 = 121.87491791012201,\\
&K_7 = 5.479845907075026,\ K_8 = -14.839135522972217
\end{aligned}$$

The first-order approximate solution 4.83 of Eqs. 4.69 and 4.70 is:

$$\begin{aligned}
&\overline{\theta}(\eta) = 1.7088972298801255\eta\text{-}2.391966408311804\eta^4+\\
&+1.6457567505899613\eta^5\text{-}3.250883605007901\eta^6+\\
&+13.242602468308318\eta^7\text{-}19.019690867884986\mu^8+\\
&+11.870456282620454\eta^9\text{-}2.5256699788779886\eta^{10}\text{-}\\
&\text{-}0.5043372580278791\eta^{11}+0.22483538671170028\eta^{12}
\end{aligned} \tag{4.101}$$

In Figs. 4.39, 4.40 and 4.41 are plotted the residual 4.84 corresponding to Eq. 4.101, the first-order approximate solution $\overline{\theta}(\eta)$ and $\overline{\theta}'(\eta)$, respectively. In Table 4.18 are presented some values of $\overline{\theta}$ and $\overline{\theta}'$ for the solution 4.101.

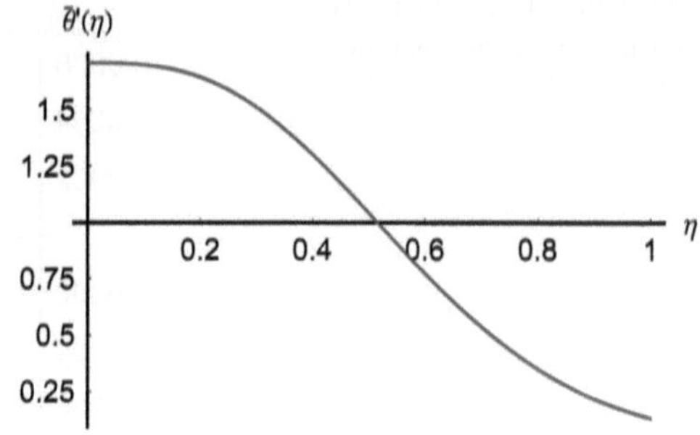

Fig. 4.41 The first-order approximate solution $\overline{\theta}'(\eta)$ obtained from 4.101 for Re = 1, S = 0.2, Pe = 5

Table 4.18 Values of $\overline{\theta}(\eta)$ and $\overline{\theta}'(\eta)$ for Re = 1, S = 0.2, Pe = 5

η	$\overline{\theta}(\eta)$	$\overline{\theta}'(\eta)$
0	0	1.7088972298801255
0.1	0.17066487870749117	1.700035714627032
0.2	0.33839750987227657	1.643523782910699
0.3	0.4967897694853447	1.5106045513799229
0.4	0.6379229689295305	1.3007987621351786
0.5	0.7552743642425533	1.0412422464389728
0.6	0.845905442576122	0.7735643303710193
0.7	0.9110054030181	0.5358345911538496
0.8	0.9548129421874232	0.3496675864009542
0.9	0.9827759493936382	0.21822118613606123
1	1	0.13292561165157402

From Figs. 4.30, 4.33 and 4.36 it is clear that the values of the residual decrease with increasing the Peclet number for the same value of cross-flow Reynolds number and elastic number (Re = 1, S = 0). Again, from Figs. 4.33 and 4.39 we observe that the values of the residual increases with increasing the elastic number S if the cross-flow Reynolds number and Peclet number are the same (Re = 1, Pe = 5).

On the other hand, the values of temperature increase if the Peclet number Pe increases in the condition that the cross-flow Reynolds number and elastic number are the same (Re = 1, S = 0). If the cross-flow Reynolds number and Peclet number are the same (Re = 1, Pe = 5), then for smaller values of elastic number, the values of the temperature are smaller (see Tables 4.16 and 4.18). Also, the derivative of the temperature increases with increasing Peclet number for Newtonian fluid, but the derivative of temperature increases with increasing the elastic number only on the domain $\eta \in [0,0.39]$ and an opposite aspect is observed on the domain $\eta \in [0.39,1]$.

In all cases, the involvement of eight convergence-control parameters in the solutions 4.83 lead to very accurate first-order approximations.

In order to estimate the effect of the elasticity of the fluid on the heat transfer on walls we can determine the values of $\overline{\theta}'(0)$ and $\overline{\theta}'(1)$ in Table 4.19 for different values of cross-flow Reynolds number, Peclet number and elastic number.

Table 4.19 The effect of elasticity of the fluid on the heat transfer on the walls

		Pe = 2		Pe = 5		Pe = 8	
S	Re	$\theta'(0)$	$\theta'(1)$	$\theta'(0)$	$\theta'(1)$	$\theta'(0)$	$\theta'(1)$
0	1	1.30200348	0.47141770	1.70208081	0.13426152	2.02511331	0.03484506
0.2	1	1.30489776	0.46982787	1.70889722	0.13292561	2.03485894	0.03418965

Table 4.20 The shear stress on the walls

S	Re	$\tau_{xy}(0)$	$\tau_{xy}(1)$	$\tau_{zy}(0)$	$\tau_{zy}(1)$
0	1	6.454267072	12.65470854	0.48949305607	−0.413008402
0.2	1	6.601882878	−6.994883915	0.468184408	−1060.264106

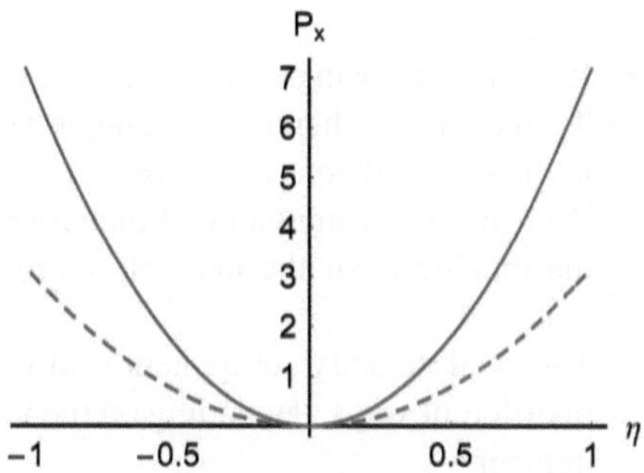

Fig. 4.42 Pressure variation in x direction: (*red solid line*) s = 0, (*blue blue dashed line*) s = 0.2 (colour figure online)

For the values of Peclet number, Pe = 2, Pe = 5 and Pe = 8, the heat transfer of the impermeable plate at η = 0 is considerably higher than that of the porous plate at η = 1 in both cases of Newtonian and non-Newtonian fluid. It is observed that for an increasing elastic number, the heat transfer increases for the impermeable plate, whereas it decreases for the porous plate. Also, for Newtonian fluid, this change in heat transfer on the walls is more pronounced for Pe = 8.

The shear stress on the walls of the channel is presented in Table 4.20, where we used Eqs. 4.32, 4.33, 4.90, 4.94, 4.96 and 4.97.

From this table we can conclude that the main effect of the elastic number on the shear stress in the x and z directions on the impermeable wall (η = 0) increases their values, whereas they decrease on the porous plate (η = 1).

At this point, we represent in Figs. 4.42 and 4.43 the pressure variation in the x and y directions, respectively, using Eqs. 4.90 and 4.94.

From Figs. 4.42 and 4.43 we notice that the elasticity of the fluid decreases the pressure variations in both directions.

The most important results of this section can be outlined as below:

- In the condition that the elastic element of the fluid increases, then the normal velocity increases

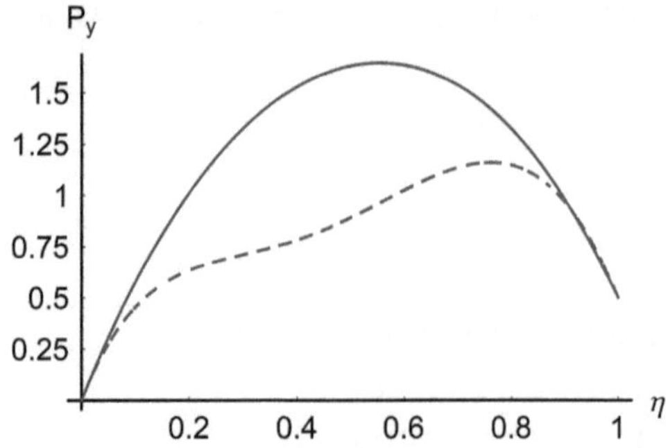

Fig. 4.43 Pressure variation in y direction: (*red solid line*) s = 0, (*blue blue dashed line*) s = 0.2 (colour figure online)

- The tangential velocity increases with increasing the elastic numbers up to approximately $\eta = 0.52$ and thereafter decreases with increasing the elastic number
- With an increasing value of the cross-flow Reynolds number, the normal velocity increases whereas the tangential velocity increases up to approximately $\eta = 0.45$ and thereafter decreases with increasing Re
- With an increasing value of the cross-flow Reynolds number, the point at which the maximum of the tangential velocity occurs moves away from the porous plate
- The axial velocity component is affected by elastic elements in the viscous fluid in different ways, depending on the values of the cross-flow Reynolds and elastic number
- The axial velocity for the viscoelstic fluid is higher than for a Newtonian fluid up to approximately $\eta = 0.452$, and thereafter the axial velocity slightly decreases with increasing the elasticity of the fluid
- The value of the temperature increases if the Peclet number increases
- The derivative of the temperature increases with increasing Peclet number for Newtonian fluid. The derivative of the temperature increases along with increasing the elastic number up to approximately $\eta = 0.39$
- The heat transfer of the impermeable plate is considerable higher than that on the porous plate for both Newtonian and non-Newtonian fluid
- The shear stresses of the impermeable wall are less for the Newtonian fluid than for viscoelastic fluid. For the porous wall is true the reverse of the situation
- The elasticity of the fluid decreases the pressure variation in both directions x and y
- The changes in tangential velocity, heat transfer and shear stress are much more noticeable for the case of a large cross-flow Reynolds number
- On the other hand, a larger number of optimal convergence-control parameters C_i in the optimal auxiliary function H lead to a better accuracy of the results
- OHAM allows us to solve directly the nonlinear differential equation 4.26 in which is involved an unknown constant C. To the best of our knowledge, it is for the first time when such a rigorous determination of this constant is reported.
- In all cases presented in this paper we obtain an excellent accuracy of the first-order approximate solutions

4.2 Thin Film Flow of an Oldroyd 6-Constant Fluid over Moving Belt

It is known that in practical applications the behavior of viscoelastic fluids cannot be represented by that of Newtonian fluids. The inadequacy of the classical Navier-Stokes model to describe the fluids which do not obey the Newtonian postulate that the stress tensor is multiple of the shear strain, has led to the development of several models of non-Newtonian fluids. During the past few decades, various constitutive equations have been proposed for these non-Newtonian fluids. The differential equations that describe non-Newtonian fluids are in general nonlinear and highly complicated, but there is not a single constitutive equation available in the literature by which one can study the flow behavior of viscoelastic fluids. The thin film flows have been attracted the attention of numerous researchers, due to their wide applications which include for example microchip production, lining of mammalian lungs and flow of surface active materials or MHD generators, accelerators, pumps and flow meters.

During the last years, many researchers have in view different problems related to such non-Newtonian fluids. The Oldroyd fluid has acquired a special status amongst the many fluids of the rate type, as it included as special cases the classical Newtonian fluid and the Maxwell fluid [15]. This fluid takes into account the properties of stress relaxation and retardation. The two-dimensional steady and flow of an incompressible Maxwell fluid in a corner formed two planes, one of which is sliding past the other at a certain angle, was first investigated by Strauss [16]. Hancock et al. have investigated the effects of inertia forces by constructing regular perturbation series for the stream function in [17]. Rajagopal and Bhatnagar [18] presented two solutions for the flow of an Oldroyd-B fluid: the flow past an infinite porous plate and the longitudinal and torsional oscillations of an infinitely long rod of finite radius. The two-dimensional steady and slow flow of an Oldroyd 6-constant fluid between intersecting planes, one of which is fixed and the other moving, has been analyzed by Bariş [19]. The effects of the non-Newtonian parameter on the flow pattern are carefully delimited. There is, unlike the case of Newtonian fluid, a secondary flow near the corner. Such flow are of considerable practical interest as they locally appear in a cylinder with a moving piston, or at the edge of a blade used to scrape up liquid from a surface.

Very recently, Hayat et al. [20] studied some steady unidimensional flow of an Oldroyd 8-constant MHD fluid in bounded domain. The MHD solutions for a Newtonian fluid as well as those corresponding to the Oldroyd 3 and 6-constant fluids, a Maxwell fluid and a second grade one, appear as limiting cases of the obtained solutions. Hayat et al. [21] and Wang et al. [22] considered Couette and Poiseuille flows of an Oldroyd 6-constant fluid with magnetic field. Sajid et al. [23] investigated the problem of wire coating by withdrawal from a bath of MHD Oldroyd 8-constant fluid, using homotopy analysis method, and Khan et al. [24] also discussed an Oldroyd 8-constant MHD fluid but between coaxial cylinders subject to partial slip at the boundaries. The Oldroyd 6-constant fluid and fluid-slip on the three nonlinear boundary value problems are studied by Hayat et al. [25]. Ellahi et al. [26] investigated an Oldroyd 8-constant fluid over a suddenly moved plate and Sajid and Hayat [27] obtained an exact solution for thin film flow of an Oldroyd 8-constant fluid. Using Fourier sine and Laplace transforms, Vieru et al. [28], established exact solutions for the unsteady flow of an incompressible generalized Oldroyd-B fluid due to an infinite constantly accelerating plate. Hayat et al. [29] obtained the exact solution of a thin film flow of an Oldroyd 6-constant fluid and Ellahi et al. [30] for flows of an Oldroyd 8-constant fluid with nonlinear slip conditions. The energetic balance for the motion of an Oldroyd-B fluid due to an impulsively moved plate is presented by Fetecău et al. [31]. 3D flow of a generalized Oldroyd-B fluid due to a constant pressure gradient between two side walls perpendicular to a plate is presented by Zheng et al. [32]. Shah et al. [33] used the optimal homotopy asymptotic method in the study of the wire coating in a pressure type die with the bath of Oldroyd 8-constant fluid with pressure gradient. In [34], the slip-effect on the rotating flows on an Oldroyd-B fluid in a porous space is examinated by Hayat et al. The fluid is permeated by a transverse magnetic field. The governing equations include the modified Darcy's law for an Oldroyd-B fluid.

4.2.1 Governing Equations

The Cauchy stress $\mathbf{T}$ is:

$$\mathbf{T} = -p\mathbf{I} + \mathbf{S}, \tag{4.102}$$

where $-p\mathbf{I}$ is the indeterminate part of the stress due to the constraint of incompressibility and the extra stress tensor $\mathbf{S}$ is defined by

$$\begin{aligned} &\mathbf{S} + \lambda_1 \frac{D\mathbf{S}}{Dt} + \frac{\lambda_3}{2}(\mathbf{S}\mathbf{A}_1 + \mathbf{A}_1\mathbf{S}) + \frac{\lambda_5}{2}(tr\mathbf{S})\mathbf{A}_1 + \frac{\lambda_6}{2}[tr(\mathbf{S}\mathbf{A}_1)]\mathbf{I} = \\ &= \mu\left[\mathbf{A}_1 + \lambda_2 \frac{D\mathbf{A}_1}{Dt} + \lambda_4\mathbf{A}_1^2 + \frac{\lambda_7}{2}\left(tr(\mathbf{A}_1)^2\right)\mathbf{I}\right], \end{aligned} \tag{4.103}$$

where λ_1, λ_2, λ_3, λ_4, λ_5, λ_6, λ_7, μ are material constants, $\mathbf{A}_1$—is the first Rivlin-Ericksen tensor defined by

$$\mathbf{A}_1 = \mathbf{L} + \mathbf{L}^T, \tag{4.104}$$

where $\mathbf{L}$ is the spatial velocity gradient, $\mathbf{L} = grad\mathbf{V}$, $\mathbf{V}$ being the velocity vector. The contravariant derivative D/Dt in terms of the material derivative d/dt is defined by

$$\frac{D\mathbf{S}}{Dt} = \frac{d\mathbf{S}}{dt} - \mathbf{L}\mathbf{S} - \mathbf{S}\mathbf{L}^T \tag{4.105}$$

It should be noted that when $\lambda_i = 0, i = 1, \ldots, 7$, the model reduces to the classical linearly viscous model of a Newtonian fluid. When $\lambda_3 = \lambda_4 = \lambda_5 = \lambda_6 = \lambda_7 = 0$ it reduce to a 3-constant model of an Oldroyd-B fluid, while for $\lambda_2 = \lambda_3 = \lambda_4 = \lambda_5 = \lambda_6 = \lambda_7 = 0$ it reduces to a Maxwell model. For $\lambda_1 = \lambda_3 = \lambda_5 = \lambda_6 = \lambda_7 = 0$ the model describes a second grade fluid and for $\lambda_5 = \lambda_6 = \lambda_7 = 0$ it reduces to a Johnson-Segalman model. For $\lambda_6 = \lambda_7 = 0$ the model reduces to the Oldroyd 6-constant.

Neglecting thermal effects and body forces, the equations of motion of an incompressible fluid are:

$$div\mathbf{V} = 0 \tag{4.106}$$

$$\rho\frac{\partial \mathbf{V}}{\partial t} + \rho[\mathbf{V}\cdot grad\mathbf{V}] = div\mathbf{T} \tag{4.107}$$

where ρ is the density and t is the time. If the extra tensor and velocity are respectively:

$$\mathbf{S}(\mathbf{x}) = \begin{pmatrix} S_{xx} & S_{xy} & S_{xz} \\ S_{yx} & S_{yy} & S_{yz} \\ S_{zx} & S_{zy} & S_{zz} \end{pmatrix}, \quad \mathbf{V}(\mathbf{x}) = \begin{pmatrix} 0 \\ v(\mathbf{x}) \\ 0 \end{pmatrix} \tag{4.108}$$

From Eq. 4.108, the continuity equation 4.106 is satisfied identically and the momentum equation 4.107 becomes:

$$\frac{d\hat{p}}{dy} = \frac{d}{dx}S_{xy} \tag{4.109}$$

where

$$S_{xy} = \frac{1}{M}\left[\mu\frac{dv}{dx} + \mu\alpha_1\left(\frac{dv}{dx}\right)^3\right] \tag{4.110}$$

$$M = 1 + \alpha_2\left(\frac{dv}{dx}\right)^2 \tag{4.111}$$

$$\alpha_1 = \lambda_1\lambda_4 - (\lambda_3 + \lambda_5)(\lambda_4 - \lambda_2) \tag{4.112}$$

$$\alpha_2 = \lambda_1\lambda_3 - (\lambda_3 + \lambda_5)(\lambda_3 - \lambda_1) \tag{4.113}$$

and modified pressure $\hat{p}$ becomes

$$\hat{p} = p_1 - \frac{1}{M}\left[\mu(\lambda_4 - \lambda_3)\left(\frac{dv}{dx}\right)^2 + \mu(\lambda_4\alpha_2 - \alpha_1\lambda_3)\left(\frac{dv}{dx}\right)^4\right]. \tag{4.114}$$

From Eqs. 4.110 and 4.109 we obtain after nondimensionalization:

$$\begin{aligned}&\frac{d^2v}{dx^2}+(3\alpha_1-\alpha_2)\left(\frac{dv}{dx}\right)^2\frac{d^2v}{dx^2}+\alpha_1\alpha_2\left(\frac{dv}{dx}\right)^4\frac{d^2v}{dx^2}-\\&-m\left[1+\alpha_2\left(\frac{dv}{dx}\right)^2\right]^2=0\end{aligned} \tag{4.115}$$

where α_1 and α_2 are given in terms of the nondimensional material constants and is a nondimensional gravity parameter. The boundary conditions are

$$v(0)=1,\qquad \left[\frac{dv}{dx}+\alpha_1\left(\frac{dv}{dx}\right)^3\right]_{x=1}=0. \tag{4.116}$$

4.2.2 *Application of OHAM to Thin Film Flow of an Oldroyd 6-Constant Fluid*

To obtain approximate solutions of Eqs. 4.115 and 4.116, we choose the linear operator in the two cases.

Case 4.2.2.1
In the first case we suppose that the linear operator has the form

$$Lv=v'' \tag{4.117}$$

where prime denotes derivative with respect to x, and the nonlinear operator becomes

$$Nv=(3\alpha_1-\alpha_2)v'^2v''+\alpha_1\alpha_2v'^4v''-m\left(1+\alpha_2v'^2\right)^2. \tag{4.118}$$

The initial conditions 4.116 become

$$v(0)=1,v'(1)+\alpha_1v'^3(1)=0. \tag{4.119}$$

From 4.119 we obtain two subcases.
In the first subcase we consider the following initial conditions

$$v(0)=1,v'(1)=0. \tag{4.120}$$

Equation 4.4 can be written as

$$v_0''-m=0,v_0(0)=0,v_0'(1)=0,g(x)=-m. \tag{4.121}$$

Its solution is

$$v_0(x) = m\left(\frac{x^2}{2} - x\right) + 1. \tag{4.122}$$

The first approximation $v_1(x)$ is obtained from Eq. 4.9:

$$v_1''(x) = H(x, C_i)N(v_0), v_1(0) = 0, v_1'(1) = 0 \tag{4.123}$$

where

$$N(v_0) = (3\alpha_1 - \alpha_2)v_0'^2 v_0'' + \alpha_1\alpha_2 v_0'^4 v_0'' - m\left(1 + \alpha_2 v_0'^2\right)^2 + m. \tag{4.124}$$

Substituting Eq. 4.122 into 4.124, the nonlinear operator N becomes

$$N(v_0) = K_1(x-1)^2 + K_2(x-1)^4 \tag{4.125}$$

where

$$K_1 = 3(\alpha_1 - \alpha_2)m^3, K_2 = \alpha_2(\alpha_1 - \alpha_2)m^5. \tag{4.126}$$

The expression 4.125 is a polynomial one and the optimal auxiliary function H from Eq. 4.123 is chosen such that the product HN be of the same form with N. Therefore we choose the optimal auxiliary function H in the form

$$\begin{aligned} H(x, C_i) &= C_1(x-1)^{-2} + C_2(x-1)^{-1} + C_3 + \\ &+ C_4(x-1) + C_5(x-1)^2 \end{aligned} \tag{4.127}$$

where C_1, C_2,..., C_5 are unknown parameters.

For example we can also consider the following possibilities:

$$\widetilde{H}(x, C_i) = C_1^*(x-1)^{-1} + C_2^* + C_3^*(x-1) + C_4^*(x-1)^3$$

or

$$\begin{aligned} \widetilde{\widetilde{H}}(x, C_i) &= C_1' + C_2'(x-1) + C_3'(x-1)^2 + C_4'(x-1)^3 + \\ & C_5'(x-1)^4 + C_6'(x-1)^5 \end{aligned}$$

and so on.

Inserting Eqs. 4.127 and 4.125 into Eq. 4.123 we obtain equation:

$$\begin{aligned}v_1'' &= K_1C_1 + K_1C_2(x-1) + (K_1C_3 + K_2C_1)(x-1)^2 + \\ &+ (K_1C_4 + K_2C_2)(x-1)^3 + (K_1C_5 + K_2C_3)(x-1)^4 + \\ &+ K_2C_4(x-1)^5 + K_2C_5(x-1)^6, v_1(0) = v_1'(1) = 0.\end{aligned} \tag{4.128}$$

By solving Eq. 4.128 we obtain

$$\begin{aligned}v_1(x) &= \frac{1}{2}K_1C_1\left[(x-1)^2 - 1\right] + \frac{1}{6}K_1C_2\left[(x-1)^3 + 1\right] + \\ &+ \frac{1}{12}(K_1C_3 + K_2C_1)\left[(x-1)^4 - 1\right] + \frac{1}{20}(K_1C_4 + K_2C_2)\left[(x-1)^5 + \right. \\ &\left. +1\right] + \frac{1}{30}(K_1C_5 + K_2C_3)\left[(x-1)^6 - 1\right] + \frac{1}{42}K_2C_4\left[(x-1)^7 + 1\right] + \\ &+ \frac{1}{56}K_2C_5\left[(x-1)^8 - 1\right].\end{aligned} \tag{4.129}$$

The first-order approximate solution of Eqs. 4.115 and 4.120 is obtained from Eq. 4.5:

$$\overline{v}(x) = v_0(x) + v_1(x) \tag{4.130}$$

where $v_0(x)$ and $v_1(x)$ are given by Eqs. 4.122 and 4.129 respectively.

The residual given by Eq. 2.35 becomes

$$\begin{aligned}R^1(x, C_1, C_2, C_3, C_4, C_5) &= \overline{v}'' + (3\alpha_1 - \alpha_2)\overline{v}'^2\overline{v}'' + \\ &+ \alpha_1\alpha_2\overline{v}'^4\overline{v}'' - m(1 + \alpha_2\overline{v}'^2)^2\end{aligned} \tag{4.131}$$

with $\overline{v}$ obtained from Eq. 4.130.

In the second subcase, from Eq. 4.120 we have

$$v(0) = 1, v'(1) = \pm\frac{1}{\sqrt{-\alpha_1}}, if \alpha_1 < 0. \tag{4.132}$$

The Eq. 4.4 becomes

$$v_0'' - m = 0, v_0(0) = 0, v_0'(1) = \pm\frac{1}{\sqrt{-\alpha_1}}, g(x) = -m \tag{4.133}$$

with the solution

$$v_0(x) = \frac{m}{2}(x-1)^2 \pm \frac{1}{\sqrt{-\alpha_1}}(x-1) - \frac{m}{2} \pm \frac{1}{\sqrt{-\alpha_1}} + 1. \tag{4.134}$$

For the expression 4.134 of $v_0(x)$, the nonlinear operator N becomes

$$N(v_0) = m(\alpha_1 - \alpha_2)\left[K_3 + K_4(x-1) + K_5(x-1)^2 + \right. \\ \left. + K_6(x-1)^3 + K_7(x-1)^4\right] \tag{4.135}$$

where

$$K_3 = -\frac{1}{\alpha_1}\left(3 - \frac{\alpha_2}{\alpha_1}\right);\ K_4 = \pm\frac{2m}{\sqrt{-\alpha_1}}\left(3 - \frac{2\alpha_2}{\alpha_1}\right); \\ K_5 = -3m^2\left(1 - \frac{2\alpha_2}{\alpha_1}\right);\ K_6 = \pm\frac{4m^3\alpha_2}{\sqrt{-\alpha_1}}; K_7 = m^4\alpha_2. \tag{4.136}$$

Equation 4.9 for the function $v_1(x)$ is

$$v_1'' = H^*(x, C_i)N(v_0), v_1(0) = 0, v_1'(1) = 0 \tag{4.137}$$

where the auxiliary function H^* is chosen in the form

$$H^*(x, C_i) = \frac{1}{m(\alpha_1 - \alpha_2)}\left[C_6 + C_7(x-1) + C_8(x-1)^2 + \right. \\ \left. + C_9(x-1)^3 + C_{10}(x-1)^4\right]. \tag{4.138}$$

Substituting Eqs. 4.135 and 4.138 into Eq. 4.137, and thereafter solving this equation, we obtain the solution

$$v_1(x) = \frac{1}{2}K_3C_6\left[(x-1)^2 - 1\right] + \frac{1}{6}(K_4C_6 + K_3C_7)\left[(x-1)^3 + 1\right] + \\ + \frac{1}{12}(K_5C_6 + K_4C_7 + K_3C_8)\left[(x-1)^4 - 1\right] + \\ + \frac{1}{20}(K_6C_6 + K_5C_7 + K_4C_8 + K_5C_9)\left[(x-1)^5 + 1\right] + \\ + \frac{1}{30}(K_7C_6 + K_6C_7 + K_5C_8 + K_4C_7 + K_3C_8)\left[(x-1)^6 - 1\right] + \\ + \frac{1}{42}(K_7C_7 + K_6C_8 + K_5C_9 + K_6C_{10})\left[(x-1)^7 + 1\right] + \\ + \frac{1}{56}(K_7C_8 + K_6C_9 + K_5C_{10})\left[(x-1)^8 - 1\right] + \\ + \frac{1}{72}(K_7C_9 + K_6C_8)\left[(x-1)^9 + 1\right] + \frac{1}{90}K_7C_{10}\left[(x-1)^{10} - 1\right]. \tag{4.139}$$

The first-order approximate solution of Eqs. 4.115 and 4.132 is

$$\overline{v}(x) = v_0(x) + v_1(x) \tag{4.140}$$

where $v_0(x)$ and $v_1(x)$ are given by Eqs. 4.134 and 4.139 respectively.

Case 4.2.2.2
In this second case, the linear operator can be chosen in the form:

$$Lv = v'' + \lambda v' \tag{4.141}$$

where λ is an unknown parameter.

From Eq. 4.119 we can consider the following two subcases.

In the first subcase, the initial conditions are

$$v(0) = 1, v'(1) = 0. \tag{4.142}$$

Equation 4.4 becomes

$$v_0'' + \lambda v_0' - \lambda^2 e^{-\lambda} = 0, v_0(0) = 1, v_0'(1) = 0, g(x) = -\lambda v_0' + \lambda^2 e^{-\lambda}. \tag{4.143}$$

The solution of Eq. 4.143 has the form

$$v_0(x) = \lambda e^{-\lambda} x + e^{-\lambda x}. \tag{4.144}$$

The nonlinear operator $N(v_0)$ becomes:

$$N(v_0) = M_0 + M_1 e^{-\lambda x} + M_2 e^{-2\lambda x} + M_3 e^{-3\lambda x} + M_4 e^{-4\lambda x} + M_5 e^{-5\lambda x} \tag{4.145}$$

where

$$\begin{aligned} &M_0 = -m\left(1 + \alpha_2 \lambda^2 e^{-2\lambda}\right)^2; \\ &M_1 = \lambda^2 + (3\alpha_1 - \alpha_2)\lambda^4 e^{-2\lambda} + \alpha_1 \alpha_2 \lambda^6 e^{-4\lambda} + 4m\left(\alpha_2 \lambda^2 e^{-\lambda} + \alpha_2^2 \lambda^4 e^{-3\lambda}\right); \\ &M_2 = -2(3\alpha_1 - \alpha_2)\lambda^4 e^{-\lambda} - 4\alpha_1 \alpha_2 \lambda^6 e^{-3\lambda} - 2m\left(\alpha_2 \lambda^2 + 3\alpha_2^2 \lambda^4 e^{-2\lambda}\right) - \\ &-4m\alpha_2^2 \lambda^4 e^{-2\lambda}; \\ &M_3 = (3\alpha_1 - \alpha_2)\lambda^4 + 6\alpha_1 \alpha_2 \lambda^6 e^{-2\lambda} + 4m\alpha_2^2 \lambda^4 e^{-\lambda}; \\ &M_4 = -4\alpha_1 \alpha_2 \lambda^6 e^{-\lambda} - m\alpha_2^2 \lambda^4; \;\; M_5 = \alpha_1 \alpha_2 \lambda^6. \end{aligned} \tag{4.146}$$

The second function $v_1(x)$ is obtained from Eq. 4.9:

$$v_1'' + \lambda v_1' = \widetilde{H}(x, C_i) N(v_0), v_1(0) = 0, v_1'(1) = 0. \tag{4.147}$$

The expression 4.145 of the nonlinear operator $N(v_0)$ is an exponential one and therefore, the optimal auxiliary function $\widetilde{H}(x, C_i)$ can be chosen in the form

$$\widetilde{H}(x, C_i) = C_1 + C_2 e^{-\lambda x} + C_3 e^{-2\lambda x}. \tag{4.148}$$

where C_1, C_2, and C_3 are unknown parameters. Also we can use the following expressions for the auxiliary function:

$$\widetilde{H}(x, C_i) = C_{1'} + C_{2'} e^{-\lambda x} + C_{3'} e^{-2\lambda x} + C_{4'} e^{-3\lambda x}$$

or

$$\widetilde{H}(x, C_i) = C_1^* + C_2^* e^{-\lambda x} + C_3^* e^{-4\lambda x} + C_4^* e^{-5\lambda x}$$

and so on.

Now, substituting Eqs. 4.148 and 4.145 into Eq. 4.147, we obtain the equation

$$\begin{aligned} & v_1'' + \lambda v_1' = M_0 C_1 + (M_1 C_1 + M_0 C_2) e^{-\lambda x} + \\ & + (M_2 C_1 + M_1 C_2 + M_0 C_3) e^{-2\lambda x} + (M_3 C_1 + M_2 C_2 + M_1 C_3) e^{-3\lambda x} + \\ & + (M_4 C_1 + M_3 C_2 + M_2 C_3) e^{-4\lambda x} + (M_5 C_1 + M_4 C_2 + M_3 C_3) e^{-5\lambda x} + \\ & + (M_5 C_2 + M_4 C_3) e^{-6\lambda x} + M_5 C_3 e^{-7\lambda x}, v_1(0) = v_1'(1) = 0. \end{aligned} \tag{4.149}$$

From Eq. 4.149 we can obtain the solution $v_1(x)$, such that the first-order approximate solution of Eqs. 4.115 and 4.142 in this subcase, can be written in the form

$$\begin{aligned} & \overline{v}(x) = v_0(x) + v_1(x) = A + \left(\frac{M_0 C_1}{\lambda} + \lambda e^{-\lambda} \right) x + \\ & + \left(B + 1 - \frac{M_1 C_1 + M_0 C_2}{\lambda} x \right) e^{-\lambda x} + \frac{M_2 C_1 + M_1 C_2 + M_0 C_3}{2\lambda^2} e^{-2\lambda x} + \\ & + \frac{M_3 C_1 + M_2 C_2 + M_1 C_3}{6\lambda^2} e^{-3\lambda x} + \frac{M_4 C_1 + M_3 C_2 + M_2 C_3}{12\lambda^2} e^{-4\lambda x} + \\ & + \frac{M_5 C_1 + M_4 C_2 + M_3 C_3}{20\lambda^2} e^{-5\lambda x} + \frac{M_5 C_2 + M_4 C_3}{30\lambda^2} e^{-6\lambda x} + \frac{M_5 C_3}{42\lambda^2} e^{-7\lambda x} \end{aligned} \tag{4.150}$$

where $v_0(x)$ is given by Eq. 4.144 and

$$
\begin{aligned}
A = & \Big(-\frac{e^{\lambda}}{\lambda^2}M_0 + \frac{1-\lambda}{\lambda^2}M_1 + \frac{2e^{-\lambda}-1}{2\lambda^2}M_2 + \frac{3e^{-2\lambda}-1}{6\lambda^2}M_3 + \\
& + \frac{4e^{-3\lambda}-1}{12\lambda^2}M_4 + \frac{5e^{-4\lambda}-1}{20\lambda^2}M_5\Big)C_1 + \Big(\frac{1-\lambda}{\lambda^2}M_0 + \frac{2e^{-\lambda}-1}{2\lambda^2}M_1 + \\
& + \frac{3e^{-2\lambda}-1}{6\lambda^2}M_2 + \frac{4e^{-3\lambda}-1}{12\lambda^2}M_3 + \frac{5e^{-4\lambda}-1}{20\lambda^2}M_4 + \frac{6e^{-5\lambda}-1}{30\lambda^2}M_5\Big)C_2 + \\
& + \Big(\frac{2e^{-\lambda}-1}{2\lambda^2}M_0 + \frac{3e^{-2\lambda}-1}{6\lambda^2}M_1 + \frac{4e^{-3\lambda}-1}{12\lambda^2}M_2 + \frac{5e^{-4\lambda}-1}{20\lambda^2}M_3 + \\
& + \frac{6e^{-5\lambda}-1}{30\lambda^2}M_4 + \frac{7e^{-6\lambda}-1}{42\lambda^2}M_5\Big)C_3; \\
B = & \Big(\frac{e^{\lambda}}{\lambda^2}M_0 + \frac{\lambda-1}{\lambda^2}M_1 - \frac{e^{-\lambda}}{\lambda^2}M_2 - \frac{e^{-2\lambda}}{2\lambda^2}M_3 - \frac{e^{-3\lambda}}{3\lambda^2}M_4 - \\
& - \frac{e^{-4\lambda}}{4\lambda^2}M_5\Big)C_1 + \Big(\frac{\lambda-1}{\lambda^2}M_0 - \frac{e^{-\lambda}}{\lambda^2}M_1 - \frac{e^{-2\lambda}}{2\lambda^2}M_2 - \frac{e^{-3\lambda}}{3\lambda^2}M_3 - \\
& - \frac{e^{-4\lambda}}{4\lambda^2}M_4 - \frac{e^{-5\lambda}}{5\lambda^2}M_5\Big)C_2 + \Big(-\frac{e^{-\lambda}}{\lambda^2}M_0 - \frac{e^{-2\lambda}}{2\lambda^2}M_1 - \frac{e^{-3\lambda}}{3\lambda^2}M_2 - \\
& - \frac{e^{-4\lambda}}{4\lambda^2}M_3 - \frac{e^{-5\lambda}}{5\lambda^2}M_4 - \frac{e^{-6\lambda}}{6\lambda^2}M_5\Big)C_3.
\end{aligned}
\tag{4.151}
$$

In the last subcase, the initial conditions are

$$
v(0) = 1, v'(1) = \pm\frac{1}{\sqrt{-\alpha_1}}, if \alpha_1 < 0. \tag{4.152}
$$

Equation 4.4 has the form

$$
\begin{aligned}
& v_0'' + \lambda v_0' - \lambda^2 e^{-\lambda} = 0, v_0(0) = 1, v_0'(1) = \pm\frac{1}{\sqrt{-\alpha_1}}, \\
& g(x) = -\lambda v_0' + \lambda^2 e^{-\lambda}
\end{aligned}
\tag{4.153}
$$

and has the solution

$$
v_0(x) = \left(\pm\frac{1}{\sqrt{-\alpha_1}} + \lambda e^{-\lambda}\right)x + e^{-\lambda x}. \tag{4.154}
$$

By means of Eq. 4.154, the nonlinear operator $N(v_0)$ becomes:

$$
N(v_0) = N_0 + N_1 e^{-\lambda x} + N_2 e^{-2\lambda x} + N_3 e^{-3\lambda x} + N_4 e^{-4\lambda x} + N_5 e^{-5\lambda x} \tag{4.155}
$$

where

$$
\begin{aligned}
N_0 &= -m\left[1+\alpha_2\left(\pm\tfrac{1}{\sqrt{-\alpha_1}}+\lambda e^{-\lambda}\right)^2\right]^2;\\
N_1 &= \lambda^2+\lambda^2(3\alpha_1-\alpha_2)\left(\pm\tfrac{1}{\sqrt{-\alpha_1}}+\lambda e^{-\lambda}\right)^2+\alpha_1\alpha_2\lambda^2\Big(\pm\frac{1}{\sqrt{-\alpha_1}}+\\
&+\lambda e^{-\lambda}\Big)^4+4m\alpha_2\lambda\left(\pm\frac{1}{\sqrt{-\alpha_1}}+\lambda e^{-\lambda}\right)+4m\alpha_2^2\lambda\left(\pm\tfrac{1}{\sqrt{-\alpha_1}}+\lambda e^{-\lambda}\right)^3;\\
N_2 &= -2(3\alpha_1-\alpha_2)\lambda^3\left(\pm\frac{1}{\sqrt{-\alpha_1}}+\lambda e^{-\lambda}\right)-4\alpha_1\alpha_2\lambda^3\Big(\pm\frac{1}{\sqrt{-\alpha_1}}+\\
&+\lambda e^{-\lambda}\Big)^3-2m\alpha_2\lambda^2-6m\alpha_2^2\lambda^2\left(\pm\tfrac{1}{\sqrt{-\alpha_1}}+\lambda e^{-\lambda}\right)^2;\\
N_3 &= (3\alpha_1-\alpha_2)\lambda^4+6\alpha_1\alpha_2\lambda^4\left(\pm\tfrac{1}{\sqrt{-\alpha_1}}+\lambda e^{-\lambda}\right)^2+\\
&+4m\alpha_2^2\lambda^3\left(\pm\frac{1}{\sqrt{-\alpha_1}}+\lambda e^{-\lambda}\right);\\
N_4 &= -4\alpha_1\alpha_2\lambda^5\left(\pm\frac{1}{\sqrt{-\alpha_1}}+\lambda e^{-\lambda}\right)-m\alpha_2^2\lambda^4; N_5=\alpha_1\alpha_2\lambda^6.
\end{aligned}
\tag{4.156}
$$

For the equation in $v_1(x)$ we choose the optimal auxiliary function $\tilde{\tilde{H}}(x, C_i)$ in the form

$$\tilde{\tilde{H}}(x, C_i) = C_4 + C_5e^{-\lambda x} + C_6e^{-2\lambda x}. \tag{4.157}$$

The Eq. 4.9 can be written, by means of Eqs. 4.157 and 4.155 as

$$
\begin{aligned}
&v_1'' + \lambda v_1' = N_0C_4 + (N_1C_4 + N_0C_5))e^{-\lambda x}+\\
&+(N_2C_4 + N_1C_5 + N_0C_6)e^{-2\lambda x} + (N_3C_4 + N_2C_5 + N_1C_6)e^{-3\lambda x}+\\
&+(N_4C_4 + N_3C_5 + N_2C_6)e^{-4\lambda x} + (N_5C_4 + N_4C_5 + N_3C_6)e^{-5\lambda x}+\\
&+(N_5C_5 + N_4C_6)e^{-6\lambda x} + N_5C_6e^{-7\lambda x}, v_1(0) = v_1'(1) = 0.
\end{aligned}
\tag{4.158}
$$

The first-order approximate solution in the last subcase has the form

$$
\begin{aligned}
\overline{v}(x) &= v_0(x) + v_1(x) = D + \left(\frac{N_0C_4}{\lambda} + \lambda e^{-\lambda} \pm \frac{1}{\sqrt{-\alpha_1}}\right)x+\\
&+\left(E + 1 - \frac{N_1C_4 + N_0C_5}{\lambda}x\right)e^{-\lambda x} + \frac{N_2C_4 + N_1C_5 + N_0C_6}{2\lambda^2}e^{-2\lambda x}+\\
&+\frac{N_3C_4 + N_2C_5 + N_1C_6}{6\lambda^2}e^{-3\lambda x} + \frac{N_4C_4 + N_3C_5 + N_2C_6}{12\lambda^2}e^{-4\lambda x}+\\
&+\frac{N_5C_4 + N_4C_5 + N_3C_6}{20\lambda^2}e^{-5\lambda x} + \frac{N_5C_5 + N_4C_6}{30\lambda^2}e^{-6\lambda x} + \frac{N_5C_6}{42\lambda^2}e^{-7\lambda x}
\end{aligned}
\tag{4.159}
$$

where

$$D = \Big(-\frac{e^{\lambda}}{\lambda^2}N_0 + \frac{1-\lambda}{\lambda^2}N_1 + \frac{2e^{-\lambda}-1}{2\lambda^2}N_2 + \frac{3e^{-2\lambda}-1}{6\lambda^2}N_3 +$$
$$+\frac{4e^{-3\lambda}-1}{12\lambda^2}N_4 + \frac{5e^{-4\lambda}-1}{20\lambda^2}N_5\Big)C_4 + \Big(\frac{1-\lambda}{\lambda^2}N_0 + \frac{2e^{-\lambda}-1}{2\lambda^2}N_1 +$$
$$+\frac{3e^{-2\lambda}-1}{6\lambda^2}N_2 + \frac{4e^{-3\lambda}-1}{12\lambda^2}N_3 + \frac{5e^{-4\lambda}-1}{20\lambda^2}N_4 + \frac{6e^{-5\lambda}-1}{30\lambda^2}N_5\Big)C_5 +$$
$$+\Big(\frac{2e^{-\lambda}-1}{2\lambda^2}N_0 + \frac{3e^{-2\lambda}-1}{6\lambda^2}N_1 + \frac{4e^{-3\lambda}-1}{12\lambda^2}N_2 + \frac{5e^{-4\lambda}-1}{20\lambda^2}N_3 +$$
$$+\frac{6e^{-5\lambda}-1}{30\lambda^2}N_4 + \frac{7e^{-6\lambda}-1}{42\lambda^2}N_5\Big)C_6;$$

$$E = \left(\frac{e^{\lambda}}{\lambda^2}N_0 + \frac{\lambda-1}{\lambda^2}N_1 - \frac{e^{-\lambda}}{\lambda^2}N_2 - \frac{e^{-2\lambda}}{2\lambda^2}N_3 - \frac{e^{-3\lambda}}{3\lambda^2}N_4 - \frac{e^{-4\lambda}}{4\lambda^2}N_5\right)C_4 +$$
$$+\left(\frac{\lambda-1}{\lambda^2}N_0 - \frac{e^{-\lambda}}{\lambda^2}N_1 - \frac{e^{-2\lambda}}{2\lambda^2}N_2 - \frac{e^{-3\lambda}}{3\lambda^2}N_3 - \frac{e^{-4\lambda}}{4\lambda^2}N_4 - \frac{e^{-5\lambda}}{5\lambda^2}N_5\right)C_5 + \quad (4.160)$$
$$+\left(-\frac{e^{-\lambda}}{\lambda^2}N_0 - \frac{e^{-2\lambda}}{2\lambda^2}N_1 - \frac{e^{-3\lambda}}{3\lambda^2}N_2 - \frac{e^{-4\lambda}}{4\lambda^2}N_3 - \frac{e^{-5\lambda}}{5\lambda^2}N_4 - \frac{e^{-6\lambda}}{6\lambda^2}N_5\right)C_6.$$

4.2.3 Numerical Results and Discussions

We illustrate the accuracy of our procedure for different values of the coefficients α_1, α_2 and and for every the optimal auxiliary functions and $H, H*, \widetilde{H}, \widetilde{\widetilde{H}}$. Also we represent graphically the variation of velocity profile of the belt problem for an Oldroyd 6-constant fluid.

Example 4.2.3.1 If we take into consideration, the solution given by Eq. 4.130, for $\alpha_1 = 0.5$, $\alpha_2 = 0.5$, $m = 1$ and $v'(1) = 0$ (Newtonian case), it is interesting to remark that from Eq. 4.126 results that $K_1 = K_2 = 0$ and therefore $v_1(x) = 0$. It follows that $\overline{v} = v_0(x) = m\left(\frac{x^2}{2} - x\right) + 1$, which is the exact solution of Eqs. 4.115 and 4.116.

Example 4.2.3.2 In this example we consider $\alpha_1 = 0.5$, $\alpha_2 = 0.75$, $m = 1$ and $v'(1) = 0$ (non-Newtonian case). The values of the optimal convergence-control parameters for the case 4.2.2.1 and the first subcase, are

$$C_1 = 0.0309007883,\ C_2 = 0.5673047654,\ C_3 = 1.7430618401,$$
$$C_4 = 4.9462074642,\ C_5 = 3.3490697012.$$

Table 4.21 Comparison between OHAM results given by Eq. 4.161 and numerical results for $\alpha_1 = 0.5$, $\alpha_2 = 0.75$, $m = 1$ and $v'(1) = 0$ and the optimal auxiliary function (4.127)

x	$v_{numeric}$	$\overline{v}_{OHAM}$, Eq. 4.161	$relative\ error = \left\|v_{numeric} - \overline{v}_{OHAM}\right\|$
0	1	1	0
0.1	0.8866085725	0.8866821774	7.3×10^{-5}
0.2	0.7875621671	0.7875887398	2.6×10^{-5}
0.3	0.7025338925	0.7025310249	2.8×10^{-5}
0.4	0.6308917264	0.6308667405	2.4×10^{-5}
0.5	0.5717826705	0.5717915984	8.9×10^{-6}
0.6	0.5245086113	0.5245288249	2.0×10^{-5}
0.7	0.4884393924	0.4884348043	4.5×10^{-6}
0.8	0.4630600752	0.4630345952	2.5×10^{-5}
0.9	0.4480006226	0.4479965346	4.0×10^{-6}
1	0.4430235508	0.4430506264	2.7×10^{-5}

The first-order approximate solution 4.130 can be written in the form

$$\begin{aligned}\overline{v}(x) &= 0.4430506264 + 0.4884122043(x-1)^2 - \\ &-0.0709130956(x-1)^3 - 0.1094241898(x-1)^4 - \\ &-0.1908012620(x-1)^5 - 0.0946208790(x-1)^6 - \\ &-0.0220812833(x-1)^7 - 0.0112134030(x-1)^8\end{aligned} \tag{4.161}$$

In Table 4.21 we present a comparison between the first-order approximate solution given by Eq. 4.161 and numerical results for some values of variable and the corresponding relative errors.

Example 4.2.3.3 For $\alpha_1 = 0.5$, $\alpha_2 = 1$, $m = 1$ and $v'(1) = 0$ we obtain for Eq. 4.130:

$$C_1 = 0.0145037003, C_2 = 0.3432811995, C_3 = 1.1572350528,$$
$$C_4 = 5.1398275696, C_5 = 3.7778914694$$

and the first-order approximate solution in the form:

$$\begin{aligned}\overline{v}(x) &= 0.3569716548 + 0.4891222247(-1+x)^2 - \\ &-0.0858202998(x-1)^3 - 0.1452587024(x-1)^4 - \\ &-0.3940690977(x-1)^5 - 0.2081818243(x-1)^6 - \\ &-0.0611884234(x-1)^7 - 0.0337311738(x-1)^8.\end{aligned} \tag{4.162}$$

In Table 4.22 is presented a comparison between the first-order approximate solution given by Eq. 4.162 and numerical results, and the corresponding relative errors.

Table 4.22 Comparison between OHAM results given by Eq. 4.162 and numerical results for $\alpha_1 = 0.5$, $\alpha_2 = 1$, $m = 1$ and $v'(1) = 0$ and the auxiliary function 4.127

x	$v_{numeric}$	$\overline{v}_{OHAM}$, Eq. 4.162	$relative\ error = \lvert v_{numeric} - \overline{v}_{OHAM} \rvert$
0	1	1	0
0.1	0.8571788783	0.8572229943	4.4×10^{-5}
0.2	0.7361604138	0.7361798440	1.9×10^{-5}
0.3	0.6360415408	0.6360346979	6.8×10^{-6}
0.4	0.5548519523	0.5548435245	8.4×10^{-6}
0.5	0.4903083144	0.4903091700	8.5×10^{-7}
0.6	0.4402488104	0.4402657870	1.6×10^{-5}
0.7	0.4029480153	0.4029501998	2.1×10^{-6}
0.8	0.3771255723	0.3771041676	2.1×10^{-5}
0.9	0.3619470473	0.3619379098	9.1×10^{-6}
1	0.3569430437	0.3569716548	2.8×10^{-5}

Example 4.2.3.4 If we consider $\alpha_1 = 0.5, \alpha_2 = 0.75, m = 1.5$ and $v'(1) = 0$, then the convergence control parameters for the case 4.2.2.1 and the first subcase are

$$C_1 = -0.0103555184, C_2 = -0.2555006995, C_3 = -2.5463142522,$$
$$C_4 = -3.1417178283, C_5 = -1.0373234408$$

and therefore the first-order approximate solution has the form:

$$\begin{aligned}\overline{v}(x) &= 0.0939116440 + 0.7631062030(x-1)^2+\\ &+0.1077893576(x-1)^3 + 0.5383418691(x-1)^4+\\ &+0.4158131167(x-1)^5 + 0.2083746269(x-1)^6+\\ &+0.1065063382(x-1)^7 + 0.0263744694(x-1)^8.\end{aligned} \tag{4.163}$$

This approximate solution is compared in Table 4.23 with the numerical results.

Example 4.2.3.5 In the case $\alpha_1 = 0.5, \alpha_2 = 1, m = 1.5$ and $v'(1) = 0$, the optimal convergence-control parameters for the case 4.2.2.1 and the first subcase are

$$C_1 = -0.0430647, C_2 = -0.997138, C_3 = -6.40181,$$
$$C_4 = -9.14949, C_5 = -3.90268.$$

The first-order approximate solution becomes:

$$\begin{aligned}\overline{v}(x) &= -0.1330803613 + 0.8590076019(x-1)^2+\\ &+0.8413350971(x-1)^3 + 2.7143908882(x-1)^4+\\ &+2.5052659024(x-1)^5 + 1.4688065504(x-1)^6+\\ &+0.8271305377(x-1)^7 + 0.2646068580(x-1)^8.\end{aligned} \tag{4.164}$$

The OHAM solution 4.164 is compared in Table 4.24 with the numerical results.

Table 4.23 Comparison between OHAM results given by Eq. 4.163 and numerical results for $\alpha_1 = 0.5$, $\alpha_2 = 0.75$, $m = 1.5$ and $v'(1) = 0$ and the auxiliary function 4.127

x	$v_{numeric}$	$\bar{v}_{OHAM}$, Eq. 4.163	*relative error* $= \lvert v_{numeric} - \bar{v}_{OHAM} \rvert$
0	1	1	0
0.1	0.8122030427	0.8122723528	6.9×10^{-5}
0.2	0.6480524303	0.6480757122	2.3×10^{-5}
0.3	0.5075106818	0.5074963533	1.4×10^{-5}
0.4	0.3899587013	0.3899662747	7.5×10^{-6}
0.5	0.2943579028	0.2943935300	3.5×10^{-5}
0.6	0.2193544932	0.2193300304	2.4×10^{-5}
0.7	0.1631889587	0.1631613755	2.7×10^{-5}
0.8	0.1243031560	0.1243139043	1.0×10^{-5}
0.9	0.1014691098	0.1014847907	1.5×10^{-5}
1	0.0939598983	0.0939116440	4.8×10^{-5}

Table 4.24 Comparison between OHAM results given by Eq. 4.164 and numerical results for $\alpha_1 = 0.5$, $\alpha_2 = 1$, $m = 1.5$ and $v'(1) = 0$ and the optimal auxiliary function 4.127

x	$v_{numeric}$	$\bar{v}_{OHAM}$, Eq. 4.164	*relative error* $= \lvert v_{numeric} - \bar{v}_{OHAM} \rvert$
0	1	0.9999999999	2.2×10^{-16}
0.1	0.7497063820	0.7498345352	1.2×10^{-4}
0.2	0.5327963111	0.5327805533	1.5×10^{-5}
0.3	0.3498945869	0.3498605615	3.4×10^{-5}
0.4	0.2010814564	0.2012282330	1.4×10^{-4}
0.5	0.0852885962	0.0853862885	9.7×10^{-5}
0.6	−0.0006296037	−0.0008156336	1.8×10^{-4}
0.7	−0.0615295648	−0.0616797274	1.5×10^{-4}
0.8	−0.1020045446	−0.1018253040	1.7×10^{-4}
0.9	−0.1251782938	−0.1250838453	9.4×10^{-5}
1	−0.1327530483	−0.133080361	3.2×10^{-4}

Example 4.2.3.6 For $\alpha_1 = -0.5$, $\alpha_2 = 0.5$, $m = 1$ and $v'(1) = \frac{1}{\sqrt{0.5}}$, for the case 4.2.2.1, and the second subcase, the optimal convergence-control parameters are

$$C_1 = -0.2487584861,\ C_2 = 0.3794941533,\ C_3 = -0.4121126222,$$
$$C_4 = 0.4150647670,\ C_5 = -0.5193428345$$

The first-order approximate solution given by Eq. 4.140 can be written as

Table 4.25 Comparison between OHAM results given by Eq. 4.165 and numerical results for $\alpha_1 = -0.5, \alpha_2 = 0.5, m = 1$ and $v'(1) = \frac{1}{\sqrt{0.5}}$ and the auxiliary function 4.138

x	$v_{numeric}$	$\overline{v}_{OHAM}$, Eq. 4.165	*relative error* $= \lvert v_{numeric} - \overline{v}_{OHAM} \rvert$
0	1	1	0
0.1	1.2229062807	1.2229013971	4.8×10^{-9}
0.2	1.4377929759	1.4377807450	1.2×10^{-5}
0.3	1.6445808407	1.6445867997	5.9×10^{-6}
0.4	1.8432159625	1.8432255865	9.6×10^{-6}
0.5	2.0336046251	2.0336047334	1.0×10^{-7}
0.6	2.2156306053	2.2156224054	8.1×10^{-6}
0.7	2.3891355914	2.3891330225	2.5×10^{-6}
0.8	2.5539053793	2.5539142068	8.8×10^{-6}
0.9	2.7096483135	2.7096510721	2.7×10^{-6}
1	2.8559539636	2.8559439901	9.9×10^{-6}

$$\begin{aligned}
&\overline{v}(x) = 2.4142135623 + 1.4142135623(x-1)- \\
&-0.4950339447\left[(x-1)^2 - 1\right] - 0.0803371704\left[(x-1)^3 + 1\right]- \\
&-0.0140724640\left[(x-1)^4 - 1\right] + 0.0102108835\left[(x-1)^5 + 1\right]- \\
&-0.0603676324\left[(x-1)^6 - 1\right] - 0.1091646985\left[(x-1)^7 + 1\right]- \\
&-0.0661814531\left[(x-1)^8 - 1\right] - 0.0175193191\left[(x-1)^9 + 1\right]- \\
&-0.0028852379\left[(x-1)^{10} - 1\right].
\end{aligned} \tag{4.165}$$

In Table 4.25 we present a comparison between the approximate solution 4.165 obtained by means of OHAM, and numerical results, and the corresponding relative errors.

Example 4.2.3.7 In this case we consider $\alpha_1 = -0.5,\ \alpha_2 = 0.75,\ m = 1$ and $v'(1) = \frac{1}{\sqrt{0.5}}$, such that for Eq. 4.140, the optimal convergence-control parameters are

$$C_1 = -0.2484286614, C_2 = 0.4310993122, C_3 = -0.4720538492,$$
$$C_4 = 0.5090078876, C_5 = -0.6742203389$$

The first-order approximate solution given by Eq. 4.140 is

$$\begin{aligned}
&\overline{v}(x) = 2.4142135623 + 1.4142135623(x-1)- \\
&-0.6179289764\left[(x-1)^2 - 1\right] - 0.0560133962\left[(x-1)^3 + 1\right]+ \\
&+0.0071974457\left[(x-1)^4 - 1\right] + 0.0344624859\left[(x-1)^5 + 1\right]- \\
&-0.0793096739\left[(x-1)^6 - 1\right] - 0.1669817690\left[(x-1)^7 + 1\right]- \\
&-0.1122347657\left[(x-1)^8 - 1\right] + 0.0344266489\left[(x-1)^9 + 1\right]- \\
&-0.0056185028\left[(x-1)^{10} - 1\right].
\end{aligned} \tag{4.166}$$

Table 4.26 Comparison between OHAM results given by Eq. 4.166 and numerical results for $\alpha_1 = -0.5$, $\alpha_2 = 0.75$, $m = 1$ and $v'(1) = \frac{1}{\sqrt{0.5}}$ and the auxiliary function 4.138

x	$v_{numeric}$	$\overline{v}_{OHAM}$, Eq. 4.166	$relative\ error = \lvert v_{numeric} - \overline{v}_{OHAM} \rvert$
0	1	1	0
0.1	1.2518351629	1.2518238888	1.1×10^{-5}
0.2	1.4920691495	1.4920531530	1.5×10^{-5}
0.3	1.7207387643	1.72074398947	5.2×10^{-6}
0.4	1.9379031842	1.93792002203	1.6×10^{-5}
0.5	2.1436214696	2.14362291135	1.4×10^{-6}
0.6	2.3379068070	2.33789447422	1.2×10^{-5}
0.7	2.5207347633	2.52072993082	4.8×10^{-6}
0.8	2.6920220811	2.69203422220	1.2×10^{-5}
0.9	2.8516000667	2.85160438618	4.3×10^{-6}
1	2.9991636824	2.99914870737	1.4×10^{-5}

In Table 4.26 we present a comparison between the present solution 4.166 and numerical results and corresponding relative errors.

Example 4.2.3.8 For $\alpha_1 = -0.5$, $\alpha_2 = 1$, $m = 1$ and $v'(1) = \frac{1}{\sqrt{0.5}}$, the optimal convergence-control parameters are

$$C_1 = -0.2481244286,\ C_2 = 0.4823327023,\ C_3 = -0.5396781473,$$
$$C_4 = 0.6132014735,\ C_5 = -0.8540900975$$

and the first-order approximate solution given by Eq. 4.140 is

$$\begin{aligned}
&\overline{v}(x) = 2.4142135623 + 1.4142135623(x-1)- \\
&-0.7406221430\left[(x-1)^2 - 1\right] - 0.0148810042\left[(x-1)^3 + 1\right] + \\
&+0.03592103224\left[(x-1)^4 - 1\right] + 0.06391596843\left[(x-1)^5 + 1\right] - \\
&-0.10448164910\left[(x-1)^6 - 1\right] - 0.24482492876\left[(x-1)^7 + 1\right] - \\
&-0.17646854016\left[(x-1)^8 - 1\right] - 0.0585869683\left[(x-1)^9 + 1\right] - \\
&-0.0094898899\left[(x-1)^{10} - 1\right].
\end{aligned} \tag{4.167}$$

In Table 4.27 we present a comparison between the solution 4.167 and numerical results, and corresponding relative errors.

In the following Tables, we present only comparisons between the present solutions obtained by means of OHAM and numerical results and corresponding relative errors (Tables 4.28, 4.29, 4.30, 4.31, 4.32, 4.33, 4.34, 4.35, 4.36, 4.37, 4.38, 4.39, 4.40, 4.41, 4.42).

Table 4.27 Comparison between OHAM results given by Eq. 4.167 and numerical results for $\alpha_1 = -0.5$, $\alpha_2 = 1$, $m = 1$ and $v'(1) = \frac{1}{\sqrt{0.5}}$ and the auxiliary function 4.138

x	$v_{numeric}$	$\bar{v}_{OHAM}$, Eq. (81)	*relative error* = $\lvert v_{numeric} - \bar{v}_{OHAM} \rvert$
0	1	1	0
0.1	1.2839691268	1.2839541027	1.5×10^{-5}
0.2	1.5522080417	1.5521893463	1.8×10^{-5}
0.3	1.8048997685	1.8049032366	3.4×10^{-6}
0.4	2.0422481654	2.0422731488	2.4×10^{-5}
0.5	2.2645158331	2.2645193174	3.4×10^{-6}
0.6	2.4718989261	2.4718821917	1.6×10^{-5}
0.7	2.6645698769	2.6645620846	7.7×10^{-6}
0.8	2.8426469875	2.8426623139	1.5×10^{-5}
0.9	3.0061618853	3.0061679939	6.1×10^{-6}
1	3.1549980906	3.1549778194	2.0×10^{-5}

Table 4.28 Comparison between OHAM results given by Eq. 4.140 and numerical results for $\alpha_1 = -0.5$, $\alpha_2 = 0.5$, $m = 1.5$ and $v'(1) = \frac{1}{\sqrt{0.5}}$ and the auxiliary function 4.138

x	$v_{numeric}$	$\bar{v}_{OHAM}$	*relative error* = $\lvert v_{numeric} - \bar{v}_{OHAM} \rvert$
0	1	1	0
0.1	1.2609509091	1.2607564771	1.9×10^{-4}
0.2	1.5099315977	1.5092509182	6.8×10^{-4}
0.3	1.7469050126	1.7471433465	2.3×10^{-4}
0.4	1.9718544689	1.9726191668	7.6×10^{-4}
0.5	2.1847229322	2.1849754601	2.5×10^{-4}
0.6	2.3853998014	2.3850003567	3.9×10^{-4}
0.7	2.5737023327	2.5735062125	1.9×10^{-4}
0.8	2.7493402910	2.7498662597	5.2×10^{-4}
0.9	2.9118599070	2.9120716519	2.1×10^{-4}
1	3.0605243186	3.0595422315	9.8×10^{-4}

Table 4.29 Comparison between OHAM results given by Eq. 4.140 and numerical results for $\alpha_1 = -0.5$, $\alpha_2 = 0.75$, $m = 1$ and $v'(1) = \frac{1}{\sqrt{0.5}}$ and the optimal auxiliary function 4.138

x	$v_{numeric}$	$\overline{v}_{OHAM}$	*relative error* $= \lvert v_{numeric} - \overline{v}_{OHAM} \rvert$
0	1	1	0
0.1	1.3085076286	1.3086293331	1.2×10^{-4}
0.2	1.5989200672	1.5982295694	6.9×10^{-4}
0.3	1.8714226404	1.8715831764	1.6×10^{-4}
0.4	2.1262167929	2.1270250021	8.0×10^{-4}
0.5	2.3635521486	2.3638995934	3.4×10^{-4}
0.6	2.5835932960	2.5832535309	3.3×10^{-4}
0.7	2.7864530758	2.7863211230	1.3×10^{-4}
0.8	2.9721424944	2.9728245480	6.8×10^{-4}
0.9	3.1405027991	3.1408732113	3.7×10^{-4}
1	3.2910606175	3.2900917909	9.6×10^{-4}

Table 4.30 Comparison between OHAM results given by Eq. 4.140 and numerical results for $\alpha_1 = -0.5$, $\alpha_2 = 1$, $m = 1$ and $v'(1) = \frac{1}{\sqrt{0.5}}$ and the optimal auxiliary function 4.138

x	$v_{numeric}$	$\overline{v}_{OHAM}$	*relative error* $= \lvert v_{numeric} - \overline{v}_{OHAM} \rvert$
0	1	1	0
0.1	1.3620021521	1.3619942603	7.8×10^{-6}
0.2	1.6989187493	1.6989013673	1.7×10^{-5}
0.3	2.0111542818	2.0111557494	1.4×10^{-6}
0.4	2.2991388016	2.2991755901	3.6×10^{-5}
0.5	2.5634672715	2.5634733579	6.0×10^{-6}
0.6	2.8046481809	2.8046272723	2.0×10^{-5}
0.7	3.0231881146	3.0231849910	3.1×10^{-6}
0.8	3.2195266334	3.2195597898	3.3×10^{-5}
0.9	3.3939541762	3.3939648992	1.0×10^{-5}
1	3.5464381070	3.5464081807	2.9×10^{-5}

Table 4.31 Comparison between OHAM results given by Eq. 4.150 and numerical results for $\alpha_1 = 0.5$, $\alpha_2 = 0.5$, $m = 1$ and $v'(1) = 0$ and the optimal auxiliary function 4.148

x	$v_{numeric}$	$\overline{v}_{OHAM}$	$relative\ error = \left\|v_{numeric} - \overline{v}_{OHAM}\right\|$
0	1	1	0
0.1	0.9053448275	0.9052579961	8.6×10^{-5}
0.2	0.8203448275	0.8203616740	1.6×10^{-5}
0.3	0.7453448275	0.7453758909	3.1×10^{-5}
0.4	0.6803448275	0.6803496596	4.8×10^{-6}
0.5	0.6253448275	0.6253174083	2.7×10^{-5}
0.6	0.5803448275	0.5803001231	4.4×10^{-5}
0.7	0.5453448275	0.5453063866	3.8×10^{-5}
0.8	0.5203448275	0.5203333245	1.1×10^{-5}
0.9	0.5053448275	0.5053674704	2.2×10^{-5}
1	0.5003448275	0.5003855577	4.0×10^{-5}

Table 4.32 Comparison between OHAM results given by Eq. 4.150 and numerical results for $\alpha_1 = 0.5$, $\alpha_2 = 0.75$, $m = 1$ and $v'(1) = 0$ and the optimal auxiliary function 4.148

x	$v_{numeric}$	$\overline{v}_{OHAM}$	$relative\ error = \left\|v_{numeric} - \overline{v}_{OHAM}\right\|$
0	1	1	0
0.1	0.8866085725	0.8863361236	2.7×10^{-4}
0.2	0.7875621671	0.7875077487	5.4×10^{-5}
0.3	0.7025338925	0.7026438847	1.0×10^{-4}
0.4	0.6308917264	0.6309536994	6.1×10^{-5}
0.5	0.5717826705	0.5717301604	5.2×10^{-5}
0.6	0.5245086113	0.5243482669	1.6×10^{-4}
0.7	0.4884393924	0.4882588929	1.8×10^{-4}
0.8	0.4630600752	0.4629799268	8.0×10^{-5}
0.9	0.4480006226	0.4480861921	8.5×10^{-5}
1	0.4430235508	0.4431992337	1.7×10^{-4}

Table 4.33 Comparison between OHAM results given by Eq. 4.150 and numerical results for $\alpha_1 = 0.5$, $\alpha_2 = 1$, $m = 1$ and $v'(1) = 0$ and the optimal auxiliary function 4.148

x	$v_{numeric}$	$\overline{v}_{OHAM}$	$relative\ error = \lvert v_{numeric} - \overline{v}_{OHAM} \rvert$
0	1	1	6.6×10^{-16}
0.1	0.8571788783	0.8574415758	2.6×10^{-4}
0.2	0.7361604138	0.7371670017	1.0×10^{-3}
0.3	0.6360415408	0.6363653381	3.2×10^{-4}
0.4	0.5548519523	0.5540906717	7.6×10^{-4}
0.5	0.4903083144	0.4886618248	1.6×10^{-3}
0.6	0.4402488104	0.4383723210	1.8×10^{-3}
0.7	0.4029480153	0.4016994772	1.2×10^{-3}
0.8	0.3771255723	0.3772291464	1.0×10^{-4}
0.9	0.3619470473	0.3635699289	1.6×10^{-3}
1	0.3569430437	0.3593259550	2.3×10^{-3}

Table 4.34 Comparison between OHAM results given by Eq. 4.150 and numerical results for $\alpha_1 = 0.5$, $\alpha_2 = 0.5$, $m = 1.5$ and $v'(1) = 0$ and the optimal auxiliary function 4.148

x	$v_{numeric}$	$\overline{v}_{OHAM}$	$relative\ error = \lvert v_{numeric} - \overline{v}_{OHAM} \rvert$
0	1	1.0000000018	1.8×10^{-9}
0.1	0.8577489626	0.8577105235	3.8×10^{-5}
0.2	0.7302489626	0.7302905954	4.1×10^{-5}
0.3	0.6177489626	0.6177941197	4.5×10^{-5}
0.4	0.5202489626	0.5202636998	1.4×10^{-5}
0.5	0.4377489626	0.4377305945	1.8×10^{-5}
0.6	0.3702489626	0.3702146522	3.4×10^{-5}
0.7	0.3177489626	0.3177242586	2.4×10^{-5}
0.8	0.2802489626	0.2802562722	7.3×10^{-5}
0.9	0.2577489626	0.2577959802	4.7×10^{-5}
1	0.2502489626	0.2503170184	6.8×10^{-5}

Table 4.35 Comparison between OHAM results given by Eq. 4.150 and numerical results for $\alpha_1 = 0.5$, $\alpha_2 = 0.75$, $m = 1.5$ and $v'(1) = 0$ and the optimal auxiliary function 4.148

x	$v_{numeric}$	$\bar{v}_{OHAM}$	$relative\ error = \lvert v_{numeric} - \bar{v}_{OHAM} \rvert$
0	1	1	0
0.1	0.8122030427	0.8121518633	5.1×10^{-5}
0.2	0.6480524030	0.6479781411	7.4×10^{-5}
0.3	0.5075106818	0.5075931624	8.2×10^{-5}
0.4	0.3899587013	0.3902155536	2.5×10^{-4}
0.5	0.2943579028	0.2945933839	2.3×10^{-4}
0.6	0.2193544932	0.2193140489	4.0×10^{-5}
0.7	0.1631889587	0.1629677570	2.2×10^{-4}
0.8	0.1243031560	0.1242157441	8.7×10^{-5}
0.9	0.1014691098	0.1018089431	3.3×10^{-4}
1	0.0939598983	0.0945851040	6.2×10^{-4}

Table 4.36 Comparison between OHAM results given by Eq. 4.150 and numerical results for $\alpha_1 = 0.5$, $\alpha_2 = 1$, $m = 1.5$ and $v'(1) = 0$ and the optimal auxiliary function 4.148

x	$v_{numeric}$	$\bar{v}_{OHAM}$	$relative\ error = \lvert v_{numeric} - \bar{v}_{OHAM} \rvert$
0	1	0.9999999999	3.5×10^{-15}
0.1	0.7497063820	0.7498561224	1.4×10^{-4}
0.2	0.5327963111	0.5319377563	8.5×10^{-4}
0.3	0.3498945869	0.3497527318	1.4×10^{-4}
0.4	0.2010814564	0.2022779050	1.1×10^{-3}
0.5	0.0852885962	0.0867149009	1.4×10^{-3}
0.6	−0.0006296037	−0.0003046742	3.2×10^{-4}
0.7	−0.0615295648	−0.0622411985	7.1×10^{-5}
0.8	−0.1020045446	−0.1024854604	4.8×10^{-4}
0.9	−0.1251782938	−0.1243191394	8.5×10^{-4}
1	−0.1327530483	−0.1309114642	1.8×10^{-3}

Table 4.37 Comparison between OHAM results given by Eq. 4.159 and numerical results for $\alpha_1 = 0.5$, $\alpha_2 = 0.5$, $m = 1$ and $v'(1) = \frac{1}{\sqrt{0.5}}$ and the optimal auxiliary function 4.157

x	$v_{numeric}$	$\overline{v}_{OHAM}$	relative error $= \lvert v_{numeric} - \overline{v}_{OHAM} \rvert$
0	1	1.0000000000	3.5×10^{-15}
0.1	1.2229384382	1.2230063532	6.7×10^{-5}
0.2	1.4378247530	1.4378525565	2.7×10^{-5}
0.3	1.6446154175	1.6445994145	1.6×10^{-5}
0.4	1.8432480259	1.8432271985	2.0×10^{-5}
0.5	2.0336362803	2.0336453830	9.1×10^{-6}
0.6	2.2156629647	2.2157024590	3.9×10^{-5}
0.7	2.3891700476	2.3891957459	2.5×10^{-5}
0.8	2.5539441869	2.5538810605	6.3×10^{-5}
0.9	2.7096939733	2.7094820881	2.1×10^{-4}
1	2.8560115627	2.8556993037	3.1×10^{-4}

Table 4.38 Comparison between OHAM results given by Eq. 4.159 and numerical results for $\alpha_1 = -0.5$, $\alpha_2 = 0.75$, $m = 1$ and $v'(1) = \frac{1}{\sqrt{0.5}}$ and the optimal auxiliary function 4.157

x	$v_{numeric}$	$\overline{v}_{OHAM}$	relative error $= \lvert v_{numeric} - \overline{v}_{OHAM} \rvert$
0	1	1	0
0.1	1.2518656079	1.2518515888	1.4×10^{-5}
0.2	1.4920877143	1.4920910613	3.3×10^{-6}
0.3	1.7207480054	1.7207531923	5.1×10^{-6}
0.4	1.9379133831	1.9379068594	6.5×10^{-6}
0.5	2.1436299399	2.1436163389	1.3×10^{-5}
0.6	2.3379146593	2.3379104050	4.2×10^{-6}
0.7	2.5207436438	2.5207591446	1.5×10^{-5}
0.8	2.6920346069	2.6920579596	2.3×10^{-5}
0.9	2.8516191935	2.8516179705	1.2×10^{-6}
1	2.9991957021	2.9991619152	3.3×10^{-5}

Table 4.39 Comparison between OHAM results given by Eq. 4.159 and numerical results for $\alpha_1 = -0.5$, $\alpha_2 = 1$, $m = 1$ and $v'(1) = \frac{1}{\sqrt{0.5}}$ and the optimal auxiliary function 4.157

x	$v_{numeric}$	$\overline{v}_{OHAM}$	$relative\ error = \lvert v_{numeric} - \overline{v}_{OHAM} \rvert$
0	1	1.0000000000	7.1×10^{-15}
0.1	1.2839954362	1.2839898927	5.5×10^{-6}
0.2	1.5522090538	1.5522083786	6.7×10^{-7}
0.3	1.8048758830	1.8048779484	2.0×10^{-6}
0.4	2.0422291469	2.0422291573	1.0×10^{-8}
0.5	2.2644931580	2.2644898117	3.3×10^{-6}
0.6	2.4718738134	2.4718705633	3.2×10^{-6}
0.7	2.6645442766	2.6645460387	1.7×10^{-6}
0.8	2.8426240012	2.8426304402	6.4×10^{-6}
0.9	3.0061452466	3.0061463212	1.0×10^{-6}
1	3.1549957250	3.1549849565	1.0×10^{-5}

Table 4.40 Comparison between OHAM results given by Eq. 4.159 and numerical results for $\alpha_1 = -0.5$, $\alpha_2 = 0.5$, $m = 1.5$ and $v'(1) = \frac{1}{\sqrt{0.5}}$ and the optimal auxiliary function 4.157

x	$v_{numeric}$	$\overline{v}_{OHAM}$	$relative\ error = \lvert v_{numeric} - \overline{v}_{OHAM} \rvert$
0	1	1	0
0.1	1.2609808120	1.2603664398	6.1×10^{-4}
0.2	1.5099516324	1.5093655735	5.8×10^{-4}
0.3	1.7469198626	1.7467254400	1.9×10^{-4}
0.4	1.9718652899	1.9721741750	3.0×10^{-4}
0.5	2.1847302867	2.1854400932	7.0×10^{-4}
0.6	2.3854057847	2.3862516582	8.4×10^{-4}
0.7	2.5737099736	2.5743374079	6.2×10^{-4}
0.8	2.7493549252	2.7494260892	7.1×10^{-5}
0.9	2.9118887501	2.9112465530	6.4×10^{-4}
1	3.0605838497	3.0595277920	1.0×10^{-3}

Table 4.41 Comparison between OHAM results given by Eq. 4.159 and numerical results for $\alpha_1 = -0.5$, $\alpha_2 = 0.75$, $m = 1.5$, $v'(1) = 1/\sqrt{0.5}$ and the optimal auxiliary function 4.157

x	$v_{numeric}$	$\overline{v}_{OHAM}$	$relative\ error = \lvert v_{numeric} - \overline{v}_{OHAM} \rvert$
0	1	0.9999999999	2.8×10^{-14}
0.1	1.3085289178	1.3084934573	3.5×10^{-5}
0.2	1.5989107009	1.5989174796	6.7×10^{-6}
0.3	1.8703842220	1.8713975406	1.3×10^{-5}
0.4	2.1261798486	2.1261654650	1.4×10^{-5}
0.5	2.3635084974	2.3634750640	3.3×10^{-5}
0.6	2.5835451658	2.5835323438	1.2×10^{-5}
0.7	2.7864041509	2.7864403839	3.6×10^{-5}
0.8	2.9720990828	2.9721581992	5.9×10^{-5}
0.9	3.1404738992	3.1404724101	1.4×10^{-6}
1	3.2910682398	3.2909802656	8.7×10^{-5}

Table 4.42 Comparison between OHAM results given by Eq. 4.159 and numerical results for $\alpha_1 = -0.5$, $\alpha_2 = 1$, $m = 1.5$, $v'(1) = 1/\sqrt{0.5}$ and the optimal auxiliary function 4.157

x	$v_{numeric}$	$\overline{v}_{OHAM}$	$relative\ error = \lvert v_{numeric} - \overline{v}_{OHAM} \rvert$
0	1	0.9999999999	2.2×10^{-13}
0.1	1.3620158757	1.3619756005	4.0×10^{-5}
0.2	1.6988832705	1.6988902831	7.0×10^{-6}
0.3	2.0110642417	2.0110793966	1.5×10^{-5}
0.4	2.2990536573	2.2990383227	1.5×10^{-5}
0.5	2.5633704925	2.5633333487	3.7×10^{-5}
0.6	2.8045422031	2.8045269744	1.5×10^{-5}
0.7	3.0230772590	3.0231164379	3.9×10^{-5}
0.8	3.2194185417	3.2194842324	6.5×10^{-5}
0.9	3.3938604328	3.3938593947	1.0×10^{-6}
1	3.5463864691	3.5462883840	9.8×10^{-5}

In Figs. 4.44, 4.45, 4.46, 4.47, 4.48, 4.49, 4.50, 4.51, 4.52, 4.53, 4.54, 4.55, 4.56, 4.57, 4.58, 4.59, 4.60, 4.61, 4.62, 4.63, 4.64, 4.65, 4.66 are plotted the velocity $\overline{v}$ against the horizontal distance x. Figures 4.44 and 4.45 show the variation of velocity field for Newtonian fluid, and Figs. 4.46, 4.47, 4.48, 4.49 show the variation of velocity field for non-Newtonian case, for different values of the parameter α_2 and $\alpha_1 = 0.5$. It is observed that the speed of belt decreases with the increase of m. Figures 4.50, 4.51, 4.52, 4.53, 4.54, 4.55 are showing the variation of the velocity field for the value $\alpha_1 = -0.5$ and different values of the parameter α_2. In these cases the speed of belt increases with the increase of m for different values of the auxiliary function. In Figs. 4.56, 4.57, 4.58, 4.59 we present the variation of velocity field for $\alpha_1 = 0.5$ and different values of the parameters m and α_2.

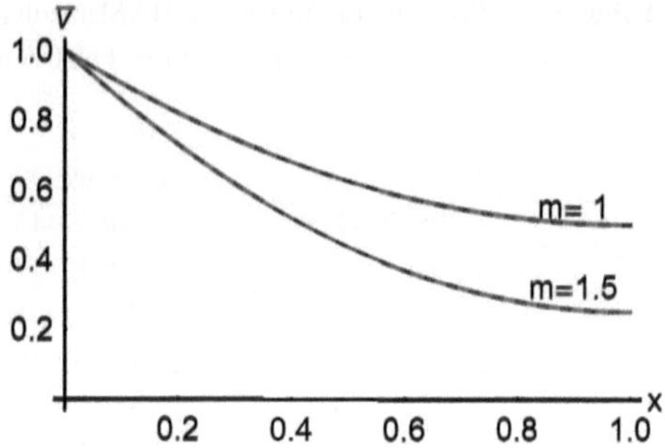

Fig. 4.44 Variation of velocity profile with increasing parameter m for $\alpha_1 = 0.5$, $\alpha_2 = 0.5$, $v'(1) = 0$ with optimal auxiliary function 4.127: (*red solid line*) numerical solution, (*blue dashed line*) OHAM solution (colour figure online)

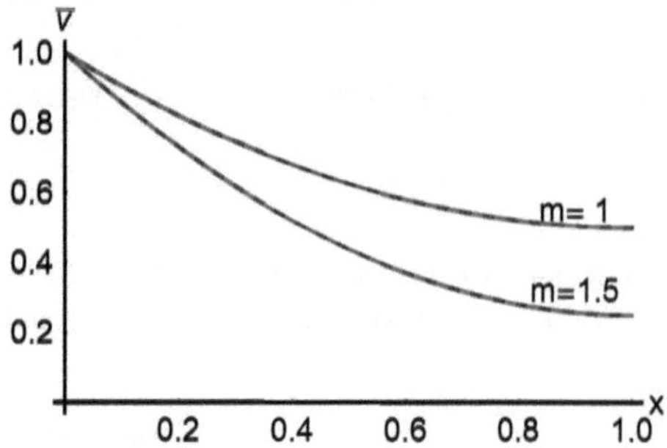

Fig. 4.45 Variation of velocity profile with increasing parameter m for $\alpha_1 = 0.5$, $\alpha_2 = 0.5$, $v'(1) = 0$ with optimal auxiliary function 4.148: (*red solid line*) numerical solution, (*blue dashed line*) OHAM solution (colour figure online)

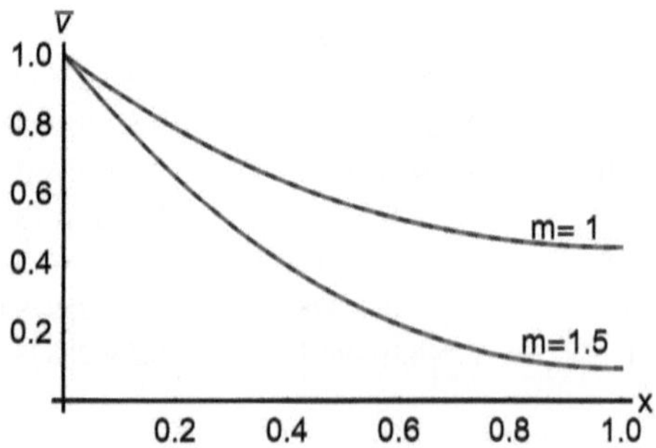

Fig. 4.46 Variation of velocity profile with increasing parameter m for $\alpha_1 = 0.5$, $\alpha_2 = 0.75$, $v'(1) = 0$ with optimal auxiliary function 4.127: (*red solid line*) numerical solution, (*blue dashed line*) OHAM solution (colour figure online)

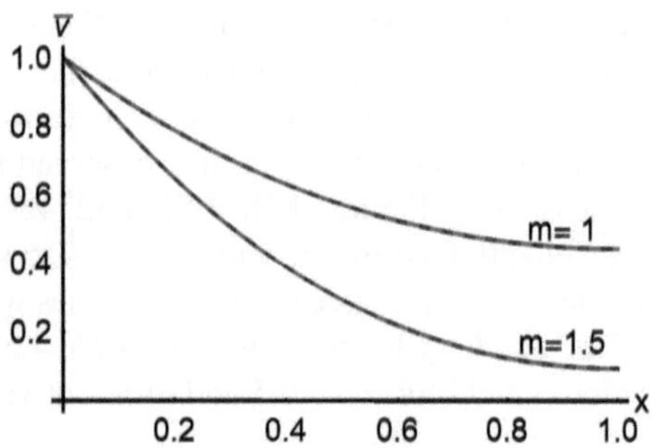

Fig. 4.47 Variation of velocity profile with increasing parameter m for $\alpha_1 = 0.5$, $\alpha_2 = 0.75$, $v'(1) = 0$ with optimal auxiliary function 4.148: ____numerical solution, (*blue dashed line*) OHAM solution (colour figure online)

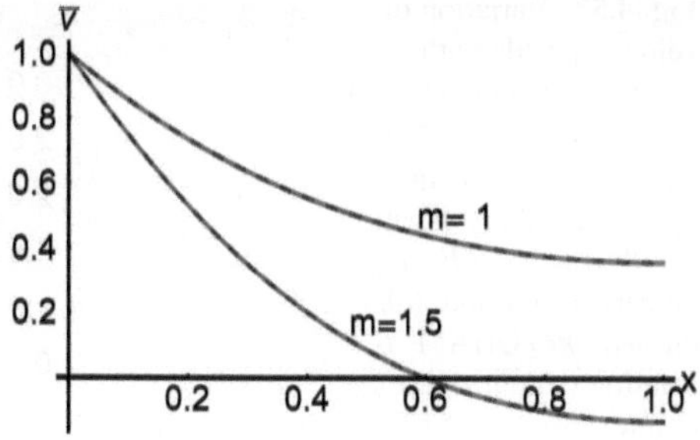

Fig. 4.48 Variation of velocity profile with increasing parameter m for $\alpha_1 = 0.5$, $\alpha_2 = 1$, $v'(1) = 0$ with optimal auxiliary function 4.127: (*red solid line*) numerical solution, (*blue dashed line*) OHAM solution (colour figure online)

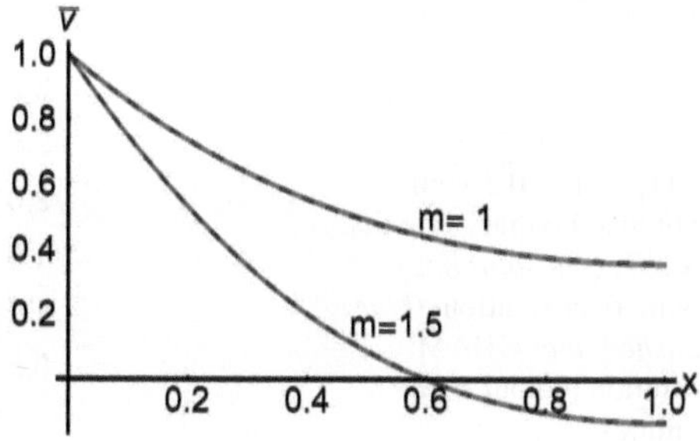

Fig. 4.49 Variation of velocity profile with increasing parameter m for $\alpha_1 = 0.5$, $\alpha_2 = 1$, $v'(1) = 0$ with optimal auxiliary function 4.148: (*red solid line*) numerical solution, (*blue dashed line*) OHAM solution (colour figure online)

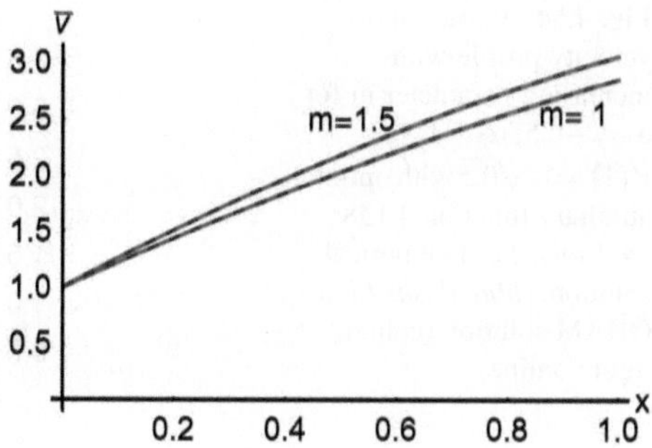

Fig. 4.50 Variation of velocity profile with increasing parameter m for $\alpha_1 = -0.5$, $\alpha_2 = 0.5$, $v'(1) = 1/\sqrt{0.5}$ with optimal auxiliary function 4.138: (*red solid line*) numerical solution, (*blue dashed line*) OHAM solution (colour figure online)

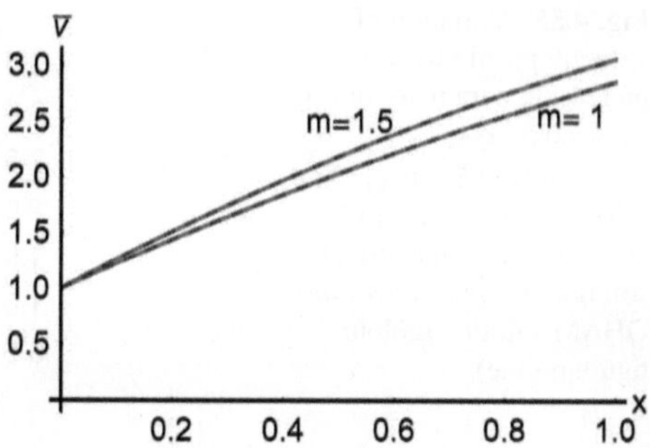

Fig. 4.51 Variation of velocity profile with increasing parameter m for $\alpha_1 = -0.5$, $\alpha_2 = 0.5$, $v'(1) = 1/\sqrt{0.5}$ with optimal auxiliary function 4.157: (*red solid line*) numerical solution, (*blue dashed line*) OHAM solution (colour figure online)

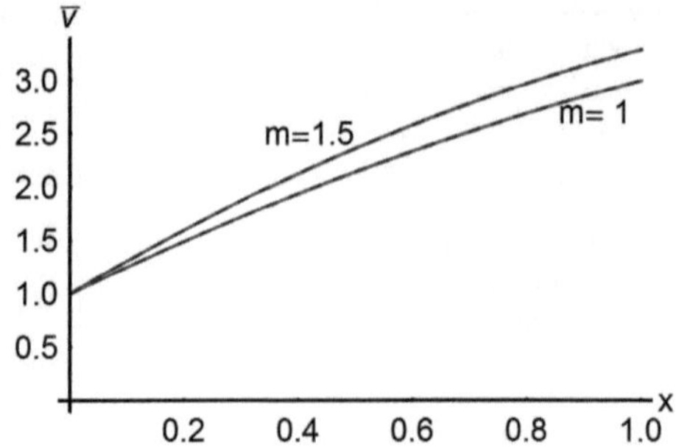

Fig. 4.52 Variation of velocity profile with increasing parameter m for $\alpha_1 = -0.5$, $\alpha_2 = 0.75$, $v'(1) = 1/\sqrt{0.5}$ with optimal auxiliary function 4.138: (*red solid line*) numerical solution, (*blue dashed line*) OHAM solution (colour figure online)

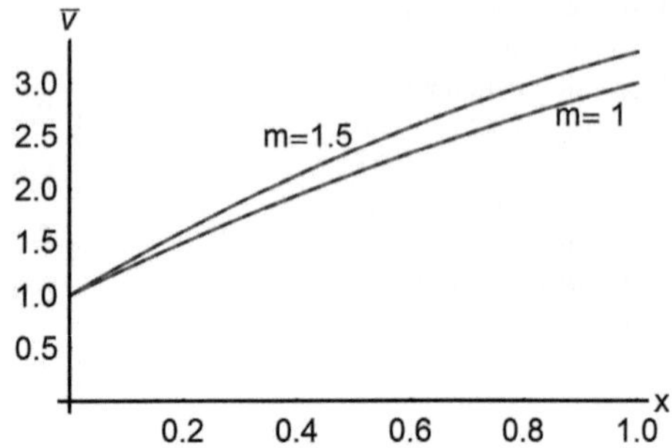

Fig. 4.53 Variation of velocity profile with increasing parameter m for $\alpha_1 = -0.5$, $\alpha_2 = 0.75$, $v'(1) = 1/\sqrt{0.5}$ with optimal auxiliary function 4.157: (*red solid line*) numerical solution, (*blue dashed line*) OHAM solution (colour figure online)

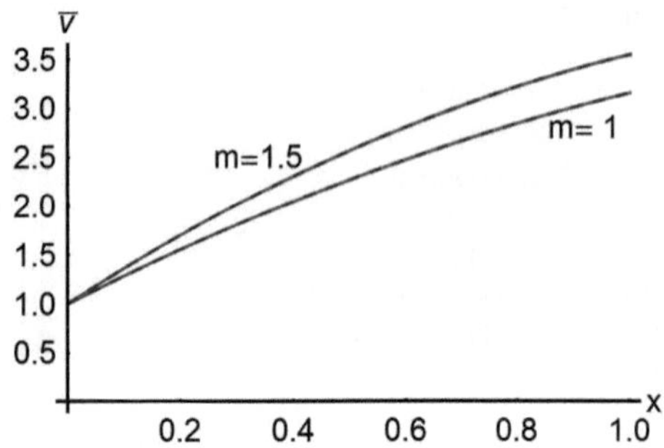

Fig. 4.54 Variation of velocity profile with increasing parameter m for $\alpha_1 = -0.5$, $\alpha_2 = 1$, $v'(1) = 1/\sqrt{0.5}$ with optimal auxiliary function 4.138: (*red solid line*) numerical solution, (*blue dashed line*) OHAM solution (colour figure online)

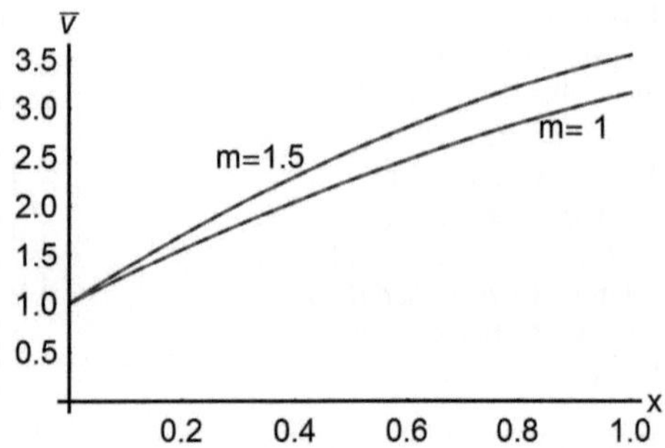

Fig. 4.55 Variation of velocity profile with increasing parameter m for $\alpha_1 = -0.5$, $\alpha_2 = 1$, $v'(1) = 1/\sqrt{0.5}$ with optimal auxiliary function 4.157: (*red solid line*) numerical solution, (*blue dashed line*) OHAM solution (colour figure online)

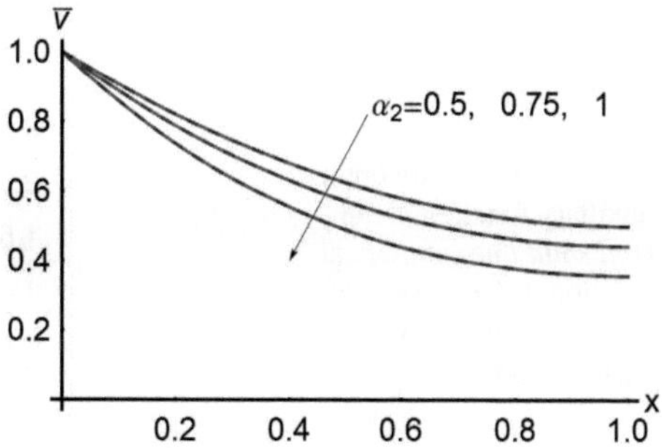

Fig. 4.56 Variation of velocity profile with increasing parameter α_2 for $\alpha_1 = 0.5$, $m = 1$, $v'(1) = 0$ with optimal auxiliary function 4.127: (*red solid line*) numerical solution, (*blue dashed line*) OHAM solution (colour figure online)

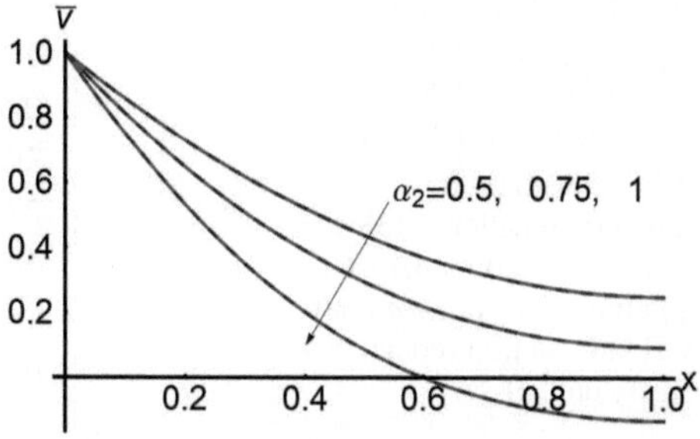

Fig. 4.57 Variation of velocity profile with increasing parameter α_2 for $\alpha_1 = 0.5$, $m = 1.5$, $v'(1) = 0$ with optimal auxiliary function 4.127: (*red solid line*) numerical solution, (*blue dashed line*) OHAM solution (colour figure online)

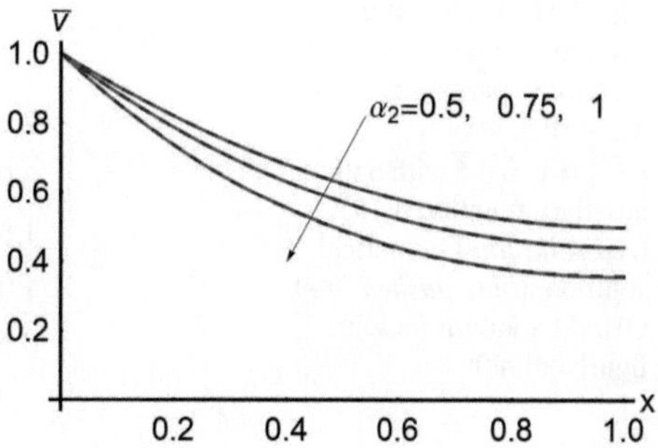

Fig. 4.58 Variation of velocity profile with increasing parameter α_2 for $\alpha_1 = 0.5$, $m = 1$, $v'(1) = 0$ with optimal auxiliary function 4.148: (*red solid line*) numerical solution, (*blue dashed line*) OHAM solution (colour figure online)

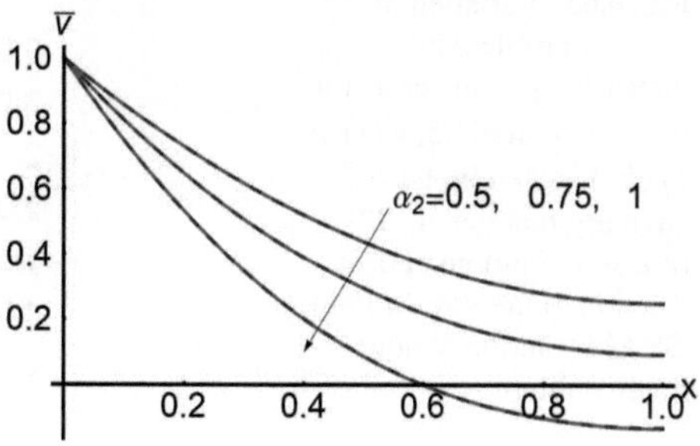

Fig. 4.59 Variation of velocity profile with increasing parameter α_2 for $\alpha_1 = 0.5$, $m = 1.5$, $v'(1) = 0$ with optimal auxiliary function 4.148: (*red solid line*) numerical solution, (*blue dashed line*) OHAM solution (colour figure online)

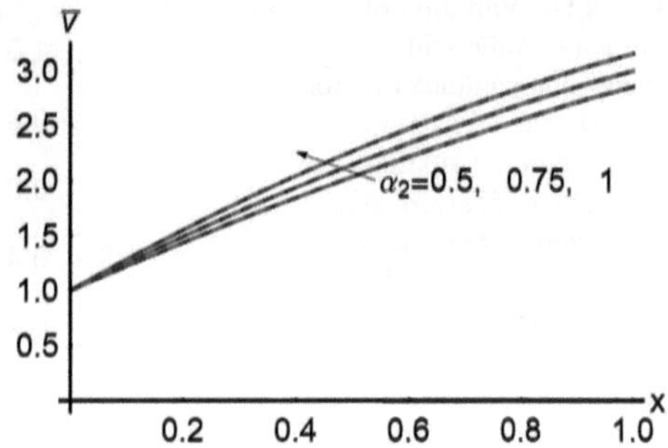

Fig. 4.60 Variation of velocity profile with increasing parameter α_2 for $\alpha_1 = -0.5$, m = 1, $v'(1) = 1/\sqrt{0.5}$ with optimal auxiliary function 4.138: (*red solid line*) numerical solution, (*blue dashed line*) OHAM solution (colour figure online)

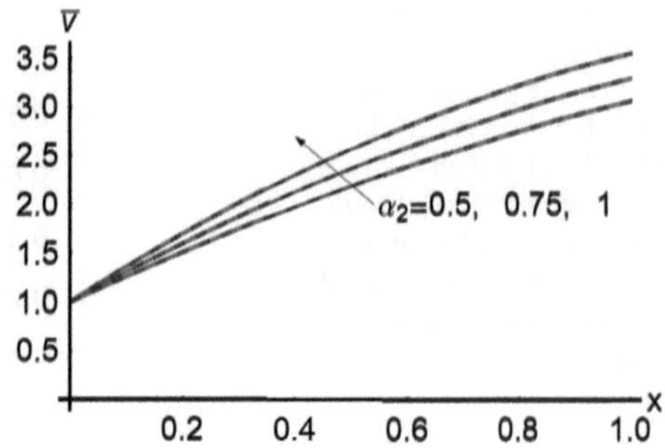

Fig. 4.61 Variation of velocity profile with increasing parameter α_2 for $\alpha_1 = -0.5$, m = 1.5, $v'(1) = 1/\sqrt{0.5}$ with optimal auxiliary function 4.138: (*red solid line*) numerical solution, (*blue dashed line*) OHAM solution (colour figure online)

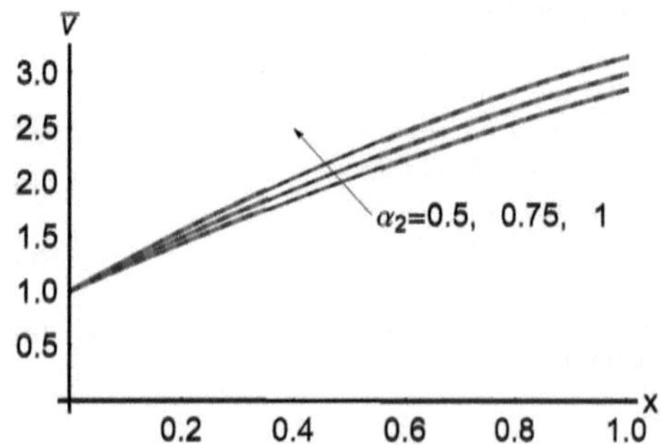

Fig. 4.62 Variation of velocity profile with increasing parameter α_2 for $\alpha_1 = -0.5$, m = 1, $v'(1) = 1/\sqrt{0.5}$ with optimal auxiliary function 4.157: (*red solid line*) numerical solution, (*blue dashed line*) OHAM solution (colour figure online)

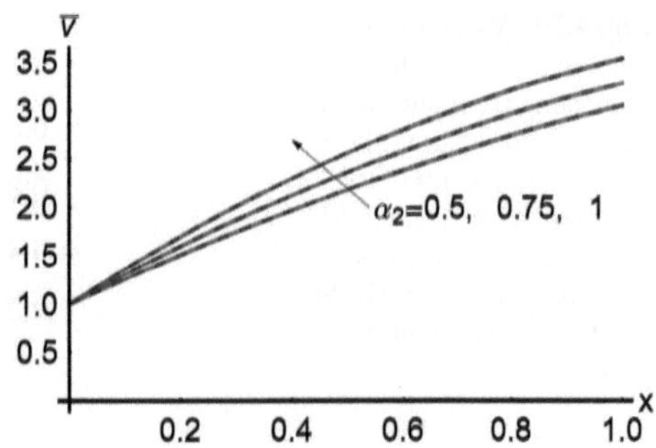

Fig. 4.63 Variation of velocity profile with increasing parameter α_2 for $\alpha_1 = -0.5$, m = 1.5, $v'(1) = 1/\sqrt{0.5}$ with optimal auxiliary function 4.157: (*red solid line*) numerical solution, (*blue dashed line*) OHAM solution (colour figure online)

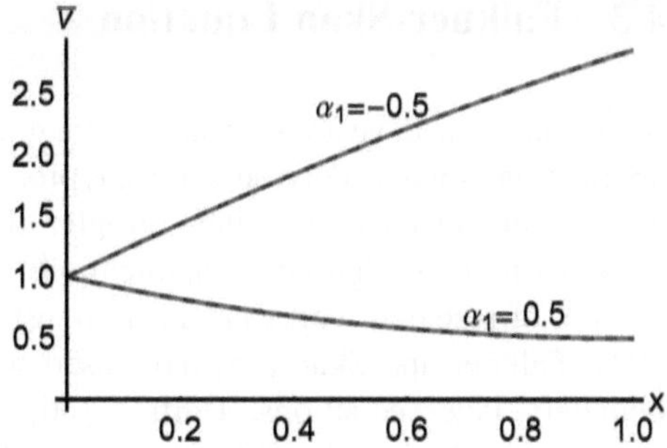

Fig. 4.64 Variation of velocity profile with increasing parameter α_1 for $\alpha_2 = 0.5$, $m = 1$, case 4.2.2.1: (*red solid line*) numerical solution, (*blue dashed line*) OHAM solution (colour figure online)

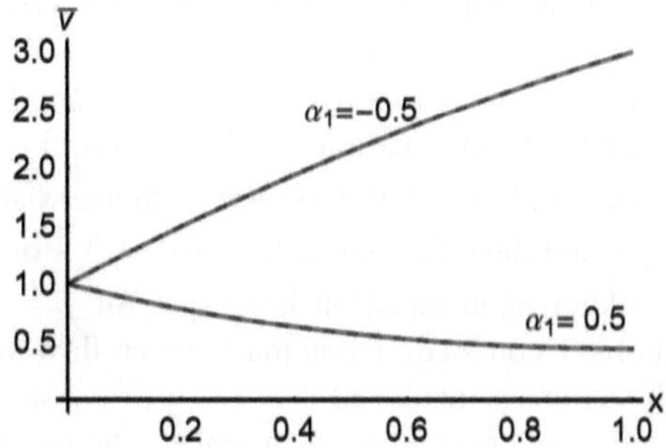

Fig. 4.65 Variation of velocity profile with increasing parameter α_1 for $\alpha_2 = 0.75$, $m = 1$, case 4.2.2.1: (*red solid line*) numerical solution, (*blue dashed line*) OHAM solution (colour figure online)

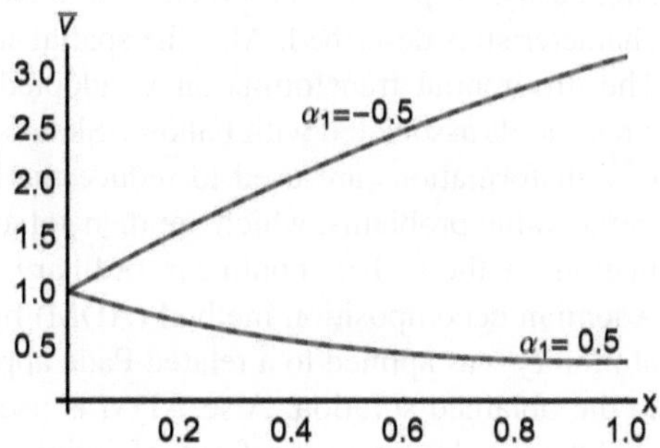

Fig. 4.66 Variation of velocity profile with increasing parameter α_1 for $\alpha_2 = 1$, $m = 1$, case 4.8.2.1: (*red solid line*) numerical solution, (*blue dashed line*) OHAM solution (colour figure online)

Here we observe that for any values of the parameter m, the speed of belt increases with the increase of α_2. On the other hand, we remark from Figs. 4.56, 4.57, 4.58, 4.59 that the speed of belt decrease for any values of $\alpha_1 > 0$. But, from Figs. 4.60, 4.61, 4.62, 4.63 we deduce that the speed of belt increase for any values of $\alpha_1 < 0$. We observe in these last cases like Figs. 4.56, 4.57, 4.58, 4.59 that the speed of belt increase for with the increase of α_2. In Figs. 4.64, 4.65, 4.66 are presented graphically the variation of velocity profile with increasing parameter α_1. It is observed that the speed of belt decreases with the increase of α_1. In all cases, we deduce that the speed of belt decreases for $\alpha_1 = 0.5$ and speed of belt increases for $\alpha_1 = -0.5$ and for any values of m.

4.3 Falkner-Skan Equation

Some materials have a viscoelastic behavior with many industrial and technological applications such as polymers, plastic processing, cosmetics, paint flow, adhesives, accelerators, electrostatic filters, droplet filters, etc.

Motivated by significant applications of viscoelastic materials, a substantial amount of research works has been invested in the study of nonlinear systems. In 1931, Falkner and Skan [35] have used some approximate procedures to solve a boundary layer equations. Hartree [36] found the numerical solution using a shooting method with $F''(0)$ (see Eq. 4.175) as free parameter. The boundary conditions which arise in the research of viscous flow past a wedge of angle $\beta\pi$, $\beta > 0$ corresponds to flow toward the wedge and $\beta < 0$ corresponds to flow away from the wedge. The special case $\beta = 0$ is called the Blasius equation where the wedge reduced to a flat plate. In [37] and [38] is proved that if $0 \leq \beta \leq 1$ then the Falkner-Skan equation 4.175 with initial conditions 4.176 admits a unique smooth solution. For $-0.1988 < \beta < 0$ there exists two solutions, that is one with $F''(0) > 0$ and the other one with $F''(0) < 0$. Botta et al. [39] showed that the solution of Falkner-Skan equation is unique for $\beta > 1$ under the restriction $0 < F'(0) < 1$. Forced convection boundary layer flow over a wedge with uniform suction or injection is analyzed by Yih [40]. Asaithambi [41] studied the Falkner-Skan equation using finite difference scheme. In [42], Zaturska and Banks presented new solution branch in function of parameter β. This solution branch is found to end singularity at $\beta = 1$; its structure is analytically investigated and the principal characteristics described. Also the spatial stability of such solutions is commented. The differential transformation is adopted to investigate the velocity and shear stress fields associated with Falkner-Skan boundary-layer problem in [43]. A group of transformations are used to reduce the boundary value problem into a pair of initial value problems, which are then solved by means of the differential transformation method. The nonlinear ordinary differential equation is solved using Adomian decomposition method (ADM) by Elgazery [44] such that the condition at infinity was applied to a related Padé approximation and Laplace transformation to the obtained solution. Also ADM is used in [45] by Alizadeh et al., to find an analytical solution in the form of infinite power series. Magnetohydrodynamics effects on the Falkner-Skan wedge flow are studied by Abbasbandy and Hayat in [46]. From a fluid mechanical point of view, the pathophysiological situation in myocardical bridges involves fluid flow in a time dependent flow geometry caused by contracting cardiac muscles overlying an intramural segment of the coronar artery. A boundary-layer model for the calculation of the pressure drop and flow separation is presented in [47] under the assumption that the idealized flow through a constriction is given by near equilibrium velocity profiles of the Falkner-Skan cooke family, the evolution of the boundary-layer is obtained by the simultaneous solution of the Falkner-Skan equation and the transient non-Karman integral momentum equation.

Pirkhedri et al. [48] developed a numerical technique transforming the governing partial differential equation into a nonlinear third-order boundary value problem by similarity variables and then solve it by the rational Legendre collocation method. It used transformed Hermite-Gauss nodes as interpolation points. The steady Falkner-Skan solution for gravity-driven film flow of micropolar fluid is investigated in [49]. The ordinary differential equations are solved numerically using an implicit finite-difference scheme known as the Keller-box method. In [50] Lakestani, truncated the semi-infinite physical domain of the problem to a finite domain and expanding the required approximate solution as the elements of Chebyshev cardinal functions. Yun proposed in [51] an iterative method for solving the Falkner-Skan equation in the form of polynomial series without requiring any differentiations or integrations of the previous iterate solutions. He suggests a correction method which is compared with the successive differences of the iterations. In [52] Hendi and Hussian considered Falkner-Skan flow over a porous surface. They have taken into account the case of uniform suction/blowing. Stream function formulation and suitable transformations reduce the arising problem to ordinary differential equation which has been solved by homotopy analysis method.

4.3.1 The Governing Equation

The two-dimensional laminar boundary-layer equations of an incompressible fluid subject to a pressure gradient are [35, 36, 42, 45, 53]:

$$u\frac{\partial u}{\partial x}+v\frac{\partial u}{\partial y}=-\frac{1}{\rho}p'+\nu\frac{\partial^2 u}{\partial y^2} \tag{4.168}$$

$$\frac{\partial u}{\partial x}+\frac{\partial v}{\partial y}=0 \tag{4.169}$$

where p' is the pressure gradient, $p'=-\rho U\frac{\partial U}{\partial x}$, u is the stream wise velocity in the direction of the fluid flow, v is the velocity in the direction normal to u, ν is the constant kinematic viscosity, $U(x)$ is the velocity at the edge of the boundary-layer which obeys the power-law relation $U(x)=ax^m$, $(x>0)$, where a is the mean stream velocity and m is a constant. The relevant boundary conditions for fixed plate, are

$$y=0: u=0, v=0; u\to a \;\; as \;\; y\to\infty. \tag{4.170}$$

A stream function $\psi(x,y)$ is introduced such that

$$u=\frac{\partial\psi}{\partial y}, v=-\frac{\partial\psi}{\partial x}. \tag{4.171}$$

The Eq. 4.169 of continuity is satisfied identically. The momentum equation 4.168 becomes:

$$\frac{\partial\psi}{\partial y}\frac{\partial^2\psi}{\partial x\partial y}-\frac{\partial\psi}{\partial x}\frac{\partial^2\psi}{\partial y^2}=U\frac{\partial U}{\partial x}+v\frac{\partial^3\psi}{\partial y^3}. \tag{4.172}$$

Integrating Eq. 4.172 and using similarity variable yields:

$$\psi=x^{(1+m)/2}\sqrt{\frac{2va}{1+m}}F(\eta) \tag{4.173}$$

$$\eta=\frac{y}{x^{(1-m)/2}}\sqrt{\frac{(1+m)a}{2v}}. \tag{4.174}$$

Substituting Eqs. 4.173 and 4.174 into Eq. 4.172 gives the equation of Falkner-Skan in the form:

$$F'''(\eta)+F(\eta)F''(\eta)+\beta\left(1-F'(\eta)^2\right)=0 \tag{4.175}$$

with the initial and boundary conditions

$$F(0)=0, F'(0)=0, F'(\infty)=0 \tag{4.176}$$

where $\beta=\frac{2m}{m+1}$ is a measure of the pressure gradient, and prime denotes derivative with respect to η.

4.3.2 Application of OHAM to Falkner-Skan Equation

To use the basic ideas of the proposed method, we choose the linear operator

$$L(\mathsf{F}(\eta))=\frac{\partial^3\mathsf{F}(\eta)}{\partial\eta^3}+K\frac{\partial^2\mathsf{F}(\eta)}{\partial\eta^2} \tag{4.177}$$

where K is the unknown parameter at this moment.

The nonlinear operator is

$$N(\mathsf{F}(\eta))=\mathsf{F}(\eta)\frac{\partial^2\mathsf{F}(\eta)}{\partial\eta^2}-K\frac{\partial^2\mathsf{F}(\eta)}{\partial\eta^2}+\beta\left[1-\left(\frac{\partial\mathsf{F}(\eta)}{\partial\eta}\right)^2\right] \tag{4.178}$$

The boundary conditions are

$$\mathsf{F}(0)=0, \frac{\partial\mathsf{F}(0)}{\partial\eta}=0, \frac{\partial\mathsf{F}(\infty)}{\partial\eta}=1. \tag{4.179}$$

Equation 4.4 can be written in the form

$$F_0'''(\eta) + KF_0''(\eta) = 0, F_0(0) = 0, F_0'(0) = 0, F_0'(\infty) = 1 \tag{4.180}$$

and has the solution

$$F_0(\eta) = \eta + \frac{e^{-K\eta} - 1}{K}. \tag{4.181}$$

From Eqs. 4.177 and 4.178 one obtain the expression

$$N(F_0(\eta)) = -kF''(\eta) + F(\eta)F''(\eta) + \beta\left[1 - F'(\eta)^2\right]. \tag{4.182}$$

Substituting Eq. 4.181 into Eq. 4.182, we obtain

$$N(F_0(\eta)) = (K\eta + 2\beta - 1 - K)e^{-K\eta} + (1 - \beta)e^{-2K\eta}. \tag{4.183}$$

If we consider the first-order approximate solution, Eq. 4.5 becomes

$$\overline{F}(\eta) = F_0(\eta) + F_1(\eta) \tag{4.184}$$

where $F_1(\eta)$ is obtained from Eq. 4.9:

$$F_1'''(\eta) + KF_1''(\eta) = H(\eta, C_i)N(F_0(\eta)). \tag{4.185}$$

Substituting Eqs. 4.181 and 4.183 into Eq. 4.185 we obtain the equation:

$$\begin{aligned} &F_1'''(\eta) + KF_1''(\eta) = H(\eta, C_i)\left[(K\eta + 2\beta - 1 - K)e^{-K\eta} + \right. \\ &\left. +(1 - \beta)e^{-2K\eta}\right], F_1(0) = 0, F_1'(0) = 0, F_1'(\infty) = 0. \end{aligned} \tag{4.186}$$

where as optimal auxiliary function H we have in view the following four cases:

Case 4.3.2.1
If the optimal auxiliary convergence-control function has the form

$$\begin{aligned} &H(\eta, C_i) = C_1 + C_2\eta + C_3\eta^2 + C_4\eta^3 + \\ &+(C_5 + C_6\eta + C_7\eta^2)e^{-K\eta} + C_8e^{-2K\eta} \end{aligned} \tag{4.187}$$

then the Eq. 4.186 can be written as

$$
\begin{aligned}
F_1'''(\eta) + KF_1''(\eta) = &\{(2\beta - 1 - K^2)C_1 + [(2\beta - 1 - K^2)C_2 + KC_1]\eta + \\
&+[(2\beta - 1 - K^2)C_3 + KC_2]\eta^2 + [(2\beta - 1 - K^2)C_4 + KC_3]\eta^3 + \\
&+KC_4\eta^4\}e^{-K\eta} + \{(1-\beta)C_1 + (2\beta - 1 - K^2)C_5 + [(1-\beta)C_2 + KC_5 + \\
&+(2\beta - 1 - K^2)C_6]\eta + [(1-\beta)C_3 + KC_6 + (2\beta - 1 - K^2)C_7]\eta^2 + \\
&+[(1-\beta)C_4 + KC_7]\eta^3\}e^{-2K\eta} + \{(1-\beta)C_5 + (2\beta - 1 - K^2)C_8 + \\
&+[(1-\beta)C_6 + KC_8]\eta + (1-\beta)C_7\eta^2\}e^{-3K\eta} + (1-\beta)C_8e^{-4K\eta}.
\end{aligned}
\tag{4.188}
$$

Finally, using Eq. 4.181 and solving Eq. 4.188, we determine the first-order approximate solution given by Eq. 4.184 in the form

$$
\begin{aligned}
\overline{F}(\eta) = &\eta - \frac{1}{K} + \frac{4K^2 - 7\beta - 5}{4K^3}C_1 + \frac{8K^2 - 15\beta - 17}{4K^4}C_2 + \\
&+\frac{48K^2 - 93\beta - 147}{8K^5}C_3 + \frac{96K^2 - 189\beta - 387}{4K^6}C_4 + \frac{9K^2 - 14\beta - 4}{36K^3}C_5 + \\
&+\frac{54K^2 - 92\beta - 43}{216K^4}C_6 + \frac{81K^2 - 146\beta - 97}{216K^5}C_7 + \frac{48K^2 - 69\beta - 11}{432K^3}C_8 + \\
&+\Big[\frac{C_4}{5K}\cdot\eta^5 + \left(\frac{C_3}{4K} + \frac{2\beta + 7 - K^2}{4K^2}C_4\right)\cdot\eta^4 + \Big(\frac{C_2}{3K} + \frac{2\beta + 5 - 2K^2}{3K^2}C_3 + \\
&+\frac{4\beta + 10 - 2K^2}{K^3}C_4\Big)\cdot\eta^3 + \Big(\frac{C_1}{2K} + \frac{2\beta + 3 - K^2}{2K^2}C_2 + \frac{4\beta + 7 - 2K^2}{K^3}C_3 + \\
&+\frac{18\beta + 39 - 9K^2}{K^4}C_4\Big)\cdot\eta^2 + \Big(\frac{2\beta + 1 - K^2}{K^2}C_1 + \frac{4\beta + 4 - 2K^2}{K^3}C_2 + \\
&+\frac{12\beta + 18 - 6K^2}{K^4}C_3 + \frac{48\beta + 96 - 24K^2}{K^5}C_4\Big)\cdot\eta + \frac{1}{K} + \frac{3\beta + 3 - 2K^2}{2K^3}C_1 + \\
&+\frac{13\beta + 19 - 8K^2}{4K^4}C_2 + \frac{41\beta + 79 - 24K^2}{4K^5}C_3 + \frac{339\beta + 813 - 192K^2}{8K^6}C_4 + \\
&+\frac{10\beta + 5 - 6K^2}{12K^3}C_5 + \frac{49\beta + 41 - 27K^2}{36K^4}C_6 + \frac{718\beta + 875 - 378K^2}{216K^5}C_7 + \\
&+\frac{9\beta + 2 - 6K^2}{36K^3}C_8\Big]\cdot e^{-K\eta} + \Big[\left(\frac{\beta - 1}{4K^3}C_4 - \frac{1}{4K^2}C_7\right)\cdot\eta^3 + \Big(\frac{\beta - 1}{4K^3}C_3 + \\
&+\frac{3\beta - 3}{2K^4}C_4 - \frac{1}{4K^2}C_6 + \frac{K^2 - 5 - 2\beta}{4K^3}C_7\Big)\cdot\eta^2 + \Big(\frac{\beta - 1}{4K^3}C_2 + \frac{\beta - 1}{K^4}C_3 + \\
&+\frac{33\beta - 33}{8K^5}C_4 - \frac{1}{4K^2}C_5 + \frac{K^2 - 2\beta - 3}{4K^3}C_6 + \frac{8K^2 - 16\beta - 25}{8K^4}C_7\Big)\cdot\eta +
\end{aligned}
$$

$$
\begin{aligned}
&+\frac{\beta-1}{4K^3}C_1+\frac{\beta-1}{2K^4}C_2+\frac{11\beta-11}{8K^5}C_3+\frac{39\beta-39}{8K^6}C_4+\frac{K^2-2\beta-1}{4K^3}C_5+\\
&+\frac{4K^2-8\beta-7}{8K^4}C_6+\frac{11K^2-22\beta-28}{8K^5}C_7]\cdot e^{-2K\eta}+[\frac{\beta-1}{18K^3}C_7\eta^2+\\
&+\left(\frac{\beta-1}{18K^3}C_6+\frac{7\beta-7}{54K^4}C_7-\frac{1}{18K^2}C_8\right)\cdot\eta+\frac{\beta-1}{18K^3}C_5+\frac{7\beta-7}{108K^4}C_6+\\
&+\frac{11\beta-11}{108K^5}C_7+\frac{6K^2-1-12\beta}{108K^3}C_8]\cdot e^{-3K\eta}+\frac{\beta-1}{48K^3}C_8e^{-4K\eta}
\end{aligned}
\tag{4.189}
$$

Case 4.3.2.2

The optimal auxiliary function $H(\eta, C_i)$ has the form

$$
\begin{aligned}
H(\eta,C_i)=C_1+C_2\eta+C_3\eta^2+C_4\eta^3+C_5\eta^4+\\
+(C_6+C_7\eta+C_8\eta^2)e^{-K\eta}.
\end{aligned}
\tag{4.190}
$$

In this case, Eq. 4.186 becomes

$$
\begin{aligned}
&F_1'''(\eta)+KF_1''(\eta)=\{(2\beta-1-K^2)C_1+[(2\beta-1-K^2)C_2+KC_1]\eta+\\
&+[(2\beta-1-K^2)C_3+KC_2]\eta^2+[(2\beta-1-K^2)C_4+KC_3]\eta^3+\\
&+[(2\beta-1-K^2)C_5+KC_4]\eta^4+KC_5\eta^5\}e^{-K\eta}+\{(1-\beta)C_1+\\
&+(2\beta-1-K^2)C_6+[(1-\beta)C_2+KC_6+(2\beta-1-K^2)C_7]\eta+\\
&+[(1-\beta)C_3+KC_7+(2\beta-1-K^2)C_8]\eta^2+[(1-\beta)C_4+KC_8]\eta^3+\\
&+(1-\beta)C_5\eta^4\}e^{-2K\eta}+[(1-\beta)C_6+(1-\beta)C_7\eta+(1-\beta)C_8\eta^2].
\end{aligned}
\tag{4.191}
$$

The first-order approximate solution 4.184 in this case is obtained from Eqs. 4.191 and 4.181 and can be written as

$$
\begin{aligned}
&\overline{F}(\eta)=\eta-\frac{1}{K}+\frac{4K^2-7\beta-5}{4K^3}C_1+\frac{8K^2-15\beta-17}{4K^4}C_2+\\
&+\frac{48K^2-93\beta-147}{8K^5}C_3+\frac{96K^2-189\beta-387}{4K^6}C_4+\\
&+\frac{960K^2-1935\beta-4785}{8K^7}C_5+\frac{9K^2-14\beta-4}{36K^3}C_6+\\
&+\frac{54K^2-92\beta-43}{216K^4}C_7+\frac{81K^2-146\beta-97}{216K^5}C_8+[\frac{C5}{6K}\eta^6+\\
&+\left(\frac{C4}{5K}+\frac{2\beta+9-K^2}{9K^2}C_5\right)\cdot\eta^5+(\frac{C_3}{4K}+\frac{2\beta+7-K^2}{4K^2}C_4+\\
&+\frac{4\beta+13-2K^2}{K^3}C_5)\cdot\eta^4+(\frac{C2}{3K}+\frac{2\beta+5-K^2}{3K^2}C_3+
\end{aligned}
$$

$$
\begin{aligned}
&+\frac{4\beta+10-2K^2}{K^3}C_4+\frac{24\beta+68-12K^2}{K^4}C_5)\cdot\eta^3+(\frac{C1}{2K}+\\
&+\frac{2\beta+3-K^2}{2K^2}C_2+\frac{4\beta+7-2K^2}{K^3}C_3+\frac{18\beta+39-9K^2}{K^4}C_4+\\
&+\frac{96\beta+252-48K^2}{K^5}C_5)\cdot\eta^2+(\frac{2\beta+1-K^2}{K^2}C_1+\frac{4\beta+4-2K^2}{K^3}C_2+\\
&+\frac{12\beta+18-6K^2}{K^4}C_3+\frac{48\beta+96-24K^2}{K^5}C_4+\\
&+\frac{240\beta+600-120K^2}{K^6}C_5)\cdot\eta+\frac{1}{K}+\frac{3\beta+3-2K^2}{2K^3}C_1+\\
&+\frac{13\beta+19-8K^2}{4K^4}C_2+\frac{41\beta+79-24K^2}{4K^5}C_3+\\
&+\frac{339\beta+813-192K^2}{8K^6}C_4+\frac{1053\beta+2307-480K^2}{4K^7}C_5+\\
&+\frac{10\beta+5-6K^2}{12K^3}C_6+\frac{49\beta+41-27K^2}{36K^4}C_7+\\
&+\frac{718\beta+875-378K^2}{216K^5}C_8]\cdot e^{-K\eta}+[\frac{\beta-1}{4K^3}C_5\eta^4+(\frac{\beta-1}{4K^3}C_4-\\
&-\frac{2\beta-2}{K^4}C_5-\frac{1}{4K^2}C_8)\cdot\eta^3+(\frac{\beta-1}{4K^3}C_3+\frac{3\beta-3}{2K^4}C_4+\frac{33\beta-33}{4K^5}C_5-\\
&-\frac{1}{4K^2}C_7+\frac{2\beta-7-K^2}{4K^3}C_8)\cdot\eta^2+(\frac{\beta-1}{4K^3}C_2+\frac{\beta-1}{K^4}C_3+\\
&+\frac{33\beta-33}{8K^5}C_4+\frac{39-39\beta}{2K^6}C_5-\frac{1}{4K^2}C_6+\frac{K^2-2\beta-3}{4K^3}C_7+\\
&+\frac{8K^2-16\beta-25}{8K^4}C_8)\cdot\eta+\frac{\beta-1}{4K^3}C_1+\frac{\beta-1}{2K^4}C_2+\frac{11\beta-11}{8K^5}C_3+\\
&+\frac{39\beta-39}{8K^6}C_4+\frac{171\beta-171}{8K^7}C_5+\frac{K^2-2\beta-1}{4K^3}C_6+\\
&+\frac{4K^2-8\beta-7}{8K^4}C_7+\frac{11K^2-22\beta-28}{8K^5}C_8]\cdot e^{-2K\eta}+[\frac{\beta-1}{18K^3}C_8\eta^2+\\
&+\left(\frac{\beta-1}{18K^3}C_7+\frac{7\beta-7}{54K^4}C_8\right)\cdot\eta+\frac{\beta-1}{18K^3}C_6+\\
&+\frac{7\beta-7}{108K^4}C_7+\frac{11\beta-11}{108K^5}C_8]\cdot e^{-3K\eta}
\end{aligned}
\tag{4.192}
$$

Case 4.3.2.3
If the optimal auxiliary function $H(\eta, C_i)$ has the form

$$
\begin{aligned}
H(\eta,C_i)&=C_1+C_2\eta+C_3\eta^2+C_4\eta^3+\\
&+(C_5+C_6\eta+C_7\eta^2+C_9\eta^3)e^{-K\eta}+C_8e^{-2K\eta}
\end{aligned}
\tag{4.193}
$$

then the first-order approximate solution Eq. 4.184 has the form

$$\overline{\overline{F}}(\eta) = \overline{F}(\eta) + C_9\left[-\frac{6804K^2+793+908\beta}{648K^6}+ \right.$$
$$+\frac{1215K^2+3872+2365\beta}{216K^6}e^{-K\eta}+\left(-\frac{1}{4K^2}\eta^4+\frac{K^2-7-2\beta}{4K^3}\eta^3+\right.$$
$$\left.+\frac{6K^2-27-12\beta}{4K^4}\eta^2+\frac{33K^2-123-66\beta}{8K^5}\eta+\frac{39K^2-132-78\beta}{8K^6}\right)e^{-2K\eta}+$$
$$\left.+\left(\frac{\beta-1}{18K^3}\eta^3+\frac{7\beta-7}{36K^4}\eta^2+\frac{11\beta-11}{36K^5}\eta+\frac{131\beta-131}{648K^6}\right)e^{-3K\eta}\right] \tag{4.194}$$

where $\overline{F}(\eta)$ is given by Eq. 4.189.

Case 4.3.2.4
In the last case, we consider

$$\begin{aligned} H(\eta, C_i) = C_1 + C_2\eta + C_3\eta^2 + C_4\eta^3 + C_5\eta^4 + \\ +(C_6 + C_7\eta + C_8\eta^2 + C_9\eta^3)e^{-K\eta} \end{aligned} \tag{4.195}$$

such that the first-order approximate solution Eq. 4.184 becomes

$$\overline{\overline{F}}(\eta) = \overline{F}(\eta) + C_9\left[-\frac{6804K^2+793+908\beta}{648K^6}+ \right.$$
$$\left(-\frac{1}{4K^2}\eta^4+\frac{1215K^2+3872+2365\beta}{216K^6}e^{-K\eta}+\frac{K^2-7-2\beta}{4K^3}\eta^3+\right.$$
$$\left.+\frac{6K^2-27-12\beta}{4K^4}\eta^2+\frac{33K^2-123-66\beta}{8K^5}\eta+\frac{39K^2-132-78\beta}{8K^6}\right)e^{-2K\eta}+$$
$$\left.+\left(\frac{\beta-1}{18K^3}\eta^3+\frac{7\beta-7}{36K^4}\eta^2+\frac{11\beta-11}{36K^5}\eta+\frac{131\beta-131}{648K^6}\right)e^{-3K\eta}\right] \tag{4.196}$$

where $\overline{F}(\eta)$ is given by Eq. (4.192).

4.3.3 *Numerical Examples*

In order to prove the accuracy of the obtained results, we will determine the convergence-control parameters which appear in Eqs. 4.189, 4.192, 4.194 and 4.196 by means of Galerkin method. Let $R(\eta, C_i)$ be the residual to within the approximate solution $\overline{F}(\eta)$ (or $\overline{\overline{F}}(\eta)$) given by Eqs. 4.189, 4.192, 4.194 and 4.196 satisfies Eq. 4.175:

$$R(\eta, C_i) = \overline{F}'''(\eta, C_i) + \overline{F}(\eta, C_i)\overline{F}''(\eta, C_i) + \beta\left(1 - \overline{F}'^2(\eta, C_i)\right). \tag{4.197}$$

Since $R(\eta, C_i)$ contains the parameters C_j, $j = 1, 2, \ldots$ we may determine them on requiring that

$$J_i(C_j) = \int_0^{\infty} R(\eta, C_j) f_i(\eta) d\eta = 0, i = 1, 2, \ldots, q, j = 1, 2, \ldots, q. \tag{4.198}$$

where f_i are linear independent functions, taken as weighting functions. Equations 4.189 and 4.192 contains nine unknown optimal convergene-control parameters: K and C_i, $i = 1, 2, \ldots, 8$ and therefore we consider the following nine weighting functions ($q = 9$)

$$\begin{aligned} &f_1 = e^{-K\eta}, f_2 = \eta e^{-K\eta}, f_3 = \eta^2 e^{-K\eta}, f_4 = \eta^3 e^{-K\eta}, f_5 = \eta^4 e^{-K\eta}, \\ &f_6 = \eta^5 e^{-K\eta}, f_7 = \eta e^{-2K\eta}, f_8 = \eta^2 e^{-2K\eta}, f_9 = \eta^3 e^{-2K\eta}. \end{aligned} \tag{4.199}$$

For Eqs. 4.194 and 4.196 which contain the unknown parameters: K and C_i, $i = 1, 2, \ldots, 9$ we consider weighting functions ($q = 11$)

$$\begin{aligned} &f_1 = e^{-K\eta}, f_2 = \eta e^{-K\eta}, f_3 = \eta^2 e^{-K\eta}, f_4 = \eta^3 e^{-K\eta}, f_5 = \eta^4 e^{-K\eta}, \\ &f_6 = \eta^4 e^{-2K\eta} + \alpha \eta e^{-4K\eta}, f_7 = \eta e^{-2K\eta}, f_8 = \eta^2 e^{-2K\eta}, f_9 = \eta^3 e^{-2K\eta}, \\ &f_{10} = \eta^5 e^{-K\eta} + \eta e^{-3K\eta}, f_{11} = \eta^3 e^{-4K\eta} + \eta^7 e^{-2K\eta}. \end{aligned} \tag{4.200}$$

where α is an unknown parameter.

Example 4.3.3.1 In the first case we consider that $\beta = 1/2$.

4.3.3.1.a) For Eq. 4.189 and from the system 4.198, following the procedure described above it is obtained the optimal convergence-control parameters:

$$\begin{aligned} &C_1 = -0.0974633576, C_2 = 0.3334342895, C_3 = -0.0846971328, \\ &C_4 = 0.0054445300, C_5 = 10.8640621331, C_6 = -8.9072774043, \\ &C_7 = 0.7917175772, C_8 = 0.4377125759, K = 0.9345058664 \end{aligned}$$

and consequently the first-order approximate solution 4.189 can be written in the form:

$$\begin{aligned} &\overline{F}(\eta) = -0.8095502989 + \eta + \big(-2.2237146927 + 0.2852293525\eta + \\ &+0.0785389555\eta^2 + 0.0216006979\eta^3 + 0.0115505513\eta^4 + \\ &+0.0011652211\eta^5\big)e^{-0.9345058664\eta} + \big(3.0954492754 + 2.0345244657\eta + \\ &+0.8905286290\eta^2 - 0.1261379297\eta^3\big)e^{-1.8690117328\eta} + \big(-0.0565973734 + \\ &+0.2080470361\eta - 0.0269476991\eta^2\big)e^{-2.8035175992\eta} - \\ &-0.0055869102 \cdot e^{-3.7380234657\eta} \end{aligned} \tag{4.201}$$

Table 4.43 Comparison between OHAM results given by Eq. 4.201 and numerical solutions for $\beta = 1/2$

η	$F_{numerical}(\eta)$	$\overline{F}_{OHAM}(\eta)$ $Eq. 4.201$	$relative\ error = \lvert F_{numerical}(\eta) - \overline{F}_{OHAM}(\eta)\rvert$
0	$-5.7757240036 \times 10^{-25}$	$5.5511151231 \times 10^{-16}$	$5.5511151289 \times 10^{-16}$
4/5	0.2543480764	0.2543149422	0.0000331341
8/5	0.8550267840	0.8550621314	0.0000353473
12/5	1.6045273996	1.6043588322	0.0001685673
16/5	2.3963133788	2.3962404320	0.0000729467
4	3.1955002598	3.1953781616	0.0001220981
24/5	3.9954529746	3.9949610905	0.0004918840
28/5	4.7954513976	4.7946972370	0.0007541605
32/5	5.5954513676	5.5946526226	0.0007987450
36/5	6.3954513670	6.3946242000	0.0008271670
8	7.1954513667	7.19442747097	0.0010238957

Table 4.44 Comparison between OHAM results for velocity $\overline{F}'(\eta)$ obtained from Eq. 4.201 and numerical results for $\beta = 1/2$

η	$F'_{numerical}(\eta)$	$\overline{F}'_{OHAM}(\eta)$ $from\ Eq. 4.201$	$relative\ error = \lvert F'_{numerical}(\eta) - \overline{F}'_{OHAM}(\eta)\rvert$
0	$1.8528845721 \times 10^{-21}$	$4.4408920985 \times 10^{-16}$	$4.4408735696 \times 10^{-16}$
4/5	0.5833048177	0.5833149514	0.0000101337
8/5	0.8760975697	0.8759651454	0.0001324243
12/5	0.9760687561	0.9759519582	0.0001167979
16/5	0.9971920750	0.9973768978	0.0001848228
4	0.9998081986	0.9994693079	0.0003388907
24/5	0.9999925769	0.9995222975	0.0004702794
28/5	0.9999998398	0.9998331589	0.0001666808
32/5	0.9999999978	1.0000050268	5.028974×10^{-6}
36/5	0.9999999995	0.9998832648	0.0001167346
8	0.9999999995	0.9996174949	0.0003825046

In the Tables 4.43 and 4.44 we present a comparison between the first-order approximate solution given by Eq. 4.201, and velocity obtained from Eq. 4.201 respectively with numerical results for some values of variable η and the corresponding relative errors.

4.3.3.1.b) From Eq. 4.192 obtained by means of the optimal auxiliary convergence-control function H given by Eq. 4.190 we obtain the following results for the parameters:

$$C_1 = 15.3365053132, C_2 = -20.7743165892, C_3 = 9.3076203696,$$
$$C_4 = -2.0941285422, C_5 = 0.2277189607, C_6 = -5.2584096507,$$
$$C_7 = -45.5148999446, C_8 = 8.3240015527, K = 2.1171968259$$

Table 4.45 Comparison between OHAM results given by Eq.4.202 and numerical solutions for $\beta = 1/2$

η	$F_{numerical}(\eta)$	$\overline{F}_{OHAM}(\eta)$ *Eq.* 4.202	*relative error* = $\lvert F_{numerical}(\eta) - \overline{F}_{OHAM}(\eta)\rvert$
0	$-5.775724 \times 10^{-25}$	$-1.33226762 \times 10^{-15}$	$1.33226762 \times 10^{-15}$
4/5	0.2543480764	0.2543333746	0.0000147017
8/5	0.8550267840	0.8549561291	0.0000706549
12/5	1.6045273996	1.6042714779	0.0002559216
16/5	2.3963133788	2.3958902142	0.0004231645
4	3.1955002598	3.1948843675	0.0006158922
24/5	3.9954529746	3.9943062991	0.0011466754
28/5	4.7954513976	4.7936944532	0.0017569444
32/5	5.5954513676	5.5932562778	0.0021950898
36/5	6.3954513670	6.3930163452	0.0024350218
8	7.1954513667	7.1929060945	0.0025452721

The first-order approximate solution 4.192 becomes

$$\begin{aligned}&\overline{F}(\eta) = -0.8071635483 + \eta + \big(2.8068611339 - 3.4953295790\eta +\\ &+4.4303812597\eta^2 - 2.5930631506\eta^3 + 0.8330363534\eta^4 -\\ &0.1666768365\eta^5 + 0.0179261368\eta^6\big)e^{-2.1171968259\eta} + \big(-2.0785335331 +\\ &+0.0310432086\eta + 0.2175675516\eta^2 - 0.4253321356\eta^3 -\\ &-0.0029993386\eta^4\big)e^{-4.2343936518\eta} + \big(0.0788359475 + 0.1063683100\eta -\\ &-0.0243638517\eta^2\big)e^{-6.3515904777\eta}\end{aligned} \tag{4.202}$$

In Tables 4.45 and 4.46 are given some values of stream function 4.202, and velocity obtained from Eq. 4.202 respectively for different values of η and the corresponding relative errors.

4.3.3.1.c) For Eq. 4.194 which depends of the optimal auxiliary convergence-control function H given by Eq. 4.193 we obtain:

$$\begin{gathered}C_1 = 17.1086190380, C_2 = -6.1886479153, C_3 = 0.7626899878,\\ C_4 = -0.0317286051, C_5 = -63.8867694459, C_6 = 0.7750251041,\\ C_7 = -7.4945476964, C_8 = -9.8195426109, C_9 = -0.4164447206,\\ K = 1.1269038309, \alpha = 0.8746910670\end{gathered}$$

Therefore, the first-order approximate solution for stream function is

Table 4.46 Comparison between OHAM results for velocity $\overline{F}'(\eta)$ obtained from Eq. 4.202 and numerical results for $\beta = 1/2$

η	$F'_{numerical}(\eta)$	$\overline{F}'_{OHAM}(\eta)$ from Eq. 4.202	relative error = $\lvert F'_{numerical}(\eta) - \overline{F}'_{OHAM}(\eta)\rvert$
0	$1.85288457 \times 10^{-21}$	$2.55351295 \times 10^{-15}$	$2.55351110 \times 10^{-15}$
4/5	0.5833048177	0.5831981849	0.0001066327
8/5	0.8760975697	0.8760769428	0.0000206268
12/5	0.9760687561	0.9757219753	0.0003467808
16/5	0.9971920750	0.9970876920	0.0001043829
4	0.9998081986	0.9993456222	0.0004625764
24/5	0.9999925769	0.9992025737	0.0007900032
28/5	0.9999998398	0.9993191919	0.0006806479
32/5	0.9999999978	0.9995868200	0.0004131778
36/5	0.9999999995	0.9997970596	0.0002029398
8	0.9999999995	0.9999136481	0.0000863514

Table 4.47 Comparison between OHAM results given by Eq. 4.203 and numerical solutions for $\beta = 1/2$

η	$F_{numerical}(\eta)$	$\overline{\overline{F}}_{OHAM}(\eta)$ Eq. 4.203	relative error = $\lvert F_{numerical}(\eta) - \overline{\overline{F}}_{OHAM}(\eta)\rvert$
0	$-7.58518159 \times 10^{-25}$	$-2.2204460492 \times 10^{-16}$	$2.22044604 \times 10^{-16}$
4/5	0.2543480764	0.2543410815	$6.9949043881 \times 10^{-6}$
8/5	0.8550267840	0.8550382932	0.0000115092
12/5	1.6045273993	1.6045272608	$1.3846035207 \times 10^{-7}$
16/5	2.3963133781	2.3963351657	0.0000217876
4	3.1955002585	3.1955812100	0.0000809514
24/5	3.9954529727	3.9955457399	0.0000927671
28/5	4.7954513949	4.7955738251	0.0001224302
32/5	5.5954513640	5.5956241393	0.0001727753
36/5	6.3954513623	6.3956765240	0.0002251617
8	7.1954513606	7.1957387803	0.0002874196

$$\begin{aligned}\overline{\overline{F}}(\eta) &= -2.4537934149 + \eta + \big(-18.9229176282 + 1.0563974022\eta + \\ &+3.6621795605\eta^2 - 1.0933800148\eta^3 + 0.1271627184\eta^4 + \\ &+0.0056311114\eta^5\big)e^{-1.1269038309\eta} + \big(19.9100776837 + 25.8158411889\eta + \\ &+9.3137451766\eta^2 + 1.9358858129\eta^3\big)e^{-2.25380766196\eta} + \big(1.3951574877 + \\ &+0.7157493617\eta + 0.1454729886\eta^2\big)e^{-3.3807114929\eta} + \\ &+0.0714758716e^{-4.5076153239\eta}\end{aligned} \tag{4.203}$$

In Tables 4.47 and 4.48 we present some values of stream function 4.203 and velocity obtained from Eq. 4.203 respectively for different values of variable η, and the corresponding relative errors.

Table 4.48 Comparison between OHAM results for velocity $\overline{\overline{F}}'(\eta)$ obtained from Eq. 4.203 and numerical results for $\beta = 1/2$

η	$F'_{numerical}(\eta)$	$\overline{\overline{F}}'_{OHAM}(\eta)\, from\, Eq.\, 4.203$	$relative\ error = \lvert F'_{numerical}(\eta) - \overline{\overline{F}}'_{OHAM}(\eta)\rvert$
0	$-2.96395573 \times 10^{-20}$	0	$2.96395357 \times 10^{-20}$
4/5	0.5833048175	0.5832372756	0.0000675419
8/5	0.8760975695	0.8761763161	0.0000787466
12/5	0.9760687557	0.9760153463	0.0000534094
16/5	0.9971920744	0.9972855939	0.0000935194
4	0.9998081978	0.9998421267	0.0000339289
24/5	0.9999925759	1.0000087749	0.0000161989
28/5	0.9999998387	1.0000557917	0.0000559529
32/5	0.9999999965	1.0000650747	0.0000650781
36/5	0.9999999980	1.0000682953	0.0000682973
8	0.9999999978	1.0000901947	0.0000901968

Comparing the results presented in Table 4.43 with the results presented in Table 4.47 and the other hand, the results presented in Table 4.44 with the results presented in Table 4.48, respectively, it is clear that the analytical solutions obtained by our procedure prove to be more accurate along with an increased number of terms in the optimal auxiliary convergence-control function.

4.3.3.1.d) If we consider Eq. 4.196 depending of the optimal auxiliary convergence-control function given by Eq. 4.195 then from system 4.198 we obtain following results:

$$\begin{gathered}C_1 = 28.8801485140, C_2 = -30.7469629234, C_3 = 10.1656115824,\\ C_4 = -1.3058736876, C_5 = 0.0629713196, C_6 = -11.2295919004,\\ C_7 = 62.3525565579, C_8 = -18.6274524032, C_9 = 13.7997786872,\\ K = 1.9932224781, \alpha = 2.7977542979\end{gathered}$$

The first-order approximate solution for stream function 4.196 becomes

$$\begin{aligned}&\overline{\overline{F}}(\eta) = -0.8042841228 + \eta + \big(5.055972832 - 7.4607384673\eta +\\ &+7.6988739549\eta^2 - 3.9526112515\eta^3 + 1.0002001924\eta^4 -\\ &-0.1204170249\eta^5 + 0.0052654533\eta^6\big)e^{-1.9932224781\eta} + \big(-4.126293156-\\ &-0.2733114331\eta - 0.151609344\eta^2 - 0.5576699112\eta^3 -\\ &-0.8693555337\eta^4\big)e^{-3.9864449562\eta} + \big(-0.0807202735 - 0.2092400883\eta -\\ &-0.0196584681\eta^2 - 0.0484063456\eta^3\big)e^{-5.9796674343\eta}\end{aligned} \tag{4.204}$$

In Tables 4.49 and 4.50 are presented a comparison between the first-order approximate solution given by Eq. 4.204 and velocity obtained from Eq. 4.204 respectively, with numerical results for some values of variable η, and the corresponding relative errors.

Table 4.49 Comparison between OHAM results given by Eq. 4.204 and numerical solutions for $\beta = 1/2$

η	$F_{numerical}(\eta)$	$\overline{\overline{F}}_{OHAM}(\eta)$ $Eq.$ 4.204	$relative\ error =$ $\lvert F_{numerical}(\eta) - \overline{\overline{F}}_{OHAM}(\eta)\rvert$
0	$-5.775724 \times 10^{-25}$	$8.88178419 \times 10^{-16}$	$8.881784204 \times 10^{-16}$
4/5	0.2543480764	0.2543611093	0.0000130328
8/5	0.8550267840	0.8550475945	0.0000208105
12/5	1.6045273993	1.6045751046	0.0000477053
16/5	2.3963133781	2.3963685295	0.0000551514
4	3.1955002585	3.1955790615	0.0000788029
24/5	3.9954529727	3.9955715667	0.0001185939
28/5	4.7954513949	4.7956217908	0.0001703959
32/5	5.5954513640	5.5956679550	0.0002165910
36/5	6.3954513623	6.3956963294	0.0002449671
8	7.1954513606	7.1957094147	0.0002580540

Table 4.50 Comparison between OHAM results for velocity $\overline{\overline{F}}'(\eta)$ obtained from Eq. 4.204 and numerical results for $\beta = 1/2$

η	$F'_{numerical}(\eta)$	$\overline{\overline{F}}'_{OHAM}(\eta)\, from\, Eq.$ 4.204	$relative\ error =$ $\lvert F'_{numerical}(\eta) - \overline{\overline{F}}'_{OHAM}(\eta)\rvert$
0	$1.85288457 \times 10^{-21}$	$-5.32907051 \times 10^{-21}$	$5.32904087 \times 10^{-21}$
4/5	0.5833048177	0.5833475838	0.0000427662
8/5	0.8760975697	0.8761071968	$9.62735898 \times 10^{-6}$
12/5	0.9760687561	0.9760898679	0.0000211121
16/5	0.9971920750	0.9972074735	0.0000153990
4	0.9998081986	0.9998485848	0.0000403869
24/5	0.9999925769	1.0000520335	0.0000594575
28/5	0.9999998398	1.0000653592	0.0000655205
32/5	0.9999999978	1.0000472907	0.0000472941
36/5	0.9999999995	1.0000244710	0.0000244730
8	0.9999999995	1.0000098270	$9.82922032 \times 10^{-6}$

If we compare the results presented in Tables 4.45 and 4.49 and then the results presented in Tables 4.46 and 4.50, respectively, we can arrive at conclusion that the analytical results obtained by OHAM are more accurate along with an increased number of terms in the optimal auxiliary convergence-control function. It is important to establish the value of the shear-stress profile $\overline{F''}(0)$. In Table 4.51 we present a comparison between the values of $\overline{F''}(0)$ obtained using OHAM from Eqs. 4.201, 4.202, 4.203 and 4.204 and numerical results for $\beta = 1/2$. Our results are in very good agreement with the numerical results.

Table 4.51 Comparison between the values of $\overline{F}''(0)$ obtained by means of OHAM and numerical results for $\beta = 1/2$

Type of equation	Eq. (4.201)	Eq. (4.202)	Eq. (4.203)	Eq. (4.204)	Numerical
$\overline{F}'_{OHAM}(0)$	0.9276773361	0.9276376088	0.9276092354	0.9277933551	0.9276800471

Table 4.52 Comparison between OHAM results given by Eq. 4.205 and numerical solutions for $\beta = 1$

η	$F_{numerical}(\eta)$	$\overline{F}_{OHAM}(\eta)$ Eq. 4.205	relative error = $\lvert F_{numerical}(\eta) - \overline{F}_{OHAM}(\eta)\rvert$
0	$2.22951024 \times 10^{-25}$	$-1.11022302 \times 10^{-16}$	$1.11022302 \times 10^{-16}$
4/5	0.3124230332	0.3124218993	1.133942×10^{-6}
8/5	0.9797795327	0.9797813157	1.783054×10^{-6}
12/5	1.7552539494	1.7552573134	3.364056×10^{-6}
16/5	2.5523254690	2.5523466484	0.0000211794
4	3.3521093590	3.3521308223	0.00002146329
24/5	4.1520998358	4.1521444024	0.0000445665
28/5	4.9520995946	4.9522018532	0.0001022586
32/5	5.7520995955	5.7522601900	0.0001605945
36/5	6.5520996012	6.5522991212	0.0001995199
8	7.3520996086	7.3523194836	0.0002198750

Example 4.3.3.2 In this second case we suppose that $\beta = 1$.

4.3.3.2.a) For Eq. 4.189, from the system 4.198 we obtained the values of the optimal convergence-control parameters:

$$C_1 = 5.9913279914, C_2 = -6.6828413743, C_3 = 2.0729320860,$$
$$C_4 = -0.1950415110, C_5 = 18.0871429890, C_6 = -84.4868066767,$$
$$C_7 = 17.6775180449, C_8 = -0.3871636300, K = 2.0064973399$$

For these values of the parameters, from Eq. 4.189 we obtained the first-order approximate solution in the form

$$\begin{aligned}\overline{F}(\eta) = &-0.6476658539 + \eta + \big(1.1858804994 - 1.0205705493\eta + \\ &+1.1912223638\eta^2 - 0.7433934009\eta^3 + 0.1980363504\eta^4 - \\ &-0.0194409937\eta^5\big)e^{-2.0064973399\eta} + \big(-0.5332638964 + 0.2249079388\eta + \\ &+0.3083821306\eta^2 + 0.0240412688\eta^3\big)e^{-4.0129946798\eta} + \\ &+(-0.004950749 + 0.0053425041\eta)e^{-6.0194920197\eta}\end{aligned} \tag{4.205}$$

In Tables 4.52 and 4.53 are presented some values of stream function given by Eq. 4.205 and velocity obtained from Eq. 4.205, respectively, for different values of variable η. Also the corresponding relative errors are given in these cases.

Table 4.53 Comparison between OHAM results for velocity $\overline{F}'(\eta)$ obtained from Eq. 4.205 and numerical results for $\beta = 1$

η	$F'_{numerical}(\eta)$	$\overline{F}'_{OHAM}(\eta)\, from\, Eq.\, 4.205$	$relative\ error = \lvert F'_{numerical}(\eta) - \overline{F}'_{OHAM}(\eta)\rvert$
0	$-4.59912420 \times 10^{-21}$	$4.44089209 \times 10^{-16}$	$4.440893808 \times 10^{-16}$
4/5	0.6859374677	0.6859347711	2.696657×10^{-6}
8/5	0.9323482529	0.9323468816	1.371273×10^{-6}
12/5	0.9905493983	0.9905683352	0.00001893
16/5	0.9991860373	0.9991967641	0.00001072
4	0.9999584304	0.9999621135	3.683007×10^{-6}
24/5	0.9999987792	1.0000547905	0.0000560112
28/5	0.9999999836	1.0000795225	0.0000795389
32/5	1.0000000061	1.0000621951	0.0000621890
36/5	1.0000000081	1.0000356813	0.0000356731
8	1.0000000101	1.0000168920	0.0000168818

4.3.3.2.b) For Eq. 4.192, the system 4.198 has the solutions

$$C_1 = 25.6413506662, C_2 = -27.7257520445, C_3 = 10.2440262047,$$
$$C_4 = -1.6311566772, C_5 = 0.1133872279, C_6 = -37.8620215531,$$
$$C_7 = 4.5810479839, C_8 = -9.7143709526, K = 2.0340821793$$

The first-order approximate solution can be written as

$$\overline{F}(\eta) = -0.6482098275 + \eta + \big(1.7192094539 - 4.5907375859\eta + \\ +5.3333876914\eta^2 - 3.0101935973\eta^3 + 0.8973516349\eta^4 - 0.1394863598\eta^5 \\ +0.0092906134\eta^6\big)e^{-2.0340821793\eta} + \big(-1.0709996263 + 2.7307483905\eta + \\ +2.3599924746\eta^2 + 0.5869724289\eta^3\big)e^{-4.0681643586\eta} \tag{4.206}$$

In Tables 4.54 and 4.55 we present some values of stream function 4.206 and velocity obtained from Eq. 4.206 respectively, for different values of variable η, and the corresponding relative errors.

4.3.3.2.c) For Eq. 4.194 with the optimal auxiliary convergence-control function H given by Eq. 4.193 we have:

$$C_1 = 15.1069993472, C_2 = -6.1116892850, C_3 = 0.8419581636,$$
$$C_4 = -0.0391594463, C_5 = -348.1410130441, C_6 = 354.6124244346,$$
$$C_7 = -78.2878044901, C_8 = -3.8910409852, C_9 = -4.3179402181,$$
$$K = 1.3940527605, \alpha = 1.2080121437$$

Table 4.54 Comparison between OHAM results given by Eq. 4.206 and numerical solutions for $\beta = 1$

η	$F_{numerical}(\eta)$	$\overline{F}_{OHAM}(\eta)$ $Eq.$ 4.206	$relative\ error =$ $\lvert F_{numerical}(\eta) - \overline{F}_{OHAM}(\eta)\rvert$
0	$2.22951024 \times 10^{-25}$	$-2.22044604 \times 10^{-16}$	$2.22044605 \times 10^{-16}$
4/5	0.3124230332	0.3124250905	2.0573489×10^{-6}
8/5	0.9797795327	0.9797816632	2.1304868×10^{-6}
12/5	1.7552539494	1.7552458857	8.063655×10^{-6}
16/5	2.5523254690	2.5523068656	0.0000186033
4	3.3521093590	3.3520844463	0.0000249127
24/5	4.1520998358	4.1520336380	0.0000661977
28/5	4.9520995946	4.9519649131	0.0001346814
32/5	5.7520995955	5.7518965404	0.0002030550
36/5	6.5520996012	6.5518462101	0.0002533911
8	7.3520996086	7.3518163433	0.0002832652

Table 4.55 Comparison between OHAM results for velocity $\overline{F}'(\eta)$ obtained from Eq. 4.206 and numerical results for $\beta = 1$

η	$F'_{numerical}(\eta)$	$\overline{F}'_{OHAM}(\eta)$ $from\ Eq.$ 4.206	$relative\ error =$ $\lvert F'_{numerical}(\eta) - \overline{F}'_{OHAM}(\eta)\rvert$
0	$-4.59912420 \times 10^{-21}$	$1.77635683 \times 10^{-15}$	$1.77636143 \times 10^{-15}$
4/5	0.6859374677	0.6859334525	4.015197×10^{-6}
8/5	0.9323482529	0.9323592586	0.0000110057
12/5	0.9905493983	0.9905212799	0.0000281183
16/5	0.9991860373	0.9991867807	7.433778×10^{-7}
4	0.9999584304	0.9999317943	0.0000266361
24/5	0.9999987792	0.9999248393	0.0000739398
28/5	0.9999999836	0.9999090229	0.0000909606
32/5	1.0000000061	0.9999238617	0.0000761443
36/5	1.0000000081	0.9999506141	0.0000493940
8	1.0000000101	0.9999733687	0.0000266414

The first-order approximate solution for stream function 4.56 becomes:

$$\begin{aligned}\overline{\overline{F}}(\eta) &= -0.6477776791 + \eta + \big(-18.001344737 + 2.2989361198\eta + \\ &+2.4130328127\eta^2 - 0.8773035418\eta^3 + 0.1154431274\eta^4 - \\ &-0.0056180723\eta^5\big)e^{-1.3940527605\eta} + \big(18.631306321 + 23.515563981\eta + \\ &+12.357602161\eta^2 + 3.3122912941\eta^3 + 0.5554669357\eta^4\big)e^{-2.7881055211\eta} + \\ &+(0.0178160913 + 0.1112333145\eta)e^{-4.1821582817\eta}\end{aligned} \tag{4.207}$$

Table 4.56 Comparison between OHAM results given by Eq. 4.207 and numerical solutions for $\beta = 1$

η	$F_{numerical}(\eta)$	$\overline{\overline{F}}_{OHAM}(\eta)\ Eq.\ 4.207$	$relative\ error = \lvert F_{numerical}(\eta) - \overline{\overline{F}}_{OHAM}(\eta)\rvert$
0	$4.69328062 \times 10^{-25}$	$6.21724893 \times 10^{-15}$	$6.21724893 \times 10^{-15}$
4/5	0.3124230332	0.3124256050	2.5717950×10^{-6}
8/5	0.9797795326	0.9797870784	7.545842×10^{-6}
12/5	1.7552539491	1.7552584440	4.494942×10^{-6}
16/5	2.5523254682	2.5523369280	0.0000114597
4	3.3521093576	3.3521193533	9.995766×10^{-6}
24/5	4.1520998332	4.1521125574	0.0000127241
28/5	4.9520995903	4.9521192859	0.0000196955
32/5	5.7520995889	5.7521233093	0.0000237203
36/5	6.5520995917	6.5521296538	0.0000300620
8	7.3520995953	7.3521431339	0.0000435385

Table 4.57 Comparison between OHAM results for velocity $\overline{\overline{F}}'(\eta)$ obtained from Eq. 4.207 and numerical results for $\beta = 1$

η	$F'_{numerical}(\eta)$	$\overline{\overline{F}}'_{OHAM}(\eta)\ from\ Eq.\ 4.207$	$relative\ error = \lvert F'_{numerical}(\eta) - \overline{\overline{F}}'_{OHAM}(\eta)\rvert$
0	$-9.39015431 \times 10^{-21}$	$-4.0856207306 \times 10^{-14}$	$4.08561979 \times 10^{-14}$
4/5	0.6859374676	0.6859366849	7.826990×10^{-7}
8/5	0.9323482527	0.9323490349	7.821535×10^{-7}
12/5	0.9905493979	0.9905533068	3.908903×10^{-6}
16/5	0.9991860366	0.9991905903	4.553709×10^{-6}
4	0.9999584293	0.9999557316	2.697690×10^{-6}
24/5	0.9999987775	1.0000074611	8.683616×10^{-6}
28/5	0.9999999812	1.0000067856	6.804426×10^{-6}
32/5	1.0000000029	1.0000048084	4.805536×10^{-6}
36/5	1.0000000039	1.0000121725	0.0000121685
8	1.0000000049	1.0000209604	0.0000209554

In Tables 4.56 and 4.57 we present some values of stream function 4.207 and velocity obtained from Eq. 4.207 respectively for different values of variable η, and the corresponding relative errors.

4.3.3.2.d) If we have in view Eq. 4.196, with the optimal auxiliary convergence-control function H given by Eq. 4.195 then we obtain

$$C_1 = 2.5755597425, C_2 = -0.5965183787, C_3 = -0.0172692621,$$
$$C_4 = 0.0136562951, C_5 = -0.0010079176, C_6 = -5.9530458601,$$
$$C_7 = -1.9930045747, C_8 = -1.9931630292, C_9 = 0.0743455822,$$
$$K = 1.1143724942, \alpha = 0.2855970092$$

Table 4.58 Comparison between OHAM results given by Eq. 4.208 and numerical solutions for $\beta = 1$

η	$F_{numerical}(\eta)$	$\overline{\overline{F}}_{OHAM}(\eta)$ Eq. 4.208	relative error = $\lvert F_{numerical}(\eta) - \overline{\overline{F}}_{OHAM}(\eta) \rvert$
0	$4.69328062 \times 10^{-25}$	0	$4.69328062 \times 10^{-25}$
4/5	0.3124230332	0.3124174580	5.575244×10^{-6}
8/5	0.9797795326	0.9797740164	5.516204×10^{-6}
12/5	1.7552539491	1.7552534223	5.267840×10^{-7}
16/5	2.5523254682	2.5523611425	0.0000356742
4	3.3521093576	3.3521105770	1.219470×10^{-6}
24/5	4.1520998332	4.1521005519	7.187418×10^{-7}
28/5	4.9520995903	4.9521462123	0.0000466219
32/5	5.7520995889	5.7521684524	0.0000688634
36/5	6.5520995917	6.5521510110	0.0000514192
8	7.3520995953	7.3521263893	0.0000267940

and therefore first-order approximate solution 4.196 is given by

$$\begin{aligned}\overline{\overline{F}}(\eta) &= -0.6475361092 + \eta + \big(-7.0772773941 + 1.5599095672\eta + \\ &+0.3833481708\eta^2 - 0.141867195\eta^3 + 0.0068820136\eta^4 + 0.0015709234\eta^5 \\ &-0.0001507451\eta^6\big)e^{-1.11437249 42\eta} + \big(7.7248135031 + 6.7700065534\eta + \\ &+2.2685016445\eta^2 + 0.2970574871\eta^3 - 0.014966987\eta^4\big)e^{-2.2287449884\eta}\end{aligned} \tag{4.208}$$

In the Tables 4.58 and 4.59 we present a comparison between the first-order approximate solution given by Eq. 4.208 and velocity obtained from Eq. 4.208 respectively, with numerical results, and the corresponding relative errors.

If we compare the results presented in Tables 4.52 and 4.56 and then the results presented in Tables 4.53 and 4.57 we deduce that the analytical results obtained by OHAM are more accurate along with an increased number of terms in the optimal auxiliary convergence-control function H. The same conclusions are deduced if we compare the results presented in Tables 4.54 and 4.58 and then the results presented in Tables 4.55 and 4.59, respectively.

In Table 4.60 we present a comparison between the values of $\overline{F}''(0)$ obtained using OHAM from Eqs. 4.205, 4.206, 4.207 and 4.208 and numerical integration results from $\beta = 1$. We can deduce that the results obtained by means of OHAM are nearly identical with those obtained through numerical integration.

Example 4.3.3.3 In the last case we consider $\beta = 1.6$.

4.3.3.3.a) For Eq. 4.189 the values of the optimal convergence-control parameters are obtained from the system 4.198:

Table 4.59 Comparison between OHAM results for velocity $\overline{\overline{F}}'(\eta)$ obtained from Eq. 4.208 and numerical results for $\beta = 1$

η	$F'_{numerical}(\eta)$	$\overline{\overline{F}}'_{OHAM}(\eta)$ *from Eq.* 4.208	*relative error* = $\lvert F'_{numerical}(\eta) - \overline{\overline{F}}'_{OHAM}(\eta)\rvert$
0	$-9.39015431 \times 10^{-21}$	0	$9.39015431 \times 10^{-21}$
4/5	0.6859374676	0.6859539897	0.0000165220
8/5	0.9323482527	0.9323092256	0.0000390271
12/5	0.9905493979	0.9906138617	0.0000644637
16/5	0.9991860366	0.9991779292	8.107355×10^{-6}
4	0.9999584293	0.9999145872	0.0000438420
24/5	0.9999987775	1.0000414108	0.0000426333
28/5	0.9999999812	1.0000539936	0.0000540124
32/5	1.0000000029	0.9999989176	1.085287×10^{-6}
36/5	1.0000000039	0.9999655183	0.0000344856
8	1.0000000049	0.9999795865	0.0000204184

Table 4.60 Comparison between the values of $\overline{\overline{F}}''(0)$ obtained by means of OHAM and numerical results for $\beta = 1$

Type of equation	Eq. 4.205	Eq. 4.206	Eq. 4.207	Eq. 4.208	Numerical
$\overline{\overline{F}}''_{OHAM}(0)$	1.23258247	1.23257895	1.23262084	1.23257391	1.232558769

$$\begin{aligned}
&C_1 = -4.4749746430, C_2 = 2.6648959478, C_3 = -0.5277449747,\\
&C_4 = 0.0344313836, C_5 = 0.9084258281, C_6 = 11.1816458907,\\
&C_7 = -0.5492649796, C_8 = 1.5695573134, K = 1.7389825314
\end{aligned}$$

The first-order approximate solution for stream function 4.189 can be written as

$$\begin{aligned}
&\overline{F}(\eta) = -0.5440478033 + \eta + (1.6577374986 - 0.4175138596\eta -\\
&-0.4722933401\eta^2 + 0.2774993818\eta^3 - 0.05544381588\eta^4 +\\
&+0.0039599458\eta^5)e^{-1.7389825314\eta} + (-1.1629370587 - 1.5183251610\eta -\\
&-0.7578665325\eta^2 - 0.1287736808\eta^3)e^{-3.4779650629\eta} +\\
&+(0.0455165672 + 0.0373698159\eta - 0.0034815689\eta^2)e^{-5.2169475944\eta} +\\
&+0.0037307962e^{-6.9559301258\eta}
\end{aligned} \tag{4.209}$$

In Tables 4.61 and 4.62 are presented some values of the stream function given by Eq. 4.209 and velocity obtained from Eq. 4.209, respectively, for different values of variable η. Also the corresponding relative errors are given in these Tables.

4.3.3.3.b) For Eq. 4.192, the convergence-control parameters obtained from the system 4.198 have the values:

Table 4.61 Comparison between OHAM results given by Eq. 4.209 and numerical solutions for $\beta = 1.6$

η	$F_{numerical}(\eta)$	$\overline{F}_{OHAM}(\eta)$ $Eq.$ 4.209	$relative\ error = \lvert F_{numerical}(\eta) - \overline{F}_{OHAM}(\eta)\rvert$
0	$-1.69232752 \times 10^{-25}$	$-1.11022302 \times 10^{-16}$	$1.11022302 \times 10^{-16}$
4/5	0.3599784956	0.3599782796	2.160525×10^{-7}
8/5	1.0696147641	1.0696156734	9.093019×10^{-7}
12/5	1.8571623264	1.8571633242	9.977941×10^{-7}
16/5	2.6560434964	2.6560463745	2.878145×10^{-6}
4	3.4559804211	3.4559769848	3.436309×10^{-6}
24/5	4.2559782714	4.2559782610	1.038556×10^{-8}
28/5	5.0559782427	5.0559829283	4.685640×10^{-6}
32/5	5.8559782660	5.8559798484	1.582358×10^{-6}
36/5	6.6559783023	6.6559717863	6.516027×10^{-6}
8	7.4559783545	7.4559638500	0.0000145044

Table 4.62 Comparison between OHAM results for velocity $\overline{F}'(\eta)$ obtained from Eq. 4.209 and numerical results for $\beta = 1.6$

η	$F'_{numerical}(\eta)$	$\overline{F}'_{OHAM}(\eta)$ $from\ Eq.$ 4.209	$relative\ error = \lvert F'_{numerical}(\eta) - \overline{F}'_{OHAM}(\eta)\rvert$
0	$4.07171384 \times 10^{-20}$	$9.99200722 \times 10^{-16}$	$9.9916000502 \times 10^{-16}$
4/5	0.7609225381	0.7609240247	1.486624×10^{-6}
8/5	0.9619780339	0.9619706813	7.352687×10^{-6}
12/5	0.9960567187	0.9960657372	9.018419×10^{-6}
16/5	0.9997439284	0.9997365151	7.413383×10^{-6}
4	0.9999899725	0.9999872482	2.724280×10^{-6}
24/5	0.9999997837	1.0000081444	8.360731×10^{-6}
28/5	1.0000000206	1.0000013591	1.338446×10^{-6}
32/5	1.0000000368	0.9999918129	8.223935×10^{-6}
36/5	1.0000000541	0.9999891905	0.0000108635
8	1.0000000764	0.9999914106	8.665844×10^{-6}

$$
\begin{aligned}
&C_1 = 12.1397319477, C_2 = -14.7300094090, C_3 = 6.0298078722,\\
&C_4 = -1.0787489057, C_5 = 0.0852147302, C_6 = -14.8695758423,\\
&C_7 = -7.8344847253, C_8 = -2.4044068709, K = 2.0881719331
\end{aligned}
$$

The first-order approximate solution for stream function 4.192 becomes

$$
\begin{aligned}
&\overline{F}(\eta) = -0.544241252 + \eta + (0.398631080 - 1.040810138\eta +\\
&+1.69949761\eta^2 - 1.248940652\eta^3 + 0.460677086\eta^4 - 0.086297165\eta^5 +\\
&+0.00680138\eta^6)e^{-2.088171933\eta} + (0.219770025 + 1.364996288\eta +\\
&+1.025546039\eta^2 + 0.114703592\eta^3 + 0.001403806\eta^4)e^{-4.176343866\eta} +\\
&+(-0.0741598535 - 0.038516290\eta - 0.008802133\eta^2)e^{-6.264515799\eta}
\end{aligned}
\tag{4.210}
$$

Table 4.63 Comparison between OHAM results given by Eq. 4.210 and numerical solutions for $\beta = 1.6$

η	$F_{numerical}(\eta)$	$\overline{F}_{OHAM}(\eta)\,Eq.\,4.210$	*relative error* = $\lvert F_{numerical}(\eta) - \overline{F}_{OHAM}(\eta)\rvert$
0	$-1.69232752 \times 10^{-25}$	$1.33226762 \times 10^{-15}$	$1.33226762 \times 10^{-15}$
4/5	0.3599784956	0.3599798414	1.345740×10^{-6}
8/5	1.0696147641	1.0696168199	2.055868×10^{-6}
12/5	1.8571623264	1.8571595718	2.754588×10^{-6}
16/5	2.6560434964	2.6560360629	7.433444×10^{-6}
4	3.4559804211	3.4559769853	3.435805×10^{-6}
24/5	4.2559782714	4.2559486488	0.0000296226
28/5	5.0559782427	5.0558929543	0.0000852884
32/5	5.8559782660	5.8558366505	0.0001416154
36/5	6.6559783023	6.6557973327	0.0001809696
8	7.4559783545	7.4557756153	0.0002027391

Table 4.64 Comparison between OHAM results for velocity $\overline{F}'(\eta)$ obtained from Eq. 4.210 and numerical results for $\beta = 1.6$

η	$F'_{numerical}(\eta)$	$\overline{F}'_{OHAM}(\eta)\,from\,Eq.\,4.210$	*relative error* = $\lvert F'_{numerical}(\eta) - \overline{F}'_{OHAM}(\eta)\rvert$
0	$4.071713847 \times 10^{-20}$	$-3.55271367 \times 10^{-15}$	$3.55275439 \times 10^{-15}$
4/5	0.7609225381	0.7609190201	3.517984×10^{-6}
8/5	0.9619780339	0.9619872725	9.238582×10^{-6}
12/5	0.9960567187	0.9960397833	0.0000169354
16/5	0.9997439284	0.9997513373	7.408827×10^{-6}
4	0.9999899725	0.9999817723	8.200210×10^{-6}
24/5	0.9999997837	0.9999433979	0.0000563857
28/5	1.0000000206	0.9999242317	0.0000757889
32/5	1.0000000368	0.9999386977	0.0000613391
36/5	1.0000000541	0.9999628614	0.0000371926
8	1.0000000764	0.9999814328	0.0000186436

In Tables 4.63 and 4.64 are given some values of the stream function 4.210 and velocity obtained from 4.210 respectively, and the corresponding relative errors in comparison with the numerical results.

4.3.3.3.c) For Eq. 4.194, with the optimal auxiliary convergence-control function H given by Eq. 4.193, the system 4.198 has the solutions:

$$C_1 = 3.7343092835, C_2 = -1.6919201004, C_3 = 0.2585003215,$$
$$C_4 = -0.0131947678, C_5 = 26.9486507241, C_6 = -47.5028568351,$$
$$C_7 = 5.3389432018, C_8 = -8.7586264382, C_9 = 0.2924008927,$$
$$K = 1.4882721635, \alpha = 1974.2364175690$$

Table 4.65 Comparison between OHAM results given by Eq. 4.211 and numerical solutions for $\beta = 1.6$

η	$F_{numerical}(\eta)$	$\overline{F}_{OHAM}(\eta)$ *Eq.* 4.211	*relative error* = $\lvert F_{numerical}(\eta) - \overline{F}_{OHAM}(\eta)\rvert$
0	$-1.69232752 \times 10^{-25}$	$-1.69309011 \times 10^{-15}$	$1.69309011 \times 10^{-15}$
4/5	0.3599784956	0.3599790235	5.278046×10^{-7}
8/5	1.0696147640	1.0696154385	6.744980×10^{-7}
12/5	1.8571623261	1.8571642418	1.915749×10^{-6}
16/5	2.6560434955	2.6560445435	1.048024×10^{-6}
4	3.4559804190	3.4559816246	1.205614×10^{-6}
24/5	4.2559782671	4.2559820737	3.806680×10^{-6}
28/5	5.0559782344	5.0559813610	3.126620×10^{-6}
32/5	5.8559782515	5.8559788900	6.385224×10^{-7}
36/5	6.6559782783	6.6559788961	6.177966×10^{-7}
8	7.4559783175	7.4559820586	$3.74109314 \times 10^{-6}$

The first-order approximate solution for stream function 4.194 can be written as

$$\begin{aligned}
\overline{\overline{F}}(\eta) &= -0.544001931 + \eta + \big(-4.354906951 + 1.3261535367\eta + \\
&+0.3072449473\eta^2 - 0.1940259496\eta^3 + 0.0315309032\eta^4 - \\
&-0.0017731659\eta^5\big)e^{-1.4882721635\eta} + \big(4.8182102767 + 6.0137273891\eta + \\
&+2.4451256821\eta^2 + 0.8109067414\eta^3 - 0.0330030433\eta^4\big)e^{-2.976544327\eta} + \\
&+\big(0.1139109044 - 0.168675728\eta + 0.060940104\eta^2 + \\
&+0.0029567211\eta^3\big)e^{-4.4648164905\eta} - 0.033212299e^{-5.9530886541\eta}
\end{aligned} \tag{4.211}$$

In Tables 4.65 and 4.66 we present some values of stream function 4.211 and velocity obtained from Eq. 4.211, respectively for different values of variable η, and the corresponding relative errors.

4.3.3.3.d) If we have in view Eq. 4.196 with the optimal auxiliary convergence-control function H given by Eq. 4.195 then the optimal convergence-control parameters obtained from system 4.198 are:

$$\begin{aligned}
&C_1 = -9.9309813061,\ C_2 = 6.8914081433,\ C_3 = -1.7496966452,\\
&C_4 = 0.2003197512,\ C_5 = -0.0103677381,\ C_6 = 10.1134115449,\\
&C_7 = 7.0897290367, C_8 = 1.9899138244, C_9 = 0.0159776712,\\
&K = 1.5329158534,\ a = -0.9799303286
\end{aligned}$$

In this case, the first-order approximate solution for stream function 4.196 becomes

Table 4.66 Comparison between OHAM results for velocity $\overline{\overline{F}}'(\eta)$ obtained from Eq. 4.211 and numerical results for $\beta = 1.6$

η	$F'_{numerical}(\eta)$	$\overline{\overline{F}}'_{OHAM}(\eta)$ *from Eq.* 4.211	*relative error* = $\lvert F'_{numerical}(\eta) - \overline{\overline{F}}'_{OHAM}(\eta)\rvert$
0	$-2.83226641 \times 10^{-21}$	$2.66453525 \times 10^{-15}$	$2.66453809 \times 10^{-15}$
4/5	0.7609225380	0.7609232141	6.760967×10^{-7}
8/5	0.9619780338	0.9619796540	1.620229×10^{-6}
12/5	0.9960567183	0.9960575197	8.014067×10^{-7}
16/5	0.9997439274	0.9997418375	2.089891×10^{-6}
4	0.9999899704	0.9999929170	2.946522×10^{-6}
24/5	0.9999997799	1.0000016372	1.857256×10^{-6}
28/5	1.0000000145	0.9999969829	3.031551×10^{-6}
32/5	1.0000000272	0.9999978766	2.150573×10^{-6}
36/5	1.0000000399	1.0000022282	2.188252×10^{-6}
8	1.0000000667	1.0000052427	5.176061×10^{-6}

Table 4.67 Comparison between OHAM results given by Eq. 4.212 and numerical solutions for $\beta = 1.6$

η	$F_{numerical}(\eta)$	$\overline{\overline{F}}_{OHAM}(\eta)$ *Eq.* 4.212	*relative error* = $\lvert F_{numerical}(\eta) - \overline{\overline{F}}_{OHAM}(\eta)\rvert$
0	$-1.69232752 \times 10^{-25}$	$-2.33146835 \times 10^{-15}$	$2.33146835 \times 10^{-15}$
4/5	0.35997849	0.3599784037	9.195208×10^{-8}
8/5	1.06961476	1.0696144385	3.254580×10^{-7}
12/5	1.85716232	1.8571634346	1.108560×10^{-6}
16/5	2.65604349	2.6560430260	4.694645×10^{-7}
4	3.455980419	3.4559810382	6.192629×10^{-7}
24/5	4.25597826	4.2559845299	6.262826×10^{-6}
28/5	5.05597823	5.0559832266	4.992205×10^{-6}
32/5	5.85597825	5.8559812915	3.039980×10^{-6}
36/5	6.65597827	6.6559890626	0.0000107842
8	7.4559783175	7.4560065554	0.0000282378

$$\begin{aligned}
&\overline{\overline{F}}(\eta) = -0.5439219913 + \eta + (4.650996506 - 2.0398248866\eta - \\
&-0.4858446135\eta^2 + 0.550362856\eta^3 - 0.1603616877\eta^4 + 0.0213068753\eta^5 \\
&-0.0011272349\eta^6)e^{-1.5329158534\eta} + (-4.2651113968 - 4.2739140037\eta - \\
&-1.660992309\eta^2 - 0.2098183934\eta^3 - 0.0021316125\eta^4)e^{-3.06583170690\eta} + \\
&+(0.158036881 + 0.0939831378\eta + 0.0187520087\eta^2 + \\
&+0.0001478554\eta^3)e^{-4.5987475603\eta}
\end{aligned} \tag{4.212}$$

In Tables 4.67 and 4.68 are presented some values of stream function given by Eq. 4.212 and velocity obtained from Eq. 4.212, respectively, for different values of variable η. Also the corresponding relative errors are given in these cases.

Table 4.68 Comparison between OHAM results for velocity $\overline{\overline{F}}'(\eta)$ obtained from Eq. 4.212 and numerical results for $\beta = 1.6$

η	$F'_{numerical}(\eta)$	$\overline{\overline{F}}'_{OHAM}(\eta)$ *from Eq.* 4.212	*relative error* = $\lvert F'_{numerical}(\eta) - \overline{\overline{F}}'_{OHAM}(\eta)\rvert$
0	$-2.83226641 \times 10^{-21}$	$3.55271367 \times 10^{-15}$	$3.55271651 \times 10^{-15}$
4/5	0.7609225380	0.7609232919	7.538631×10^{-7}
8/5	0.9619780338	0.9619761488	1.885029×10^{-6}
12/5	0.9960567183	0.9960594353	2.717000×10^{-6}
16/5	0.9997439274	0.9997398342	4.093197×10^{-6}
4	0.9999899704	0.9999971751	7.204662×10^{-6}
24/5	0.9999997799	1.0000032843	3.504355×10^{-6}
28/5	1.0000000145	0.9999951986	4.815848×10^{-6}
32/5	1.0000000272	1.0000024525	2.425327×10^{-6}
36/5	1.0000000399	1.0000168296	0.0000167896
8	1.0000000667	1.0000253832	0.0000253165

Table 4.69 Comparison between the values of $\overline{\overline{F}}''(0)$ obtained by means of OHAM and numerical results for $\beta = 1.6$

Type of equation	Eq. 4.209	Eq. 4.210	Eq. 4.211	Eq. 4.212	Numerical
$\overline{\overline{F}}''_{OHAM}(0)$	1.52151589	1.52151820	1.52151245	1.52151403	1.52151402

In Table 4.69 we present a comparison between the values of $\overline{\overline{F}}''(0)$ obtained through OHAM from Eq. 4.209, 4.210, 4.211, 4.212 and numerical results for $\beta = 1.6$. We can observe that the analytical solutions obtained by OHAM are very accurate, being nearly identical with the numerical results.

The accuracy of the obtained results are verified graphically in Figs. 4.67, 4.68, 4.69, 4.70, 4.71, 4.72, 4.73, 4.74. The approximate solutions obtained by means of OHAM and with the optimal auxiliary convergence-control functions H given by Eqs. 4.187, 4.190, 4.193 and 4.195 are compared with numerical integration results for different values of β in the Figs. 4.65, 4.67, 4.69 and 4.71 respectively.

On the other hand, the velocity profile obtained from the corresponding approximate solutions are compared with numerical integration results in Figs. 4.68, 4.70, 4.72 and 4.74, respectively. In all cases presented in Figs. 4.67,4.68, 4.69, 4.70, 4.71, 4.72, 4.73, 4.74, it is shown that the first-order approximate solutions and the velocity increase with an increase in the parameter β.

By comparing the results presented in Tables 4.60 and 4.64 and then the results presented in Tables 4.61 and 4.65 we can write that the results obtained by OHAM are more accurate along with increasing the number of terms in the optimal auxiliary convergence-control function H. Now, comparing the results from Tables 4.62 and 4.66 and then the results from Tables 4.63 and 4.67 the conclusions are the same.

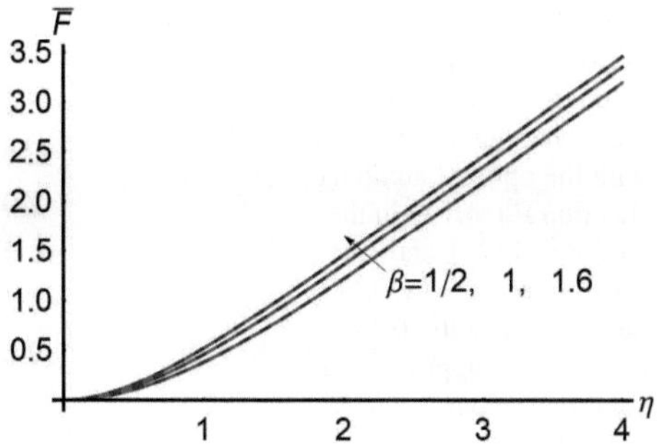

Fig. 4.67 Comparison between the approximate solutions 4.201, 4.205 and 4.209 and numerical results with the optimal auxiliary function Eq. 4.187 in the case $\beta = 1/2$, 1 and 1.6: (*red solid line*) numerical results; (*blue dashed line*) approximate solution (colour figure online)

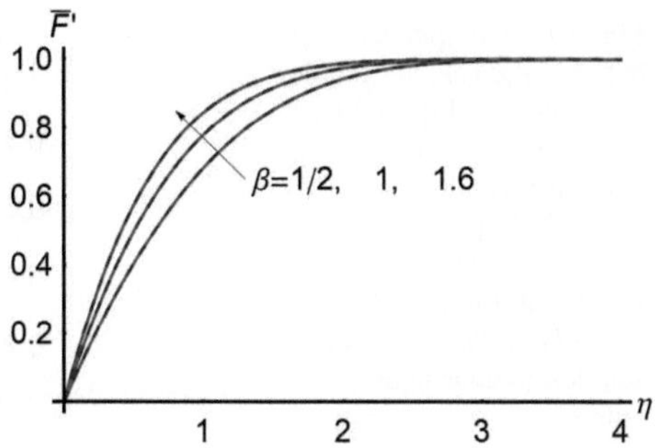

Fig. 4.68 Comparison between the velocity profile obtained from Eqs. 4.201, 4.205 and 4.209 and numerical results for $\beta = 1/2$, 1 and 1.6 respectively: (*red solid line*) numerical results; (*blue dashed line*) approximate solution (colour figure online)

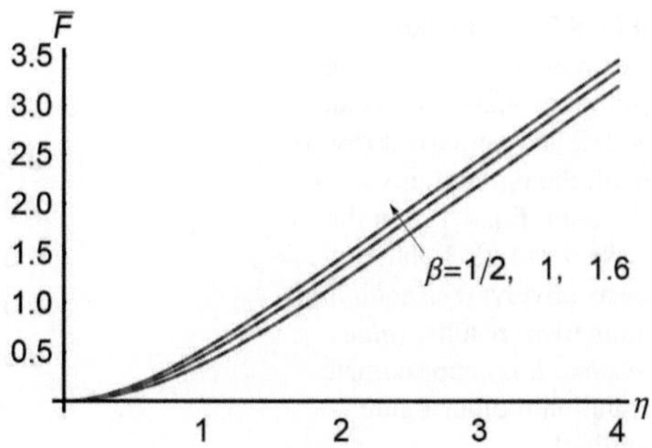

Fig. 4.69 Comparison between the approximate solutions 4.202, 4.206 and 4.210 and numerical results with the optimal auxiliary function Eq. 4.190 in the case $\beta = 1/2$, 1 and 1.6 respectively: (*red solid line*) numerical results (*blue dashed line*) approximate solution (colour figure online)

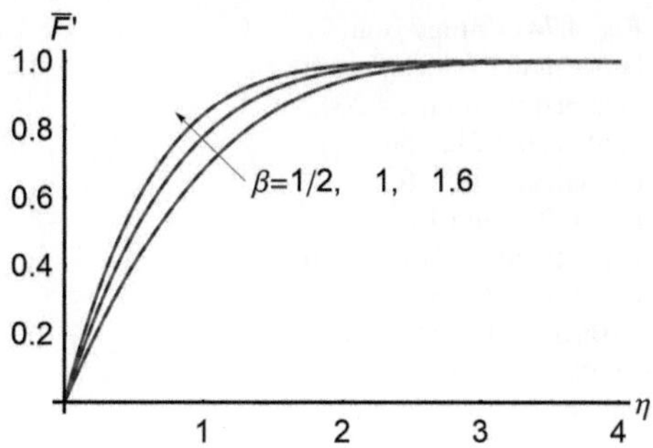

Fig. 4.70 Comparison between the velocity profile obtained from Eqs. 4.202, 4.206 and 4.210 and numerical results for $\beta = 1/2$, 1 and 1.6 respectively: (*red solid line*) numerical results; (*blue dashed line*) approximate solution (colour figure online)

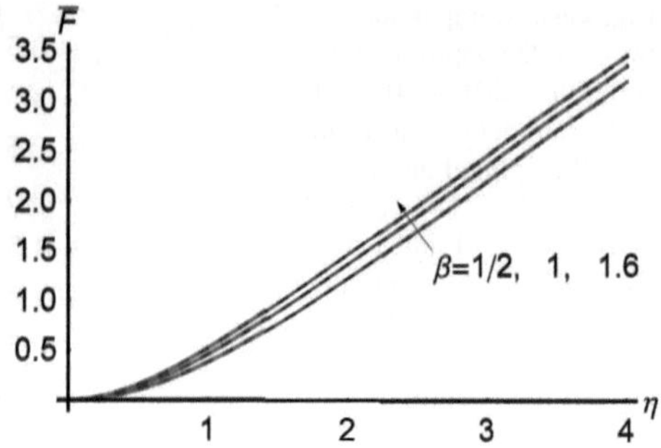

Fig. 4.71 Comparison between the approximate solutions 4.203, 4.207 and 4.211 and numerical results with the optimal auxiliary function Eq. 4.193 in the case $\beta = 1/2$, 1 and 1.6 respectively: (*red solid line*) numerical results; (*blue dashed line*) approximate solution (colour figure online)

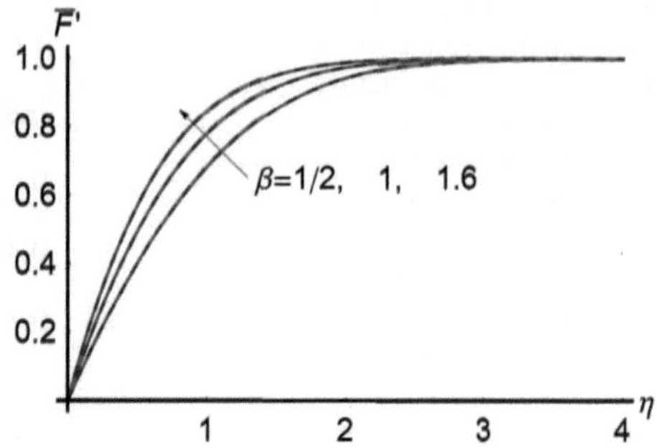

Fig. 4.72 Comparison between the velocity profile obtained from Eqs. 4.203, 4.207 and 4.211 and numerical results for $\beta = 1/2$, 1 and 1.6 respectively: (*red solid line*) numerical results; (*blue dashed line*) approximate solution (colour figure online)

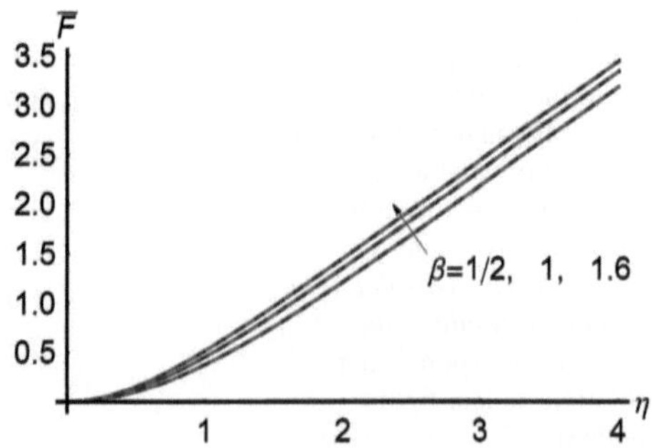

Fig. 4.73 Comparison between the approximate solutions 4.204, 4.208 and 4.212 and numerical results with the optimal auxiliary function Eq. 4.195 in the case $\beta = 1/2$, 1 and 1.6 respectively: (*red solid line*) numerical results; (*blue dashed line*) approximate solution (colour figure online)

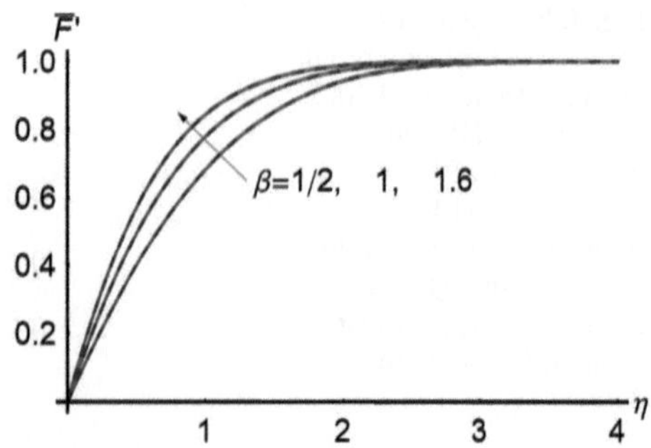

Fig. 4.74 Comparison between the velocity profile obtained from Eqs. 4.204, 4.208 and 4.212 and numerical results for $\beta = 1/2$, 1 and 1.6 respectively: (*red solid line*) numerical results; (*blue dashed line*) approximate solution (colour figure online)

4.4 Viscous Flow Due to a Stretching Surface with Partial Slip

The study of boundary layer viscous flow due to a stretching surface is very significant because of its several applications in engineering or industrial processes for example: materials manufactured by extrusion of polymer sheets, paper production, drawing of elastic films, glass fiber, crystal growing, the heat treated materials traveling between a feed roll and the wind-up roll on a conveyor belt poses the features of a moving continuous surface. Also, there exist some situations where there may be a partial slip between the fluid and the boundary, e.g., the fluid may be modeled as a rarefied gas. The non-slip boundary condition is know as an important tenets of the Navier-Stokes theory. But there are some situations wherein such a condition can not be appropriate. Sakiadis [54, 55] first studied the boundary layer flow over a continuously stretching surface. Following these works, many studies have been carried out. Crane [56] found a closed form exponential solution for the planar viscous flow of a linear stretching case. Sparrow and his collaborators [57, 58] analyzed a number of flow problems taking the velocity slip into account. Ackroyd [59] has given an algorithm for the moving boundaries and for the flow of a fluid past a rotating disk, that eliminates the boundary condition at infinity. Rajagopal et al. [60] obtained the numerical solution for the flow of viscoelastic second order fluid past a stretching sheet. Ariel [61] and Anderson [62] have reported the analytical closed form solutions of the second grade fluid and of the fourth order nonlinear differential equations arising due to the MHD flow, respectively.

More recently, Wang [63] has considered both the two-dimensional and the axisymmetric cases for the viscous flow due to a stretching sheet with slip and suction. For the case two-dimensional, an exact solution is found and its uniqueness is proved. For the axisymmetric stretching sheet, both existence and uniqueness are shown and since there is no exact form solution, numerical solution is obtained. A number of boundary layer flows induced by the axisymmetric stretching of a sheet are studied in [64]. The sheet is stretched with a velocity proportional to the distance from the vertical axis. The unsteady axisymmetric flow and heat transfer of a viscous fluid has been considered by Sajid et al. [65]. The analytic solution for the axisymmetric flow of a viscous fluid over a nonliniarly stretching sheet has been considered in [66].

Sahoo in [67] investigated the effects of partial slip on the MHD flow and heat transfer of an electrically conducting second grade fluid past an axisymmetric stretching sheet. Miklavčič and Wang [68] first studied properties of the flow to a shrinking sheet with suction. Suction occurs when the fluid condenses on the surface. Considerable attention has been given in different field including magnetic flows [69], different thermal and flow boundary conditions [70], nanofluids and non-Newtonian fluids [71], the slip flow to a stretching cylinder [72], or the axisymmetric stagnation-point flow of a viscous fluid over a lubricated surface [73] and so on.

4.4.1 Equation of Motion

Consider the three dimensional flow of a viscous fluid bounded by a stretching surface. If (u, v, w) is the velocity components in the Cartesian directions (x, y, z) respectively, then the continuity and steady constant property Navier-Stokes equations for viscous fluid flow are [60, 63, 66]:

$$u_x + v_y + w_z = 0 \tag{4.213}$$

$$uu_x + vu_y + wu_z = -\frac{p_x}{\rho} + \nu\nabla^2 u \tag{4.214}$$

$$uv_x + vv_y + wv_z = -\frac{p_y}{\rho} + \nu\nabla^2 v \tag{4.215}$$

$$uw_x + vw_y + ww_z = -\frac{p_z}{\rho} + \nu\nabla^2 w \tag{4.216}$$

where ρ is the density, p is the pressure and ν is the kinematic viscosity. If a is the stretching constant, $a > 0$, W is the suction velocity and m is a parameter describing the type of stretching, then the velocity components on the stretching surface are:

$$u = ax, v = a(m-1)y, w = -W \tag{4.217}$$

For $m = 1$, we have planar stretching case, while for $m = 2$ we have axisymmetric stretching case. In order to simplify the governing equations we use the similarity transform:

$$u = ax\Phi'(\eta), v = a(m-1)y\Phi'(\eta), w = -m\sqrt{av}\Phi(\eta) \tag{4.218}$$

where $\eta = z\sqrt{\frac{a}{v}}$ and prime denotes the derivation with respect to η. Equation 4.213 is automatically satisfied and Eq. 4.216 can be written in the form

$$p = v\rho w_z - \frac{1}{2}\rho W^2 + C \tag{4.219}$$

where C is a constant. Equations 4.214 and 4.215 reduce to the ordinary differential equation

$$\Phi'''(\eta) + m\Phi(\eta)\Phi''(\eta) - \Phi'(\eta)^2 = 0 \tag{4.220}$$

as there is no lateral pressure gradient at infinity. If N denotes a slip constant, then on the surface of the stretching sheet, the velocity slip is assumed to be proportional to the local sheer stress

$$u - ax = \rho vNu_z < 0, v - a(m-1)y = \rho vNv_z < 0 \tag{4.221}$$

From the similarity transform 4.218 we obtain

$$\Phi'(0) = 1 + \lambda\Phi''(0) \tag{4.222}$$

with $\lambda = \rho N\sqrt{av} > 0$ a non-dimensional parameter indicating the relative importance of partial slip. If $\lambda = 0$ there is no slip. Given a suction velocity of $-W$ on the stretching surface, we have the boundary condition

$$\Phi(0) = \alpha \tag{4.223}$$

where $\alpha = \frac{W}{m\sqrt{av}}$ is a non-dimensional constant which determines the transpiration rate at the surface and $\alpha < 0$ if injection from the surface occurs, $\alpha > 0$ for suction and $\alpha = 0$ for an impermeable sheet. Also, since there is no lateral velocity at infinity, we have yet the condition

$$\Phi'(\infty) = 0 \tag{4.224}$$

4.4.2 Application of OHAM to Viscous Fluid Given by Eq. 4.220

We introduce the basic ideas of the OHAM by considering Eq. 4.220 with the boundary conditions given by Eqs. 4.222, 4.223 and 4.224. We choose the linear operator L by the form

$$L(\Phi(\eta)) = \Phi''' + K\Phi''. \tag{4.225}$$

Another possibility to choose the linear operator is $L(\Phi(\eta)) = \Phi''' + K^2\Phi'$, and so on, where $K > 0$ is an unknown parameter.

Equation 4.4 becomes:

$$\Phi_0'''(\eta) + K\Phi_0''(\eta) = 0, \Phi_0(0) = \alpha, \Phi_0'(0) = 1 + \lambda\Phi_0''(0), \Phi_0'(\infty) = 0 \tag{4.226}$$

which has the following solution

$$\Phi_0(\eta) = \alpha + \frac{1 - e^{-K\eta}}{K(1 + \lambda K)} \tag{4.227}$$

The nonlinear operator $N(\Phi(\eta))$ is given from Eq. 4.220

$$N(\Phi(\eta)) = -k\Phi''(\eta) + m\Phi(\eta)\Phi''(\eta) - \Phi'(\eta)^2 \tag{4.228}$$

such that substituting Eq. 4.227 into 4.228, we obtain

$$N(\Phi_0(\eta)) = \frac{\beta e^{-K\eta} + (m-1)e^{-2K\eta}}{(1 + \lambda K)^2} \tag{4.229}$$

where

$$\beta = K(K - m\alpha)(1 + \lambda K) - m. \tag{4.230}$$

Having in view Eq. 4.9, the optimal auxiliary function $H(\eta, C_i)$ must follow the terms appearing in Eq. 4.229, such that we can choose the function $H(\eta, C_i)$ in the following forms:

$$H(\eta, C_i) = \frac{1}{(1+\lambda K)^2}\left[C_1 + C_2\eta + (C_3 + C_4\eta)e^{-K\eta} + (C_5 + C_6\eta)e^{-2K\eta}\right] \tag{4.231}$$

or

$$H^*(\eta, C_i) = \frac{1}{(1+\lambda K)^2}\left(C_1 + C_2\eta + C_3\eta^2 + C_4 e^{-K\eta} + C_5 e^{-2K\eta}\right) \tag{4.232}$$

or yet

$$\begin{aligned} H^{**}(\eta, C_i) = &\frac{1}{(1+\lambda K)^2}\left[C_1 + (C_2 + C_3\eta)e^{-K\eta} + \right.\\ &\left. + (C_4 + C_5\eta)e^{-2K\eta} + C_6 e^{-3K\eta}\right] \end{aligned} \tag{4.233}$$

and so on, where C_1, C_2, ... are unknown parameters at this moment.

If we choose only the expression 4.231 for $H(\eta, C_i)$, then by using Eqs. 4.229, 4.231 and 4.6 we obtain the equation in $\Phi_1(\eta, C_i)$

$$\begin{aligned} &\Phi_1''' + K\Phi_1'' = (\beta C_2\eta + \beta C_1)e^{-K\eta} + \\ &+ [(mC_2 - C_2 + \beta C_4)\eta + (m-1)C_1 + \beta C_3]e^{-2K\eta} + \\ &+ [(mC_4 - C_4 + \beta C_6)\eta + (m-1)C_3 + \beta C_5]e^{-3K\eta} + \\ &+ [(m-1)C_6\eta + (m-1)C_5]e^{-4K\eta} \\ &\Phi_1(0) = 0, \Phi_1'(0) = \lambda\Phi_1''(0), \Phi_1'(\infty) = 0 \end{aligned} \tag{4.234}$$

It is interesting to remark that if we consider $m = 1$, and $\beta = 0$ (the planar stretching case for impermeable sheet) into Eq. 4.234 then we have

$$\Phi_1''' + K\Phi_1'' = 0, \Phi_1(0) = 0, \Phi_1'(0) = \lambda\Phi_1''(0), \Phi_1'(\infty) = 0 \tag{4.235}$$

such that $\Phi_1(\eta) = 0$, and therefore we can obtain the exact solution of the equation

$$\Phi''' + \Phi\Phi'' - \Phi'^2 = 0, \Phi(0) = 0, \Phi'(0) = 1 + \lambda\Phi''(0), \Phi'(\infty) = 0 \tag{4.236}$$

in the form given by Eq. 4.225:

$$\Phi(\eta) = \frac{1 - e^{-K\eta}}{K(1 + \lambda K)} \tag{4.237}$$

where K is obtained from the condition $\beta = 0$ into Eq. 4.230

$$\lambda K^3 + (1 - \alpha\lambda)K^2 - \alpha K - 1 = 0. \tag{4.238}$$

Beyond of this remarkable case, solving Eq. 4.234 and then substituting this result and Eq. 4.227 into Eq. 4.5, we obtain the approximate solution of the first-order in the form:

$$\begin{aligned}
\overline{\Phi}(\eta, C_i) &= A + \alpha + \frac{1}{K(1+\lambda K)} + [\frac{\beta C_2}{2K^2}\eta^2 + \left(\frac{\beta C_1}{K^2} + \frac{2\beta C_2}{K^3}\right)\eta + \\
&+ B - \frac{1}{K(1+\lambda K)}]e^{-K\eta} + [\frac{(1-m)C_2 - \beta C_4}{4K^3}\eta + \\
&+ \frac{1-m}{4K^3}C_1 + \frac{1-m}{2K^4}C_2 - \frac{\beta}{4K^3}C_3 - \frac{\beta}{2K^4}C_4]e^{-2K\eta} + \\
&+ [\frac{(1-m)C_4 - \beta C_6}{18K^3}\eta + \frac{1-m}{18K^3}C_3 + \frac{7(1-m)}{108K^4}C_4 - \frac{\beta}{18K^3}C_5 - \\
&- \frac{7\beta}{108K^4}C_6]e^{-3K\eta} + \left[\frac{1-m}{48K^3}C_6\eta + \frac{1-m}{48K^3}C_5 + \frac{5(1-m)}{288K^4}C_6\right]e^{-4K\eta}
\end{aligned} \tag{4.239}$$

where

$$\begin{aligned}
A &= \frac{1 - m - 4\beta + \lambda K(3 - 3m - 8\beta)}{4K^3(1+\lambda K)}C_1 + \frac{1 - m - 8\beta + \lambda K(2 - 2m - 12\beta)}{4K^4(1+\lambda K)}C_2 + \\
&+ \frac{4 - 4m + \lambda K(16 - 16m - 27\beta)}{36K^3(1+\lambda K)}C_3 + \frac{8 - 8m - 27\beta + \lambda K(20 - 20m - 54\beta)}{108K^4(1+\lambda K)}C_4 + \\
&+ \frac{9 - 9m - 16\beta + \lambda K(45 - 45m - 64\beta)}{144K^3(1+\lambda K)}C_5 + \\
&+ \frac{27 - 27m - 64\beta + \lambda K(81 - 81m - 160\beta)}{864K^4(1+\lambda K)}C_6 \\
B &= \frac{(2\beta + m - 1)(1 + 2\lambda K)}{2K^3(1+\lambda K)}C_1 + \frac{8\beta + 3m - 3 + 4\lambda K(m - 1 + 3\beta)}{9K^4(1+\lambda K)}C_2 + \\
&+ \frac{3\beta + m - 1 + 3\lambda K(m + 2\beta - 1)}{6K^3(1+\lambda K)}C_3 + \frac{27\beta + 5m - 5 + 9\lambda K(m - 1 + 4\beta)}{36K^4(1+\lambda K)}C_4 + \\
&+ \frac{2\beta + m - 1 + \lambda K(6\beta + 4m - 4)}{12K^3(1+\lambda K)}C_5 + \\
&+ \frac{20\beta + 7m - 7 + \lambda K(36\beta + 16m - 16)}{144K^4(1+\lambda K)}C_6
\end{aligned} \tag{4.240}$$

4.4.3 Numerical Examples

We illustrate the accuracy of OHAM by comparing obtained approximate solutions with the numerical integration results obtained by means of a fourth-order Runge-Kutta method in combination with the shooting method. In all cases, the unknown parameters are optimally identified via Galerkin method. For this, we use the following nine weighted functions f_i given by

$$\begin{aligned}
&f_1(\eta) = \gamma e^{-3K\eta} + \eta^2 e^{-2K\eta} + \delta\eta e^{-K\eta}, f_2(\eta) = \eta e^{-K\eta},\\
&f_3(\eta) = \eta^2 e^{-K\eta}, f_4(\eta) = 1 + \gamma\eta e^{-K\eta},\\
&f_5(\eta) = e^{-2K\eta}, f_6(\eta) = e^{-K\eta}, f_7(\eta) = \eta e^{-4K\eta},\\
&f_8(\eta) = e^{-4K\eta}, f_9(\eta) = \delta\eta e^{-2K\eta} + \eta^3 e^{-2K\eta},
\end{aligned} \tag{4.241}$$

where γ and δ are unknown parameters.

The optimal convergence-control parameters $K, \gamma, \delta, C_1, \ldots, C_6$ are determined from equations:

$$J_j = \int_0^\infty R(\eta, C_i) f_j(\eta) d\eta = 0, i = 1, \ldots, 6, j = 1, \ldots, 9 \tag{4.242}$$

where the residual $R(\eta, C_i)$ is given by Eq. 2.35

$$R(\eta, C_i) = \overline{\Phi}'''(\eta) + m\overline{\Phi}(\eta)\overline{\Phi}''(\eta) - \overline{\Phi}'(\eta)^2. \tag{4.243}$$

and $\overline{\Phi}(\eta)$ is given by Eq. 4.239.

Example 4.4.3.1 Consider planar stretching case with impermeable sheet, $m = 1$, $\alpha = 0$ and $\lambda = 1$. In this case, from Eqs. 4.242 and 4.243 we obtain

$$\begin{aligned}
&K = 0.4631238249, \gamma = 0.6858214854, \delta = -0.1052314539,\\
&C_1 = -0.0119658357, C_2 = 0.0007396024, C_3 = -0.0936132872,\\
&C_4 = -0.0475837579, C_5 = -0.0772995955, C_6 = -0.0149226174.
\end{aligned}$$

The first-order approximate solution given by Eq. 4.239 can be written in the form

$$\begin{aligned}
&\overline{\Phi}(\eta) = 0.7549045180 + \big(- 0.0011830813\eta^2 + 0.0280632559\eta -\\
&-0.1942637382\big) e^{-0.4631238249\eta} - \big(0.0821765397\eta +\\
&+0.5165482911\big) e^{-0.9262476499\eta} - 0.0440924886 e^{-1.3893714748\eta}.
\end{aligned} \tag{4.244}$$

It is easy to verify the accuracy of the obtained solution if we compare analytical approximate solution 4.244 with the numerical ones. It can be seen from Tables 4.70 and 4.71 that the first-order approximate solution $\overline{\Phi}$ given by Eq. 4.244 and $\overline{\Phi}'(\eta)$ obtained from Eq. 4.244 respectively are very accurate.

Table 4.70 The results of the first-order approximate solution $\overline{\Phi}$ given by Eq. 4.244 and numerical solution for $m = 1$, $\alpha = 0$, $\lambda = 1$

η	$\Phi_{numerical}$	$\overline{\Phi}_{OHAM}$ *given by Eq.* 4.244	*relative error* $\varepsilon = \lvert\Phi_{numerical} - \overline{\Phi}_{OHAM}\rvert$
0	-1.4690727610^{-21}	$\cdot 3.33066907 \times 10^{-16}$	3.330×10^{-16}
4/5	0.3422054517	0.3422055158	6.412×10^{-8}
8/5	0.5292803844	0.5292802974	8.696×10^{-8}
12/5	0.6315494379	0.6315484434	9.945×10^{-7}
16/5	0.6874573050	0.6874558597	1.445×10^{-6}
4	0.7180206952	0.7180203943	3.008×10^{-7}
24/5	0.7347289032	0.7347306790	1.775×10^{-6}
28/5	0.7438628555	0.7438659956	3.140×10^{-6}
32/5	0.7488561521	0.7488587475	2.595×10^{-6}
36/5	0.7515858591	0.7515858161	4.305×10^{-8}
8	0.7530781197	0.7530740167	4.102×10^{-6}

Table 4.71 The results of $\overline{\Phi}'(\eta)$ obtained from Eq. 4.244 and numerical solution for $m = 1$, $\alpha = 0$, $\lambda = 1$

η	$\Phi'_{numerical}$	$\overline{\Phi}'_{OHAM}$ *from Eq.* 4.244	*relative error* $\varepsilon = \lvert\Phi'_{numerical} - \overline{\Phi}'_{OHAM}\rvert$
0	0.5698402909	0.5698404419	1.509×10^{-7}
4/5	0.3115170361	0.3115171244	8.831×10^{-8}
8/5	0.1702983601	0.1702976402	7.19×10^{-7}
12/5	0.0930977404	0.0930965030	1.237×10^{-6}
16/5	0.0508941277	0.0508945198	3.920×10^{-7}
4	0.0278225054	0.0278248245	2.319×10^{-6}
24/5	0.0152098506	0.0152123356	2.484×10^{-6}
28/5	0.0083148326	0.0083154964	6.638×10^{-7}
32/5	0.0045455059	0.0045434479	2.057×10^{-6}
36/5	0.0024849112	0.0024805243	4.386×10^{-6}
8	0.0013584343	0.0013529018	5.532×10^{-6}

Example 4.4.3.2 In the second case for the same planar stretching case with impermeable sheet $m = 1$, $\alpha = 0$ but $\lambda = 5$, from Eqs. 4.242 the values of the parameters are

$$K = 0.5252370049, \gamma = 4.8796805607, \delta = -1.2890550305,$$
$$C_1 = -0.0279185716, C_2 = -1.9918945423 \times 10^{-6},$$
$$C_3 = 1.6897196188 \times 10^{-8}, C_4 = -1.0550664807 \times 10^{-8},$$
$$C_5 = -2.601365111 \times 10^{-8}, C_6 = -6.4375057414 \times 10^{-9}.$$

Table 4.72 The results of the first-order approximate solution $\overline{\Phi}$ given by Eq. 4.245 and numerical solution for $m = 1$, $\alpha = 0$, $\lambda = 5$

η	$\Phi_{numerical}$	$\overline{\Phi}_{OHAM}$ *given by Eq.* 4.245	*relative error* $\varepsilon = \lvert \Phi_{numerical} - \overline{\Phi}_{OHAM} \rvert$
0	$-3.69915169 \times 10^{-21}$	$1.11022302 \times 10^{-16}$	1.110×10^{-16}
4/5	0.1801529748	0.1801529944	1.956×10^{-8}
8/5	0.2985062397	0.2985062551	1.531×10^{-8}
12/5	0.3762595705	0.3762595800	9.478×10^{-9}
16/5	0.4273403785	0.4273403825	3.999×10^{-9}
4	0.4608984141	0.4608984122	1.941×10^{-9}
24/5	0.4829446930	0.4829446859	7.085×10^{-9}
28/5	0.4974282071	0.4974281997	7.399×10^{-9}
32/5	0.5069432900	0.5069432849	5.118×10^{-9}
36/5	0.5131943125	0.5131943126	1.071×10^{-10}
8	0.5173009861	0.5173009860	1.022×10^{-10}

Table 4.73 The results of $\overline{\Phi}'(\eta)$ obtained from Eq. 4.245 and numerical solution for $m = 1$, $\alpha = 0$, $\lambda = 5$

η	$\Phi'_{numerical}$	$\overline{\Phi}'_{OHAM}$ *from Eq.* 4.245	*relative error* $\varepsilon = \lvert \Phi'_{numerical} - \overline{\Phi}'_{OHAM} \rvert$
0	0.2757990287	0.2757990287	2.220×10^{-16}
4/5	0.1811889384	0.1811888525	8.594×10^{-8}
8/5	0.1190337497	0.1190337776	2.788×10^{-8}
12/5	0.0782004298	0.0782003971	3.268×10^{-8}
16/5	0.0513745131	0.0513745109	2.133×10^{-9}
4	0.0337509879	0.0337509843	3.600×10^{-9}
24/5	0.0221730406	0.0221730371	3.513×10^{-9}
28/5	0.0145667928	0.0145667922	5.860×10^{-10}
32/5	0.0095697621	0.0095697957	3.360×10^{-8}
36/5	0.0062869728	0.0062869703	2.482×10^{-9}
8	0.0041302880	0.0041302862	1.824×10^{-9}

The approximate solution 4.239 becomes:

$$\begin{aligned}\overline{\Phi}(\eta) &= 0.5251657155 + \big(-1.3352013705 \times 10^{-9}\eta^2 - \\ &-0.0000374387\eta - 0.5251657155\big)e^{-0.5252370049\eta} + \\ &+\big(6.7324779815 \times 10^{-12}\eta + 1.4853703052 \times 10^{-11}\big)e^{-1.0504740099\eta} + \\ &+\big(9.1285170601 \times 10^{-13}\eta + 5.7164338355 \times 10^{-12}\big)e^{-1.5757110148\eta}\end{aligned} \tag{4.245}$$

The analytical approximate solution $\overline{\Phi}$ given by Eq. 4.245 and $\overline{\Phi}'(\eta)$ obtained from Eq. 4.245 are compared in Tables 4.72 and 4.73 respectively with the numerical results.

Table 4.74 The results of the first-order approximate solution $\overline{\Phi}$ given by Eq. 4.246 and numerical solution for $m = 1$, $\alpha = 0$, $\lambda = 10$

η	$\Phi_{numerical}$	$\overline{\Phi}_{OHAM}$ given by Eq. 4.246	relative error $\varepsilon = \left\|\Phi_{numerical} - \overline{\Phi}_{OHAM}\right\|$
0	$1.50215999 \times 10^{-21}$	0	1.502×10^{-21}
4/5	0.1268262363	0.1268262411	4.720×10^{-9}
8/5	0.2165139331	0.2165139494	1.628×10^{-8}
12/5	0.2799383807	0.2799384028	2.205×10^{-8}
16/5	0.3247902719	0.3247902712	7.518×10^{-10}
4	0.3565081685	0.3565081653	3.148×10^{-9}
24/5	0.3789381153	0.3789381074	7.870×10^{-9}
28/5	0.3947998981	0.3947998887	9.438×10^{-9}
32/5	0.4060168733	0.4060168640	9.342×10^{-9}
36/5	0.4139491814	0.4139491720	9.425×10^{-9}
8	0.4195586644	0.4195586620	2.446×10^{-9}

Example 4.4.3.3 In the third case for the planar stretching case with impermeable sheet $m = 1$, $\alpha = 0$ and with sleep parameter $\lambda = 10$, from Eqs. 4.242 we obtain

$$\begin{aligned}&K = 0.4331821853, \gamma = 3.0639340595, \delta = -0.5722383481,\\&C_1 = -0.0125089730, C_2 = -9.6275057188 \times 10^{-7},\\&C_3 = 1.3219663971 \times 10^{-8}, C_4 = -6.7307907487 \times 10^{-9},\\&C_5 = -2.0218988984 \times 10^{-8}, C_6 = -4.1394827886 \times 10^{-9}.\end{aligned}$$

The first-order approximate solution 4.239 is

$$\begin{aligned}&\overline{\Phi}(\eta) = 0.4331053011 + \left(-1.2809679965 \times 10^{-9}\eta^2 -\right.\\&\left.-0.0000332989\eta - 0.4331053011\right)e^{-0.433182185\eta} +\\&+\left(1.0336892501 \times 10^{-11}\eta + 2.7423124117 \times 10^{-11}\right)e^{-0.8663643706\eta} +\\&+\left(1.4127245039 \times 10^{-12}\eta + 1.070516162 \times 10^{-11}\right)e^{-1.2995465559\eta}\end{aligned} \tag{4.246}$$

In Tables 4.74 and 4.75 we present some values of $\overline{\Phi}$ given by Eq. 4.246 and $\overline{\Phi}'(\eta)$ obtained from Eq. 4.246, respectively in comparison with the numerical results.

The values of $\Phi''(0)$ and $\Phi(\infty)$ are given in Table 4.76 for the same planar stretching case with impermeable sheet, calculated by means OHAM, and by numerical integration.

From Tables 4.70, 4.71, 4.72, 4.73, 4.74, 4.75, 4.76, it can be seen that the first-order approximate solution $\overline{\Phi}$ and $\overline{\Phi}'$ given by OHAM in comparison with the numerical results are very accurate.

Example 4.4.3.4 In this case, we consider planar stretching case but suction sheet, $m = 1$, $\alpha = 3$ and with slip parameter $\lambda = 1$. The parameters obtained by means of Eqs. 4.242 are:

Table 4.75 The results of $\overline{\Phi}'(\eta)$ obtained from Eq. 4.246 and numerical solution for $m = 1$, $\alpha = 0$, $\lambda = 10$

η	$\Phi'_{numerical}$	$\overline{\Phi}'_{OHAM}$ *from Eq.* 4.246	*relative error* $\varepsilon = \lvert\Phi'_{numerical} - \overline{\Phi}'_{OHAM}\rvert$
0	0.1875802018	0.1875802018	1.110×10^{-16}
4/5	0.1326509468	0.1326510845	1.377×10^{-7}
8/5	0.0938068684	0.0938068626	5.887×10^{-9}
12/5	0.0663374780	0.0663373956	8.247×10^{-8}
16/5	0.0469118131	0.0469118136	5.295×10^{-10}
4	0.0331746279	0.0331746255	2.410×10^{-9}
24/5	0.0234601074	0.0234600987	8.697×10^{-9}
28/5	0.0165902831	0.0165902771	5.954×10^{-9}
32/5	0.0117321466	0.0117321456	9.209×10^{-10}
36/5	0.0082966254	0.0082966210	4.450×10^{-9}
8	0.0058671110	0.0058671212	1.015×10^{-8}

Table 4.76 Values of $\Phi''(0)$ and $\Phi(\infty)$ in the case of planar stretching case with impermeable sheet $m = 1$, $\alpha = 0$

λ	1	5	10
$\Phi''_{numeric}(0)$	−0.4301597090	−0.1448401942	−0.0812419798
$\Phi''_{OHAM}(0)$	−0.4301595580	−0.1448401942	−0.0812419798
$\Phi_{numeric}(\infty)$	0.7548776662	0.5251657154	0.4331053011
$\Phi_{OHAM}(\infty)$	0.7549045180	0.5251657155	0.4331053011

$$K = 1.0265261151, \gamma = 3.011278740, \delta = -0.9721507438,$$
$$C_1 = 2.7091920761 \times 10^{-11}, C_2 = -4.3645109956 \times 10^{-12},$$
$$C_3 = 9.1196480013 \times 10^{-10}, C_4 = -3.3227889086 \times 10^{-10},$$
$$C_5 = 1.3649192039, C_6 = -0.0000235838$$

and the first-order approximate solution 4.239 becomes

$$\begin{aligned}\overline{\Phi}(\eta) &= 3.0795956234 + \big(-2.3514669361 \times 10^{-12}\eta^2 + \\ &+ 2.0029806302 \times 10^{-11}\eta - 4.9211079655 \times 10^{-11}\big)e^{-1.0265261151\eta} + \\ &+ \big(8.7197902556 \times 10^{-11}\eta - 6.9432003042 \times 10^{-11}\big)e^{-2.0530522302\eta} + \\ &+ \big(1.3753249143 \times 10^{-6}\eta - 0.0795956233\big)e^{-3.0795783454\eta}\end{aligned} \tag{4.247}$$

In Tables 4.77 and 4.78 are given the values of $\overline{\Phi}(\eta)$ and $\overline{\Phi}'(\eta)$ respectively, from Eq. 4.247 in comparison with the numerical results.

Example 4.4.3.5 For planar stretching case and suction sheet, $m = 1$, $\alpha = 3$ and slip parameter $\lambda = 5$, we have

Table 4.77 The results of the first-order approximate solution $\overline{\Phi}$ given by Eq. 4.247 and numerical solution for $m = 1$, $\alpha = 3$, $\lambda = 1$

η	$\Phi_{numerical}$	$\overline{\Phi}_{OHAM}$ givenby Eq. 4.247	relative error $\varepsilon = \lvert\Phi_{numerical} - \overline{\Phi}_{OHAM}\rvert$
0	3	3	0
4/5	3.0728203311	3.0728203305	6.548×10^{-10}
8/5	3.0790189008	3.0790189008	7.097×10^{-11}
12/5	3.0795465321	3.0795465320	6.988×10^{-11}
16/5	3.0795914447	3.0795914447	1.960×10^{-11}
4	3.0795952677	3.0795952677	6.594×10^{-13}
24/5	3.0795955932	3.0795955932	7.611×10^{-12}
28/5	3.0795956209	3.0795956209	1.132×10^{-11}
32/5	3.0795956232	3.0795956232	8.887×10^{-12}
36/5	3.0795956234	3.0795956234	9.428×10^{-13}
8	3.0795956235	3.0795956234	1.596×10^{-11}

Table 4.78 The results of $\overline{\Phi}'(\eta)$ obtained from Eq. 4.247 and numerical solution for $m = 1$, $\alpha = 3$, $\lambda = 1$

η	$\Phi'_{numerical}$	$\overline{\Phi}'_{OHAM}$ fromEq.4.247	relative error $\varepsilon = \lvert\Phi'_{numerical} - \overline{\Phi}'_{OHAM}\rvert$
0	0.2451223337	0.2451223337	3.330×10^{-16}
4/5	0.0208651636	0.0208651626	1.062×10^{-9}
8/5	0.0017760720	0.0017760724	3.183×10^{-10}
12/5	0.0001511816	0.0001511818	2.060×10^{-10}
16/5	0.0000128678	0.0000128688	9.511×10^{-10}
4	$1.0954211531 \times 10^{-6}$	$1.0954116047 \times 10^{-6}$	9.548×10^{-12}
24/5	$9.3042799740 \times 10^{-8}$	$9.3243000146 \times 10^{-8}$	2.002×10^{-10}
28/5	$8.3530523334 \times 10^{-9}$	$7.9369842720 \times 10^{-9}$	4.160×10^{-10}
32/5	$6.7554276598 \times 10^{-10}$	$6.7561582843 \times 10^{-10}$	7.306×10^{-14}
36/5	$1.0911464277 \times 10^{-10}$	$5.7516709294 \times 10^{-11}$	5.159×10^{-11}
8	$-3.2255392572 \times 10^{-11}$	$4.9013452922 \times 10^{-12}$	3.715×10^{-11}

$$\begin{aligned}
&K = 1.0016112565, \gamma = 2.9079023283, \delta = -0.9038557714,\\
&C_1 = 1.3658235079 \times 10^{-6}, C_2 = -2.1653508962 \times 10^{-7},\\
&C_3 = 0.0000445395, C_4 = -0.0000160260,\\
&C_5 = 0.0752835577, C_6 = -0.0012155499
\end{aligned}$$

and therefore, the first-order approximate solution 4.239 has the form

$$\begin{aligned}
&\overline{\Phi}(\eta) = 3.0205594824 + \big(-5.4255801019 \times 10^{-7}\eta^2 +\\
&+4.6777709040 \times 10^{-6}\eta - 0.0000118225\big)e^{-1.0016112565\eta} +\\
&+(0.0000200454\eta - 0.0000156837)e^{-2.0032225131\eta} +\\
&+(0.0003378695\eta - 0.0205319761)e^{-3.0048337696\eta}
\end{aligned} \tag{4.248}$$

Table 4.79 The results of the first-order approximate solution $\overline{\Phi}$ given by Eq. 4.248 and numerical solution for $m = 1$, $\alpha = 3$, $\lambda = 5$

η	$\Phi_{numerical}$	$\overline{\Phi}_{OHAM}$ *given by Eq.* 4.248	*relative error* $\varepsilon = \lvert\Phi_{numerical} - \overline{\Phi}_{OHAM}\rvert$
0	3	3	8.881×10^{-16}
4/5	3.0187247563	3.0187247676	1.124×10^{-8}
8/5	3.0203957172	3.0203957372	2.004×10^{-8}
12/5	3.0205448301	3.0205448573	2.721×10^{-8}
16/5	3.0205581365	3.0205581673	3.072×10^{-8}
4	3.0205593241	3.0205593556	3.151×10^{-8}
24/5	3.0205594299	3.0205594622	3.231×10^{-8}
28/5	3.0205594395	3.0205594731	3.365×10^{-8}
32/5	3.0205594403	3.0205594759	3.561×10^{-8}
36/5	3.0205594404	3.0205594779	3.747×10^{-8}
8	3.0205594404	3.0205594794	3.905×10^{-8}

Table 4.80 The results of $\overline{\Phi}'(\eta)$ obtained from Eq. 4.248 and numerical solution for $m = 1$, $\alpha = 3$, $\lambda = 5$

η	$\Phi'_{numerical}$	$\overline{\Phi}'_{OHAM}$ *from Eq.* 4.248	*relative error* $\varepsilon = \lvert\Phi'_{numerical} - \overline{\Phi}'_{OHAM}\rvert$
0	0.0621010118	0.0621010277	1.587×10^{-8}
4/5	0.0055417735	0.0055417870	1.357×10^{-8}
8/5	0.0004945360	0.0004945468	1.088×10^{-8}
12/5	0.0000441313	0.0000441381	6.752×10^{-9}
16/5	$3.93832266 \times 10^{-6}$	$3.94034066 \times 10^{-6}$	2.017×10^{-9}
4	$3.51850149 \times 10^{-7}$	$3.52031130 \times 10^{-7}$	1.809×10^{-10}
24/5	$3.12529413 \times 10^{-8}$	$3.26669626 \times 10^{-8}$	1.414×10^{-9}
28/5	$2.70446249 \times 10^{-9}$	$5.01427659 \times 10^{-9}$	2.309×10^{-9}
32/5	$2.58845911 \times 10^{-10}$	$2.72961298 \times 10^{-9}$	2.470×10^{-9}
36/5	$1.08218302 \times 10^{-10}$	$2.21264141 \times 10^{-9}$	2.104×10^{-9}
8	$-3.95822832 \times 10^{-11}$	$1.67312148 \times 10^{-9}$	1.712×10^{-9}

In Tables 4.79 and 4.80 we present some values of $\overline{\Phi}(\eta)$ given by Eq. 4.248 and $\overline{\Phi}'(\eta)$ obtained from Eq. 4.248, respectively in comparison with the numerical results.

Example 4.4.3.6 The planar stretching case and suction sheet, $m = 1$, $\alpha = 3$ but slip parameter $\lambda = 10$ we obtain from Eqs. 4.242:

$$K = 1.1395297367, \gamma = 2.9650196892, \delta = -0.9588570845,$$
$$C_1 = 0.0000646091, C_2 = -0.0000113762, C_3 = 0.0031645975,$$
$$C_4 = -0.0010792973, C_5 = 0.0087672199, C_6 = 0.0022371918$$

Table 4.81 The results of the first-order approximate solution $\overline{\Phi}$ given by Eq. 4.249 and numerical solution for $m = 1$, $\alpha = 3$, $\lambda = 10$

η	$\Phi_{numerical}$	$\overline{\Phi}_{OHAM}$ *given by Eq.* 4.249	*relative error* $\varepsilon = \lvert\Phi_{numerical} - \overline{\Phi}_{OHAM}\rvert$
0	3	2.9999999999	4.440×10^{-16}
4/5	3.0097173475	3.0097177871	4.395×10^{-7}
8/5	3.0105913868	3.0105920909	7.040×10^{-7}
12/5	3.0106700032	3.0106708927	8.895×10^{-7}
16/5	3.0106770744	3.0106779056	8.312×10^{-7}
4	3.0106777105	3.0106784439	7.334×10^{-7}
24/5	3.0106777676	3.0106785350	7.673×10^{-7}
28/5	3.0106777728	3.0106786810	9.082×10^{-7}
32/5	3.0106777732	3.0106788473	1.074×10^{-6}
36/5	3.0106777733	3.0106789863	1.213×10^{-6}
8	3.0106777733	3.0106790838	1.310×10^{-6}

Table 4.82 The results of $\overline{\Phi}'(\eta)$ obtained from Eq. 4.249 and numerical solution for $m = 1$, $\alpha = 3$, $\lambda = 10$

η	$\Phi'_{numerical}$	$\overline{\Phi}'_{OHAM}$ *from Eq.* 4.249	*relative error* $\varepsilon = \lvert\Phi'_{numerical} - \overline{\Phi}'_{OHAM}\rvert$
0	0.0321473348	0.0321479832	6.484×10^{-7}
4/5	0.0028915342	0.0028919142	3.799×10^{-7}
8/5	0.0002600819	0.0002604166	3.347×10^{-7}
12/5	0.0000233938	0.0000234630	6.916×10^{-8}
16/5	$2.10451274 \times 10^{-6}$	$1.94797538 \times 10^{-6}$	1.565×10^{-7}
4	$1.89179753 \times 10^{-7}$	$1.38497197 \times 10^{-7}$	5.068×10^{-8}
24/5	$1.75549222 \times 10^{-8}$	$1.43292069 \times 10^{-7}$	1.257×10^{-7}
28/5	$1.33562917 \times 10^{-9}$	$2.08664068 \times 10^{-7}$	2.073×10^{-7}
32/5	$1.47310477 \times 10^{-10}$	$1.96955313 \times 10^{-7}$	1.968×10^{-7}
36/5	$-6.14876692 \times 10^{-11}$	$1.47972374 \times 10^{-7}$	1.480×10^{-7}
8	$-2.11307071 \times 10^{-11}$	$9.70994588 \times 10^{-8}$	9.712×10^{-8}

and the first-order approximate solution 4.239 can be written as

$$\begin{aligned}\overline{\Phi}(\eta) = 3.0106792212 + (&-0.0000661253\eta^2 + \\ +0.0005189796\eta &- 0.0011722736)e^{-1.1395297367\eta} + \\ +(0.0027526798\eta &- 0.0032398505)e^{-2.2790594735\eta} + \\ +(-0.0012679592\eta &- 0.0062670970)e^{-3.4185892103\eta}\end{aligned} \tag{4.249}$$

In Tables 4.81 and 4.82, the numerical solution for $m = 1$, $\alpha = 3$, $\lambda = 10$ is compared with the first-order approximate solution $\overline{\Phi}(\eta)$ and $\overline{\Phi}'(\eta)$ respectively, obtained from Eq. 4.249.

Table 4.83 Values of $\Phi''(0)$ and $\Phi(\infty)$ in the case of planar stretching case with suction sheet $m = 1, \alpha = 3$

λ	1	5	10
$\Phi''_{numeric}(0)$	−0.7548776662	−0.1875797976	−0.0967852665
$\Phi''_{OHAM}(0)$	−0.7548776662	−0.1875797944	−0.0967852016
$\Phi_{numeric}(\infty)$	3.0795956235	3.0205594824	3.0106792212
$\Phi_{OHAM}(\infty)$	3.0795956235	3.0205594404	3.0106777733

In Table 4.83, we present the values of $\Phi''(0)$ and $\Phi(\infty)$ in the case of planar stretching case with suction sheet.

The results presented in Tables 4.77, 4.78, 4.79, 4.80, 4.81, 4.82, 4.83 indicates a very good agreement between the numerical results and the results obtained by means of OHAM.

Example 4.4.3.7 Now, we consider the axisymmetric flow with impermeable sheet, $m = 2$, $\alpha = 0$ and with slip parameter $\lambda = 1$. From Eqs. 4.242 we obtain the following set of parameters

$$K = 1.2284861855, \gamma = -1.2107679860, \delta = 5.9462560050,$$
$$C_1 = -0.0272717347, C_2 = -0.0098884303, C_3 = -0.0096082725,$$
$$C_4 = -0.1430856756, C_5 = -0.0461779063, C_6 = 0.0013806913.$$

In this case, the first-order approximate solution 4.239 is

$$\begin{aligned}\overline{\Phi}(\eta) &= 0.5509446955 + \big(-0.0077420177\eta^2-\\ &-0.0679124177\eta - 0.6421304231\big)e^{-1.2284861855\eta}+\\ &+(0.0469289624\eta + 0.0831404603)e^{-2.4569723711\eta}+\\ &+(0.0041898146\eta + 0.0075368933)e^{-3.6854585567\eta}+\\ &+(0.0000155147\eta + 0.0005083739)e^{-4.9139447423\eta}\end{aligned} \tag{4.250}$$

In Tables 4.84 and 4.85 we present a comparison between the first-order approximate solution $\overline{\Phi}(\eta)$ given by Eq. 4.250 and $\overline{\Phi}'(\eta)$ obtained from Eq. 4.250, respectively and the numerical solution in the case $m = 2, \alpha = 0, \lambda = 10$.

Example 4.4.3.8 For the case of the axisymmetric flow with impermeable sheet, $m = 2$, $\alpha = 0$ and with slip parameter $\lambda = 5$, from Eqs. 4.242 we have

$$K = 0.8408529662, \gamma = -2.5936764722, \delta = 12.6879512489,$$
$$C_1 = -0.0053548861, C_2 = -0.0012448162, C_3 = -0.0048905544,$$
$$C_4 = -0.0180483421, C_5 = -0.0064961222, C_6 = 0.0010523763$$

Table 4.84 The results of the first-order approximate solution $\overline{\Phi}$ given by Eq. 4.250 and numerical solution for $m = 2$, $\alpha = 0$, $\lambda = 1$

η	$\Phi_{numerical}$	$\overline{\Phi}_{OHAM}$ *given by Eq.* 4.250	*relative error* $\varepsilon = \lvert \Phi_{numerical} - \overline{\Phi}_{OHAM} \rvert$
0	$4.53294975 \times 10^{-21}$	0	4.532×10^{-21}
4/5	0.3059142785	0.3059142583	2.024×10^{-8}
8/5	0.4461452851	0.4461454761	1.910×10^{-7}
12/5	0.5069389666	0.5069388037	1.629×10^{-7}
16/5	0.5326162622	0.5326159023	3.599×10^{-7}
4	0.5433388098	0.5433393496	5.398×10^{-7}
24/5	0.54779490007	0.5477960148	1.114×10^{-6}
28/5	0.5496430435	0.5496436022	5.587×10^{-7}
32/5	0.5504089078	0.5504081313	7.765×10^{-7}
36/5	0.5507261667	0.5507239003	2.266×10^{-6}
8	0.5508575704	0.5508540509	3.519×10^{-6}

Table 4.85 The results of $\overline{\Phi}'(\eta)$ obtained from Eq. 4.250 and numerical solution for $m = 2$, $\alpha = 0$, $\lambda = 1$

η	$\Phi'_{numerical}$	$\overline{\Phi}'_{OHAM}$ *from Eq.* 4.250	*relative error* $\varepsilon = \lvert \Phi'_{numerical} - \overline{\Phi}'_{OHAM} \rvert$
0	0.5374903499	0.5374903555	5.591×10^{-7}
4/5	0.2555506238	0.2555503452	2.785×10^{-7}
8/5	0.1127824708	0.1127828241	3.533×10^{-7}
12/5	0.0480153265	0.0480144420	8.845×10^{-7}
16/5	0.0201187063	0.0201193276	6.213×10^{-7}
4	0.0083728499	0.083740894	1.239×10^{-6}
24/5	0.0034746602	0.0034746801	1.988×10^{-8}
28/5	0.0014402426	0.0014389244	1.318×10^{-6}
32/5	0.0005966811	0.0005948028	1.878×10^{-6}
36/5	0.0002471495	0.0002453854	1.764×10^{-6}
8	0.0001023595	0.0001010119	1.347×10^{-6}

and therefore, the first-order approximate solution 4.238 becomes

$$\begin{aligned}
&\overline{\Phi}(\eta) = 0.3780173693 + \big(- 0.0023588678\eta^2 - \\
&-0.0315158198\eta - 0.4396463038\big)e^{-0.8408529662\eta} + \\
&+(0.0208604015\eta + 0.0573797361)e^{-1.6817059324\eta} + \\
&+(0.001423055078\eta + 0.0040581045)e^{-2.5225588986\eta} + \\
&+(-0.0000368781\eta + 0.0001910937)e^{-3.3634118648\eta}
\end{aligned} \tag{4.251}$$

In Tables 4.86 and 4.87 the numerical solution for $m = 2$, $\alpha = 0$, $\lambda = 5$ is compared with the first-order approximate solution $\overline{\Phi}(\eta)$ and $\overline{\Phi}'(\eta)$ respectively, obtained from Eq. 4.251.

Table 4.86 The results of the first-order approximate solution $\overline{\Phi}$ given by Eq. 4.251 and numerical solution for $m = 2$, $\alpha = 0$, $\lambda = 5$

η	$\Phi_{numerical}$	$\overline{\Phi}_{OHAM}$ *givenby Eq.* 4.251	*relative error* $\varepsilon = \lvert\Phi_{numerical} - \overline{\Phi}_{OHAM}\rvert$
0	$-5.65791538 \times 10^{-21}$	$5.55111512 \times 10^{-17}$	5.551×10^{-17}
4/5	0.1600040676	0.1600041067	3.910×10^{-8}
8/5	0.2550767447	0.2550767290	1.567×10^{-8}
12/5	0.3096381374	0.3096382859	1.485×10^{-7}
16/5	0.3402900384	0.3402900681	2.968×10^{-8}
4	0.3572971242	0.3572968761	2.480×10^{-7}
24/5	0.3666672040	0.3666670201	1.838×10^{-7}
28/5	0.3718094321	0.3718096527	2.206×10^{-7}
32/5	0.3746253326	0.3746259498	6.172×10^{-7}
36/5	0.3761654885	0.3761662159	7.273×10^{-7}
8	0.3770073263	0.3770078077	4.813×10^{-7}

Table 4.87 The results of $\overline{\Phi}'(\eta)$ obtained from Eq. 4.251 and numerical solution for $m = 2$, $\alpha = 0$, $\lambda = 5$

η	$\Phi'_{numerical}$	$\overline{\Phi}'_{OHAM}$ *fromEq.*4.251	*relative error* $\varepsilon = \lvert\Phi'_{numerical} - \overline{\Phi}'_{OHAM}\rvert$
0	0.2530332735	0.2530332797	6.295×10^{-9}
4/5	0.1535171630	0.1535171106	5.239×10^{-8}
8/5	0.0892746569	0.0892747527	9.579×10^{-8}
12/5	0.0505489758	0.0505490965	1.207×10^{-7}
16/5	0.0281733373	0.0281729632	3.740×10^{-7}
4	0.0155613273	0.0155611338	1.934×10^{-7}
24/5	0.0085518839	0.0085522312	3.472×10^{-7}
28/5	0.0046866399	0.0046872209	5.810×10^{-7}
32/5	0.0025644425	0.0025647958	3.533×10^{-7}
36/5	0.0014020278	0.0014019372	9.055×10^{-8}
8	0.0007661537	0.0007656582	4.955×10^{-7}

Example 4.4.3.9 In this case we consider the axisymmetric flow with impermeable sheet, $m = 2$, $\alpha = 0$ and with slip parameter $\lambda = 10$. From Eqs. 4.242 yields

$$K = 0.5239724222, \gamma = 1.6648535124, \delta = -1.8794890289,$$
$$C_1 = 0.0112870692, C_2 = -0.0004699006, C_3 = -0.1276778030,$$
$$C_4 = 0.0206177280, C_5 = 0.1080628083, C_6 = 0.02844119967$$

Table 4.88 The results of the first-order approximate solution $\overline{\Phi}$ given by Eq. 4.252 and numerical solution for $m = 2$, $\alpha = 0$, $\lambda = 10$

η	$\Phi_{numerical}$	$\overline{\Phi}_{OHAM}$ *given by Eq.* 4.252	*relative error* $\varepsilon = \lvert\Phi_{numerical} - \overline{\Phi}_{OHAM}\rvert$
0	$-6.46524366 \times 10^{-21}$	$-2.77555756 \times 10^{-16}$	2.775×10^{-16}
4/5	0.1126417938	0.1126404827	1.311×10^{-6}
8/5	0.1864912649	0.1864889581	2.306×10^{-6}
12/5	0.2336569168	0.2336539425	2.974×10^{-6}
16/5	0.2632492851	0.2632470296	2.255×10^{-6}
4	0.2816018688	0.2816009735	8.952×10^{-7}
24/5	0.2929002361	0.2929000387	1.973×10^{-7}
28/5	0.2998238959	0.2998231514	7.444×10^{-7}
32/5	0.3040546698	0.3040522683	2.401×10^{-6}
36/5	0.3066354098	0.3066306698	4.739×10^{-6}
8	0.3082079556	0.3082006277	7.327×10^{-6}

Table 4.89 The results of $\overline{\Phi}'(\eta)$ obtained from Eq. 4.252 and numerical solution for $m = 2$, $\alpha = 0$, $\lambda = 10$

η	$\Phi'_{numerical}$	$\overline{\Phi}'_{OHAM}$ *from Eq.* 4.252	*relative error* $\varepsilon = \lvert\Phi'_{numerical} - \overline{\Phi}'_{OHAM}\rvert$
0	0.1708836323	0.1708818873	1.745×10^{-6}
4/5	0.1137440701	0.1137428294	1.240×10^{-6}
8/5	0.0734177895	0.0734164390	1.350×10^{-6}
12/5	0.0463876465	0.0463876308	1.563×10^{-8}
16/5	0.0288985239	0.0289001166	1.592×10^{-6}
4	0.0178411952	0.0178426995	1.504×10^{-6}
24/5	0.0109523552	0.0109524790	1.237×10^{-7}
28/5	0.0066998042	0.0066983503	1.453×10^{-6}
32/5	0.0040895458	0.0040869503	2.595×10^{-6}
36/5	0.0024929328	0.0024897729	3.159×10^{-6}
8	0.0015184181	0.0015151809	3.237×10^{-6}

The first-order approximate solution can be written in the form

$$\begin{aligned}\overline{\Phi}(\eta) &= 0.3106373548 + \big(-0.0006102510\eta^2 + \\ &+0.0246579539\eta - 0.3170175340\big)e^{-0.5239724222\eta} + \\ &+(-0.0247341994\eta + 0.0442009306)e^{-1.0479448445\eta} + \\ &+(-0.0157948648\eta - 0.0156201575)e^{-1.5719172668\eta} + \\ &+(-0.0041189012\eta - 0.0222005938)e^{-2.0958896689\eta}\end{aligned} \tag{4.252}$$

In Tables 4.88 and 4.89 we present a comparison between the first-order approximate solution $\overline{\Phi}(\eta)$ given by Eq. 4.252 and $\overline{\Phi}'(\eta)$ obtained from Eq. 4.252, respectively and the numerical solutions in the case $m = 2$, $\alpha = 0$, $\lambda = 10$.

Table 4.90 Values of $\Phi''(0)$ and $\Phi(\infty)$ in the case of stretching flow with impermeable sheet $m = 2$, $\alpha = 0$

λ	1	5	10
$\Phi''_{numeric}(0)$	−0.4625096500	−1.1493933452	−0.0829116367
$\Phi''_{OHAM}(0)$	−0.4625096444	−0.1493933440	−0.0829118112
$\Phi_{numeric}(\infty)$	0.5509502913	0.37801840311	0.31065126857
$\Phi_{OHAM}(\infty)$	0.5509446955	0.3780173693	0.31063735488

In Table 4.90 we present the values of $\Phi''(0)$ and $\Phi(\infty)$ for the stretching flow with impermeable sheet.

Example 4.4.3.10 For the axisymmetric flow with suction sheet $m = 2, \alpha = 3$ and slip parameter $\lambda = 1$ we obtain the following values of the parameters:

$$\begin{aligned}
&K = 1.5574274146, \gamma = -0.7860056759, \delta = 3.6690514687,\\
&C_1 = 0.0107133300, C_2 = -0.0024557002, C_3 = 0.4688574884,\\
&C_4 = -0.2507161620, C_5 = 0.1381947406, C_6 = -1.1772137282
\end{aligned}$$

with the first-order approximate solution obtained from 4.238:

$$\begin{aligned}
&\overline{\Phi}(\eta) = 3.0235364880 + \big(0.0026339287\eta^2 +\\
&+0.0162169332\eta - 0.0279371463\big)e^{-1.5574274146\eta} +\\
&+(0.0864947892\eta - 0.0510827083)e^{-3.1148548292\eta} +\\
&+(0.0937682076\eta + 0.0527717186)e^{-4.6722822438\eta} +\\
&+(0.0064921812\eta + 0.0027116480)e^{-6.2297096585\eta}
\end{aligned} \tag{4.253}$$

In Tables 4.91 and 4.92 we present a comparison between the first-order approximate solution $\overline{\Phi}(\eta)$ and $\overline{\Phi}'(\eta)$, respectively obtained from Eq. 4.253, respectively and the numerical solutions in the case $m = 2$, $\alpha = 3$, $\lambda = 1$.

Example 4.4.3.11 For the case of the axisymmetric flow with suction sheet $m = 2$, $\alpha = 3$ and slip parameter $\lambda = 5$, the parameters are

$$\begin{aligned}
&K = 1.7780169491, \gamma = -0.5957717503, \delta = 2.8222930255,\\
&C_1 = 0.0000862400, C_2 = -0.0000222658, C_3 = 0.0046073879,\\
&C_4 = -0.0026158399, C_5 = 0.0214772965, C_6 = -0.0217910850
\end{aligned}$$

and therefore, the first-order approximate solution 4.238 has the form

$$\begin{aligned}
&\overline{\Phi}(\eta) = 3.0053612258 + \big(- 0.0001065840\eta^2 +\\
&+0.0005858601\eta - 0.0008994531\big)e^{-1.7780169491\eta} +\\
&+(0.0035222493\eta - 0.0022439786)e^{-3.55603389829\eta} +\\
&+(0.0065444365\eta - 0.0021760449)e^{-5.3340508474\eta} +\\
&+(0.0000807662\eta - 0.0000417491)e^{-7.1120677965\eta}
\end{aligned} \tag{4.254}$$

Table 4.91 The results of the first-order approximate solution $\overline{\Phi}$ given by Eq. 4.253 and numerical solution for $m = 2$, $\alpha = 3$, $\lambda = 1$

η	$\Phi_{numerical}$	$\overline{\Phi}_{OHAM}$ *given by Eq.* 4.253	*relative error* $\varepsilon = \lvert\Phi_{numerical} - \overline{\Phi}_{OHAM}\rvert$
0	3	3	4.440×10^{-16}
4/5	3.0233419039	3.0233421081	2.041×10^{-7}
8/5	3.0235270847	3.0235272605	1.757×10^{-7}
12/5	3.0235285524	3.0235292386	6.861×10^{-7}
16/5	3.0235285641	3.0235265431	2.020×10^{-6}
4	3.0235285642	3.0235273675	1.196×10^{-6}
24/5	3.0235285642	3.0235304949	1.930×10^{-6}
28/5	3.0235285642	3.0235332843	4.720×10^{-6}
32/5	3.0235285642	3.0235349868	6.422×10^{-6}
36/5	3.0235285642	3.0235358443	7.280×10^{-6}
8	3.0235285642	3.0235362289	7.664×10^{-6}

Table 4.92 The results of $\overline{\Phi}'(\eta)$ obtained from Eq. 4.253 and numerical solution for $m = 2$, $\alpha = 3$, $\lambda = 1$

η	$\Phi'_{numerical}$	$\overline{\Phi}'_{OHAM}$ *from Eq.* 4.253	*relative error* $\varepsilon = \lvert\Phi'_{numerical} - \overline{\Phi}'_{OHAM}\rvert$
0	0.1421402629	0.1421402659	3.066×10^{-9}
4/5	0.0011287363	0.0011270165	1.719×10^{-6}
8/5	$8.94612288 \times 10^{-6}$	0.0000126036	3.657×10^{-6}
12/5	$7.06128678 \times 10^{-8}$	$-3.17690153 \times 10^{-6}$	3.247×10^{-6}
16/5	$4.84453399 \times 10^{-10}$	$-1.64770366 \times 10^{-6}$	1.648×10^{-6}
4	$-4.96135311 \times 10^{-11}$	$3.18485332 \times 10^{-6}$	3.184×10^{-6}
24/5	$-1.52925333 \times 10^{-12}$	$4.03974509 \times 10^{-6}$	4.039×10^{-6}
28/5	$-8.25565615 \times 10^{-13}$	$2.80847387 \times 10^{-6}$	2.808×10^{-6}
32/5	$9.36359087 \times 10^{-14}$	$1.51596351 \times 10^{-6}$	1.515×10^{-6}
36/5	$3.44504044 \times 10^{-14}$	$7.09404172 \times 10^{-7}$	7.094×10^{-7}
8	$-2.56256584 \times 10^{-14}$	$3.02954051 \times 10^{-7}$	3.029×10^{-7}

In Tables 4.93 and 4.94, the numerical solution for $m = 2$, $\alpha = 3$, $\lambda = 5$ is compared with the first-order approximate solution $\overline{\Phi}(\eta)$ and $\overline{\Phi}'(\eta)$ respectively, obtained from Eq. 4.254.

Example 4.4.3.12 In the last case we consider the axisymmetric flow with suction sheet $m = 2$, $\alpha = 3$ and slip parameter $\lambda = 10$, such that the parameters are

$$K = 1.8412120648, \gamma = -0.554362126, \delta = 2.6330245292,$$
$$C_1 = 9.5831454966 \times 10^{-6}, C_2 = -2.5525898638 \times 10^{-6}, C_3 = 0.0005454355,$$
$$C_4 = -0.0003132222, C_5 = 0.0056651734, C_6 = -0.0035106130.$$

Table 4.93 The results of the first-order approximate solution $\overline{\Phi}$ given by Eq. 4.254 and numerical solution for $m = 2$, $\alpha = 3$, $\lambda = 5$

η	$\Phi_{numerical}$	$\overline{\Phi}_{OHAM}$ *given by Eq.* 4.254	*relative error* $\varepsilon = \lvert\Phi_{numerical} - \overline{\Phi}_{OHAM}\rvert$
0	3	2.9999999999	4.440×10^{-16}
4/5	3.0053172442	3.0053172436	6.174×10^{-10}
8/5	3.0053606396	3.0053606637	2.405×10^{-8}
12/5	3.0053609937	3.0053609790	1.457×10^{-8}
16/5	3.0053609960	3.0053609371	5.948×10^{-8}
4	3.0053609966	3.0053610206	2.398×10^{-8}
24/5	3.0053609966	3.0053611190	1.229×10^{-7}
28/5	3.0053609966	3.0053611803	1.836×10^{-7}
32/5	3.0053609966	3.0053612080	2.118×10^{-7}
36/5	3.0053609960	3.0053612197	2.230×10^{-7}
8	3.0053609966	3.0053612238	2.271×10^{-7}

Table 4.94 The results of $\overline{\Phi}'(\eta)$ obtained from Eq. 4.254 and numerical solution for $m = 2$, $\alpha = 3$, $\lambda = 5$

η	$\Phi'_{numerical}$	$\overline{\Phi}'_{OHAM}$ *from Eq.* 4.254	*relative error* $\varepsilon = \lvert\Phi'_{numerical} - \overline{\Phi}'_{OHAM}\rvert$
0	0.0322162764	0.0322162764	3.591×10^{-11}
4/5	0.0002629857	0.0002629038	8.184×10^{-8}
8/5	$2.14586145 \times 10^{-6}$	$2.23702432 \times 10^{-6}$	9.116×10^{-8}
12/5	$1.73166136 \times 10^{-8}$	$-1.12765143 \times 10^{-7}$	1.300×10^{-7}
16/5	$1.51392598 \times 10^{-10}$	$4.22596213 \times 10^{-8}$	4.210×10^{-8}
4	$3.45389945 \times 10^{-11}$	$1.35600169 \times 10^{-7}$	1.355×10^{-7}
24/5	$-6.39472134 \times 10^{-12}$	$1.01940372 \times 10^{-7}$	1.019×10^{-7}
28/5	$-1.77525312 \times 10^{-12}$	$5.20549040 \times 10^{-8}$	5.205×10^{-8}
32/5	$-3.14834890 \times 10^{-13}$	$2.18946309 \times 10^{-8}$	2.189×10^{-8}
36/5	$-2.76742719 \times 10^{-14}$	$8.19686257 \times 10^{-9}$	8.196×10^{-9}
8	$-1.18816854 \times 10^{-16}$	$2.84092336 \times 10^{-9}$	2.840×10^{-9}

The first-order approximate solution 4.239 can be written as

$$\begin{aligned}\overline{\Phi}(\eta) &= 3.0027282856 + \big(-0.0000243991\eta^2 + \\ &+0.0001301957\eta - 0.0001939768\big)e^{-1.8412120648\eta} + \\ &+(0.0008131418\eta - 0.0005329181)e^{-3.6824241297\eta} + \\ &+(0.0020278091\eta - 0.0019877853)e^{-5.5236361946\eta} + \\ &+(0.0000117173\eta - 0.0000136053)e^{-7.3648482595\eta}\end{aligned} \tag{4.255}$$

In Tables 4.95 and 4.96 we present the numerical solution for $m = 2$, $\alpha = 3$, $\lambda = 10$ which is compared with the first-order approximate solution $\overline{\Phi}$ and $\overline{\Phi}'(\eta)$ respectively, obtained from Eq. 4.255.

Table 4.95 The results of the first-order approximate solution $\overline{\Phi}(\eta)$ given by Eq. 4.255 and numerical solution for $m = 2$, $\alpha = 3$, $\lambda = 10$

η	$\Phi_{numerical}$	$\overline{\Phi}_{OHAM}$ given by Eq. 4.255	relative error $\varepsilon = \lvert\Phi_{numerical} - \overline{\Phi}_{OHAM}\rvert$
0	3	3	0
4/5	3.0027058802	3.0027058796	6.211×10^{-10}
8/5	3.0027280544	3.0027280603	5.873×10^{-9}
12/5	3.0027282361	3.0027282309	5.146×10^{-9}
16/5	3.0027282376	3.0027282263	1.122×10^{-8}
4	3.0027282376	3.0027282464	8.833×10^{-9}
24/5	3.0027282376	3.0027282666	2.903×10^{-8}
28/5	3.0027282376	3.0027282779	4.035×10^{-8}
32/5	3.0027282376	3.0027282828	4.525×10^{-8}
36/5	3.0027282376	3.0027282847	4.709×10^{-8}
8	3.0027282376	3.0027282853	4.771×10^{-8}

Table 4.96 The results of $\overline{\Phi}'(\eta)$ obtained from Eq. 4.255 and numerical solution for $m = 2$, $\alpha = 3$, $\lambda = 10$

η	$\Phi'_{numerical}$	$\overline{\Phi}'_{OHAM}$ from Eq. 4.255	relative error $\varepsilon = \lvert\Phi'_{numerical} - \overline{\Phi}'_{OHAM}\rvert$
0	0.0163824516	0.0163824515	5.792×10^{-12}
4/5	0.0001342659	0.0001342487	1.717×10^{-8}
8/5	$1.10030088 \times 10^{-6}$	$1.11623704 \times 10^{-6}$	1.593×10^{-8}
12/5	$9.22925006 \times 10^{-9}$	$-1.78482556 \times 10^{-8}$	2.707×10^{-8}
16/5	$7.67306969 \times 10^{-11}$	$1.42118265 \times 10^{-8}$	1.413×10^{-8}
4	$2.22399659 \times 10^{-11}$	$2.92679094 \times 10^{-8}$	2.924×10^{-8}
24/5	$-4.15280439 \times 10^{-12}$	$1.97148332 \times 10^{-8}$	1.971×10^{-8}
28/5	$-1.20337821 \times 10^{-12}$	$9.31644717 \times 10^{-9}$	9.317×10^{-9}
32/5	$-2.52380813 \times 10^{-13}$	$3.66730924 \times 10^{-9}$	3.667×10^{-9}
36/5	$-2.76511080 \times 10^{-14}$	$1.29192233 \times 10^{-9}$	1.291×10^{-9}
8	$1.97308246 \times 10^{-15}$	$4.22625545 \times 10^{-10}$	4.226×10^{-10}

In Table 4.97 are presented the values of $\Phi''(0)$ and $\Phi(\infty)$ for stretching flow with suction sheet.

From Tables 4.91, 4.92, 4.93, 4.94, 4.95, 4.96, 4.97 results that there exist an excellent agreement between the numerical results and the results obtained by means of OHAM.

On the other hand, considering the effect of slip parameter on the velocity $\overline{\Phi}(\eta)$ in both flows, Figs. 4.75, 4.76, 4.77, 4.78, 4.79, 4.80, 4.81, 4.82 have been displayed. It is shown in Figs. 4.75 and 4.76, the variation of $\overline{\Phi}$ and $\overline{\Phi'}$ respectively for planar flow and impermeable sheet. In Figs. 4.77 and 4.78 have been plotted the variation of $\overline{\Phi}$ and $\overline{\Phi'}$ respectively for planar flow and suction sheet.

Table 4.97 Values of $\Phi''(0)$ and $\Phi(\infty)$ in the case of stretching flow with suction sheet $m = 2$, $\alpha = 3$

λ	1	5	10
$\Phi''_{numeric}(0)$	−0.8578597370	−0.1935567447	0.0163824516
$\Phi''_{OHAM}(0)$	−0.8578597340	−0.1935567447	0.0163824515
$\Phi_{numeric}(\infty)$	3.0235285642	3.005309966	3.0027282376
$\Phi_{OHAM}(\infty)$	3.0235364880	3.0053612258	3.0027282856

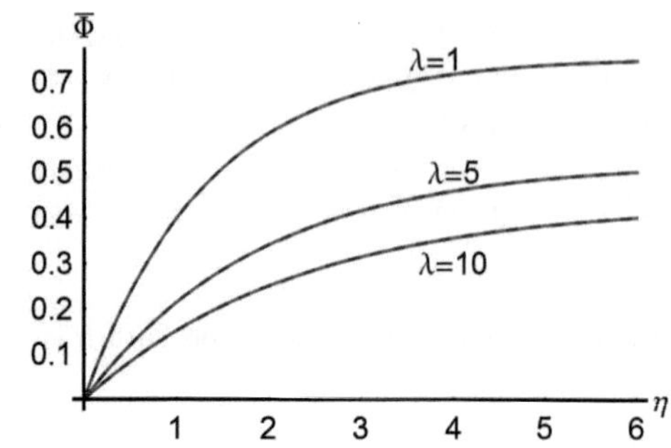

Fig. 4.75 Variation of $\overline{\Phi}$ by increasing the slip parameter λ for planar flow and impermeable sheet ($m = 1$, $\alpha = 0$): (*red solid line*) numerical results; (*blue dashed line*) approximate solution (colour figure online)

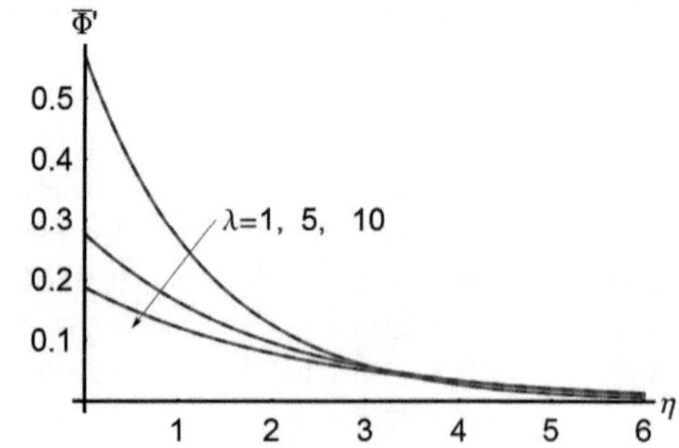

Fig. 4.76 Variation of $\overline{\Phi}'$ by increasing the slip parameter λ for planar flow and impermeable sheet ($m = 1$, $\alpha = 0$): (*red solid line*) numerical results; (*blue dashed line*) approximate solution (colour figure online)

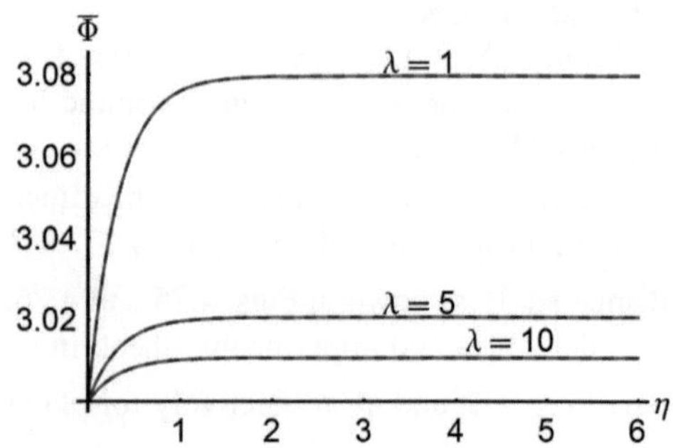

Fig. 4.77 Variation of $\overline{\Phi}$ by increasing the slip parameter λ for planar flow and impermeable sheet ($m = 1$, $\alpha = 3$): (*red solid line*) numerical results; (*blue dashed line*) approximate solution (colour figure online)

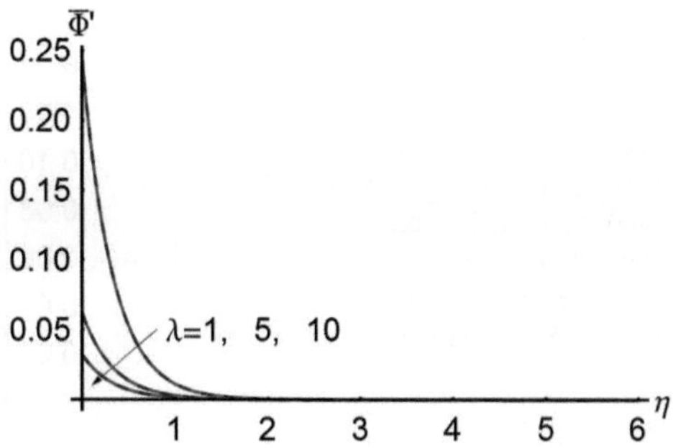

Fig. 4.78 Variation of $\overline{\Phi}'$ by increasing the slip parameter λ for planar flow and impermeable sheet ($m = 1$, $\alpha = 3$): (*red solid line*) numerical results; (*blue dashed line*) approximate solution (colour figure online)

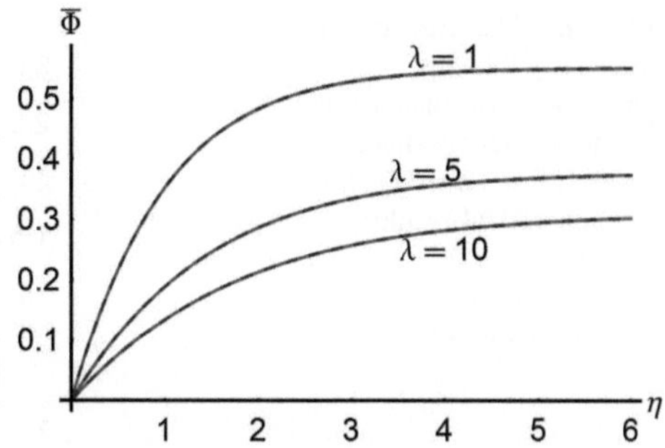

Fig. 4.79 Variation of $\overline{\Phi}$ by increasing the slip parameter λ for planar flow and impermeable sheet ($m = 2$, $\alpha = 0$): (*red solid line*) numerical results; (*blue dashed line*) approximate solution (colour figure online)

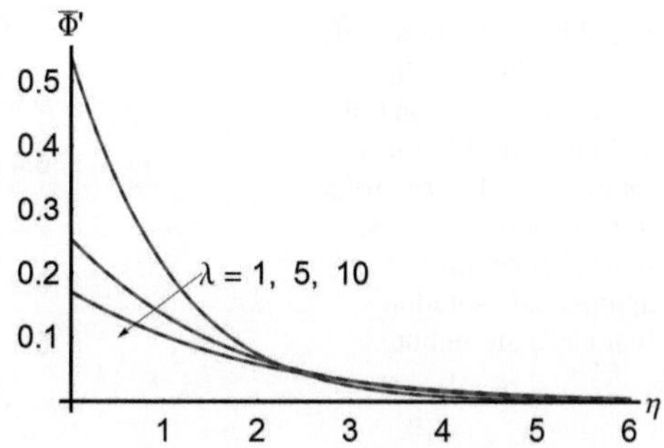

Fig. 4.80 Variation of $\overline{\Phi}'$ by increasing the slip parameter λ for planar flow and impermeable sheet ($m = 2$, $\alpha = 0$): (*red solid line*) numerical results; (*blue dashed line*) approximate solution (colour figure online)

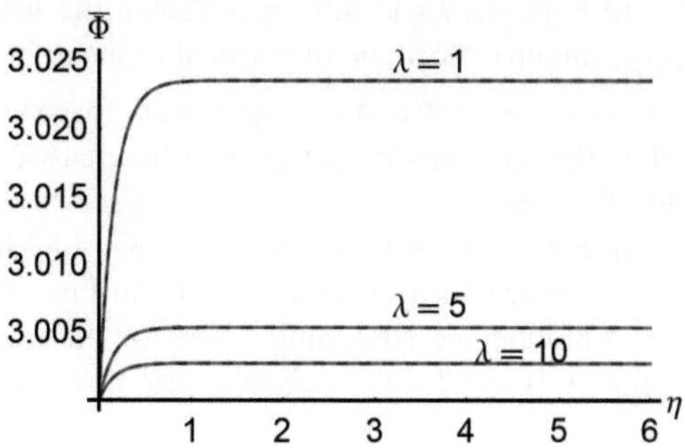

Fig. 4.81 Variation of $\overline{\Phi}$ by increasing the slip parameter λ for planar flow and impermeable sheet ($m = 2$, $\alpha = 3$): (*red solid line*) numerical results; (*blue dashed line*) approximate solution (colour figure online)

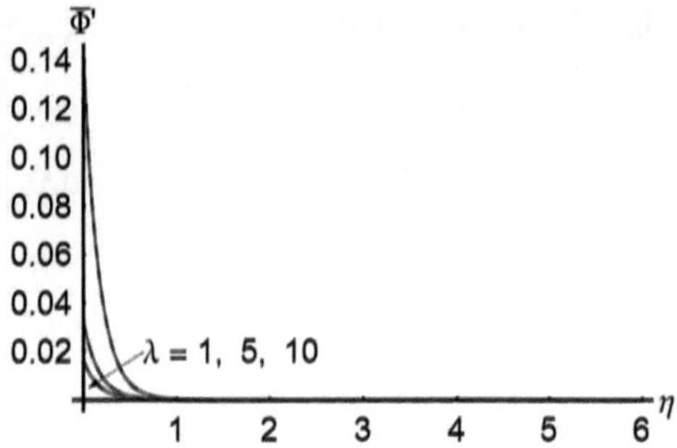

Fig. 4.82 Variation of $\overline{\Phi}'$ by increasing the slip parameter λ for planar flow and impermeable sheet ($m = 2$, $\alpha = 3$): (*red solid line*) numerical results; (*blue dashed line*) approximate solution (colour figure online)

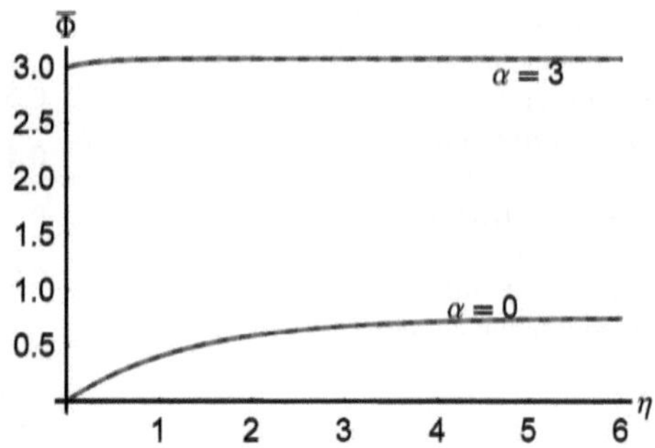

Fig. 4.83 Variation of $\overline{\Phi}$ by increasing the slip parameter λ for planar flow and impermeable sheet ($m = 1$, $\lambda = 1$): (*red solid line*) numerical results; (*blue dashed line*) approximate solution (colour figure online)

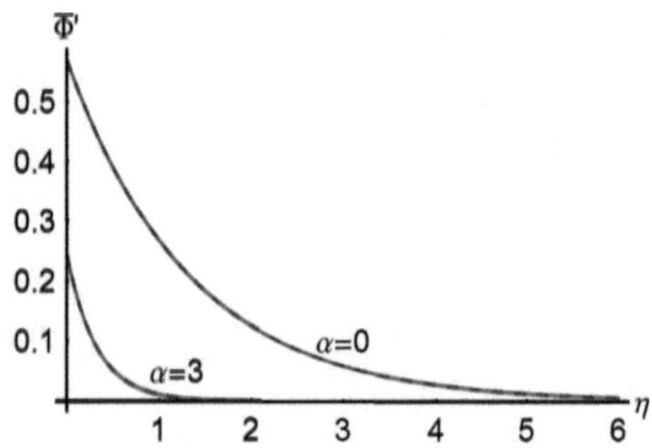

Fig. 4.84 Variation of $\overline{\Phi}'$ by increasing the slip parameter λ for planar flow and impermeable sheet ($m = 1$, $\lambda = 1$): (*red solid line*) numerical results; (*blue dashed line*) approximate solution (colour figure online)

In Figs. 4.79 and 4.80 it is shown the variation of $\overline{\Phi}$ and $\overline{\Phi'}$ respectively for axysimmetric flow and impermeable sheet. In Figs. 4.81 and 4.82 have been plotted the variation of $\overline{\Phi}$ and $\overline{\Phi'}$ respectively for axysimmetric flow and suction sheet. It is clear that the velocity components decreases with an increase in the slip parameter for all cases.

In Figs. 4.83, 4.84, 4.85, 4.86, 4.87, 4.88 have been plotted the planar cases for every value of slip parameter λ and in Figs. 4.89, 4.90, 4.91, 4.92, 4.93, 4.94 have been plotted the stretching cases for different values of λ. It is evident that the velocity is less for the axisymmetric flow when compared with the planar case.

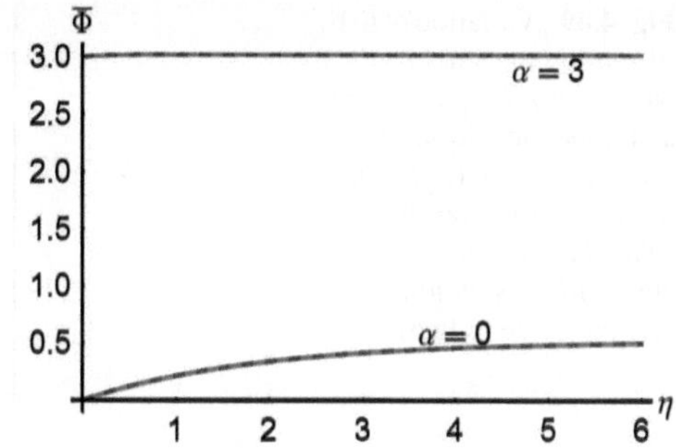

Fig. 4.85 Variation of $\overline{\Phi}$ by increasing the slip parameter λ for planar flow and impermeable sheet ($m = 1, \lambda = 5$): (*red solid line*) numerical results; (*blue dashed line*) approximate solution (colour figure online)

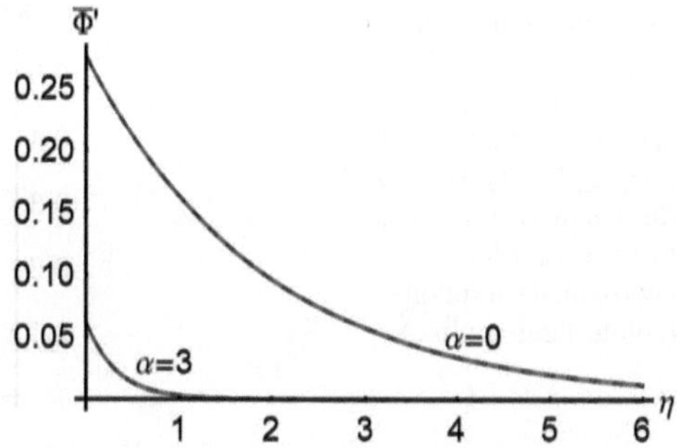

Fig. 4.86 Variation of $\overline{\Phi}'$ by increasing the slip parameter λ for planar flow and impermeable sheet ($m = 1, \lambda = 5$): (*red solid line*) numerical results; (*blue dashed line*) approximate solution (colour figure online)

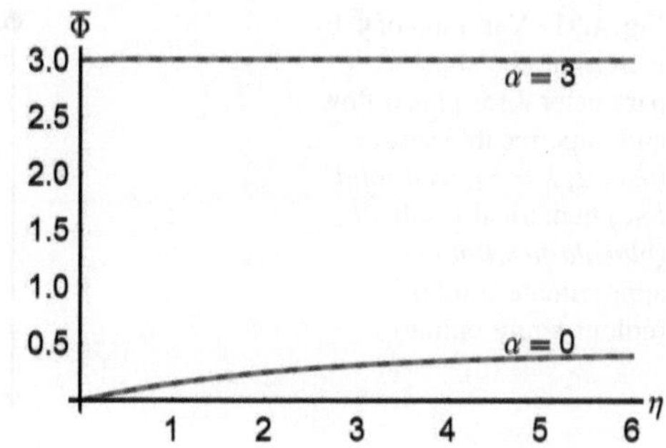

Fig. 4.87 Variation of $\overline{\Phi}$ by increasing the slip parameter λ for planar flow and impermeable sheet ($m = 1, \lambda = 10$): (*red solid line*) numerical results; (*blue dashed line*) approximate solution (colour figure online)

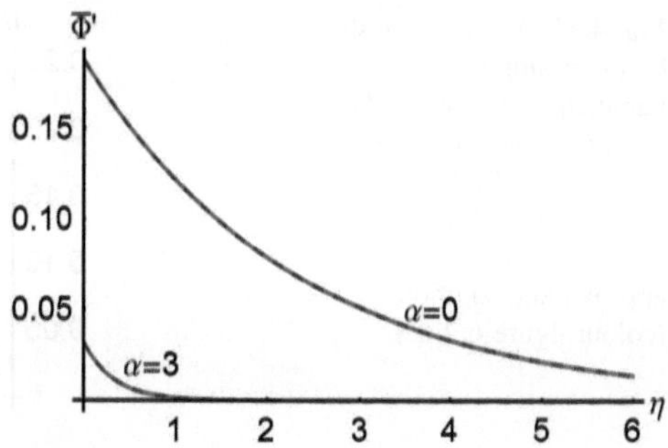

Fig. 4.88 Variation of $\overline{\Phi}'$ by increasing the slip parameter λ for planar flow and impermeable sheet ($m = 1, \lambda = 10$): (*red solid line*) numerical results; (*blue dashed line*) approximate solution (colour figure online)

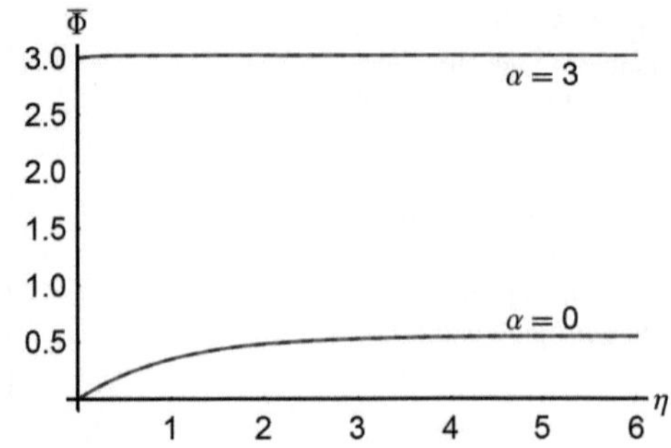

Fig. 4.89 Variation of $\overline{\Phi}$ by increasing the slip parameter λ for planar flow and impermeable sheet ($m = 2$, $\lambda = 1$): (*red solid line*) numerical results; (*blue dashed line*) approximate solution (colour figure online)

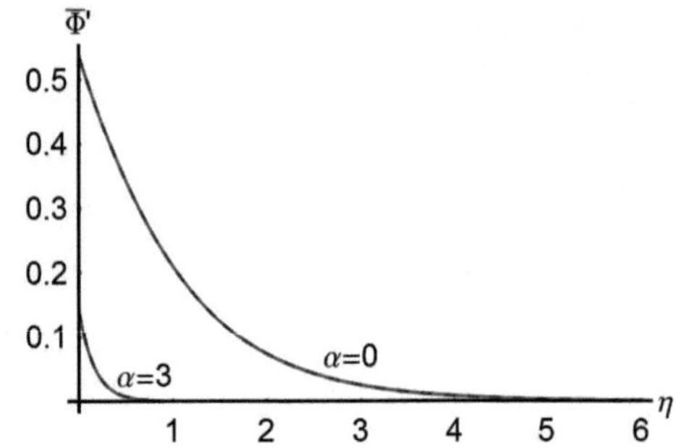

Fig. 4.90 Variation of $\overline{\Phi}'$ by increasing the slip parameter λ for planar flow and impermeable sheet ($m = 2$, $\lambda = 1$): (*red solid line*) numerical results; (*blue dashed line*) approximate solution (colour figure online)

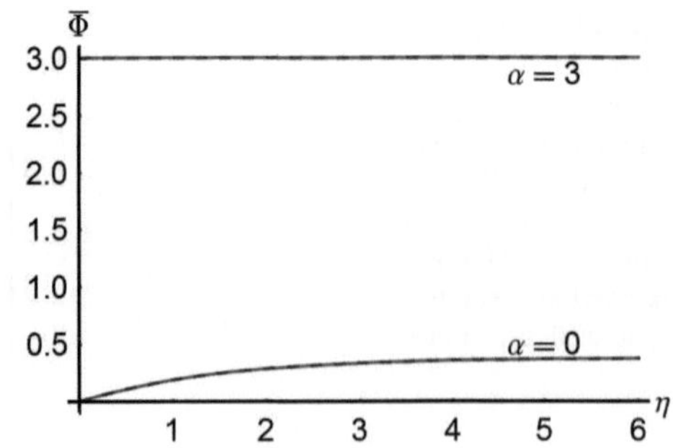

Fig. 4.91 Variation of $\overline{\Phi}$ by increasing the slip parameter λ for planar flow and impermeable sheet ($m = 2$, $\lambda = 5$): (*red solid line*) numerical results: (*blue dashed line*) approximate solution (colour figure online)

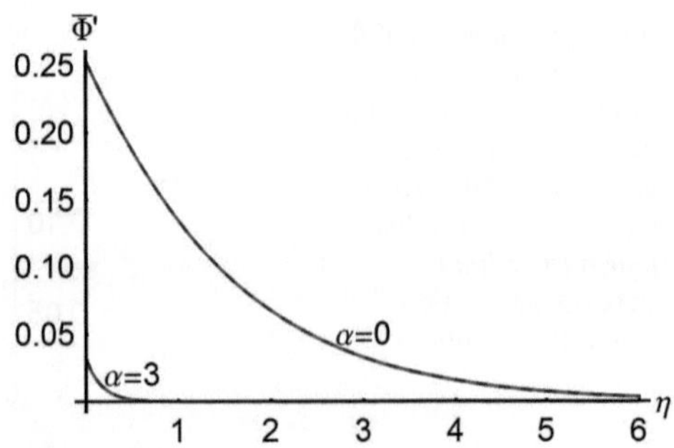

Fig. 4.92 Variation of $\overline{\Phi}'$ by increasing the slip parameter λ for planar flow and impermeable sheet ($m = 2$, $\lambda = 5$): (*red solid line*) numerical results: (*blue dashed line*) approximate solution (colour figure online)

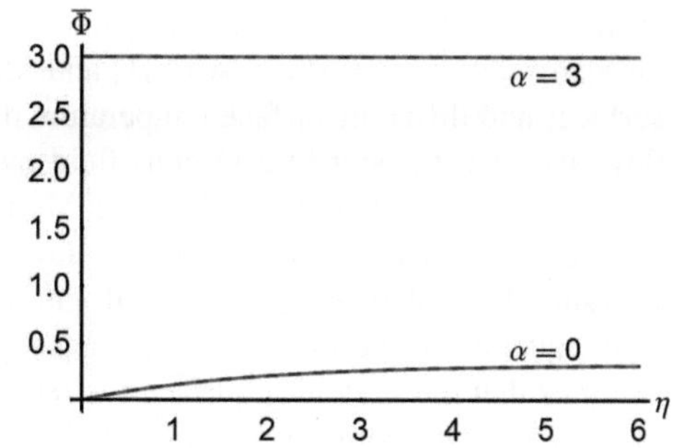

Fig. 4.93 Variation of $\overline{\Phi}$ by increasing the slip parameter λ for planar flow and impermeable sheet ($m = 2$, $\lambda = 10$): (*red solid line*) numerical results: (*blue dashed line*) approximate solution (colour figure online)

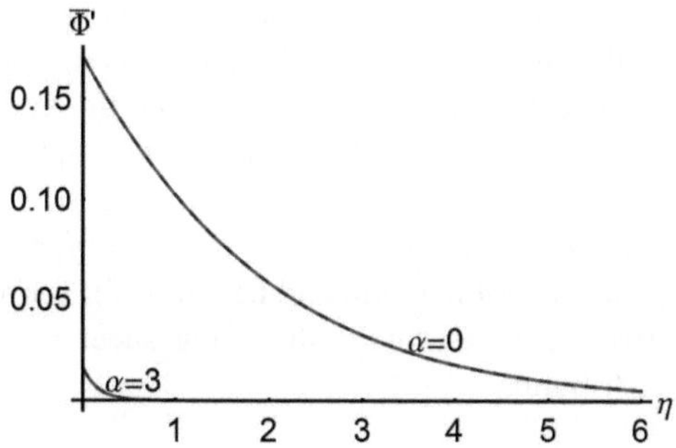

Fig. 4.94 Variation of $\overline{\Phi}'$ by increasing the slip parameter λ for planar flow and impermeable sheet ($m = 2$, $\lambda = 10$): (*red solid line*) numerical results; (*blue dashed line*) approximate solution (colour figure online)

4.5 The Flow and Heat Transfer in a Viscous Fluid Over an Unsteady Stretching Surface

The flow and heat transfer in a viscous fluid over a stretching surface is a relevant problem in many industrial and engineering processes. Examples are manufacture and drawing of plastics and rubber sheets, polymer extrusion, wire drawing, glass-fiber and paper production, crystal growing, continuous casting, and so on. Cooling of stretching surface requires control of the temperature and consequently knowledge of flow and heat transfer in a viscous fluid. Sakiadis [54, 55], studied the boundary layer flow over a continuous solid surface moving with constant speed. Crane [56] analyzed the stretching problem having in view the fluid flow over a linearly stretching surface. Tsou et al [74] studied constant surface velocity and temperature. Gupta and Gupta [75] and Maneschy et al [76] extended the Crane' work to the stretching problem with a constant surface temperature including suction or blowing and to fluids exhibiting a non-Newtonian behavior, respectively. Grubka and Bobba [77] studied the stretching problem for a surface moving with a linear velocity and with a variable surface temperature. Wang [78] introduced a similarity transformation to reduce time-dependent momentum equation to a third-order nonlinear differential equation. He analyzed the hydrodynamic behavior of a finite fluid body driven by an unsteady stretching surface. The same problem was considered by Usha and Rukmani [79] for the axisymmetric case. Anderson et al [80] analyzed the accompanying heat transfer in the liquid film driven by unsteady stretching surface. Ali [81] and Magyari et al [82] considered permeable surfaces and different surface temperature distributions. Vajravelu [83] studied the flow and heat transfer in a viscous fluid over a planar nonlinear stretching sheet. Magyari and Keller [84] applied the Merkin transformation method to the heat transfer problems of steady boundary layer flows induced by stretching surfaces. Elbashbeshy and Bazid [85] studied similarity solution of the laminar boundary layer equations corresponding to an unsteady stretching surface. Dandapat et al [86] assumed that the stretching surface is stretched impulsively from rest and the effect of inertia of the liquid is considered. The unsteady heat and fluid flow has been investigated by Ali and Magyari [87]. Liu and Anderson [88] explored the thermal characteristics of a viscous film on an unsteady stretching surface. Chen [89] analyzed the problem of MHD mixed convective flow and heat transfer of an electrically conducting, power-low fluid past a stretching surface in the presence of heat generation/absortion and thermal radiation. Dandapat et al [90] studied a thin viscous liquid film flow over a stretching sheet under different non-linear stretching velocities in presence of uniform transverse magnetic field. Cortell [91] presented momentum and heat transfer for the flow induced in a quiscent fluid by a permeable non-linear stretching sheet with a prescribed power-low temperature distribution.

4.5.1 Equations of Motion

Consider an unsteady, two dimensional flow on a continuous stretching surface, with the governing time-dependent equations for the continuity, momentum and thermal energy [78, 80, 85, 87, 88]:

$$\frac{\partial u}{\partial x}+\frac{\partial v}{\partial y}=0 \tag{4.256}$$

$$\frac{\partial u}{\partial t}+u\frac{\partial u}{\partial x}+v\frac{\partial v}{\partial y}=\nu\frac{\partial^2 u}{\partial y^2} \tag{4.257}$$

$$\frac{\partial T}{\partial t}+u\frac{\partial T}{\partial x}+v\frac{\partial T}{\partial y}=k\frac{\partial^2 T}{\partial y^2} \tag{4.258}$$

where u and v are velocity components in the x and y directions, respectively, T is the temperature and k is the thermal conductivity of the incompressible fluid. The appropriate boundary conditions are:

$$u=\frac{u_0\frac{x}{l}}{1+\gamma t}, v=0, T=T_\infty+\frac{T_0}{(1+\gamma t)^c}\left(\frac{x}{l}\right)^n at\, y=0 \tag{4.259}$$

$$u\to 0, T\to T_\infty\, at\, y\to\infty \tag{4.260}$$

where $u_0, T_0, T_\infty, \gamma$ are positive constants, c and n are arbitrary and l is a reference length.

If $Re = \frac{u_0 l}{\nu}$ and $Pr = \frac{\nu}{k}$ are the Reynolds number and the Prandl number respectively and if we choose a stream function $\Psi(x, y)$ such that

$$u = \frac{\partial \Psi}{\partial y}, v = -\frac{\partial \Psi}{\partial x} \tag{4.261}$$

then the equation 4.256 of continuity is satisfied and the mathematical analysis of the problems 4.257 and 4.258 is simplified by introducing the following similarity transformation:

$$\Psi = \frac{x}{l} \frac{f(\eta)}{\sqrt{Re}(1+\gamma t)^{1/2}} \tag{4.262}$$

$$\eta = \sqrt{Re} \frac{y}{l(1+\gamma t)^{1/2}} \tag{4.263}$$

$$T = T_\infty + T_0 \left(\frac{x}{l}\right)^n \frac{\theta(\eta)}{(1+\gamma t)^c} \tag{4.264}$$

T_0 being a reference temperature. In this way Eq. 4.261 can be written in the form:

$$u(x, y, t) = \frac{u_0}{l} \frac{x}{(1+\gamma t)} f'(\eta) \tag{4.265}$$

$$v(x, y, t) = -\frac{u_0}{\sqrt{Re}(1+\gamma t)^{1/2}} f(\eta) \tag{4.266}$$

where prime denotes differentiation with respect to η.

Substituting Eqs. 4.262, 4.263, 4.264, 4.265 and 4.266 into Eqs. 4.257 and 4.258, we obtain

$$f''' + ff'' - f'^2 + \Lambda\left(f' + \frac{1}{2}\eta f''\right) = 0 \tag{4.267}$$

$$\frac{1}{Pr}\theta'' + f\theta' - nf'\theta + \Lambda\left(c\theta + \frac{1}{2}\eta\theta'\right) = 0. \tag{4.268}$$

Here $\Lambda = \frac{\gamma l}{u_0}$ is dimensionless measure of the unsteadiness. The dimensional boundary conditions 4.259 and 4.260 become

$$u = \frac{u_0}{l} \frac{x}{(1+\gamma t)} f'(0) \; at \; y = 0 \tag{4.269}$$

$$v = -\frac{u_0}{\sqrt{Re}(1+\gamma t)^{1/2}} f(0) \; at \; y = 0 \tag{4.270}$$

$$T = T_\infty + T_0 \left(\frac{x}{l}\right)^n \frac{\theta(0)}{(1+\gamma t)^c} \; at \; y = 0 \tag{4.271}$$

such that for the dimensionless functions f and θ, the boundary/initial conditions become

$$f(0) = f_w, f'(0) = 1, f'(\infty) = 0 \tag{4.272}$$

$$\theta(0) = 1, \theta(\infty) = 0. \tag{4.273}$$

In addition to the boundary conditions 4.272 and 4.273, the requirements

$$f'(\eta) \geq 0, \theta(\eta) \geq 0, \forall \eta \geq 0 \tag{4.274}$$

must also satisfied [87].

4.5.2 *Application of OHAM to Flow and Heat Transfer*

We use the basic ideas of the OHAM by considering Eq. 4.267 with the boundary conditions given by Eq. 4.272. We can choose the linear operator in two form.

Case 4.5.2.1
If $K > 0$ is an unknown parameter at this moment, then we suppose that the linear operator has the expression:

$$L_f(\Phi(\eta)) = \Phi''' - K^2\Phi'. \tag{4.275}$$

The equation 4.4 becomes

$$\Phi_0''' - K^2\Phi_0' = 0, \Phi_0(0) = f_w, \Phi_0'(0) = 1, \Phi_0'(\infty) = 0,$$

which has the following solution

$$\Phi_0(\eta) = f_w + \frac{1 - e^{-K\eta}}{K}. \tag{4.276}$$

The nonlinear operator $N_f(\Phi(\eta))$ is obtained from Eqs. 4.267 and 4.275:

$$N_f(\Phi(\eta)) = K^2\Phi'(\eta) + \Phi(\eta)\Phi''(\eta) - \Phi'(\eta)^2 + \Lambda\left(\Phi'(\eta) + \frac{1}{2}\eta\Phi''(\eta)\right) = 0 \tag{4.277}$$

such that substituting Eq. 4.276 into Eq. 4.277, we obtain

$$N_f(\Phi_0(\eta)) = (\alpha\eta + \beta)e^{-K\eta} \tag{4.278}$$

where

$$\alpha = \frac{1}{2}K\Lambda;\ \beta = K^2 - 1 - Kf_w - \Lambda. \tag{4.279}$$

Having in view that in Eq. 4.278 appears an exponential function and that the optimal auxiliary function $H_f(\eta, C_i)$ must follow the terms appearing in Eq. 4.278, then we can choose the function $H_f(\eta, C_i)$ in the following forms:

$$\begin{gathered}H_f(\eta, C_i) = C_1 + C_2\eta + C_3\eta^2 + (C_4 + C_5\eta + C_6\eta^2)e^{-K\eta} + \\ +(C_7 + C_8\eta + C_9\eta^2)e^{-2K\eta}\end{gathered} \tag{4.280}$$

or

$$H_f^*(\eta, C_i) = C_1 + \left(C_2 + C_3\eta + C_4\eta^2\right)e^{-K\eta} \tag{4.281}$$

or yet

$$\begin{gathered}H_f^{**}(\eta, C_i) = C_1 + C_2\eta + C_3\eta^2 + (C_4 + C_5\eta)e^{-K\eta} + \\ +(C_6 + C_7\eta + C_8\eta^2)e^{-2K\eta}\end{gathered} \tag{4.282}$$

and so on, where C_1, C_2, ... are unknown parameters at this moment.

If we choose only the expression 4.280 for $H_f(\eta, C_i)$, then by using Eqs. 4.278, 4.280 and 4.9, we can obtain the equation in $\Phi_1(\eta, C_i)$:

$$\begin{gathered}\Phi_1''' - K^2\Phi_1' = [\beta C_1 + (\alpha C_1 + \beta C_2)\eta + (\alpha C_2 + \beta C_3)\eta^2 + \alpha C_3\eta^3]e^{-K\eta} + \\ +[\beta C_4 + (\alpha C_4 + \beta C_5)\eta + (\alpha C_5 + \beta C_6)\eta^2 + \alpha C_6\eta^3]e^{-2K\eta} + [\beta C_7 + \\ +(\alpha C_7 + \beta C_8)\eta + (\alpha C_8 + \beta C_9)\eta^2 + \alpha C_9\eta^3]e^{-3K\eta}, \\ \Phi_1(0) = \Phi_1'(0) = \Phi_1'(\infty) = 0.\end{gathered} \tag{4.283}$$

The solution of Eq. 4.283 can be found as

$$\begin{gathered}\Phi_1(\eta) = M_1 + \Big[N_1 + \left(\frac{7\alpha C_2}{4K^4} + \frac{3\alpha C_1}{4K^3} + \frac{3\beta C_2}{4K^3} + \frac{\beta C_1}{2K^2} + \frac{45\alpha C_3}{8K^5}\right)\eta + \\ +\left(\frac{3\alpha C_2}{4K^3} + \frac{\alpha C_1}{4K^2} + \frac{\beta C_2}{4K^2} + \frac{21\alpha\beta}{8K^4}\right)\eta^2 + \left(\frac{\alpha C_2}{6K^2} + \frac{3\alpha C_3}{4K^3} + \frac{\beta C_3}{6K^2}\right)\eta^3 + \\ +\frac{\alpha C_3}{8K^2}\eta^4\Big]e^{-K\eta} + \Big[-\frac{85\alpha C_5}{108K^5} - \frac{11\alpha C_4}{36K^4} - \frac{11\beta C_5}{36K^4} - \frac{\beta C_4}{6K^3} - \frac{575\alpha C_6}{216K^6} - \\ -\frac{85\beta C_6}{108K^5} - \left(\frac{11\alpha C_5}{18K^4} + \frac{\alpha C_4}{6K^3} + \frac{\beta C_5}{6K^3} + \frac{85\alpha C_6}{36K^5} + \frac{11\beta C_6}{18K^4}\right)\eta - \Big(\frac{\alpha C_5}{6K^3} +\end{gathered}$$

$$+\frac{11\alpha C_6}{12K^4}+\frac{\beta C_6}{6K^3})\eta^2-\frac{\alpha C_6}{6K^3}\eta^3]e^{-2K\eta}+[-\frac{115\alpha C_8}{1728K^5}-\frac{13\alpha C_7}{288K^4}-\frac{13\beta C_8}{288K^4}-$$
$$-\frac{\beta C_7}{24K^3}-\frac{865\alpha C_9}{6912K^6}-\frac{115\beta C_9}{1728K^5}-(\frac{\alpha C_7}{24K^3}+\frac{13\alpha C_8}{144K^4}+\frac{\beta C_8}{24K^3}+\frac{115\alpha C_9}{576K^5}+ \quad (4.284)$$
$$+\frac{13\beta C_9}{144K^4})\eta-\left(\frac{\alpha C_8}{24K^3}+\frac{13\alpha C_9}{96K^4}+\frac{\beta C_9}{24K^3}\right)\eta^2-\frac{\alpha C_9}{24K^3}]e^{-3K\eta}$$

where

$$M_1=-\frac{3\alpha+2K\beta}{4K^4}C_1-\frac{7\alpha+3K\beta}{4K^5}C_2-\frac{45\alpha C_3}{8K^6}-\frac{5\alpha+6K\beta}{36K^4}C_4-$$
$$-\frac{19\alpha+15K\beta}{108K^4}C_5-\frac{65\alpha+38K\beta}{216K^6}C_6-\frac{7\alpha+12K\beta}{144K^4}C_7-$$
$$-\frac{37\alpha+44K\beta}{864K^5}C_8-\frac{175\alpha+148K\beta}{3456K^6}C_9 \quad (4.285)$$
$$N_1=\frac{3\alpha+2K\beta}{4K^4}C_1+\frac{7\alpha+3K\beta}{4K^5}C_2+\frac{45\alpha C_3}{8K^6}+\frac{4\alpha+3K\beta}{9K^4}C_4+$$
$$+\frac{26\alpha+12K\beta}{27K^5}C_5+\frac{80\alpha+26K\beta}{27K^6}C_6+\frac{3\alpha+4K\beta}{32K^4}C_7+$$
$$+\frac{7\alpha+6K\beta}{64K^5}C_8+\frac{45\alpha+28K\beta}{256K^6}.$$

The first-order approximate solution 4.5 for Eqs. 4.267 and 4.272 is obtained from Eqs. 4.276 and 4.285:

$$\overline{f}(\eta)=\overline{\Phi}(\eta)=\Phi_0(\eta)+\Phi_1(\eta). \quad (4.286)$$

In what follows, we consider Eqs. 4.268 and 4.273. In this case, we choose the linear operator in the form

$$L_\theta(\theta(\eta))=\theta''+K\theta' \quad (4.287)$$

where the parameter K is defined in Eq. 4.275.

The equation 4.4 becomes

$$\theta_0''+K\theta_0'=0, \theta_0(0)=1, \theta_0(\infty)=0. \quad (4.288)$$

The equation 4.288 has the solution

$$\theta_0(\eta)=e^{-K\eta}. \quad (4.289)$$

The nonlinear operator $N_\theta(\theta(\eta))$ is obtained from Eq. 4.267:

$$N_\theta(\theta(\eta))=-\frac{k}{Pr}\theta'+\Phi\theta-n\Phi'\theta+\Lambda\left(c\theta+\frac{1}{2}\eta\theta'\right). \quad (4.290)$$

Substituting Eq. 4.289 into Eq. 4.290, we obtain

$$N_\theta(\theta_0(\eta)) = (m_1\eta + m_2)e^{-K\eta} + m_3e^{-2K\eta} \tag{4.291}$$

where

$$m_1 = -\frac{1}{2}K\Lambda; m_2 = \frac{K^2}{Pr} - Kf_w - 1 + c\Lambda; m_3 = 1 - n. \tag{4.292}$$

The optimal auxiliary function $H_\theta(\eta, C_i)$ can be choosen in the following forms:

$$\begin{aligned} H_\theta(\eta, C_i) = C_{10} + C_{11}\eta + C_{12}\eta^2 + (C_{13} + C_{14}\eta + C_{15}\eta^2)e^{-K\eta} + \\ +(C_{16} + C_{17}\eta)e^{-2K\eta} \end{aligned} \tag{4.293}$$

or

$$H_\theta^*(\eta, C_i) = C_7 + C_8\eta + C_9\eta^2 + (C_{10} + C_{11}\eta)e^{-K\eta} + C_{13}e^{-2K\eta} \tag{4.294}$$

or yet

$$H_\theta^{**}(\eta, C_i) = C_7 + (C_8 + C_9\eta)e^{-K\eta} + (C_{10} + C_{11}\eta)e^{-2K\eta} \tag{4.295}$$

and so on, where C_7, C_8, ... are unknown parameters.

If we choose the Eq. 4.293 for H_θ, then from Eqs. 4.291, 4.293 and 4.9 we obtain the equation in $\theta_1(\eta, C_i)$ as

$$\begin{aligned} \theta_1'' + K\theta_1' = \left[m_2C_{10} + (m_1C_{10} + m_2C_{11})\eta + (m_1C_{11} + m_2C_{12})\eta^2 + \right. \\ \left. +m_1C_{12}\eta^3\right]e^{-K\eta} + \left[m_2C_{13} + m_3C_{10} + (m_1C_{13} + m_2C_{14} + m_3C_{11})\eta + \right. \\ \left. +m_1C_{14} + m_2C_{15} + m_3C_{12})\eta^2 + m_1C_{15}\eta^3\right]e^{-2K\eta} + \left[m_3C_{13} + m_2C_{16} + \right. \\ \left. +(m_3C_{14} + m_1C_{16} + m_2C_{17})\eta + (m_3C_{15} + m_1C_{17})\eta^2\right]e^{-3K\eta} + \\ +(m_3C_{16} + m_3C_{17}\eta)e^{-4K\eta}, \theta_1(0) = \theta_1(\infty) = 0. \end{aligned} \tag{4.296}$$

Solving Eq. 4.296, we obtain

$$\begin{aligned} \theta_1(\eta) = \Big[P_1 - \left(\frac{2m_1C_{11}}{K^3} + \frac{m_1C_{10}}{K^2} + \frac{m_2C_{11}}{K^2} + \frac{m_2C_{10}}{K} - \frac{2m_2C_{12}}{K^3} - \frac{6m_1C_{12}}{K^4}\right)\eta - \\ -\left(\frac{m_1C_{11}}{K^2} + \frac{m_1C_{10}}{2K} + \frac{m_2C_{11}}{2K} - \frac{m_2C_{12}}{K^2} - \frac{3m_1C_{12}}{K^3}\right)\eta^2 - (\frac{m_1C_{11}}{3K} + \frac{m_2C_{12}}{3K} + \\ +\frac{m_1C_{12}}{K^2})\eta^3 - \frac{m_1C_{12}}{4K}\eta^4\Big]e^{-K\eta} + \Big[\frac{45m_1C_{15}}{8K^5} + \frac{7m_2C_{15}}{4K^4} + \frac{7m_1C_{12}}{4K^4} + \\ +\frac{7m_1C_{14}}{4K^4} + \frac{3m_1C_{13}}{4K^3} + \frac{3m_2C_{14}}{4K^3} + \frac{3m_3C_{11}}{4K^3} + \frac{m_2C_{13}}{2K^2} + \frac{m_3C_{10}}{2K^2} + (\frac{21m_1C_{15}}{4K^4} + \\ +\frac{3m_2C_{15}}{2K^3} + \frac{3m_3C_{12}}{2K^3} + \frac{3m_1C_{11}}{2K^3} + \frac{m_1C_{13}}{2K^2} + \frac{m_3C_{11}}{2K^2})\eta + (\frac{9m_1C_{15}}{4K^3} + \frac{m_2C_{15}}{2K^2} + \end{aligned}$$

$$+\frac{m_3C_{12}}{2K^3}+\frac{m_1C_{14}}{2K^2})\eta^2+\frac{m_1C_{15}}{2K^2}\eta^3]e^{-2K\eta}+[\frac{5(m_3C_{14}+m_1C_{16}+m_2C_{17})}{36K^3}+$$
$$+\frac{19m_3C_{15}}{108K^4}+\frac{6m_3C_{13}+m_2C_{16}}{36K^2}+(\frac{m_3C_{14}+m_1C_{16}+m_2C_{17}}{6K^2}+$$
$$+\frac{5(m_1C_{17}+m_3C_{15})}{18K^3})\eta+\frac{m_1C_{17}+m_3C_{15}}{6K^2}\eta^2]e^{-3K\eta}+$$
$$+\left(\frac{m_3C_{17}}{12K^2}+\frac{m_3C_{16}}{12K^3}+\frac{7m_3C_{17}}{144K^3}\eta\right)e^{-4K\eta}, \tag{4.297}$$

where

$$P_1=-\frac{m_3C_{10}}{2K^2}-\frac{3m_3C_{11}}{4K^3}-\frac{7m_3C_{12}}{4K^4}-\frac{9m_1+6Km_2+2Km_3}{12K^3}C_{13}-$$
$$-\frac{63m_1+27Km_2+5Km_3}{36K^4}C_{14}-\frac{1215m_1+378Km_2+38Km_3}{216K^5}C_{15}- \tag{4.298}$$
$$-\frac{5m_1+Km_2}{36K^3}C_{16}-\frac{20m_2+7m_3}{144K^3}C_{17}.$$

In this way, the first-order approximate solution 4.5 for Eqs. 4.268 and 4.273 becomes

$$\overline{\theta}(\eta)=\theta_0(\eta)+\theta_1(\eta,C_i). \tag{4.299}$$

where θ_0 and θ_1 are given by Eqs. 4.289 and 4.297, respectively.

Case 4.5.2.2
The second alternative for the choice of the linear operator is

$$L_f(\Phi(\eta))=\Phi'''+\frac{3K}{K\eta+1}\Phi''. \tag{4.300}$$

The initial approximate solution Φ_0 can be obtained from equation

$$\Phi_0'''+\frac{3K}{K\eta+1}\Phi_0''=0;\Phi_0(0)=f_w,\Phi_0'(0)=1,\Phi_0'(\infty)=0 \tag{4.301}$$

and therefore we have

$$\Phi_0(\eta)=\frac{1}{K}+f_w-\frac{1}{K(K\eta+1)}. \tag{4.302}$$

The nonlinear operator N_f for the expression given by Eq. 4.302 can be written as

$$N_f(\Phi_0(\eta))=\frac{\Lambda}{(K\eta+1)^2}-\frac{K\Lambda\eta+2+2Kf_w}{(K\eta+1)^3}+\frac{1+6K^2}{(K\eta+1)^4} \tag{4.303}$$

such that we choose the optimal auxiliary function H_f in the form

$$H_f(\eta, C_i) = \frac{C_1}{(K\eta+1)^2} + \frac{C_2}{(K\eta+1)^3} + \frac{C_3}{(K\eta+1)^4} + \frac{C_4}{(K\eta+1)^5} + \\ + \frac{C_5}{(K\eta+1)^6} + \frac{C_6}{(K\eta+1)^7} + \frac{C_7}{(K\eta+1)^8}. \tag{4.304}$$

The equation 4.9 becomes

$$\Phi_1''' + \frac{3K}{K\eta+1}\Phi_1'' = \frac{\Lambda C_1}{(K\eta+1)^4} - \frac{K\Lambda\eta C_1}{(K\eta+1)^5} + \frac{\Lambda C_2 - (2+2Kf_w)C_1}{(K\eta+1)^5} - \\ - \frac{K\Lambda\eta C_2}{(K\eta+1)^6} + \frac{(1+6K^2)C_1 - (2+2Kf_w)C_2 + \Lambda C_3}{(K\eta+1)^6} - \frac{K\Lambda\eta C_3}{(K\eta+1)^7} + \\ + \frac{(1+6K^2)C_2 - (2+2Kf_w)C_3 + \Lambda C_4}{(K\eta+1)^7} - \frac{K\Lambda\eta C_4}{(K\eta+1)^8} + \\ + \frac{(1+6K^2)C_3 - (2+2Kf_w)C_4 + \Lambda C_5}{(K\eta+1)^8} - \frac{K\Lambda\eta C_5}{(K\eta+1)^9} + \\ + \frac{(1+6K^2)C_4 - (2+2Kf_w)C_5 + \Lambda C_6}{(K\eta+1)^9} - \frac{K\Lambda\eta C_6}{(K\eta+1)^{10}} + \\ + \frac{(1+6K^2)C_5 - (2+2Kf_w)C_6 + \Lambda C_7}{(K\eta+1)^{10}} - \frac{K\Lambda\eta C_7}{(K\eta+1)^{11}} + \\ + \frac{(1+6K^2)C_6 - (2+2Kf_w)C_7}{(K\eta+1)^{11}} + \frac{(1+6K^2)C_7}{(K\eta+1)^{12}}, \\ \Phi_1(0) = \Phi_1'(0) = \Phi_1'(\infty) = 0. \tag{4.305}$$

The solution of the Eq. 4.305 has the form

$$\Phi_1(\eta) = M_2 + \frac{N_2}{K\eta+1} - \frac{(2-\Lambda+2Kf_w)C_1}{12K^3(K\eta+1)^2} - \frac{(1+6K^2)C_1 + (2-\Lambda+2Kf_w)C_2}{36(K\eta+1)^3} - \\ - \frac{(1+6K^2)C_2 + (2-\Lambda+2Kf_w)C_3}{80(K\eta+1)^4} - \frac{(1+6K^2)C_2 - (2-\Lambda+2Kf_w)C_4}{150(K\eta+1)^5} - \\ - \frac{(1+6K^2)C_4 + (2-\Lambda+2Kf_w)C_5}{252(K\eta+1)^6} - \frac{(1+6K^2)C_5 - (2-\Lambda+2Kf_w)C_6}{392(K\eta+1)^7} - \\ - \frac{(1+6K^2)C_6 + (2-\Lambda+2Kf_w)C_7}{576(K\eta+1)^8} - \frac{(1+6K^2)C_7}{810(K\eta+1)^9} \tag{4.306}$$

where

$$
\begin{aligned}
M_2 = & -\left(\frac{2-\Lambda+2Kf_w}{12K^3}+\frac{1+6K^2}{18K^3}\right)C_1-\left(\frac{2-\Lambda+2Kf_w}{18K^3}+\frac{3+18K^2}{80K^3}\right)C_2- \\
& -\left(\frac{3(2-\Lambda+2Kf_w)}{80K^3}+\frac{2+12K^2}{75K^3}\right)C_3-\left(\frac{2(2-\Lambda+2Kf_w)}{75K^3}+\frac{5+30K^2}{252K^3}\right)C_4- \\
& -\left(\frac{5(2-\Lambda+2Kf_w)}{252K^3}+\frac{3+18K^2}{196K^3}\right)C_5-\left(\frac{3(2-\Lambda+2Kf_w)}{196K^3}+\frac{7+42K^2}{576K^3}\right)C_6- \\
& -\left(\frac{7(2-\Lambda+2Kf_w)}{576K^3}+\frac{4+42K^2}{405K^3}\right)C_7; \\
N_2 = & \left(\frac{2-\Lambda+2Kf_w}{6K^3}+\frac{1+6K^2}{12K^3}\right)C_1+\left(\frac{2-\Lambda+2Kf_w}{12K^3}+\frac{1+6K^2}{20K^3}\right)C_2+ \\
& +\left(\frac{2-\Lambda+2Kf_w}{20K^3}+\frac{1+6K^2}{30K^3}\right)C_3+\left(\frac{2-\Lambda+2Kf_w}{30K^3}+\frac{1+6K^2}{42K^3}\right)C_4+ \\
& +\left(\frac{2-\Lambda+2Kf_w}{42K^3}+\frac{1+6K^2}{56K^3}\right)C_5+\left(\frac{2-\Lambda+2Kf_w}{56K^3}+\frac{1+6K^2}{72K^3}\right)C_6+ \\
& +\left(\frac{2-\Lambda+2Kf_w}{72K^3}+\frac{1+6K^2}{90K^3}\right)C_7
\end{aligned}
\tag{4.307}
$$

The first-order approximate solution 4.5 in the second alternative for Eqs. 4.267 and 4.272 is obtained from Eqs. 4.302 and 4.306:

$$\overline{f}(\eta)=\overline{\Phi}(\eta)=\Phi_0(\eta)+\Phi_1(\eta). \tag{4.308}$$

For Eqs. 4.268 and 4.273, the linear operator in this second alternative, we choose the expression:

$$L_\theta(\theta(\eta))=\theta''+\frac{2K}{K\eta+1}\theta'. \tag{4.309}$$

The equation 4.4 can be written in the form

$$\theta_0''+\frac{2K}{K\eta+1}\theta_0'=0, \theta_0(0)=1, \theta_0(\infty)=0 \tag{4.310}$$

and has the solution

$$\theta_0(\eta)=\frac{1}{K\eta+1}. \tag{4.311}$$

For the expression 4.311, the nonlinear operator becomes

$$N_\theta(\theta_0(\eta))=\frac{c\Lambda}{K\eta+1}-\frac{K\eta\Lambda}{2(K\eta+1)^2}+\frac{\left(\frac{2K^2}{Pr}+1-n\right)}{(K\eta+1)^3}+\frac{1+Kf_w}{(K\eta+1)^2}. \tag{4.312}$$

Taking into account Eq. 4.312, we choose the optimal auxiliary function H_θ as

$$H_\theta(\eta, C_i) = \frac{C_8}{(K\eta+1)^2} + \frac{C_9}{(K\eta+1)^3} + \frac{C_{10}}{(K\eta+1)^4} + \frac{C_{11}}{(K\eta+1)^5} + \\ + \frac{C_{12}}{(K\eta+1)^6} + \frac{C_{13}}{(K\eta+1)^7} + \frac{C_{14}}{(K\eta+1)^8} + \frac{C_{15}}{(K\eta+1)^9} + \frac{C_{16}}{(K\eta+1)^{10}} \tag{4.313}$$

such that Eq. 4.9 can be written in the form:

$$\theta_1'' + \frac{2K}{K\eta+1}\theta_1' = \frac{c\Lambda C_8}{(K\eta+1)^3} + \frac{-(1+Kf_w)C_8 + c\Lambda C_9}{(K\eta+1)^4} - \frac{K\Lambda\eta C_8}{2(K\eta+1)^4} + \\
+ \frac{\left(\frac{2K^2}{Pr}+1-n\right)C_8 - (1+Kf_w)C_9 + c\Lambda C_{10}}{(K\eta+1)^5} + \frac{K\Lambda\eta C_9}{2(K\eta+1)^5} + \\
+ \frac{\left(\frac{2K^2}{Pr}+1-n\right)C_9 - (1+Kf_w)C_{10} + c\Lambda C_{11}}{(K\eta+1)^6} - \frac{K\Lambda\eta C_{10}}{2(K\eta+1)^6} + \\
+ \frac{\left(\frac{2K^2}{Pr}+1-n\right)C_{10} - (1+Kf_w)C_{11} + c\Lambda C_{12}}{(K\eta+1)^7} - \frac{K\Lambda\eta C_{11}}{2(K\eta+1)^7} + \\
+ \frac{\left(\frac{2K^2}{Pr}+1-n\right)C_{11} - (1+Kf_w)C_{12} + c\Lambda C_{13}}{(K\eta+1)^8} - \frac{K\Lambda\eta C_{12}}{2(K\eta+1)^8} + \\
+ \frac{\left(\frac{2K^2}{Pr}+1-n\right)C_{12} - (1+Kf_w)C_{13} + c\Lambda C_{14}}{(K\eta+1)^9} - \frac{K\Lambda\eta C_{13}}{2(K\eta+1)^9} + \\
+ \frac{\left(\frac{2K^2}{Pr}+1-n\right)C_{13} - (1+Kf_w)C_{14} + c\Lambda C_{15}}{(K\eta+1)^{10}} - \frac{K\Lambda\eta C_{14}}{2(K\eta+1)^{10}} + \\
+ \frac{\left(\frac{2K^2}{Pr}+1-n\right)C_{14} - (1+Kf_w)C_{15} + c\Lambda C_{16}}{(K\eta+1)^{11}} - \frac{K\Lambda\eta C_{15}}{2(K\eta+1)^{11}} + \\
+ \frac{\left(\frac{2K^2}{Pr}+1-n\right)C_{15} - (1+Kf_w)C_{16}}{(K\eta+1)^{12}} - \frac{K\Lambda\eta C_{16}}{2(K\eta+1)^{12}} + \\
+ \frac{\left(\frac{2K^2}{Pr}+1-n\right)C_{16}}{(K\eta+1)^{13}}, \theta_1(0) = \theta_1(\infty) = 0. \tag{4.314}$$

The solution of Eq. 4.314 with the solution given by Eq. 4.312, lead at the first-order approximate solution of Eqs. 4.268 and 4.273 in the form:

$$\begin{aligned}\overline{\theta}(\eta) = \theta_0(\eta) + \theta_1(\eta) = \frac{1+P_2}{K\eta+1} + \frac{q_3C_8}{2(K\eta+1)^2} + \frac{q_2C_8+q_3C_9}{6(K\eta+1)^3} + \\ + \frac{q_1C_8+q_2C_9+q_3C_{10}}{12(K\eta+1)^4} + \frac{q_1C_9+q_2C_{10}+q_3C_{11}}{20(K\eta+1)^5} + \frac{q_1C_{10}+q_2C_{11}+q_3C_{12}}{30(K\eta+1)^6} + \\ + \frac{q_1C_{11}+q_2C_{12}+q_3C_{13}}{42(K\eta+1)^7} + \frac{q_1C_{12}+q_2C_{13}+q_3C_{14}}{56(K\eta+1)^8} + \frac{q_1C_{13}+q_2C_{14}+q_3C_{15}}{72(K\eta+1)^9} + \\ + \frac{q_1C_{14}+q_2C_{15}+q_3C_{16}}{90(K\eta+1)^{10}} + \frac{q_1C_{15}+q_2C_{16}}{110(K\eta+1)^{11}} + \frac{q_1C_{16}}{132(K\eta+1)^{12}}\end{aligned} \tag{4.315}$$

where

$$\begin{aligned} q_1 &= \frac{2}{\Pr} + \frac{1-n}{K^2}; q_2 = \frac{\Lambda}{2K^2} - \frac{1}{K^3} - \frac{f_w}{K^2}; q_3 = \frac{\Lambda}{K^2}\left(c - \frac{1}{2}\right) \\ P_2 &= -\left(\frac{q_1}{12} + \frac{q_2}{6} + \frac{q_3}{2}\right)C_8 - \left(\frac{q_1}{20} + \frac{q_2}{12} + \frac{q_3}{6}\right)C_9 - \left(\frac{q_1}{30} + \frac{q_2}{20} + \frac{q_3}{12}\right)C_{10} - \\ &- \left(\frac{q_1}{42} + \frac{q_2}{30} + \frac{q_3}{20}\right)C_{11} - \left(\frac{q_1}{56} + \frac{q_2}{42} + \frac{q_3}{30}\right)C_{12} - \left(\frac{q_1}{72} + \frac{q_2}{56} + \frac{q_3}{42}\right)C_{13} - \\ &- \left(\frac{q_1}{90} + \frac{q_2}{72} + \frac{q_3}{56}\right)C_{14} - \left(\frac{q_1}{110} + \frac{q_2}{90} + \frac{q_3}{72}\right)C_{15} - \left(\frac{q_1}{132} + \frac{q_2}{110} + \frac{q_3}{90}\right)C_{16}\end{aligned} \tag{4.316}$$

4.5.3 *Numerical Examples*

We will determine the convergence-control parameters K and C_i which appear in Eqs. 4.286, 4.299, 4.308 and 4.315 by means of the least square method. In this way, the optimal convergence-control parameters are optimally determined and the first-order approximate solutions known for different values of the known parameters f_w, Λ, Pr, n and c. In what follows, we illustrate the accuracy of the OHAM comparing previously obtained approximate solutions with the numerical integration results computed by means of the shooting method combined with fourth-order Runge-Kutta method using Wolfram Mathematica 6.0 software. For some values of the parameters f_w, Λ, Pr, n and c we will determine the approximate solutions.

Example 4.5.3.1 For the first alternative given in the subsection 4.5.2.1, we consider $f_w = -1$, $\Lambda = 1$, $c = 1/2$, $n = 1$, $Pr = 0.7$. For Eq. 4.286, following the procedure described above are obtained the optimal convergence-control parameters:

$$\begin{aligned} &C_1 = 9.6452968755, C_2 = 0.1097422956, C_3 = -0.1193121348, \\ &C_4 = -20.9241100889, C_5 = 94.9807962191, C_6 = -69.7422167093, \\ &C_7 = 10.9458650144, C_8 = -93.7824327712, C_9 = -78.5221460579, \\ &K = 1.2635440003\end{aligned}$$

Table 4.98 Comparison between OHAM results given by Eq. 4.317 and numerical results for $f_w = -1, \Lambda = 1$

η	$f_{numeric}$	$\overline{f}_{OHAM}$, Eq. 4.317	*relative error* $= \lvert f_{numeric} - \overline{f}_{OHAM} \rvert$
0	−1	−1	3.55×10^{-15}
1	−0.1497942276	−0.1498344102	4.01×10^{-5}
2	0.3108643384	0.3109544567	9.01×10^{-5}
3	0.4604991620	0.4604195294	7.96×10^{-5}
4	0.4887865463	0.4888639524	7.74×10^{-5}
5	0.4919308455	0.4919139907	1.68×10^{-5}
6	0.4921388939	0.4920918624	4.70×10^{-5}
7	0.4921471111	0.4921672825	2.01×10^{-5}
8	0.4921472622	0.4922055552	5.82×10^{-5}
9	0.4921472290	0.4922010303	5.38×10^{-5}
10	0.4921472001	0.4921818790	3.46×10^{-5}

and consequently the first-order approximate solution 4.286 can be written in the form:

$$\begin{aligned}
\overline{f}(\eta) &= 0.4921441442 + \big(- 13.0032535811 + 6.574641676\eta - \\
&-0.8531916063\eta^2 - 0.0148365254\eta^3 + 0.005901661\eta^4\big)e^{-1.263544003\eta} + \\
&+\big(5.3801670145 + 6.3720335263\eta + 5.5919507063\eta^2 - \\
&-3.6402713785\eta^3\big)e^{-2.5270880007\eta} + \big(6.1309424223 + 8.459443797\eta + \\
&+0.7793533299\eta^2 - 1.0246373515\eta^3\big)e^{-3.7906320011\eta}
\end{aligned} \tag{4.317}$$

Now, for Eq. 4.299, the optimal convergence-control parameters are:

$$C_{10} = -6.6921310048, C_{11} = 1.5495404722, C_{12} = -0.0585397574, C_{13} = -2.892719$$
$$C_{14} = 16.5956433751, C_{15} = -1.8970024371, C_{16} = 90.566874605, C_{17} = 22.38755109$$

and therefore the first-order approximate solution 4.299 becomes:

$$\begin{aligned}
\overline{\theta}(\eta) &= \big(- 12.1849128512 + 11.4017666446\eta - 2.9831783642\eta^2 + \\
&+0.2821061030\eta^3 - 0.0073174696\eta^4\big)e^{-1.2635440003\eta} + \big(7.6348206753 + \\
&+13.8411701476\eta - 3.7554538548\eta^2 + 0.3753336719\eta^3\big)e^{-2.5270880007\eta} + \\
&+\big(5.5500921758 - 0.8058065356\eta - 1.4765051770\eta^2\big)e^{-3.7906320011\eta}
\end{aligned} \tag{4.318}$$

In Tables 4.98 and 4.99 we present a comparison between the first-order approximate solutions given by Eqs. 4.317 and 4.318 respectively, with numerical results for some values of variable η and the corresponding relative errors.

Table 4.99 Comparison between OHAM results given by Eq. 4.318 and numerical results for $f_w = -1, \Lambda = 1, c = 1/2, n = 1, Pr = 0.7$

η	$\theta_{numeric}$	$\overline{\theta}_{OHAM}$, Eq. 4.318	$relative\ error = \lvert\theta_{numeric} - \overline{\theta}_{OHAM}\rvert$
0	1	1	5.32×10^{-15}
1	0.5326742751	0.5326094013	6.48×10^{-5}
2	0.2137126209	0.2136635578	4.90×10^{-5}
3	0.0624455901	0.0624695312	2.39×10^{-5}
4	0.0129760976	0.0130451746	6.90×10^{-5}
5	0.0019028731	0.0017825465	1.20×10^{-4}
6	0.0001964797	0.0001536740	4.28×10^{-5}
7	0.0000141245	0.0000939655	7.98×10^{-5}
8	5.56×10^{-7}	0.0001048851	1.04×10^{-4}
9	-1.33×10^{-7}	0.0000741555	7.42×10^{-5}
10	-1.43×10^{-7}	0.0000405093	4.06×10^{-5}

Example 4.5.3.2 In this case, we consider $f_w = -1, \Lambda = 1, c = \frac{1}{2}, n = 1, Pr = 2$. The solution $\overline{f}(\eta)$ is given by Eq. 4.317. The optimal convergence-control parameters for Eq. 4.299 are:

$$C_{10} = 30.0451241235, C_{11} = -9.3295076070, C_{12} = 0.7087908262, C_{13} = -56.8560242711$$
$$C_{14} = 14.9052101750, C_{15} = -35.7551937615, C_{16} = 202.9733744147, C_{17} = 73.8738456424$$

such that the first-order approximate solution 4.299 becomes:

$$\begin{aligned}\overline{\theta}(\eta) &= \left(7.5223382835 - 21.9218912323\eta + 9.6128365232\eta^2 - \right.\\ &\left.-1.5664765389\eta^3 + 0.0885988532\eta^4\right)e^{-1.2635440003\eta} + \left(-11.1527575799 + \right.\\ &\left.+27.9256280939\eta + 4.7569854384\eta^2 + 7.0743863591\eta^3\right)e^{-2.5270880007\eta} + \\ &+(4.6304192964 - 7.7685639892\eta - 4.8721325111\eta^2)e^{-3.7906320011\eta}\end{aligned} \tag{4.319}$$

In Table 4.100 we present a comparison between the first-order approximate solutions given by Eq. 4.319 with numerical results and corresponding relative errors.

Example 4.5.3.3 For $f_w = 0,\ \Lambda = 1,\ c = 1/2,\ n = 1,\ Pr = 0.7$, the optimal convergence-control parameters for Eq. 4.286 are:

$$C_1 = -25.0117201741, C_2 = 6.6137319587, C_3 = -19.6198516695,$$
$$C_4 = 10.2458750670, C_5 = -14.5125817256, C_6 = 110.2766434456,$$
$$C_7 = 16.4700689255, C_8 = 98.4254858353, C_9 = 175.2340779848, K = 2.74296086$$

Table 4.100 Comparison between OHAM results given by Eq. 4.319 and numerical results for $f_w = -1, \Lambda = 1, c = 1/2, n = 1, Pr = 2$

η	$\theta_{numeric}$	$\overline{\theta}_{OHAM}$, Eq. 4.319	relative error $= \lvert\theta_{numeric} - \overline{\theta}_{OHAM}\rvert$
0	1	1	4.4×10^{-15}
1	0.3342437188	0.3336572672	5.86×10^{-4}
2	0.0347397038	0.0346981189	4.15×10^{-5}
3	0.0011292869	0.0011292869	1.83×10^{-4}
4	0.0000127051	0.0005111089	4.98×10^{-4}
5	-7.03×10^{-8}	−0.0002892869	2.89×10^{-4}
6	-1.05×10^{-7}	−0.0002730640	2.72×10^{-4}
7	-9.14×10^{-8}	0.0001343804	1.34×10^{-4}
8	-8.15×10^{-8}	0.0003422905	3.42×10^{-4}
9	-7.54×10^{-8}	0.0003254867	3.25×10^{-4}
10	-6.89×10^{-8}	0.0002249729	2.25×10^{-4}

and therefore, the first-order approximate solution 4.286 can be written in the form:

$$\begin{aligned}\overline{f}(\eta) &= 0.9662741803 + (3.6114466434 - 8.7564860947\eta +\\ &+3.7113826615\eta^2 - 2.4930207584\eta^3 + 0.4470500279\eta^4)e^{-2.7429608601\eta} +\\ &+(-3.3184342100 - 5.8768121011\eta - 4.4122265025\eta^2 +\\ &+1.2214145001\eta^3)e^{-5.4859217203\eta} + (-1.2592866137 - 3.0278370028\eta -\\ &-1.8144246850\eta^2 + 0.4852193471\eta^3)e^{-8.2288825805\eta}\end{aligned} \tag{4.320}$$

For Eq. 4.299, the optimal convergence-control parameters are:

$$C_{10} = -162.5504084166, C_{11} = 85.2172492101, C_{12} = -46.4287699827, C_{13} = 508.1703952920$$
$$C_{14} = -354.5685124251, C_{15} = 786.5035291540, C_{16} = -2220.5827908553, C_{17} = -577.6054765061$$

and the first-order approximate solution 4.299 is:

$$\begin{aligned}\overline{\theta}(\eta) &= (-309.4860987610 + 433.5833040526\eta - 238.2844781153\eta^2 +\\ &+63.5624829213\eta^3 - 5.8035962478\eta^4)e^{-2.7429608601\eta} + (413.8470916973 +\\ &+189.5182352991\eta + 450.3689197225\eta^2 - 71.6838089606\eta^3)e^{-5.4859217203\eta} +\\ &+(-3.3609929363 - 53.0020202602\eta + 17.5481139404\eta^2)e^{-8.2288825805\eta}\end{aligned} \tag{4.321}$$

In Tables 4.101 and 4.102 we present a comparison between the first-order approximate solutions given by Eqs. 4.320 and 4.321 respectively, with numerical results and corresponding relative errors.

Table 4.101 Comparison between OHAM results given by Eq. 4.320 and numerical results for $f_w = 0, \Lambda = 1$

η	$f_{numeric}$	$\overline{f}_{OHAM}$, Eq. 4.320	*relative error* = $\left\|f_{numeric} - \overline{f}_{OHAM}\right\|$
0	-2.72×10^{-21}	-2.22×10^{-16}	2.22×10^{-16}
1	0.6894348681	0.6894237664	1.11×10^{-5}
2	0.9167697025	0.9167686777	1.02×10^{-6}
3	0.9608822313	0.9608821622	6.91×10^{-8}
4	0.9659198252	0.9659240893	4.26×10^{-6}
5	0.9662621701	0.9662467814	1.53×10^{-5}
6	0.9662761187	0.9662763785	2.59×10^{-7}
7	0.9662764567	0.9662882274	1.17×10^{-5}
8	0.9662764581	0.9662872359	1.07×10^{-5}
9	0.9662764551	0.9662825969	6.14×10^{-6}
10	0.9662764527	0.9662790666	2.61×10^{-6}

Table 4.102 Comparison between OHAM results given by Eq. 4.321 and numerical results for $f_w = 0, \Lambda = 1, c = 1/2, n = 1, Pr = 0.7$

η	$\theta_{numeric}$	$\overline{\theta}_{OHAM}$, Eq. 4.321	*relative error* = $\left\|\theta_{numeric} - \overline{\theta}_{OHAM}\right\|$
0	1	1	2.84×10^{-14}
1	0.4003247620	0.4004484539	1.23×10^{-4}
2	0.1184316476	0.1183669267	6.47×10^{-5}
3	0.0250381505	0.0249827647	5.53×10^{-5}
4	0.0037388119	0.0033436318	3.95×10^{-4}
5	0.0003936490	0.0002426294	1.51×10^{-4}
6	0.0000293256	-5.56×10^{-6}	3.48×10^{-5}
7	1.64×10^{-6}	-4.96×10^{-6}	6.61×10^{-6}
8	1.66×10^{-7}	-9.79×10^{-7}	1.14×10^{-6}
9	1.01×10^{-7}	-1.41×10^{-7}	2.42×10^{-7}
10	9.12×10^{-8}	-1.74×10^{-8}	1.08×10^{-7}

Example 4.5.3.4 For $f_w = 0,\ \Lambda = 1,\ c = 1/2,\ n = 1,\ Pr = 2$ the first-order approximate solution 4.286 is given by Eq. 4.320. The optimal convergence-control parameters for Eq. 4.299 are determined as:

$$C_{10} = -6.13977897514, C_{11} = 4.5056075965, C_{12} = -0.5220736728, C_{13} = -35.5868730308$$
$$C_{14} = 113.2854944599, C_{15} = -114.1037778967, C_{16} = 350.9820431138, C_{17} = 162.9773924832$$

Table 4.103 Comparison between OHAM results given by Eq. 4.322 and numerical results for $f_w = 0$, $\Lambda = 1$, $c = 1/2$, $n = 1$, $Pr = 2$

η	$\theta_{numeric}$	$\overline{\theta}_{OHAM}$, Eq. 4.322	$relative\ error = \lvert\theta_{numeric} - \overline{\theta}_{OHAM}\rvert$
0	1	0.9999999999	8.88×10^{-16}
1	0.1197331463	0.1197256799	7.46×10^{-6}
2	0.0042245165	0.0042165993	7.91×10^{-6}
3	0.0000508826	0.0000581491	7.26×10^{-6}
4	2.05×10^{-7}	0.0000140335	1.38×10^{-5}
5	-1.40×10^{-8}	2.72×10^{-6}	2.73×10^{-6}
6	-1.25×10^{-8}	2.31×10^{-7}	2.44×10^{-7}
7	-1.10×10^{-8}	-1.14×10^{-9}	9.88×10^{-9}
8	-1.01×10^{-8}	-3.85×10^{-9}	6.32×10^{-9}
9	-9.72×10^{-9}	-7.92×10^{-10}	8.93×10^{-9}
10	-9.23×10^{-9}	-1.15×10^{-10}	9.12×10^{-9}

such that the first-order approximate solution 4.299 may be written as:

$$\begin{aligned}\overline{\theta}(\eta) = &\left(-0.4134528457 + 4.7385301362\eta - 3.5149237889\eta^2 + \right.\\ &\left.+0.8627177660\eta^3 - 0.0652592091\eta^4\right)e^{-2.7429608601\,\eta} + \left(-3.1515927315 + \right.\\ &\left.+14.8125629408\eta - 17.9982969580\eta^2 + 10.3996906730\eta^3\right)e^{-5.4859217203\eta} + \\ &+(4.5650455772 - 1.8953267703\eta - 4.9513828544\eta^2)e^{-8.2288825805\eta}\end{aligned} \tag{4.322}$$

In Table 4.103 we compare the first-order approximate solutions given by Eq. 4.322 with numerical results. The corresponding relative errors are also presented.

Example 4.5.3.5 We consider $f_w = 1$, $\Lambda = 1$, $c = 1/2$, $n = 1$, $Pr = 0.7$. The optimal convergence-control parameters for Eq. 4.286 are given by:

$$\begin{aligned}&C_1 = 0.0896409781, C_2 = -6.7409754539, C_3 = -8.2094550260,\\ &C_4 = -0.4938997135, C_5 = -2.5507847956, C_6 = -17.8757706647,\\ &C_7 = -0.0794196188, C_8 = -3.2167864927, C_9 = 4.3212597652, K = 3.2735461689\end{aligned}$$

The first-order approximate solution 4.286 reads:

$$\begin{aligned}\overline{f}(\eta) = &1.6119227643 + \left(-0.8884126339 - 0.6754353143\eta - \right.\\ &\left.-0.6308104346\eta^2 - 0.4913885608\eta^3 + 0.1567385680\eta^4\right)e^{-3.2735461689\eta} + \\ &+\left(0.2723347361 + 0.5883664238\eta + 0.3787001472\eta^2 - \right.\\ &\left.-0.1390101741\eta^3\right)e^{-6.5470923379\eta} + \left(0.0041551334 + 0.0026158441\eta - \right.\\ &\left.-0.0361134342\eta^2 + 0.0084010234\eta^3\right)e^{-9.8206385068\,\eta}\end{aligned} \tag{4.323}$$

Table 4.104 Comparison between OHAM results given by Eq. 4.323 and numerical results for $f_w = 1, \Lambda = 1$

η	$f_{numeric}$	$\overline{f}_{OHAM}$, Eq. 4.323	$relative\ error = \left\|f_{numeric} - \overline{f}_{OHAM}\right\|$
0	1	1	0
1	1.5177074192	1.5177102108	2.79×10^{-6}
2	1.6030516967	1.6030543811	2.68×10^{-6}
3	1.6114161917	1.6114250084	8.81×10^{-6}
4	1.6119056438	1.6119124642	6.82×10^{-6}
5	1.6119228465	1.6119240499	1.20×10^{-6}
6	1.6119232066	1.6119229689	2.37×10^{-7}
7	1.6119232084	1.6119227834	4.25×10^{-7}
8	1.6119232063	1.6119227657	4.40×10^{-7}
9	1.6119232045	1.6119227644	4.40×10^{-7}
10	1.6119232031	1.6119227643	4.38×10^{-7}

The optimal convergence-control parameters for Eq. 4.299, are:

$$C_{10} = -109.3048316983, C_{11} = 67.7129511272, C_{12} = -41.6850053604, C_{13} = 381.1698243654$$
$$C_{14} = -360.4698971005, C_{15} = 816.0658192188, C_{16} = -1741.3618639076, C_{17} = -524.9835767955$$

and the first-order approximate solution 4.299 becomes:

$$\begin{aligned}\overline{\theta}(\eta) = &\left(-169.6126739547 + 278.0490758824\eta - 175.3219141601\eta^2 + \right.\\ &\left. +53.8811240663\eta^3 - 5.2106256700\eta^4\right)e^{-3.2735461689\eta} + \left(235.3726180577 + \right.\\ &\left. +113.5920574435\eta + 381.0769578373\eta^2 - 62.3227668940\eta^3\right)e^{-6.5470923379\eta} + \\ &+(-64.7599441030 - 43.0516620812\eta + 13.3642933814\eta^2)e^{-9.8206385068\eta}\end{aligned} \tag{4.324}$$

In Tables 4.104 and 4.105 we present a comparison between the first-order approximate solutions given by Eqs. 4.323 and 4.324 respectively, with numerical results and corresponding relative errors.

Example 4.5.3.6 For $f_w = 1,\ \Lambda = 1,\ c = 1/2,\ n = 1,\ Pr = 2$ the first-order approximate solution for $\overline{f}(\eta)$ is given by Eq. 4.323.

For Eq. 4.299 the optimal convergence-control parameters are given by:

$$C_{10} = -2.8367036538, C_{11} = 1.6885612882, C_{12} = -0.2429637782, C_{13} = 1.0483056239$$
$$C_{14} = 66.7794092539, C_{15} = -107.0774020659, C_{16} = 116.5916922691, C_{17} = 87.5482503888$$

Table 4.105 Comparison between OHAM results given by Eq. 4.324 and numerical results for $f_w = 1, \Lambda = 1, c = 1/2, n = 1, Pr = 0.7$

η	$\theta_{numeric}$	$\overline{\theta}_{OHAM}$, Eq. 4.324	relative error $= \lvert \theta_{numeric} - \overline{\theta}_{OHAM} \rvert$
0	1	1	0
1	0.2625846355	0.2627285301	1.43×10^{-4}
2	0.0500297082	0.0502158852	1.86×10^{-4}
3	0.0067456382	0.0064907372	2.54×10^{-4}
4	0.0006411974	0.0005182199	1.22×10^{-4}
5	0.0000429733	0.0000246256	1.83×10^{-5}
6	2.05×10^{-6}	2.13×10^{-7}	1.83×10^{-6}
7	8.85×10^{-8}	-9.42×10^{-8}	1.82×10^{-7}
8	2.04×10^{-8}	-1.23×10^{-8}	3.28×10^{-8}
9	1.72×10^{-8}	-1.08×10^{-9}	1.83×10^{-8}
10	1.58×10^{-8}	-7.97×10^{-11}	1.59×10^{-8}

Table 4.106 Comparison between OHAM results given by Eq. 4.325 and numerical results for $f_w = 1, \Lambda = 1, c = 1/2, n = 1, Pr = 2$

η	$\theta_{numeric}$	$\overline{\theta}_{OHAM}$, Eq. 4.325	relative error $= \lvert \theta_{numeric} - \overline{\theta}_{OHAM} \rvert$
0	1	1	2.22×10^{-16}
1	0.0288438340	0.0288405127	3.32×10^{-6}
2	0.0002626174	0.0002601131	2.50×10^{-6}
3	8.78×10^{-7}	2.34×10^{-6}	1.46×10^{-6}
4	1.68×10^{-8}	4.33×10^{-8}	2.64×10^{-8}
5	1.38×10^{-8}	-6.24×10^{-8}	7.63×10^{-8}
6	1.23×10^{-8}	-1.18×10^{-8}	2.42×10^{-8}
7	1.12×10^{-8}	-1.33×10^{-9}	1.25×10^{-8}
8	1.06×10^{-8}	-1.16×10^{-10}	1.07×10^{-8}
9	1.02×10^{-8}	-8.76×10^{-12}	1.02×10^{-8}
10	9.70×10^{-9}	-5.93×10^{-13}	9.70×10^{-9}

The first-order approximate solution 4.299 one retrieves as:

$$\begin{aligned}\overline{\theta}(\eta) = {} & (0.0531334758 + 0.8465102419\eta - 0.8618419294\eta^2 + \\ & +0.2835175722\eta^3 - 0.0303704722\eta^4)e^{-3.2735461689\eta} + (0.6743212346 + \\ & +5.4965670108\eta - 1.7750564463\eta^2 + 8.1774776144\eta^3)e^{-6.5470923379\eta} + \\ & +(0.2725452894 - 1.9452075278\eta - 2.2286802005\eta^2)e^{-9.8206385068\eta}\end{aligned} \tag{4.325}$$

In Table 4.106 we present a comparison between the first-order approximate solutions given by Eqs. 4.325 with numerical results. The corresponding relative errors are presented.

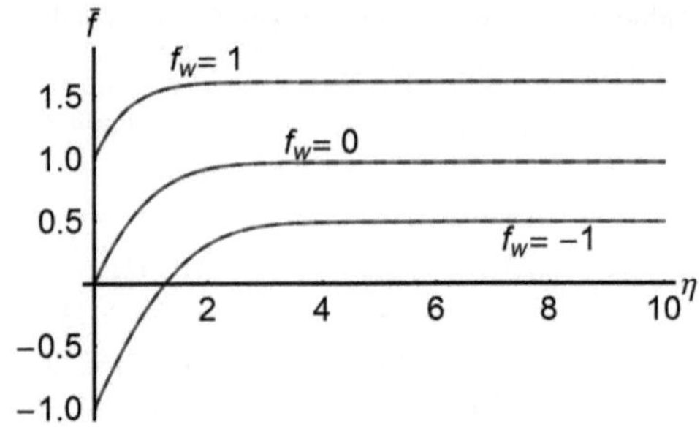

Fig. 4.95 Solutions $\overline{f}_{OHAM}(\eta)$ given by 4.317, 4.320 and 4.323 for different values of f_w: (*red solid line*) numerical results; (*blue dashed line*) approximate solution (colour figure online)

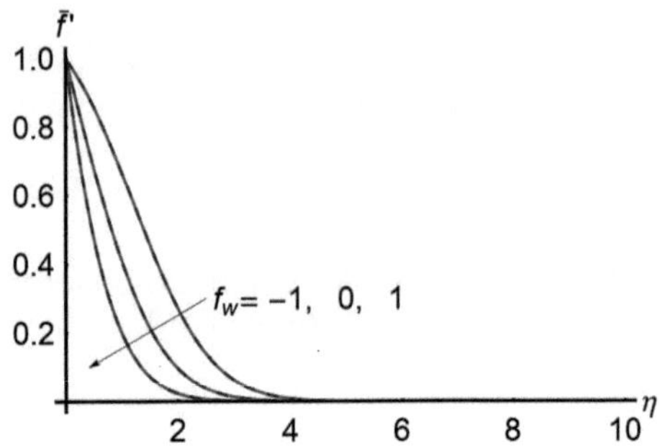

Fig. 4.96 Solutions $\overline{f}_{OHAM}(\eta)$ obtained from 4.317, 4.320 and 4.323 for different values of f_w: (*red solid line*) numerical results; (*blue dashed line*) approximate solution (colour figure online)

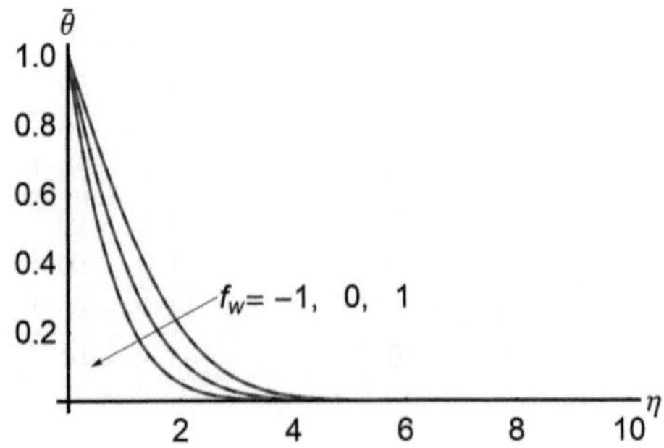

Fig. 4.97 Plots of $\overline{\theta}_{OHAM}(\eta)$ given by 4.318, 4.321 and 4.324 for $\Lambda = 1$, $c = 1/2$, $n = 1$, $\mathrm{Pr} = 0.7$ and three values of f_w: (*red solid line*) numerical results; (*blue dashed line*) approximate solution (colour figure online)

In Figs. 4.95 and 4.96 are plotted the profiles of $\overline{f}(\eta)$ and velocity profile $\overline{f'}(\eta)$ respectively for different values of f_w. It is clear that the solution $\overline{f}(\eta)$ increases with an increase of f_w and the velocity decreases with an increase of f_w. The condition $\overline{f'}(\eta) > 0$ for $\eta > 0$ is satisfied.

In Figs. 4.97, 4.98, 4.99, 4.100, 4.101 are plotted the temperature profiles given for two values of Prandl number $Pr = 0.7$ and $Pr = 2$ respectively and different values of f_w. From Figs 4.97 and 4.98 it is observe that the temperature $\overline{\theta}(\eta)$ decreases with an increase of the f_w for any values of parameter Pr.

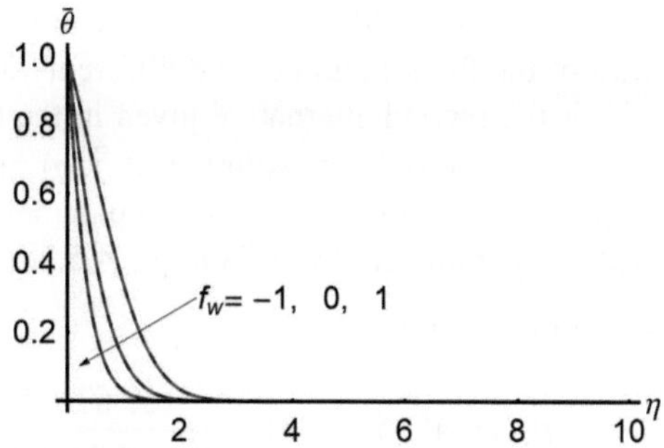

Fig. 4.98 Plots of $\overline{\theta}_{OHAM}(\eta)$ given by 4.319, 4.322 and 4.325 for $\Lambda = 1$, $c = 1/2$, $n = 1$, $\Pr = 2$ and three values of f_w: (*red solid line*) numerical results; (*blue dashed line*) approximate solution (colour figure online)

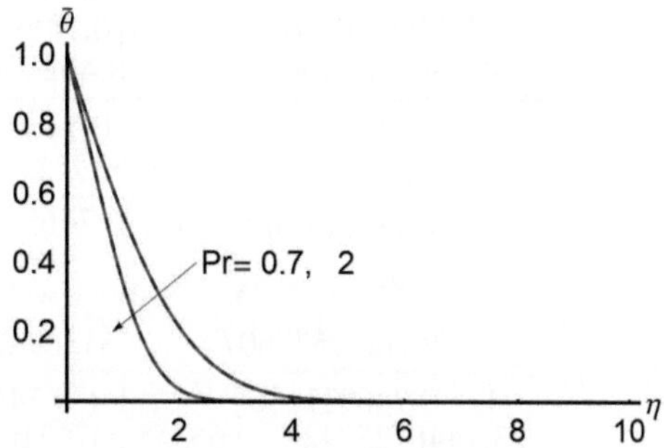

Fig. 4.99 Plots of $\overline{\theta}_{OHAM}(\eta)$ given by 4.318, and 4.319 for $\Lambda = 1$, $c = 1/2$, $n = 1$, $f_w = -1$ and two values of Pr: (*red solid line*) numerical results; (*blue dashed line*) approximate solution (colour figure online)

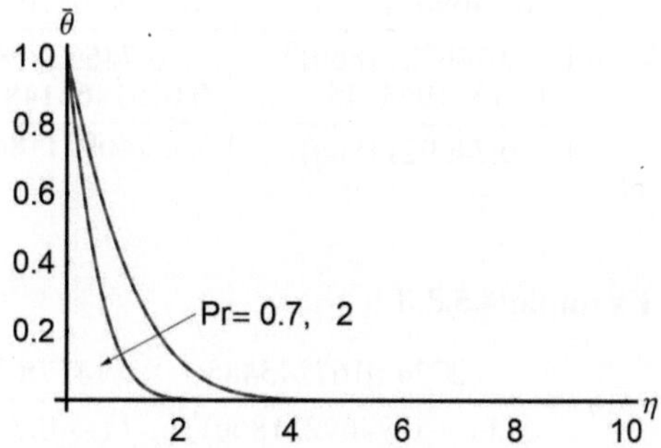

Fig. 4.100 Plots of $\overline{\theta}_{OHAM}$ (η) given by 4.321 and 4.322 for $\Lambda = 1$, $c = 1/2$, $n = 1$, $f_w = 0$ and two values of Pr: (*red solid line*) numerical results; (*blue dashed line*) approximate solution (colour figure online)

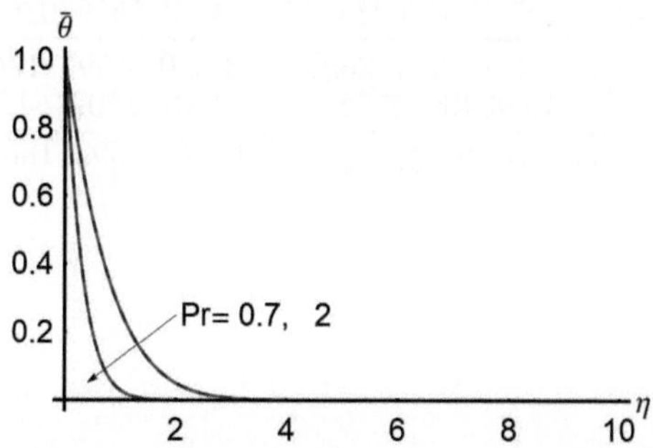

Fig. 4.101 Plots of $\overline{\theta}_{OHAM}$ (η) given by 4.324 and 4.325 for $\Lambda = 1$, $c = 1/2$, $n = 1$, $f_w = 1$ and two values of Pr: (*red solid line*) numerical results; (*blue dashed line*) approximate solution (colour figure online)

From Figs. 4.99, 4.100, 4.101 we can conclude that the temperature decrease with of the Prandl number and different values of f_w.

For the second alternative given in the Subsection 4.5.2.2, we present only the first-order approximate solution of $\overline{f}(\eta)$ and $\overline{\theta}(\eta)$ given by Eqs. 4.308 and 4.315 respectively. We mention that the parameters from Example 4.5.3. α are the same with the parameters from Example 4.5.3.α + 6 where $\alpha = 1, 2 \ldots .6$.

Example 4.5.3.7

$$\begin{aligned}\overline{f}(\eta) = 0.4835407781 - \frac{160.6727530823}{(1+0.746921186\eta)^9} + \frac{633.1433635563}{(1+0.746921186\eta)^8} - \\ - \frac{892.4875001383}{(1+0.746921186\eta)^7} + \frac{426.4144218646}{(1+0.746921186\eta)^6} + \frac{159.3296084991}{(1+0.746921186\eta)^5} - \\ - \frac{235.7972895388}{(1+0.746921186\eta)^4} + \frac{78.4024369095}{(1+0.746921186\eta)^3} - \\ - \frac{10.3553564922}{(1+0.746921186\eta)^2} + \frac{0.5395276438}{1+0.746921186\eta};\end{aligned} \tag{4.326}$$

$$\begin{aligned}\overline{\theta}(\eta) = \frac{5776.4812527607}{(1+0.746921186\eta)^{12}} - \frac{31369.266442779}{(1+0.746921186\eta)^{11}} + \frac{71958.881460483}{(1+0.746921186\eta)^{10}} - \\ - \frac{89923.1448725742}{(1+0.746921186\eta)^9} + \frac{65355.5376310228}{(1+0.746921186\eta)^8} - \frac{26746.1538968436}{(1+0.746921186\eta)^7} + \\ + \frac{4965.4696216472}{(1+0.746921186\eta)^6} + \frac{132.5980830091}{(1+0.746921186\eta)^5} - \frac{164.5468864556}{(1+0.746921186\eta)^4} + \\ + \frac{15.1631963445}{(1+0.746921186\eta)^3} - \frac{0.0191466148}{1+0.746921186\eta};\end{aligned} \tag{4.327}$$

Example 4.5.3.8

$$\begin{aligned}\overline{\theta}(\eta) = -\frac{2774.0167245485}{(1+0.746921186\eta)^{12}} + \frac{14175.2434254252}{(1+0.746921186\eta)^{11}} - \frac{29579.9446633193}{(1+0.746921186\eta)^{10}} + \\ + \frac{31500.0785269703}{(1+0.746921186\eta)^9} - \frac{16564.3693045166}{(1+0.746921186\eta)^8} + \frac{1958.3672090004}{(1+0.746921186\eta)^7} + \\ + \frac{2217.7863982347}{(1+0.746921186\eta)^6} - \frac{1125.5665403671}{(1+0.746921186\eta)^5} + \frac{207.390156582}{(1+0.746921186\eta)^4} - \\ - \frac{13.9840036775}{(1+0.746921186\eta)^3} + \frac{0.0155202164}{1+0.746921186\eta};\end{aligned} \tag{4.328}$$

Example 4.5.3.9

$$\begin{aligned}\overline{f}(\eta) = 0.9621568509 &- \frac{40.4435319395}{(1+0.2350916817\eta)^9} + \frac{116.0406686825}{(1+0.2350916817\eta)^8} - \\ &- \frac{66.6539187472}{(1+0.2350916817\eta)^7} - \frac{101.0600483755}{(1+0.2350916817\eta)^6} + \frac{157.2214635275}{(1+0.2350916817\eta)^5} - \\ &- \frac{86.2725049816}{(1+0.2350916817\eta)^4} + \frac{23.1122830793}{(1+0.2350916817\eta)^3} - \\ &- \frac{3.0970790202}{(1+0.2350916817\eta)^2} + \frac{0.1905109238}{1+0.2350916817\eta};\end{aligned} \tag{4.329}$$

$$\begin{aligned}\overline{\theta}(\eta) = &\frac{8.6480822639}{(1+0.2350916817\eta)^{12}} - \frac{278.4205849661}{(1+0.2350916817\eta)^{11}} + \frac{960.0799007261}{(1+0.2350916817\eta)^{10}} - \\ &- \frac{1139.8489060253}{(1+0.2350916817\eta)^9} + \frac{157.102753844}{(1+0.2350916817\eta)^8} + \frac{802.5519134898}{(1+0.2350916817\eta)^7} - \\ &- \frac{758.308651043}{(1+0.2350916817\eta)^6} + \frac{302.9379777076}{(1+0.2350916817\eta)^5} - \frac{58.3046748949}{(1+0.2350916817\eta)^4} + \\ &+ \frac{4.5726605819}{(1+0.2350916817\eta)^3} - \frac{0.010471684}{1+0.2350916817\eta};\end{aligned} \tag{4.330}$$

Example 4.5.3.10

$$\begin{aligned}\overline{\theta}(\eta) = &- \frac{15.3585464656}{(1+0.2350916817\eta)^{12}} + \frac{1324.9629387994}{(1+0.2350916817\eta)^{11}} - \\ &- \frac{6224.7965545045}{(1+0.2350916817\eta)^{10}} + \frac{12551.9849904292}{(1+0.2350916817\eta)^9} - \frac{13938.734218336}{(1+0.2350916817\eta)^8} + \\ &+ \frac{9296.1388960946}{(1+0.2350916817\eta)^7} - \frac{3815.8964759625}{(1+0.2350916817\eta)^6} + \frac{945.6792910964}{(1+0.2350916817\eta)^5} - \\ &- \frac{131.0527100223}{(1+0.2350916817\eta)^4} + \frac{8.0860093938}{(1+0.2350916817\eta)^3} - \frac{0.0136205223}{1+0.2350916817\eta};\end{aligned} \tag{4.331}$$

Example 4.5.3.11

$$\begin{aligned}\overline{f}(\eta) = 1.6105433181 &- \frac{11.7580011934}{(1+0.3223046368\eta)^9} + \frac{29.8301473184}{(1+0.3223046368\eta)^8} - \\ &- \frac{1.6143096961}{(1+0.3223046368\eta)^7} - \frac{59.3976259403}{(1+0.3223046368\eta)^6} + \frac{69.0721922017}{(1+0.3223046368\eta)^5} - \\ &- \frac{34.3372361715}{(1+0.3223046368\eta)^4} + \frac{8.6285265528}{(1+0.3223046368\eta)^3} - \\ &- \frac{1.0992585934}{(1+0.3223046368\eta)^2} + \frac{0.0650222037}{1+0.3223046368\eta}\end{aligned} \tag{4.332}$$

Table 4.107 Comparison between OHAM results given by Eq. 4.326 and numerical results for $f_w = -1, \Lambda = 1$

η	$f_{numeric}$	$\overline{f}_{OHAM}$ Eq. 4.326	*relative error* $= \left\lvert f_{numeric} - \overline{f}_{OHAM} \right\rvert$
0	−1	−1	2.19×10^{-12}
1	−0.1497942044	−0.1529007368	3.10×10^{-3}
2	0.3108644420	0.3104344026	4.30×10^{-4}
3	0.4604993765	0.4538634998	6.63×10^{-3}
4	0.4887868562	0.4892067993	4.19×10^{-4}
5	0.4919312215	0.4954735834	3.54×10^{-3}
6	0.4921393144	0.4948271475	2.68×10^{-3}
7	0.4921475634	0.4931440674	9.96×10^{-4}
8	0.4921477384	0.4918139645	3.33×10^{-4}
9	0.4921477238	0.4910192311	1.12×10^{-3}
10	0.4921477099	0.4906553086	1.49×10^{-3}

$$\begin{aligned}\overline{\theta}(\eta) = &-\frac{3.5244772845}{(1+0.3223046368\eta)^{12}} + \frac{82.4835925018}{(1+0.3223046368\eta)^{11}} - \\ &-\frac{560.0298081734}{(1+0.3223046368\eta)^{10}} + \frac{1639.70570286}{(1+0.3223046368\eta)^{9}} - \frac{2528.275401232}{(1+0.3223046368\eta)^{8}} + \\ &+\frac{2222.6091071034}{(1+0.3223046368\eta)^{7}} - \frac{1135.2615273237}{(1+0.3223046368\eta)^{6}} + \frac{331.3512102608}{(1+0.3223046368\eta)^{5}} - \\ &-\frac{51.4442236255}{(1+0.3223046368\eta)^{4}} + \frac{3.3913372651}{(1+0.3223046368\eta)^{3}} - \frac{0.0055123518}{1+0.3223046368\eta};\end{aligned} \tag{4.333}$$

Example 4.5.3.12

$$\begin{aligned}\overline{\theta}(\eta) = &-\frac{10.3673796978}{(1+0.3223046368\eta)^{12}} + \frac{482.8585838577}{(1+0.3223046368\eta)^{11}} - \\ &-\frac{2284.5038236813}{(1+0.3223046368\eta)^{10}} + \frac{4704.3920875677}{(1+0.3223046368\eta)^{9}} - \frac{5322.4805989766}{(1+0.3223046368\eta)^{8}} + \\ &+\frac{3596.4529013927}{(1+0.3223046368\eta)^{7}} - \frac{1485.3963216692}{(1+0.3223046368\eta)^{6}} + \frac{367.2429052041}{(1+0.3223046368\eta)^{5}} - \\ &-\frac{50.200024372}{(1+0.3223046368\eta)^{4}} + \frac{3.0055038826}{(1+0.3223046368\eta)^{3}} - \frac{0.0038335076}{1+0.3223046368\eta};\end{aligned} \tag{4.334}$$

From Tables 4.98, 4.99, 4.100, 4.101, 4.102, 4.103, 4.104, 4.105, 4.106, 4.107, 4.108, 4.109, 4.110, 4.111, 4.112, 4.113, 4.114, 4.115 for both of alternatives we can summarize that the results obtained by means of OHAM are very accurate in comparison with the numerical results. Moreover, if we represent graphically $\overline{f}$ and

Table 4.108 Comparison between OHAM results given by Eq. 4.327 and numerical results for $f_w = -1, \Lambda = 1, c = 1/2, n = 1, Pr = 0.7$

η	$\theta_{numeric}$	$\overline{\theta}_{OHAM}$ Eq. 4.327	$relative\ error = \lvert\theta_{numeric} - \overline{\theta}_{OHAM}\rvert$
0	1	1	6.36×10^{-12}
1	0.5330730009	0.5322232473	8.49×10^{-4}
2	0.2146584465	0.2129729862	1.68×10^{-3}
3	0.0628614705	0.0636111016	7.49×10^{-4}
4	0.0130346763	0.0147892863	1.75×10^{-3}
5	0.00190668436	0.0016130301	2.93×10^{-4}
6	0.0001967000	−0.0009503492	1.14×10^{-3}
7	0.0000142565	−0.0007669977	7.81×10^{-4}
8	6.74×10^{-7}	−0.0001148534	1.15×10^{-4}
9	-2.78×10^{-8}	0.0004251870	4.25×10^{-4}
10	-4.86×10^{-8}	0.0007622847	7.62×10^{-4}

Table 4.109 Comparison between OHAM results given by Eq. 4.328 and numerical results for $f_w = -1, \Lambda = 1, c = 1/2, n = 1, Pr = 2$

η	$\theta_{numeric}$	$\overline{\theta}_{OHAM}$ Eq. 4.328	$relative\ error = \lvert\theta_{numeric} - \overline{\theta}_{OHAM}\rvert$
0	1	1	8.41×10^{-12}
1	0.3347045498	0.3345263562	1.78×10^{-4}
2	0.0348756972	0.0363459695	1.47×10^{-3}
3	0.0011477450	−0.0004370307	1.58×10^{-3}
4	0.0000128032	−0.0003197189	3.32×10^{-4}
5	-1.7×10^{-7}	0.0007745297	7.74×10^{-4}
6	-1.90×10^{-7}	0.0007470530	7.47×10^{-4}
7	-1.66×10^{-7}	0.0003234150	3.23×10^{-4}
8	-1.46×10^{-7}	−0.0000957322	9.55×10^{-5}
9	-1.30×10^{-7}	0.0003952057	3.95×10^{-4}
10	-1.17×10^{-7}	−0.0005724549	5.72×10^{-4}

Table 4.110 Comparison between OHAM results given by Eq. 4.329 and numerical results for $f_w = 0, \Lambda = 1$

η	$f_{numeric}$	$\overline{f}_{OHAM}$ Eq. 4.329	$relative\ error = \lvert f_{numeric} - \overline{f}_{OHAM}\rvert$
0	-5.50×10^{-21}	-1.27×10^{-12}	1.27×10^{-12}
1	0.6894348341	0.6920308645	2.59×10^{-3}
2	0.9167696529	0.9143867087	2.38×10^{-3}
3	0.9608821303	0.9621083391	1.22×10^{-3}
4	0.9659196704	0.9679260271	2.00×10^{-3}
5	0.9662619960	0.9668394760	5.77×10^{-4}
6	0.9662759513	0.9658454054	4.30×10^{-4}
7	0.9662762950	0.9656226339	6.53×10^{-4}
8	0.9662763018	0.9658253914	4.50×10^{-4}
9	0.9662763032	0.9661465828	1.29×10^{-4}
10	0.9662763043	0.9664313764	1.55×10^{-4}

Table 4.111 Comparison between OHAM results given by Eq. 4.330 and numerical results for $f_w = 0, \Lambda = 1, c = 1/2, n = 1, Pr = 0.7$

η	$\theta_{numeric}$	$\bar{\theta}_{OHAM}$ Eq. 4.330	relative error $= \lvert\theta_{numeric} - \bar{\theta}_{OHAM}\rvert$
0	1	1	1.13×10^{-13}
1	0.4004439081	0.4010649748	6.21×10^{-4}
2	0.1185598839	0.1184236150	1.36×10^{-4}
3	0.0250249964	0.0257450396	7.20×10^{-4}
4	0.0037325749	0.0033592672	3.73×10^{-4}
5	0.0003927240	−0.0001646366	5.57×10^{-4}
6	0.0000290673	−0.0000366134	6.56×10^{-5}
7	1.45×10^{-6}	0.0002647668	2.63×10^{-4}
8	-4.85×10^{-9}	0.0002800726	2.80×10^{-4}
9	-5.37×10^{-8}	0.0001278578	1.27×10^{-4}
10	-5.03×10^{-8}	−0.0000578886	5.78×10^{-5}

Table 4.112 Comparison between OHAM results given by Eq. 4.331 and numerical results for $f_w = 0, \Lambda = 1, c = 1/2, n = 1, Pr = 2$

η	$\theta_{numeric}$	$\bar{\theta}_{OHAM}$ Eq. 4.331	relative error $= \lvert\theta_{numeric} - \bar{\theta}_{OHAM}\rvert$
0	1	0.9999999999	1.36×10^{-12}
1	0.1193941521	0.1195658111	1.71×10^{-4}
2	0.0042304240	0.0036984294	5.31×10^{-4}
3	0.0000508647	0.0003939409	5.08×10^{-5}
4	2.04×10^{-7}	0.0001801267	1.79×10^{-4}
5	-1.40×10^{-8}	0.0002168196	2.16×10^{-4}
6	-1.25×10^{-8}	−0.0001880096	1.87×10^{-4}
7	-1.10×10^{-8}	1.94×10^{-6}	1.95×10^{-6}
8	-1.02×10^{-8}	0.0001329460	1.32×10^{-4}
9	-9.83×10^{-9}	0.0001566250	1.56×10^{-4}
10	-9.25×10^{-9}	0.0001038247	1.03×10^{-4}

Table 4.113 Comparison between OHAM results given by Eq. 4.332 and numerical results for $f_w = 1, \Lambda = 1$

η	$f_{numeric}$	$\bar{f}_{OHAM}$ Eq. 4.332	relative error $= \lvert f_{numeric} - \bar{f}_{OHAM}\rvert$
0	1	1	7.10×10^{-14}
1	1.5177074192	1.5178179504	1.10×10^{-4}
2	1.6030516967	1.6030704647	1.87×10^{-5}
3	1.6114161917	1.6119936425	5.77×10^{-4}
4	1.6119056438	1.6119213150	1.56×10^{-5}
5	1.6119228465	1.6116417023	2.81×10^{-4}
6	1.6119232066	1.6116883119	2.34×10^{-4}
7	1.6119232084	1.6118378302	8.53×10^{-5}
8	1.6119232063	1.6119648304	4.16×10^{-5}
9	1.6119232045	1.6120377026	1.14×10^{-4}
10	1.6119232031	1.6120627833	1.39×10^{-4}

Table 4.114 Comparison between OHAM results given by Eq. 4.333 and numerical results for $f_w = 1, \Lambda = 1, c = 1/2, n = 1, Pr = 0.7$

η	$\theta_{numeric}$	$\overline{\theta}_{OHAM}$ Eq.4.333	*relative error* = $\left\|\theta_{numeric} - \overline{\theta}_{OHAM}\right\|$
0	1	1	7.95×10^{-13}
1	0.2626142132	0.2625812869	3.29×10^{-5}
2	0.0500241709	0.0503057086	2.81×10^{-4}
3	0.0067427056	0.0067065432	3.61×10^{-5}
4	0.0006409254	0.0003360891	3.04×10^{-4}
5	0.0000429288	0.0000198651	2.30×10^{-5}
6	2.01×10^{-6}	0.0001596003	1.57×10^{-4}
7	5.78×10^{-8}	0.0001312550	1.31×10^{-4}
8	-7.26×10^{-9}	0.0000224552	2.24×10^{-5}
9	-8.05×10^{-9}	−0.0000781005	7.80×10^{-5}
10	-7.46×10^{-9}	−0.0001393169	1.39×10^{-4}

Table 4.115 Comparison between OHAM results given by Eq. 4.334 and numerical results for $f_w = 1, \Lambda = 1, c = 1/2, n = 1, Pr = 2$

η	$\theta_{numeric}$	$\overline{\theta}_{OHAM}$ Eq.4.334	*relative error* = $\left\|\theta_{numeric} - \overline{\theta}_{OHAM}\right\|$
0	1	1	1.81×10^{-12}
1	0.0292003919	0.0292096747	9.28×10^{-6}
2	0.0005730502	0.0007562468	1.83×10^{-4}
3	0.0002597282	0.0002271428	3.25×10^{-5}
4	0.0002214295	0.0000566272	1.64×10^{-4}
5	0.0001935582	0.0001758501	1.77×10^{-5}
6	0.0001720066	0.0002527193	8.07×10^{-5}
7	0.0001548229	0.0002317234	7.69×10^{-5}
8	0.0001407890	0.0001639037	2.31×10^{-5}
9	0.0001291050	0.0000941465	3.49×10^{-5}
10	0.0001192225	0.0000427725	7.64×10^{-5}

$\overline{\theta}$ in the second alternative, the obtained results are unmodified in comparison with Figs. 4.93, 4.94, 4.95, 4.96, 4.97, 4.98, 4.99.

4.6 Blasius' Problem

Blasius equation is one of the basic equations of fluid dynamics. Blasius' equation describes the velocity profile of the fluid in the boundary layer theory on a half-infinite interval. A broad class of analytical solutions method and numerical solutions methods were used to handle this problem. The Blasius equation is the mother of all boundary layer equations in fluid mechanics. Two forms of Blasius' equation appear in the fluid mechanics theory, where each is subject to specific physical conditions. For instance, consider the two-dimensional laminar viscous flow governed by a nonlinear ordinary differential equation [92–97]

$$u'''(\eta) + \frac{1}{2}u(\eta)u''(\eta) = 0, \eta \in [0, \infty) \tag{4.335}$$

subject to the boundary conditions

$$u(0) = 0, u'(0) = 0, u'(\infty) = 1 \tag{4.336}$$

or

$$u(0) = 0, u'(0) = 1, u'(\infty) = 0 \tag{4.337}$$

It is obvious that the differential equations are the same, but differ in boundary conditions. Within Eqs.4.335–4.337, the prime denotes the derivative with respect to the similarity variable which is defined as $\eta = y\sqrt{\frac{U}{vx}}$, the dimensionless function $u(\eta)$ is related to the stream function $\psi(x, y)$ by $u(\eta) = \frac{\psi(x,y)}{\sqrt{vxU_\infty}}$. Here U_∞ is the constant velocity of the mainstream at infinity, v is the kinematic viscosity coefficient and x and y are two independent variables.

Note that Blasius equation 4.335 with initial conditions 4.336 is a special case of the Falkner-Skan equation 4.175 and 4.176 with $F(\eta) = \frac{1}{2}u(\eta)$ and $\beta = 0$.

In 1908 Blasius [98] provides a power series solution:

$$u(\eta) = \sum_{k=0}^{\infty} \left(-\frac{1}{2}\right)^k \frac{A_k \sigma^{k+1}}{(3k+2)!} \eta^{3k+2} \tag{4.338}$$

where

$$A_0 = A_1 = 1, A_k = \sum_{i=0}^{k-1} \binom{3k-1}{3i} A_i A_{k-i-1} (k \geq 2) \tag{4.339}$$

with $\binom{m}{n} = \frac{m!}{n!(m-n)!}$. Note that the expression 4.338 is not closed, because $\sigma = u''(0)$ is unknown. By means of matching two different approximations at a proper point, Blasius obtained the numerical result $\sigma = 0.332$. In 1938, Howarth [99] gained a more accurate value $\sigma = 0.33206$ by means of a numerical technique. However, by means of $\sigma = 0.33206$, $u(\eta)$ given by Eq.4.338 is valid in a small region $0 \leq \eta \leq \rho_0 \approx 5.690$. Blasius' power series 4.338 is fundamentally an analytic-numerical solution, because the value of σ is gained by numerical techniques.

Until now lots of analytical methods were proposed to solve Blasius equation. He [100] proposed a perturbation technique coupled with an iteration technique. Comparison with Howarth's numerical solution reveals that the proposed method leads to the approximate value $\sigma = 0.3296$ with 0.73 % accuracy. Asaithambi [101] found this number correct to nine decimal points as $\sigma = 0.332057336$. The variational iteration method is applied for a reliable treatment of two forms of Blasius equation which comes from boundary layer equation by Wazwaz [102]. The same author proposed earlier a modified form of the Adomian decomposition method, which is found to be fast and accurate [103]. Wang [104] employed an algorithm based mainly on applying the Adomian decomposition method (ADM) to the transformation of the Blasius equation and later, Hashim [105] improved the numerical solution of Wang using the ADM–Padé approach. Recently, Fazio [106] solved numerically the Blasius problem, and its variants and extensions, by initial value methods derived within scaling invariance theory. Sinc-collocation method, which is a procedure converging to the solution at an exponential rate, is applied in [107] and the Homotopy Analysis Method (HAM) is successfully applied by Yao and Chen in [108]. Also, Yun [109] proposed an intuitive approach to the approximate analytical solution for the Blasius problem in the form of a logarithm of the hyperbolic cosine function.

All these prove that although the Blasius problem is a century old, it is still a topic of active current research.

By the homotopy analysis method [94], Liao obtained a series solution similar to 4.338, which is valid in the whole region $[0,\infty)$. The solution reads

$$u(\eta) = \lim_{m\to\infty} \sum_{k=0}^{+\infty} \left[\left(-\frac{1}{2}\right)^k \frac{A_k \sigma^{k+1}}{(3k+2)!} \eta^{3k+2} \right] \Phi_{m,k}(h), \eta \in [0,\infty), h \in (-2,0) \quad (4.340)$$

where

$$\Phi_{m,n}(h) = \begin{cases} 0 & n > m \\ (-h)^n \sum_{k=0}^{m-n} \binom{m}{m-n-k} (n+k-1) h^k & 1 \le n \le m \\ 1 & n \le 0 \end{cases} \quad (4.341)$$

An explicit, totally analytic approximate solution for Blasius equation, by means of Homotopy Analysis Method [95] is obtained by Liao:

$$u(\eta) = \eta + \lim_{M\to\infty} \left[\sum_{m=0}^{M} b_0^{m,0} + \sum_{n=1}^{M+1} \exp(-n\lambda\eta) \left(\sum_{m=n-1}^{M} \sum_{k=0}^{2(m-n+1)} b_k^{m,n} \eta^k \right) \right] \quad (4.342)$$

where the coefficients $b_k^{m,n}$ contain the auxiliary parameter h and the so-called spatial scale parameter λ.

4.6.1 Solution of Blasius' Problem by Optimal Homotopy Asymptotic Method

Using the initial conditions 4.336, we choose the initial approximate solution as

$$u_0(\eta) = \eta - \frac{1}{k} + \frac{1}{k}e^{-k\eta} \tag{4.343}$$

where k is a positive unknown parameter. Then, from Eqs. 4.4, 4.335 and 4.343 we can choose the linear operator, the function $g(\eta)$ and the initial approximate solution in the forms:

$$L(u(\eta)) = u''' - k^2 u', g(\eta) = k^2 \tag{4.344}$$

$$L(u(\eta)) = u''' + ku'', u_0(\eta) = \eta - \frac{1}{k} + \frac{1}{k}e^{-k\eta}, g(\eta) = 0 \tag{4.345}$$

$$L(u(\eta)) = u''' + \frac{3ku''(\eta)}{(k\eta+1)}, u_0(\eta) = \eta - \frac{1}{k} + \frac{1}{k(k\eta+1)}, g(\eta) = 0 \tag{4.346}$$

$$L(u(\eta)) = u''' + \frac{6k^2u'(\eta)}{(k\eta+1)^2} - \frac{6k^2}{(k\eta+1)^4}, u_0(\eta) = \eta - \frac{1}{k} + \frac{1}{k(k\eta+1)}, g(\eta) = 0 \tag{4.347}$$

For Eq. 4.344, the nonlinear operator N given by Eq. 4.1 becomes:

$$N(u(\eta)) = k^2u' + \frac{1}{2}uu'' - k^2 \tag{4.348}$$

From Eqs. 4.343 and 4.348 we obtain

$$N(u_0(\eta)) = \left(\frac{1}{2}k\eta - k^2 - \frac{1}{2}\right)e^{-k\eta} + \frac{1}{2}e^{-2k\eta} \tag{4.349}$$

Substituting Eq. 4.349 into Eq. 4.9 it is obtained the equation in $u_1(\eta)$:

$$u_1'''(\eta) - k^2u_1'(\eta) = H(\eta, C_i)\left[\left(\frac{1}{2}k\eta - k^2 - \frac{1}{2}\right)e^{-k\eta} + \frac{1}{2}e^{-2k\eta}\right] \tag{4.350}$$

The optimal auxiliary function $H(\eta, C_i)$ can be chosen under the forms:

$$H(\eta, C_i) = 2\left[C_1\eta + C_2 + (C_3\eta + C_4)e^{-k\eta} + (C_5\eta + C_6)e^{-2k\eta}\right] \tag{4.351}$$

or

$$H^*(\eta, C_i) = C_1 + C_2e^{-k\eta} + C_3e^{-2k\eta} \tag{4.352}$$

or yet

$$H^{**}(\eta, C_i) = C_1 + (C_2 + C_3\eta)e^{-k\eta} + \left(C_4 + C_5\eta + C_6\eta^2 + C_7\eta^3\right)e^{-2k\eta} \tag{4.353}$$

and so on.

If the optimal auxiliary function is given for example by Eq. 4.351, then Eq. 4.350 becomes

$$\begin{aligned} u_1''' - k^2 u_1 = &\left[kC_1\eta^2 + (kC_2 - (2k^2+1)C_1)\eta - (2k^2+1)C_2\right]e^{-k\eta} + \\ &+\left[kC_3\eta^2 + (kC_4 - (2k^2+1)C_3 + C_1)\eta + C_2 - (2k^2+1)C_4\right]e^{-2k\eta} + \\ &+\left[kC_5\eta^2 + (kC_6 - (2k^2+1)C_5 + C_3)\eta + C_4 - (2k^2+1)C_6\right]e^{-3k\eta} + \\ &+(C_5\eta + C_6)e^{-4k\eta} \end{aligned} \tag{4.354}$$

The above equation with the boundary conditions 4.350 has the solution:

$$\begin{aligned} u_1(\eta) = A &+ \left[\frac{C_1}{6k}\eta^3 + \left(\frac{1-k^2}{2k^2}C_1 + \frac{C_2}{4k}\right)\eta^2 + \left(\frac{2-3k^2}{2k^3}C_1 + \right.\right. \\ &\left.\left.+\frac{1-4k^2}{4k^2}C_2\right)\eta + B\right]e^{-k\eta} + \left[-\frac{C_3}{6k^2}\eta^2 + \left(-\frac{C_1}{6k^3} + \frac{3k^2-4}{9k^3}C_3 - \frac{C_4}{6k^2}\right)\eta - \right. \\ &\left.-\frac{11C_1}{36k^4} - \frac{C_2}{6k^3} + \frac{33k^2-26}{54k^4}C_3 + \frac{12k^2-5}{36k^3}C_4\right]e^{-2k\eta} + \left[-\frac{C_5}{24k^2}\eta^2 + \right. \\ &+\left(-\frac{C_3}{24k^3} + \frac{12k^2-7}{144k^3}C_5 - \frac{C_6}{24k^2}\right)\eta - \frac{13}{288k^4}C_3 - \frac{C_4}{24k^3} + \\ &\left.+\frac{156k^2-37}{1728k^4}C_5 + \frac{24k^2-1}{288k^3}C_6\right]e^{-3k\eta} + \left(-\frac{C_5\eta}{60k^3} - \frac{47C_5}{3600k^4} - \frac{C_6}{60k^3}\right)e^{-4k\eta} \end{aligned} \tag{4.355}$$

where

$$\begin{aligned} A = &\left(\frac{3}{2k^2} - \frac{41}{36k^4}\right)C_1 + \left(\frac{1}{k} - \frac{5}{12k^3}\right)C_2 + \left(\frac{5}{18k^2} - \frac{37}{432k^4}\right)C_3 + \\ &+\left(\frac{1}{3k} - \frac{1}{18k^3}\right)C_4 + \left(\frac{7}{72k^2} - \frac{361}{21600k^4}\right)C_5 + \left(\frac{1}{6k} - \frac{11}{720k^3}\right)C_6 \\ B = &\left(\frac{13}{9k^4} - \frac{3}{2k^2}\right)C_1 + \left(\frac{7}{12k^3} - \frac{1}{k}\right)C_2 + \left(\frac{529}{864k^4} - \frac{8}{9k^2}\right)C_3 + \\ &+\left(\frac{17}{72k^3} - \frac{2}{3k}\right)C_4 + \left(\frac{2211}{43200k^4} - \frac{3}{16k^2}\right)C_5 + \left(\frac{17}{480k^3} - \frac{1}{4k}\right)C_6 \end{aligned}$$

The first-order approximate solution is

$$\overline{u}(\eta, C_i) = u_0(\eta) + u_1(\eta, C_i) \tag{4.356}$$

where $u_0(\eta)$ and $u_1(\eta, C_i)$ are given by Eqs. 4.343 and 4.355, respectively.

Fig. 4.102 Residual obtained using the approximate solution given by Eq. 4.356

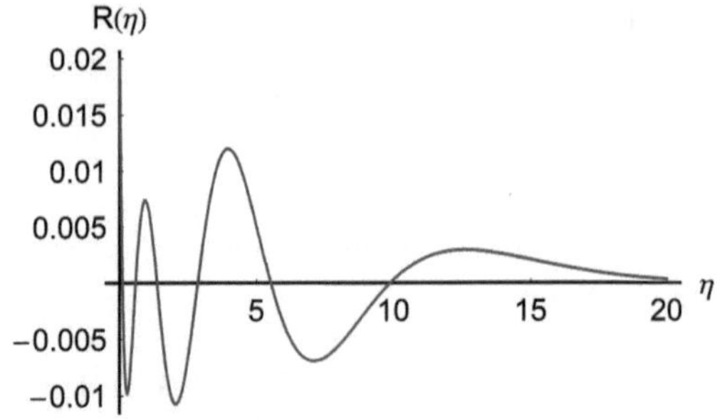

Using the least square method, the optimal convergence-control parameters C_i and k are:

$$C_1 = 0.15113501421650694, C_2 = -1.0516905081033998,$$
$$C_3 = -2.8143479872643007, C_4 = 6.702817799085998,$$
$$C_5 = -1.1737588368580167, C_6 = -6.171736371003325,$$
$$k = 0.6944867422071196$$

From the Eq. 4.356 one can get:

$$\bar{u}''(0) = 0.3320290457441719 \tag{4.357}$$

A highly accurate numerical solution of Blasius equation has been provided by Howarth [99], who gives the initial slope $u''(0) = 0.33206$. Obviously, our first-order approximate result 4.356 assures an absolute error $\varepsilon = 0.00932$ %, which is remarkable good. Our results are also comparable with Liao's results obtained by HAM [95]. One can see that HAM ensures an error of 15.38 % for the initial slope in the 5-th order of approximation, 0.64 % for the 10-th order of approximation and 0.126 % after the 15-th order of approximation, respectively, which means that the error obtained through OHAM in the first-order approximation is better than the error obtained through HAM.

From Fig. 4.102 and. Eq. 4.357 we can observe that the first-order approximate solution 4.356 obtained through OHAM is highly accurate.

4.7 Thermal Radiation on MHD Flow over a Stretching Porous Sheet

The study of magnetohydrodynamic flow (MHD) and heat transfer of an electrically conducted fluid finds useful applications in many engineering problems such as nuclear power plants, gas turbines and various propulsion devices for aircraft, missiles, satellites, space vehicles, and so on. It serves as the basis for understanding some of the important phenomena occurring in heat exchanger devices. Magneto-hydrodynamics can be regarded as a combination of fluid mechanics and electro-magnetism. The effects of thermal radiation on MHD flow and heat transfer problems have become more important industrially. At high operating temperature, thermal radiation effects could be quite significant.

In this section we determine the approximate solution of thermal radiation effects on MHD steady asymmetric flow of an electrically conducting fluid past a stretching porous sheet.

Consider a steady two-dimensional incompressible flow caused by a moving sheet, which is placed in a quiescent, electrically conducting fluid. A magnetic field of uniform strength is applied perpendicular to the stretching sheet. The magnetic Reynolds number is taken to be small enough so that the induced magnetic field can be neglected. The equation governing the dimensionless temperature $\theta(\eta)$ is given by [110]:

$$\theta''(\eta) + \beta f(\eta)\theta'(\eta) - \beta S f'(\eta)\theta(\eta) = 0 \tag{4.358}$$

where η is the similarity variable, $\beta = \frac{3RP_r}{4+3R}$, R is the radiation parameter, P_r is the modified Prandl number, S is the wall temperature parameter and $f(\eta)$ is a dimensionless stream given by the solution of nonlinear differential equation:

$$f'''(\eta) + f(\eta)f''(\eta) - f'^2(\eta) - Mf'(\eta) = 0 \tag{4.359}$$

where M is the magnetic parameter and prime denotes differentiation with respect to η. The boundary conditions for Eq. 4.359 are:

$$f(0) = \lambda, f'(0) = 1, f'(\infty) = 0 \tag{4.360}$$

where λ is the injection/suction parameter.

4.7.1 *Solution of the Problem with Optimal Homotopy Asymptotic Method*

First, we have in view Eqs. 4.359 and 4.360. For the initial/boundary conditions 4.360, the initial approximate solution is given by

$$f_0(\eta) = \lambda + \frac{1 - e^{-\gamma\eta}}{\gamma} \tag{4.361}$$

where γ is a positive unknown parameter. The linear operator L is defined by

$$L[f(\eta)] = f'''(\eta) + \gamma f''(\eta) \tag{4.362}$$

where prime denotes differentiation with respect to η.

The nonlinear operator N corresponding to Eq. 4.359 may be written as

$$N[f(\eta)] = -\gamma f''(\eta) + f(\eta)f''(\eta) - f'^2(\eta) - Mf'(\eta) \tag{4.363}$$

From Eqs. 4.363 and 4.361 it can be shown that

$$N[f_0(\eta)] = (\gamma^2 - \gamma\lambda - 1 - M)e^{-\gamma\eta} \tag{4.364}$$

The first approximation $f_1(\eta)$ has the governing equation given by Eq. 4.9:

$$f_1''' + \gamma f_1'' = H(\eta, C_i)(\gamma^2 - \gamma\lambda - 1 - M)e^{-\gamma\eta}, \quad f_1(0) = f_1'(0) = f_1'(\infty) = 0 \tag{4.365}$$

It is very interesting to remark that if

$$\gamma^2 - \gamma\lambda - 1 - M = 0 \tag{4.366}$$

then from Eq. 4.365 it follows that

$$f_1(\eta) = 0 \tag{4.367}$$

and consequently, the first-order approximate solution obtained from Eq. 4.5 can be written as

$$\bar{f}(\eta, C_i) = f_0(\eta) \tag{4.368}$$

In this case, the initial approximation 4.361 becomes the exact solution of Eq. 4.359. From the Eq. 4.366 one gets:

$$\gamma = \frac{1}{2}\left(\lambda + \sqrt{\lambda^2 + 4M + 4}\right) \tag{4.369}$$

Therefore, the exact solution of Eq. 4.359 obtained using the Optimal Homotopy Asymptotic Method is sought in the form

$$f(\eta) = \lambda + \frac{1 - e^{-\gamma\eta}}{\gamma} \tag{4.370}$$

where the parameter γ is given by Eq. 4.369.

Substituting Eq. 4.370 into Eq. 4.358 we obtain:

$$\theta''(\eta) + (a - be^{-\gamma\eta})\theta'(\eta) - ce^{-\gamma\eta}\theta(\eta) = 0 \tag{4.371}$$

where

$$a = \beta\lambda + \frac{\beta}{\gamma}, b = \frac{\beta}{\lambda}, c = \beta S \tag{4.372}$$

The initial/boundary conditions are given by:

$$\theta(0) = 1, \theta(\infty) = 0 \tag{4.373}$$

The exact solution of Eq. 4.371 with boundary conditions 4.373 is graphically known (in terms of Kummer's confluent hyper-geometric function) [111].

For Eq. 4.371, which is a differential equation with variable coefficients, with the conditions 4.373 we will determine an approximate analytical solution through the Optimal Homotopy Asymptotic Method.

For this, it is natural to choose the initial approximate solution as

$$\theta_0(\eta) = e^{-k\eta} \tag{4.374}$$

where k is an unknown positive parameter and will be determined later. The linear operator is given by

$$L[\theta(\eta)] = \theta''(\eta) - k^2\theta(\eta) \tag{4.375}$$

and therefore the nonlinear operator for Eq. 4.371 becomes

$$N[\theta(\eta)] = (k^2 - ce^{-\gamma\eta})\theta(\eta) + (a - be^{-\gamma\eta})\theta'(\eta) \tag{4.376}$$

By means of Eq. 4.374 and 4.376, we find

$$N[\theta_0(\eta)] = (k^2 - ka)e^{-k\eta} + (bk - c)e^{-(k+\gamma)\eta} \tag{4.377}$$

The first approximation is obtained from the equation

$$\theta_1'' - k^2\theta_1 = H(\eta, C_i)\left[(k^2 - ka)e^{-k\eta} + (bk - c)e^{-(k+\gamma)\eta}\right] \tag{4.378}$$

where the optimal auxiliary function H can be chosen as follows:

$$H(\eta, C_i) = C_1 + C_2\eta + (C_3 + C_4\eta)e^{(k-\gamma)\eta} \tag{4.379}$$

where C_1, C_2, C_3 and C_4 are unknown parameters.

Substituting Eq. 4.379 into Eq. 4.378 and thereafter solving this equation one can get

$$\theta_1(\eta, C_i) = \left[-\frac{AC_2}{4k}\eta^2 - \left(\frac{AC_2}{4k} + \frac{AC_1}{2k}\right)\eta + P\right]e^{-k\eta} + \left[\frac{BC_2}{\gamma^2 + 2\gamma k}\eta + \right.$$
$$\left. + \frac{BC_1}{\gamma^2 + 2\gamma k} + \frac{2(k+\gamma)BC_2}{(\gamma^2 + 2\gamma k)^2}\right]e^{-(k+\gamma)\eta} + \left[\frac{AC_4}{\gamma^2 - k^2}\eta + \frac{AC_3}{\gamma^2 - k^2} + \right. \tag{4.380}$$
$$\left. + \frac{2\gamma AC_4}{(\gamma^2 - k^2)^2}\right]e^{-\gamma\eta} + \left[\frac{BC_3}{4\gamma^2 - k^2}\eta + \frac{BC_3}{4\gamma^2 - k^2} + \frac{4\gamma BC_4}{(4\gamma^2 - k^2)^2}\right]e^{-2\gamma\eta}$$

where

$$A = k^2 - ka, \quad B = bk - c,$$
$$P = -\frac{BC_1}{\gamma^2 + 2\gamma k} - \frac{2(k+\gamma)BC_2}{(\gamma^2 + 2\gamma k)^2} - \frac{AC_3}{\gamma^2 - k^2} - \frac{2\gamma AC_4}{(\gamma^2 - k^2)^2} - \frac{BC_3}{4\gamma^2 - k^2} - \frac{4kBC_4}{(4\gamma^2 - k^2)^2}$$

In this way, the first-order approximate solution of Eqs. 4.371 and 4.373 becomes

$$\bar{\theta}(\eta, C_i) = \theta_0(\eta) + \theta_1(\eta, C_i) \tag{4.381}$$

where θ_0 and θ_1 are given by Eqs. 4.371 and 4.380, respectively.

The exact solution of Eq. 4.371 satisfying Eq. 4.373 in terms of Kummer's confluent hyper-geometric function ${}_1F_1(\cdot, \cdot, \cdot)$ is [111]

$$\theta(\eta) = e^{-\delta\gamma\eta}\frac{{}_1F_1\left(\delta - s, \delta + 1, -\frac{\beta}{\gamma^2}e^{-\gamma\eta}\right)}{{}_1F_1\left(\delta - s, \delta + 1, -\frac{\beta}{\gamma^2}\right)} \tag{4.382}$$

where $\delta = \dfrac{\beta(1 + \gamma\lambda)}{\gamma^2}$.

4.7.2 *Numerical Examples*

In what follows, we consider $P_r = 0.71$, $R = 1$, $M = 1$, $S = 1$ and two cases for the parameter λ.

Fig. 4.103 Dimensionless temperature for $R=1$, $P_r=0.71$, $\lambda=0.5$, $S=1$ and $M=1$ (*red solid line*) numerical solution; (*blue dashed line*) approximate solution (colour figure online)

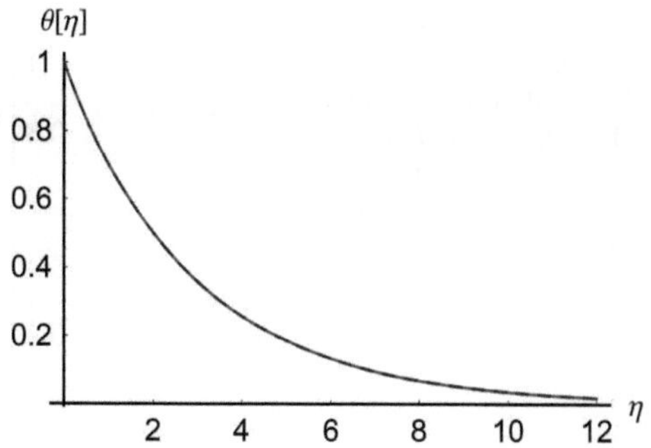

Table 4.116 Comparison between the approximate solution given by Eq. 4.381 and numerical solution for $\lambda=0.5$, $P_r=0.71$, $R=S=M=1$

η	θ_{num}	θ_{ap}	$\varepsilon=\theta_{num}-\theta_{ap}$
0	1	1	0
1	0.701430316974874	0.701437997778492	7.6808×10^{-6}
2	0.500739423444406	0.500744968103362	5.54466×10^{-6}
3	0.358757431373132	0.358761402770191	3.9714×10^{-6}
4	0.257208204973048	0.257211059173984	2.8542×10^{-6}
5	0.184426587775341	0.184428637727979	2.04995×10^{-6}
6	0.132242911773251	0.132244383072397	1.4713×10^{-6}
7	0.094825067148036	0.094826122629021	1.05548×10^{-6}
8	0.067994576103991	0.067995332126912	7.56023×10^{-7}
9	0.048755708691306	0.048756247225616	5.38534×10^{-7}
10	0.034960419481953	0.034960807376464	3.87895×10^{-7}
11	0.025068467600790	0.025068747662087	2.80061×10^{-7}
12	0.017975416152122	0.017975617758826	2.01607×10^{-7}

Case 4.7.2.1 For $\lambda=0.5$ for the above specified parameters it is obtained:

$$\begin{aligned}&C_1 = -1.1326886088324044,\ C_2 = -0.00007139101469509287,\\&C_3 = -0.05739167460388602,\ C_4 = 0.4666428622172303,\\&k = 0.3326056863769377\end{aligned} \tag{4.383}$$

and the approximate solution 4.381 is well-determined.

In Fig. 4.103 are compared the numerical and approximate solution 4.381 in the considered particular case. In Table 4.116 is presented a comparison between the present approximate solution and numerical solution.

Case 4.7.2.2 For $\lambda=1$ one gets:

$$\begin{aligned}&C_1 = -1.1475715598232705,\ C_2 = -0.00010269422287064595,\\&C_3 = -0.026421431593204012,\ C_4 = 0.19383170457375248,\\&k = 0.4564285731779241\end{aligned} \tag{4.384}$$

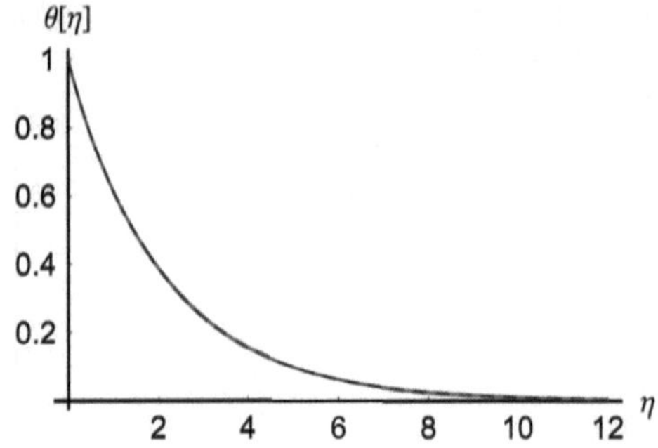

Fig. 4.104 Dimensionless temperature for $R = 1$, $P_r = 0.71$, $\lambda = 1$, $S = 1$ and $M = 1$ (*red solid line*) numerical solution; (*blue dashed line*) approximate solution (colour figure online)

Table 4.117 Comparison between the approximate solution given by Eq. 4381 and numerical solution for $\lambda = 1$, $P_r = 0.71$, $R = S = M = 1$

η	θ_{num}	θ_{ap}	$\varepsilon = \theta_{num} - \theta_{ap}$
0	1	1	0
1	0.616176494014264	0.616160621165197	1.58728×10^{-5}
2	0.388848675238033	0.388838699447494	9.97579×10^{-6}
3	0.246220865144691	0.246214550107048	6.31504×10^{-6}
4	0.155980069406119	0.155976068641317	4.00076×10^{-6}
5	0.098819001603281	0.098816466077325	2.53553×10^{-6}
6	0.062605929751198	0.062604324531284	1.60522×10^{-6}
7	0.039663494392365	0.039662478175583	1.01622×10^{-6}
8	0.025128498801881	0.025127855465832	6.43336×10^{-7}
9	0.015919962926977	0.015919558297277	4.0463×10^{-7}
10	0.010085967029792	0.010085712942880	2.54087×10^{-7}
11	0.006389884924434	0.006389725374641	1.5955×10^{-7}
12	0.004048259591152	0.004048161056802	9.85343×10^{-8}

In Fig. 4.104 are compared the numerical and approximate solution 4.381 for $\lambda = 1$. In Table 4.117 is presented a comparison between the present approximate solution and numerical solution.

In all cases we obtained an excellent agreement between the first-order approximate solutions and numerical ones.

4.8 Nonlinear Equations Arising in Heat Transfer

Some properties of different materials used in practice are dependent on temperature. The pursuit of approximate analytical solutions for heat transfer equations are of intrinsic scientific interest. A lot of methods have been used to obtain analytical solutions of ordinary differential equation, partial differential equations or integro-differential equations, such as perturbation methods, differential transform method, variational iteration method, homotopy perturbation method, optimal homotopy asymptotic method, optimal homotopy perturbation method, and so on [112–123].

4.8.1 Cooling of a Lumped System with Variable Specific Heat

Consider the cooling of a lumped system [112, 113, 115, 116, 120, 122, 123]. Let the system have volume V, surface area A, density ρ, specific heat c and initial temperature T_i. At time $t=0$, the system is exposed to a convective environment at temperature T_a with convective heat transfer coefficient h. Assume that the specific heat c is a linear function of temperature of the form

$$c = c_a[1+\beta(T-T_a)] \tag{4.385}$$

where c_a is the specific heat at temperature T_a and β is a constant. The cooling equation and the initial conditions are as follows:

$$\rho V c\frac{dT}{dt} + hA(T-T_a) = 0, T_0 = T_i \tag{4.386}$$

where by using

$$u = \frac{T-T_a}{T_i-T_a}, x = \frac{t(hA)}{\rho V c_a}, \varepsilon = \beta(T-T_a)$$

we have

$$(1+\varepsilon u)\frac{du}{dx} + u = 0, u(0) = 1, x \in [0,\infty) \tag{4.387}$$

According to Eq. 4.1 we have:

$$L(u(x)) = u'(x) + ku(x), g(x) = 0, \tag{4.388}$$

$$N(u(x)) = \varepsilon u(x)u'(x) + (1-k)u(x), u(0) = 1 \tag{4.389}$$

where k is an unknown parameter.

Eq. 4.4 becomes

$$u_0'(x) + ku_0(x) = 0, u_0(0) = 1 \tag{4.390}$$

from which we obtain

$$u_0(x) = e^{-kx} \tag{4.391}$$

Taking into account Eq. 4.390 and the nonlinear operator given by Eq. 4.389, one can get

$$N(u_0(x)) = (1-k)e^{-kx} - \varepsilon k e^{-2kx} \tag{4.392}$$

The equation 4.9 may be written as

$$u_1' + ku_1 = H(x, C_i)\left[(1-k)e^{-kx} - \varepsilon k e^{-2kx}\right], \quad u_1(0) = 0 \tag{4.393}$$

It is easy to choose the optimal auxiliary function:

$$H(x, C_i) = C_1 + \left(C_2x^2 + C_3x + C_4\right)e^{-kx} \tag{4.394}$$

Substituting Eq. 4.394 into Eq. 4.393 yields:

$$\begin{aligned} u_1' + ku_1 &= (1-k)C_1e^{-kx} + [(1-k)C_2x^2 + (1-k)C_3x - \varepsilon kC_1 + \\ &+ (1-k)C_4]e^{-2kx} - (\varepsilon kC_2x^2 + \varepsilon kC_3x + \varepsilon kC_4)e^{-3kx} \end{aligned} \tag{4.395}$$

whose solution is

$$\begin{aligned} u_1(x, C_i) &= [(1-k)C_1x + A]e^{-kx} + \left[\frac{k-1}{k}C_2x^2 + \left(\frac{2k-2}{k^2}C_2 + \frac{k-1}{k}C_3\right)x + \right. \\ &\left. + \frac{2k-2}{k^3}C_2 + \frac{k-1}{k^2}C_3 + \varepsilon C_1 + \frac{k-1}{k}C_4\right]e^{-2kx} + \\ &+ \left[\frac{\varepsilon C_2}{2}x^2 + \left(\frac{\varepsilon C_2}{2k} + \frac{\varepsilon C_3}{2}\right)x + \frac{\varepsilon C_2}{4k^2} + \frac{\varepsilon C_3}{4k} + \frac{\varepsilon C_4}{2}\right]e^{-3kx} \end{aligned} \tag{4.396}$$

where

$$A = -\varepsilon C_1 + \frac{8 - 8k - k\varepsilon}{4k^3}C_2 + \frac{4 - 4k - k\varepsilon}{4k^2}C_3 + \frac{2 - 2k - k\varepsilon}{2k}C_4$$

The first-order approximate solution of Eq. 4.387 can be written as

$$\overline{u}(x, C_i) = u_0(x) + u_1(x, C_i) \tag{4.397}$$

where $u_0(x)$ and $u_1(x, C_i)$ are known from Eqs. 4.391 and 4.396, respectively.

4.8.1.1 Numerical Examples

Using the least square method, for different values of the parameter ε, we consider the following cases:

a) For $\varepsilon = 1$ we find:

$$C_1 = -0.7728027778643591, C_2 = -0.0003985056426484715,$$
$$C_3 = 0.1975237649283658,\ C_4 = 0.22440932488190918,$$
$$k = 0.09992988662637363$$

b) In the case $\varepsilon = 2$ one can get:

$$C_1 = -0.6064285730774795, C_2 = 0.01489305000491624,$$
$$C_3 = 0.03572102820159439, C_4 = 0.256311598243528,$$
$$k = 0.09781322716741489$$

c) If $\varepsilon = 3$, then:

$$C_1 = -0.9611363904740828,\ C_2 = 0.021813438553156747,$$
$$C_3 = 0.009769248221736677,\ C_4 = 0.7212518466044552,$$
$$k = 0.10004404904649201$$

d) For $\varepsilon = 4$, we obtain:

$$C_1 = -1.0543316901843238,\ C_2 = 0.018614855191734343,$$
$$C_3 = 0.014746530650732781,\ C_4 = 0.8715830726453405,$$
$$k = 0.10038976543362527$$

e) In the last case we consider $\varepsilon = 7$, such that:

$$C_1 = 0.6396458557501352,\ C_2 = -0.005752844071843791,$$
$$C_3 = -0.173013003418252,\ C_4 = -0.6793918629147411,$$
$$k = 0.12819464120233695$$

In Figs. 4.105, 4.106, 4.107, 4.108, 4.109 are compared the numerical solutions and approximate solutions 4.397 obtained by means of the Optimal Homotopy Asymptotic Method for different values of the parameter ε.

It can be seen that the obtained approximate solutions are nearly identical with numerical results, even if the parameter ε is not small.

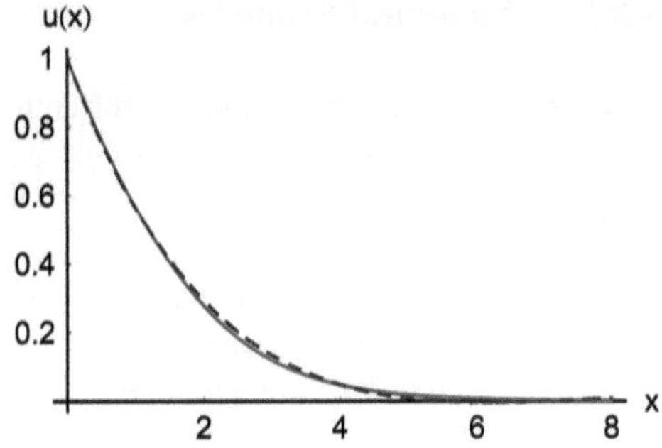

Fig. 4.105 Comparison between the approximate solution given by Eq. 4.397 for $\varepsilon = 1$ and numerical solution: (*red solid line*) numerical results; (*blue dashed line*) approximate solution (colour figure online)

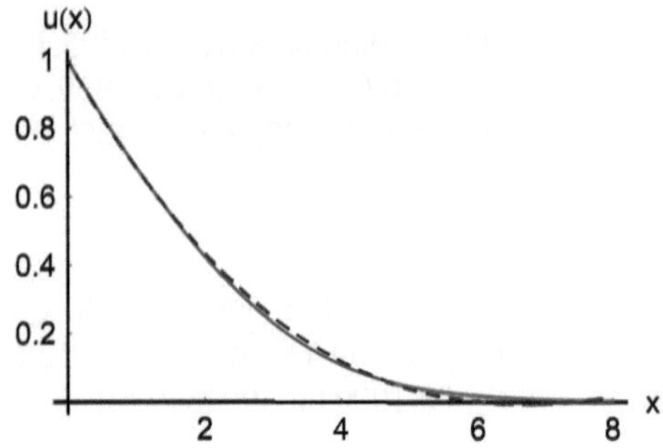

Fig. 4.106 Comparison between the approximate solution given by Eq. 4.397 for $\varepsilon = 2$ and numerical solution: (*red solid line*) numerical results; (*blue dashed line*) approximate solution

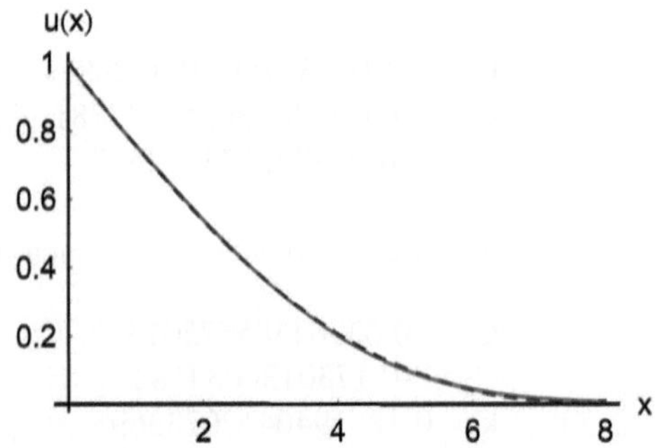

Fig. 4.107 Comparison between the approximate solution given by Eq. 4.397 for $\varepsilon = 3$ and numerical solution: (*red solid line*) numerical results; (*blue dashed line*) (colour figure online)

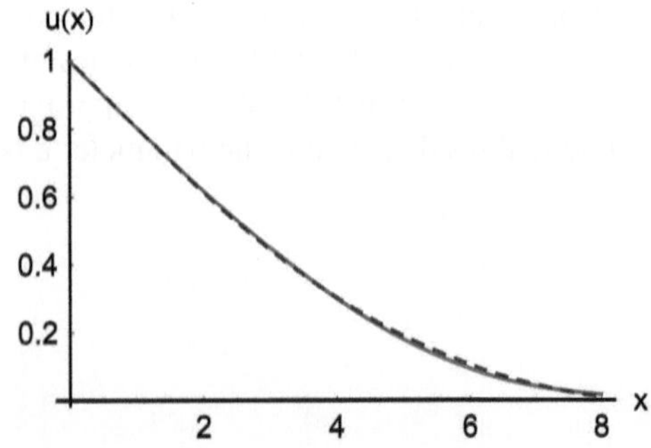

Fig. 4.108 Comparison between the approximate solution given by Eq. 4.397 for $\varepsilon = 4$ and numerical solution: (*red solid line*) numerical results; (*blue dashed line*) approximate solution (colour figure online)

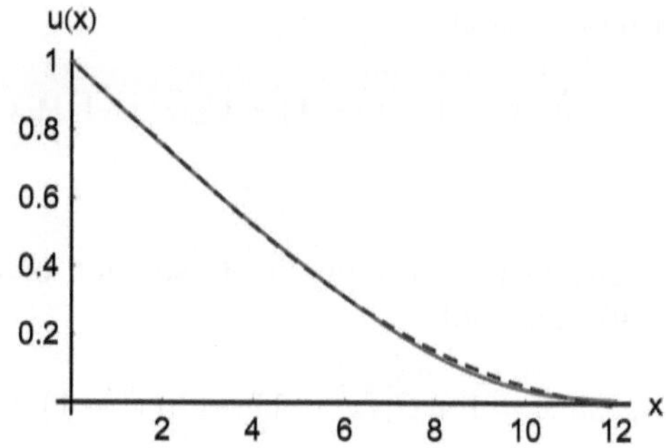

Fig. 4.109 Comparison between the approximate solution given by Eq. 4.397 for ε = 7 and numerical solution: (*red solid line*) numerical results; (*blue dashed line*) approximate solution (colour figure online)

4.8.2 *The Temperature Distribution Equation in a Thick Rectangular Fin Radiation to Free Space*

Now we will consider a nonlinear equation, the temperature distribution equation in a uniformly thick rectangular fin radiation to free space with non-linearity of high order [112, 116, 117, 120]:

$$u''(x) - \varepsilon u^4(x) = 0, u(1) = 1, u'(0) = 0, x \in [0, 1] \tag{4.398}$$

According to Eq. 4.1, we define the operators:

$$L(u(x)) = u''(x), N(u(x)) = -\varepsilon u^4(x), g(x) = 0, \tag{4.399}$$

In this case Eq. 4.4 becomes

$$u_0''(x) = 0, u_0(1) = 1, u_0'(0) = 0 \tag{4.400}$$

It is obtained:

$$u_0(x) = 1 \tag{4.401}$$

Equation 4.9 can be written as

$$u_1''(x) = H(x, C_i)\left[-\varepsilon u_0^4(x)\right], u_1(1) = 0, u_1'(0) = 0 \tag{4.402}$$

with

$$H(x, C_i) = -\frac{1}{\varepsilon}(2C_1 + 6C_2x + 12C_3x^2 + 20C_4x^3 + 30C_5x^4) \tag{4.403}$$

whose solution is

$$u_1(x) = C_1\left(x^2 - 1\right) + C_2\left(x^3 - 1\right) + C_3\left(x^4 - 1\right) + C_4\left(x^5 - 1\right) + C_5\left(x^6 - 1\right) \tag{4.404}$$

The first-order approximate solution of Eq. 4.398 is obtained from Eqs. 4.5, 4.401 and 4.404:

$$\bar{u}(x) = 1 + C_1\left(x^2 - 1\right) + C_2\left(x^3 - 1\right) + C_3\left(x^4 - 1\right) + C_4\left(x^5 - 1\right) + C_5\left(x^6 - 1\right) \tag{4.405}$$

4.8.2.1 Numerical Examples

The optimal values of the convergence-control parameters are found as follows for three considered cases:

a) For $\varepsilon = 1$, the convergence-control parameters are

$$C_1 = 0.18473120953310185, C_2 = -0.008814648369832024,$$
$$C_3 = 0.05668647446865589,\ C_4 = -0.03626644553997487,$$
$$C_5 = 0.02418180502922892$$

b) If $\varepsilon = 2$, we obtain

$$C_1 = 0.23750186964663583, C_2 = -0.04791859342827699,$$
$$C_3 = 0.1978022232692659,\ C_4 = -0.18747305814376602,$$
$$C_5 = 0.10563603613019007$$

c) For $\varepsilon = 3$, the results are

$$C_1 = 0.26785764790249195,\ C_2 = -0.11682450677783947,$$
$$C_3 = 0.41818220559839925, C_4 = -0.4431122477131337,$$
$$C_5 = 0.23191871015290566$$

Figures 4.110, 4.111, 4.112 show the comparison between the present solution 4.405 and the numerical integration results for $\varepsilon = 1$, $\varepsilon = 2$ and $\varepsilon = 3$, respectively. From Figs. 4.110, 4.111, 4.112 it can be seen that the obtained results are in excellent agreement with the results obtained by numerical integration.

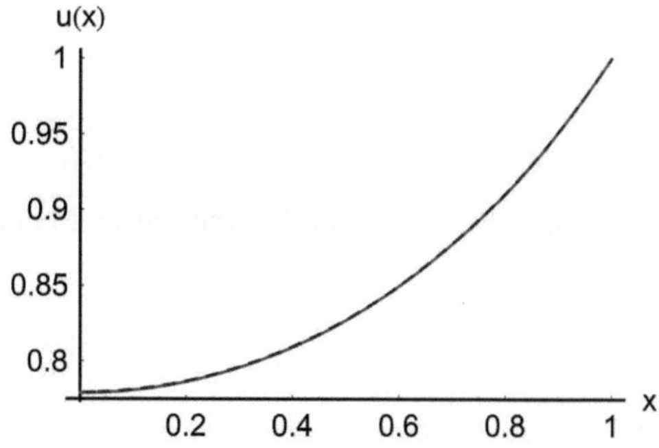

Fig. 4.110 Comparison between the approximate solution 4.405 for $\varepsilon = 1$ and numerical results: (*red solid line*) numerical results; (*blue dashed line*) approximate solution (colour figure online)

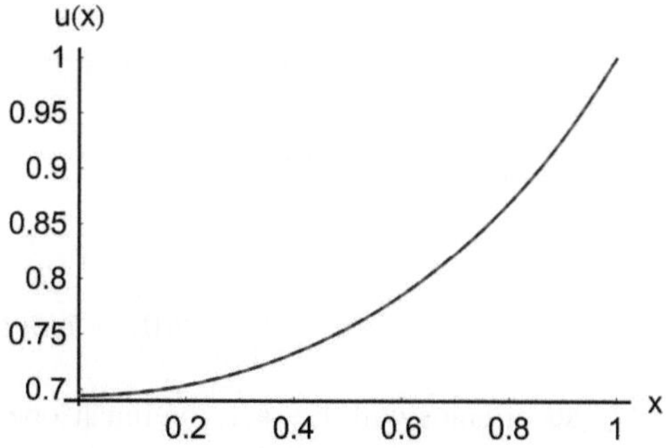

Fig. 4.111 Comparison between the approximate solution 4.405 for $\varepsilon = 2$ and numerical results: (*red solid line*) numerical results; (*blue dashed line*) approximate solution (colour figure online)

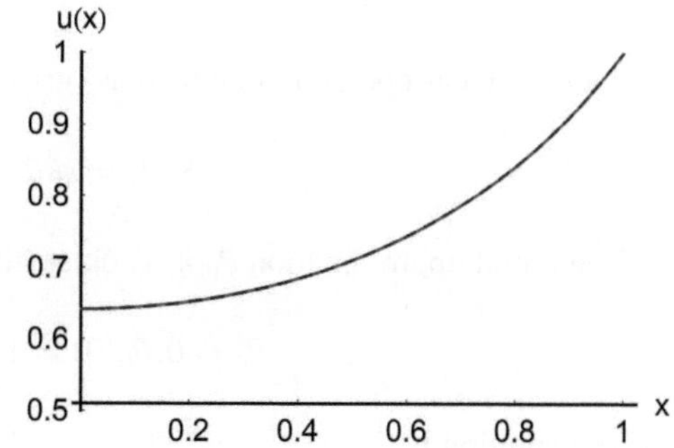

Fig. 4.112 Comparison between the approximate solution 4.405 for $\varepsilon = 3$ and numerical results: (*red solid line*) numerical results; (*blue dashed line*) approximate solution (colour figure online)

4.8.3 A Heat Transfer Problem

Consider the one-dimensional conduction of heat in a slab of thickness L that is made of a material with temperature-dependent thermal conductivity k [123]. If the temperatures of the two opposite faces of the slab are uniformly maintained at T_1 and T_2, $T_1 > T_2$, then the governing equation and boundary conditions are [115]:

$$\frac{d}{dx}\left(k\frac{dT}{dx}\right) = 0 \tag{4.406}$$

$$x = 0, T = T_1; x = L, T = T_2 \tag{4.407}$$

If we further assume that the thermal conductivity varies linearly with temperature, i.e.

$$k = k_2[1 + \beta(T - T_2)] \tag{4.408}$$

where k_2 is the thermal conductivity at temperature T_2 for some constant β, then by introducing the dimensionless quantities

$$\theta = \frac{T - T_2}{T_1 - T_2}, \eta = \frac{x}{L}, \varepsilon = \beta(T_1 - T_2) = \frac{k_1 - k_2}{k_2} \tag{4.409}$$

into Eqs. 4.406 and 4.407, we obtain

$$\theta''(\eta) + \varepsilon\theta(\eta)\theta''(\eta) + \varepsilon\theta'^2(\eta) = 0 \tag{4.410}$$

$$\theta(0) = 1, \theta(1) = 0 \tag{4.411}$$

In accordance with Eq. 4.1, the linear operator is chosen as

$$L(\theta) = \theta'' \tag{4.412}$$

and a non-linear operator is defined as $(g(\eta) = 0)$:

$$N(\theta) = \varepsilon\theta\theta'' + \varepsilon\theta'^2 \tag{4.413}$$

The initial approximation $\theta_0(\eta)$ is obtained from Eq. 4.414 for $L(u_0) = 0$ as

$$\theta_0'' = 0, \theta_0(0) = 1, \theta_0(1) = 0 \tag{4.414}$$

whose solution is

$$\theta_0(\eta) = 1 - \eta \tag{4.415}$$

From Eqs. 4.413 and 4.415 it hold that

$$N[\theta_0(\eta)] = \varepsilon \tag{4.416}$$

The governing equation for the first approximation θ_1 is obtained from Eq. 4.9:

$$\theta_1''(\eta, C_i) = H(\eta, C_i)\varepsilon, \theta_1(0) = 0, \theta_1(1) = 0 \tag{4.417}$$

where we choose the optimal auxiliary function as follows:

$$H(\eta, C_i) = \frac{1}{\varepsilon}(2C_1 + 6C_2\eta + 12C_3\eta^2 + 20C_4\eta^3 + 30C_5\eta^4 + 42C_6\eta^5) \quad (4.418)$$

such that, substituting Eq. 4.418 into Eq. 4.417, we have:

$$\begin{aligned}\theta_1(\eta, C_i) = -(C_1 + C_2 + C_3 + C_4 + C_5 + C_6)\eta + C_1\eta^2 + C_2\eta^3 + \\ + C_3\eta^4 + C_4\eta^5 + C_5\eta^6 + C_6\eta^7\end{aligned} \quad (4.419)$$

The first-order approximate solution of Eqs. 4.410 and 4.411 is given by

$$\begin{aligned}\overline{\theta}(\eta, C_i) = 1 - (1 + C_1 + C_2 + C_3 + C_4 + C_5 + C_6)\eta + C_1\eta^2 + C_2\eta^3 + \\ + C_3\eta^4 + C_4\eta^5 + C_5\eta^6 + C_6\eta^7\end{aligned} \quad (4.420)$$

4.8.3.1 Numerical Examples

In this subsection, the values of the convergence-control parameters are determined by means of Galerkin method. We present three cases corresponding to different values of the parameter ε. As weighting functions we choose $f_1 = 1 + \alpha\eta$, $f_2 = \eta^2$, $f_3 = \eta^3$, $f_4 = \eta^4$, $f_5 = \eta^5$, $f_6 = \eta^6$, $f_7 = \eta^7$, where α is an unknown parameter.

a) In the case ε = 1, we have:

$$\begin{aligned}&C_1 = 2.6023097327592746, C_2 = -12.541681406612316,\\&C_3 = 28.692418626907255,\ C_4 = -35.1752654200865,\\&C_5 = 21.91682201309714, C_6 = -5.49460354606486,\\&\alpha = -43.99259715147711\end{aligned}$$

b) For ε = 2, we obtain:

$$\begin{aligned}&C_1 = 3.7078301081758847,\ C_2 = -18.645859593405874,\\&C_3 = 45.25943591684021,\ C_4 = -58.98223777925954,\\&C_5 = 39.208506482235094,\ C_6 = -10.547675134585765,\\&\alpha = -46.559833189724436\end{aligned}$$

c) In the last case ε = 3, and the convergence-control parameters are:

$$\begin{aligned}&C_1 = 4.714005137629025,\ C_2 = -25.999513442122986,\\&C_3 = 69.12338932722459,\ C_4 = -97.51692534812231,\\&C_5 = 69.56295430440936,\ C_6 = -19.88390997901767,\\&\alpha = -53.16213709714975\end{aligned}$$

In Figs. 4.113, 4.114, 4.115 we compare the approximate solutions 4.420 obtained through the Optimal Homotopy Asymptotic Method and numerical results for different values of the parameter ε, which are not small. From the above figures, we can read that the results obtained by means of OHAM are in very good agreement with numerical results.

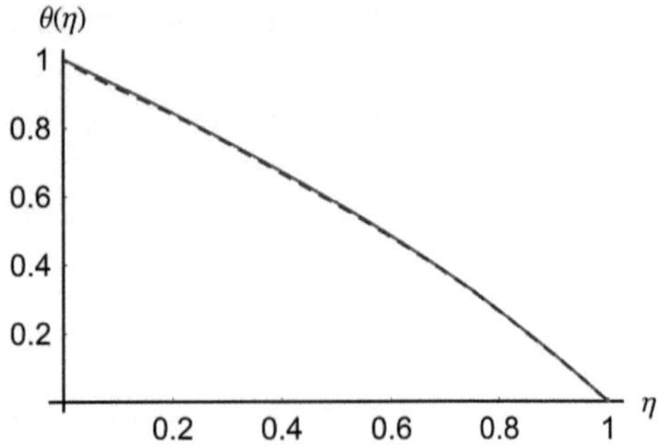

Fig. 4.113 Comparison between the approximate solution 4.420 for $\varepsilon = 1$ and numerical results: (*red solid line*) numerical results; (*blue dashed line*) approximate solution (colour figure online)

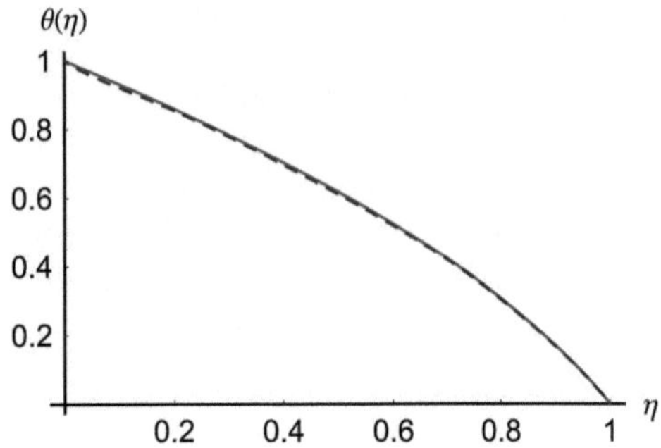

Fig. 4.114 Comparison between the approximate solution 4.420 for $\varepsilon = 2$ and numerical results: (*red solid line*) numerical results; (*blue dashed line*) approximate solution (colour figure online)

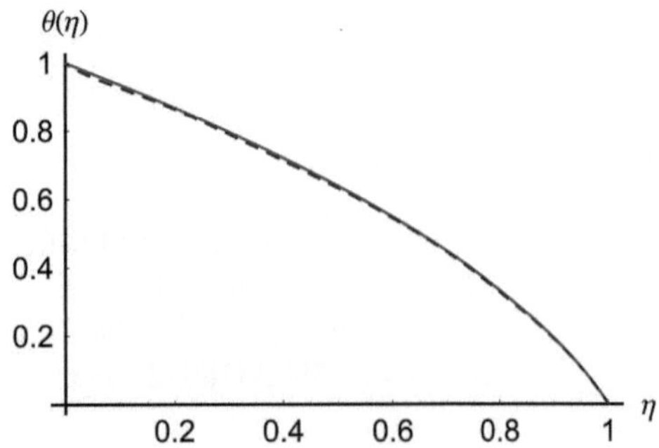

Fig. 4.115 Comparison between the approximate solution 4.420 for $\varepsilon = 3$ and numerical results: (*red solid line*) numerical results; (*blue dashed line*) approximate solution (colour figure online)

4.9 The Nonlinear Age-Structured Population Models

Recently, global climate change and warming in the atmosphere have been generally recognized. Modeling biological population dynamics plays an important role in understanding and controlling the population growth of biological species in their current living environment. Mathematical models describing this evolution have attracted a considerable amount of interest among researchers as a tool for modeling the interaction of different population communities in such diverse fields as demography, epidemiology, ecology, cell kinetics, tumor growth. The size-structured population models considering other physiological characteristics in the individuals of the population, such as length, weight, maturity level, biomass etc. mentioned in [124], appeared in the sixties as a more meaningful way to model the dynamics of species where these characteristics play an essential role in the capacity to survive, grow and produce. The first work in which the population was structured in terms of age was carried out by Sharpe and Lotka [125]. Many models about the age-structured population, simplify the mathematical analysis, by assuming that the mortality rate—as well as other modeling parameters—is bounded, automatically leading to the possibility of immortality. McKendick [126] and Lotka [127] are credited with the first age-structured model, a linear one that supports exponential solutions, just as the unstructured Malthus model [128]. A nonlinear form of that model was first proposed by Gurtin and McCamy [129] by making the fertility and mortality rates dependent on the total population size that is the integral of the age density. Also, there are models that describe the interaction among several species, such as the one introduced by Venturino [130] that models different situations as predator-prey, symbiosis, parasitism etc.

Numerical methods to approximate the solutions of such population models have been proposed in the last years. The first work in which the numerical solution of age-structured population models was dealt with was made by Douglas and Milner in 1987 [131]. First they studied a model with a Dirichlet boundary conditions, i.e. the birth law was given explicitly. Later, they studied a model with a nonlocal and linear boundary condition. On the other hand, the global existence and uniqueness of the solution of the nonlinear approximation scheme are established by using the theory of variation methods, Schauder's fixed point theorem, and the technique of prior estimates. Numerical methods of arbitrary order based on Runge-Kutta techniques were analyzed by Abia and Lopez-Marcos [132] to solve the Gurtin-McCamy's nonlinear model with a more general boundary condition. Iannelli and Milner [133] point out that standard numerical methods based on uniform meshes degenerate and do not converge near the age of unbounded mortality. Kim and Kwon [134] introduced a numerical method that reaches its optimal order of convergence when the tail of the mortality function has some specific analytic forms.

Cai and Chen [135] give a numerical algorithm based on the concept of so-called ε-approximate solution. Li [136] applied the variational iteration method.

In the last decade, active research efforts were focused on nonlinear dynamical systems that emerge in various fields such as biology, physics, mechanics, chemistry, and so on. These real problems are described by nonlinear partial differential equations. In general, it is difficult to seek exact solutions and therefore there appears an ever-increasing interest of scientists in the analytical approximate techniques for these nonlinear problems.

In this section we to construct highly accurate approximations by means of the OHAM for the age-structured population model. These equations were proposed by Gurtin and McCamy [129] and Angulo et al [137] in the form:

$$
\begin{aligned}
&\frac{\partial p(t,x)}{\partial t}+\frac{\partial p(t,x)}{\partial x}+[a_1(x)+a_2(x)P(t)]p(t,x)=0\\
&p(0,x)=p_0(x)\\
&p(t,0)=\int_0^A[b_1(\eta)-b_2(\eta)P(t)]p(t,\eta)d\eta \qquad (4.421)\\
&P(t)=\int_0^A p(t,\eta)d\eta \qquad t\geq 0,\quad 0\leq x\leq A
\end{aligned}
$$

where t is the time, x is the age, P(t) is the total population number at time t, p(t,x) is the age-specific density of individuals of age x at time t, i.e. $\int_{\xi}^{\xi+d\xi} p(t,\eta)d\eta$ is the number of individuals at time t that are aged between ξ and $\xi+d\xi$, $a_1(x)$ is the natural death rate without considering competition, $a_2(x)$ is the increase of death rate considering competition, $b_1(\eta)$ is the natural fertility rate, without considering competition, $b_2(\eta)$ is the decrease of fertility rate considering competition, ξ denotes the lowest age when an individual can bear and A is the maximum age that an individual of the population may reach.

4.9.1 Analytical Solution for Nonlinear Age-Structured Population Models Using OHAM

4.9.1.1 In the first example, we consider the nonlinear age-structured model 4.421 characterized by

$$a_1(x)=1, a_2(x)=1, p(0,x)=\frac{1}{2}e^{-x} \qquad (4.422)$$

The model 4.421 becomes [135, 136, 138]:

$$
\begin{aligned}
&\frac{\partial p(t,x)}{\partial t}+\frac{\partial p(t,x)}{\partial x}+p(t,x)+P(t)\,p(t,x)=0\\
&p(0,x)=\frac{1}{2}e^{-x} \qquad (4.423)\\
&P(t)=p(t,0)=\int_0^A p(t,\eta)d\eta
\end{aligned}
$$

where $A\to\infty$.

According to Eq. 4.1, we define the operators L and N as

$$L(p(t,x)) = \frac{\partial p(t,x)}{\partial t}$$
$$N(p(t,x;q)) = \frac{\partial p(t,x;q)}{\partial x} + p(t,x;q) + p(t,x)\int_0^\infty p(t,\eta;q)d\eta \tag{4.424}$$

and g(t,x) = 0.

The initial approximation $p_0(t,x)$ is obtained from Eqs. 4.4 and 4.423 which become

$$\frac{\partial p_0(t,x)}{\partial t} = 0,\ p_0(0,x) = \frac{1}{2}e^{-x} \tag{4.425}$$

The equation 4.425 has the solution

$$p_0(t,x) = \frac{1}{2}e^{-x} \tag{4.426}$$

The nonlinear term $N(p_0(t,x))$ obtained from Eqs. 4.424 and 4.426 can be written in the form

$$N(p_0(t,x)) = \frac{1}{4}e^{-x} \tag{4.427}$$

The equation 4.9 can be defined as

$$\frac{\partial p_1(t,x)}{\partial t} = H(t,C_i)\frac{1}{4}e^{-x},\ p_1(0,x) = 0 \tag{4.428}$$

If we choose the optimal auxiliary function H in the form

$$H(t,C_i) = 4(C_1 + C_2t + C_3t^2 + C_4t^3) \tag{4.429}$$

where C_i, i = 1,2,3,4 are unknown parameters at this moment, then from Eqs. 4.428 and 4.429, we obtain

$$p_1(t,x) = \left(C_1t + \frac{1}{2}t^2 + \frac{1}{3}C_3t^3 + \frac{1}{4}C_4t^4\right)e^{-x} \tag{4.430}$$

The first-order approximate solution is obtained from Eqs. 4.426, 4.430 and 4.5:

$$\overline{p}(t,x) = \left(\frac{1}{2} + C_1t + \frac{1}{2}C_2t^2 + \frac{1}{3}C_3t^3 + \frac{1}{4}C_4t^4\right)e^{-x} \tag{4.431}$$

From Eqs. 2.37 for s = 4, we obtain the values of the optimal convergence-control parameters:

$$C_1 = -0.249529779, C_2 = 0.241588294,\\ C_3 = -0.147126438, C_4 = 0.044479988 \tag{4.432}$$

Substituting Eq. 4.432 into Eq. 4.431 we find the approximate solution of Eq. 4.423:

$$\overline{p}(t,x) = \left(\frac{1}{2} - 0.249529779t + 0.120794147t^2 - \right.\\ \left. -0.049042146t^3 + 0.0111199972t^4\right)e^{-x} \tag{4.433}$$

By means of the homotopy analysis method (HAM) [138], we obtain the first three terms in the form:

$$p_0(t,x) = \frac{1}{2}e^{-x};\ p_1(t,x) = \frac{1}{4}hte^{-x}\\ p_2(t,x) = \left[\frac{1}{4}h(h+1)t + \frac{1}{8}h^2t^2\right]e^{-x} \tag{4.434}$$

such that the approximate solution of the second-order using HAM can be written as

$$\overline{p}_{HAM}(t,x) = p_0(t,x) + p_1(t,x) + p_2(t,x)\\ = \left[\frac{1}{2} + \frac{1}{4}h(h+2)t + \frac{1}{8}h^2t^2\right]e^{-x} \tag{4.435}$$

The square residual error of the second-order approximate solution 4.435 is defined by:

$$J(h) = \int_0^{\infty}\int_0^1 \left[\frac{\partial \overline{p}_{HAM}}{\partial t} + \frac{\partial \overline{p}_{HAM}}{\partial x} + \overline{p}_{HAM} + \overline{p}_{HAM}\int_0^{\infty} \overline{p}_{HAM}(t,\eta)d\eta\right]^2 dtdx \tag{4.436}$$

From Eqs. 4.435 and 4.436 it is found that

$$J(h) = 0.00159583h^8 + 0.0089495h^7 + 0.0472656h^6 + 0.134115h^5 +\\ +0.338542h^4 + 0.53385h^3 + 0.447917h^2 + 0.1875h + 0.03125 \tag{4.437}$$

The equation 2.37 becomes

$$\frac{\partial J}{\partial h} = 0 \tag{4.438}$$

Table 4.118 Comparison between the absolute error obtained by VIM, HAM and OHAM for Eq. 4.423

t	$\overline{p}_{VIM}(t)$ 4.440	$\overline{p}_{HAM}(t)$ 4.439	$\overline{p}_{OHAM}(t)$ 4.433	$p_{ex}(t)$ 4.441	ε_{VIM}	ε_{HAM}	ε_{OHAM}
0	0.5	0.5	0.5	0.5	0	0	0
0.2	0.454545	0.4332085	0.4545512	0.45454540	0	2.13369E-02	5.825E-06
0.4	0.41664	0.4072680	0.4166613	0.41666666	2.6666E-05	9.39856E-03	5.284E-06
0.6	0.384335	0.3438844	0.3846173	0.38461538	2.8038E-04	4.07309E-02	1.985E-06
0.8	0.35568	0.3276745	0.3571285	0.35714285	1.4628E-03	2.94683E-02	1.428E-05
1	0.328125	0.3346478	0.3333522	0.33333333	5.2083E-03	1.31452E-03	1.888E-05

From Eq. 4.438 we find h = −0.7258731 and therefore the second-order HAM solution is written as

$$\overline{p}_{HAM}(t,x) = \left(\frac{1}{2} - 0.231213578t + 0.065861451t^2\right)e^{-x} \tag{4.439}$$

Using the variational iteration method (VIM), in [136] is given the approximate solution with six terms for Eq. 4.423 in the form:

$$\overline{p}_{VIM}(t,x) = \left(\frac{1}{2} - \frac{t}{4} + \frac{t^2}{8} - \frac{t^3}{16} + \frac{t^4}{32} - \frac{t^5}{64}\right)e^{-x} \tag{4.440}$$

The problem 4.423 has the following exact solution [135, 136]:

$$p_{ex}(t,x) = \frac{e^{-x}}{t+2} \tag{4.441}$$

We remark that the Eqs. 4.433, 4.439, 4.440 and 4.441 can be written in the form:

$$\overline{p}(t,x) = e^{-x}\overline{p}(t) \tag{4.442}$$

The comparison of the relative error $\varepsilon_{app} = \left|\overline{p}_{app}(t) - p_{ex}(t)\right|$ is given in Table 4.118, where $\overline{p}_{app}$ means the approximate solution given by Eqs. 4.433, 4.439, 4.440 and 4.441 respectively. From Table 4.118 it can be seen that all methods show a higher degree of accuracy and OHAM is more accurate than the others.

Figure 4.116 shows the graphical comparison between the solutions obtained by VIM, HAM, OHAM and exact solution of Eq. 4.423. It is to observe that the solution obtained by OHAM is nearly identical with the exact solution and comparing to other methods, our procedure gives better results.

4.9.1.2 For the second example, the age-structured model 4.421 for $a_1(x) = 0$, $a_2(x) = 1$, $p(0,x) = \frac{1}{2}e^{-x}$, can be written in the form [135, 136, 138]:

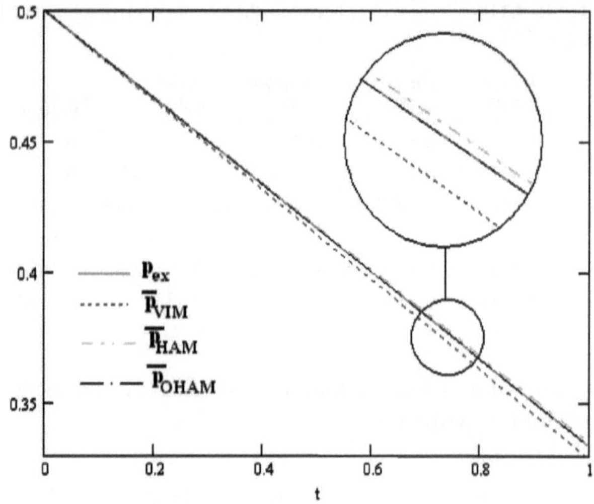

Fig. 4.116 Comparison between exact results and approximate results given by VIM, HAM and OHAM for the example 4.9.1.1

$$
\begin{aligned}
&\frac{\partial p(t,x)}{\partial t}+\frac{\partial p(t,x)}{\partial x}+P(t)p(t,x)=0\\
&p(0,x)=\frac{1}{2}e^{-x}\\
&p(t,0)=P(t)=\int_0^A p(t,\eta)d\eta
\end{aligned}
\tag{4.443}
$$

where $A \to \infty$.

In this case, the nonlinear operator N is

$$
N(p(t,x))=\frac{\partial p(t,x)}{\partial x}+p(t,x)\int_0^\infty p(t,\eta)d\eta \tag{4.444}
$$

and $g(t,x)=0$.

The initial approximation $p_0(t,x)$ is obtained from the following equation:

$$
\frac{\partial p_0(t,x)}{\partial t}=0,\ p_0(0,x)=\frac{1}{2}e^{-x} \tag{4.445}
$$

and therefore

$$
p_0(t,x)=\frac{1}{2}e^{-x} \tag{4.446}
$$

The nonlinear term $N(p_0(t,x))$ becomes

$$
N(p_0(t,x))=-\frac{1}{4}e^{-x} \tag{4.447}
$$

The equation 4.9 can be written in the form

$$\frac{\partial p_1(t,x)}{\partial t} = H(t,C_i)\left(-\frac{1}{4}e^{-x}\right), \; p_1(0,x) = 0 \tag{4.448}$$

We consider the optimal auxiliary function H in the form

$$H(t,C_i) = -4(C_1 + C_2 t + C_3 t^2 + C_4 t^3) \tag{4.449}$$

with C_i, i = 1,2,3,4 unknown parameters. From Eqs. 4.448 and 4.449 we obtain

$$p_1(t,x) = \left(C_1 t + \frac{1}{2}C_2 t^2 + \frac{1}{3}C_3 t^3 + \frac{1}{4}C_4 t^4\right)e^{-x} \tag{4.450}$$

From Eqs. 4.446, 4.450 and 4.5, the first-order approximate solution becomes

$$\overline{p}(t,x) = \left(\frac{1}{2} + C_1 t + \frac{1}{2}C_2 t^2 + \frac{1}{3}C_3 t^3 + \frac{1}{4}C_4 t^4\right)e^{-x} \tag{4.451}$$

The parameters C_i can be determined from Eqs. 2.37

$$\begin{aligned} &C_1 = 0.249881591, C_2 = 0.002069876,\\ &C_3 = -0.07193052, C_4 = 0.01644192 \end{aligned} \tag{4.452}$$

We remark that choosing H in the form 4.449 is not unique. For example, alternatively, we can choose H in the forms:

$$H'(t,C_i) = -4(C_1 + C_2 t + C_3 t^2 + C_4 t^3 + C_5 t^4) \tag{4.453}$$

$$H''(t,C_i) = -4(C_1 + C_2 t^2 + C_3 t^4 + C_4 t^6) \tag{4.454}$$

$$H'''(t,C_i) = -4(C_1 + C_2 t^2 + C_3 t^3 + C_4 t^5 + C_5 t^7) \tag{4.455}$$

and so on.

If we choose the form 4.453 for $H'(t,C_1)$ we obtain:

$$\begin{aligned} &C_1 = 0.249985695, C_2 = 0.000342152, C_3 = -0.064631739,\\ &C_4 = 0.005139228, C_5 = 0.0058203 \end{aligned} \tag{4.456}$$

For the optimal auxiliary function H given by Eq. 4.449, the first-order approximate solution 4.451 can be written as

$$\begin{aligned} &\overline{p}_{OHAM}(t,x) = (0.5 + 0.24988159t + 0.001034938t^2 -\\ &-0.02397684t^3 + 0.00411048t^4)e^{-x} \end{aligned} \tag{4.457}$$

and for the auxiliary function H' given by Eq. 4.453, the first-order approximate solution becomes

$$\begin{aligned}\overline{p}_{OHAM}(t,x) = &\big(0.5 + 0.249985695t + 0.000171076t^2 - \\ &-0.021543913t^3 + 0.001284807t^4 + 0.001161106t^5\big)e^{-x}\end{aligned} \tag{4.458}$$

By means of the homotopy analysis method (HAM), the first three terms are

$$p_0(t,x) = \frac{1}{2}e^{-x},\ p_1(t,x) = -\frac{1}{4}hte^{-x},\ p_2(t,x) = -\frac{1}{4}h(h+1)te^{-x} \tag{4.459}$$

Using HAM, the approximate solution of the second-order becomes:

$$\overline{p}_{HAM}(t,x) = p_0(t,x) + p_1(t,x) + p_2(t,x) = \left[\frac{1}{2} - \frac{h(h+2)}{4}t\right]e^{-x} \tag{4.460}$$

The square residual error of the second-order approximate solution 4.460 is given by

$$J(h) = \int_0^\infty\int_0^1 \left[\frac{\partial \overline{p}_{HAM}}{\partial t} + \frac{\partial \overline{p}_{HAM}}{\partial x} + \overline{p}_{HAM}\int_0^\infty \overline{p}_{HAM}(t,\eta)d\eta\right]^2 dtd\eta \tag{4.461}$$

or

$$\begin{aligned}J(h) = &\frac{1}{3840}\big(3h^8 + 24h^7 + 32h^6 - 144h^5 - \\ &-232h^4 + 480h^3 + 1280h^2 + 960h + 240\big)\end{aligned} \tag{4.462}$$

From the condition 4.438 we obtain h = −0.7201245 and therefore Eq. 4.460 becomes

$$\overline{p}_{HAM}(t,x) = (0.5 + 0.230417426t)e^{-x} \tag{4.463}$$

In [136] the approximate solution by variational iteration method has the form

$$\overline{p}_{VIM}(t,x) = \left(\frac{1}{2} + \frac{t}{4} - \frac{t^3}{48} + \frac{t^5}{480} - \frac{17t^7}{80640} + \frac{31t^9}{1451520}\right)e^{-x} \tag{4.464}$$

The exact solution of Eq. 4.443 is given in [135, 136]:

$$p_{ex}(t,x) = \frac{e^{-x}}{1+e^{-t}} \tag{4.465}$$

Using the so-called concept of ε-approximate solution, Cai and Chen [135] obtained the solution which is notated in Table 4.119 with $p_\varepsilon(t,x)$. In Table 4.119,

Table 4.119 The solutions of Eq. 4.443 obtained by ε-approximate solution [135], VIM (4.464), HAM (4.463), OHAM (4.457), OHAM (4.458) and exact solution (4.465).

t	x	$\overline{p}_\varepsilon(x,t)$ [135]	$\overline{p}_{VIM}(x,t)$ (4.464)	$\overline{p}_{HAM}(t,x)$ (4.463)	$\overline{p}_{OHAM}(x,t)$ (4.457)	$\overline{p}_{OHAM}(x,t)$ (4.458)	$p_{ex}(x,t)$ (4.465)
0	0	0.5	0.5	0.5	0.5	0.5	0.5
0.2	1	0.195806	0.202272623	0.200892887	0.202272064	0.202272646	0.202272623
0.2	2	0.0688321	0.074411939	0.073904363	0.074411734	0.074411947	0.074411939
0.4	2	0.0714055	0.081023564	0.080141084	0.081023736	0.081023559	0.081023564
0.4	3	0.0222456	0.029806903	0.029482257	0.029806966	0.029806903	0.029806903
0.6	4	0.0080506	0.011825607	0.011689964	0.011825593	0.011825607	0.011825607
0.6	6	0.0016002	0.001600421	0.001582064	0.00160042	0.001600421	0.001600421
0.8	8	0.0002314	0.00023146	0.000229568	0.00023146	0.00023146	0.00023146
0.8	10	0.0000293	0.000031324	0.000031068	0.000031324	0.000031324	0.000031324

Table 4.120 Comparison between the errors obtained using ε-approximate solution [135], VIM (4.464), HAM (4.463), OHAM (4.457), OHAM (4.458)

t	x	$\varepsilon^*_\varepsilon$ [135]	ε^*_{VIM} (4.464)	ε^*_{HAM} (4.463)	ε^*_{OHAM} (4.457)	ε^*_{OHAM} (4.458)
0	0	0	0	0	0	0
0.2	1	6.466623E-03	0	1.379759E-03	5.59E-07	2.3E-09
0.2	2	5.579839E-03	0	5.07576E-04	2.05E-07	8E-09
0.4	2	9.813014E-03	0	8.8248E-04	1.72E-07	6E-09
0.4	3	7.561303E-03	0	3.24646E-04	6.3E-08	0
0.6	4	3.774947E-03	0	1.35643E-04	1.4E-08	0
0.6	6	1.81E-07	0	1.8357E-05	0	0
0.8	8	1E-09	0	1.892E-06	0	0
0.8	10	2.011E-06	0	2.56E-07	0	0

we present a comparison between the results obtained by VIM, HAM, OHAM, ε-approximate solution and exact solution. In Table 4.120 we present a comparison between the errors $\varepsilon^*_{app} = \left|\overline{p}_{app}(t,x) - p_{ex}(t,x)\right|$ obtained using the ε-approximate solution [135], VIM (4.464), HAM (4.463), OHAM (4.457) and OHAM (4.448).

Figure 4.117 shows a graphical representation of the approximate solutions obtained through VIM, HAM, OHAM and the exact solution:

$$\overline{p}_{VIM}(t) = \frac{1}{2} + \frac{t}{4} - \frac{t^3}{48} + \frac{t^5}{480} - \frac{17t^7}{80640} + \frac{31t^9}{1451520} \tag{4.466}$$

$$\overline{p}_{HAM}(t) = \frac{1}{2} + 0.230417426t \tag{4.467}$$

$$\begin{gathered}\overline{p}_{OHAM}(t) = \frac{1}{2} + 0.24988159t + \\ +0.001034938t^2 - 0.02397684t^3 + 0.00411048t^4\end{gathered} \tag{4.468}$$

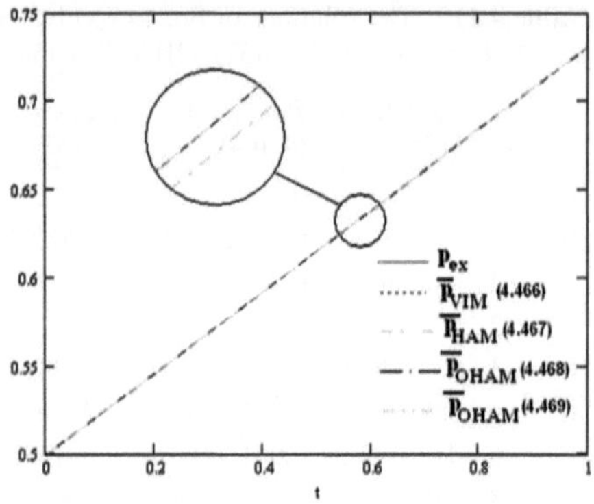

Fig. 4.117 Comparison between exact results and approximate results given by VIM, HAM and OHAM for the example 4.9.1.2

$$\overline{p}'_{OHAM}(t) = \frac{1}{2} + 0.249985695t + 0.000171076t^2 - \\ -0.021543913t^3 + 0.001284807t^4 + 0.001161406t^5 \tag{4.469}$$

From Tables 4.119 and 4.120 and Fig. 4.117, we observe that all methods provide very accurate results, especially VIM and OHAM. On the other hand, from Tables 4.119 and 4.120 we can arrive to the conclusion that the analytical results obtained by OHAM are more accurate along with an increased number of terms in the auxiliary function $H(t,C_i)$ (namely the auxiliary convergence-control function).

4.9.1.3 In the last example, we consider the nonlinear age-structured model 4.421 with $a_1(x) = \frac{1}{1-x}$, $a_2(x) = 1$ and $p(0,x) = 4(1-x)e^{-\lambda x}$. The system 4.421 becomes

$$\begin{aligned} &\frac{\partial p(t,x)}{\partial t} + \frac{\partial p(t,x)}{\partial x} + \frac{p(t,x)}{1-x} + P(t)p(t,x) = 0 \\ &p(0,x) = 4(1-x)e^{-\lambda x} \\ &p(t,0) = 4P(t) = 4\int_0^A p(t,\eta)d\eta, t \geq 0, 0 \leq x \leq A \end{aligned} \tag{4.470}$$

where A = 1 and λ = 2.5569290855.

The nonlinear operator N is given by

$$N(p(t,x)) = \frac{\partial p(t,x)}{\partial x} + \frac{p(t,x)}{1-x} + p(t,x)\int_0^1 p(t,\eta)d\eta \tag{4.471}$$

and g(t,x) = 0.

The term $p_0(t,x)$ is obtained from

$$\frac{\partial p_0(t,x)}{\partial t} = 0,\ p_0(o,x) = 4(1-x)e^{-\lambda x} \tag{4.472}$$

The equation 4.472 has the solution

$$p_0(t,x) = 4(1-x)e^{-\lambda x} \tag{4.473}$$

such that the first nonlinear term $N(p_0(t,x))$ can be written as

$$N(p_0(t,x)) = 4(1-\lambda)(1-x)e^{-\lambda x} \tag{4.474}$$

The equation 4.9 becomes

$$\frac{\partial p_1(t,x)}{\partial t} = H_1(t,C_i)4(1-\lambda)(1-x)e^{-\lambda x},\ p_1(0,x) = 0 \tag{4.475}$$

Choosing the optimal auxiliary function H_1 in the form

$$H_1(t,C_i) = \frac{1}{1-\lambda}(C_1 + C_2t + C_3t^2 + C_4t^3)$$

then from Eq. 4.475 we obtain

$$p_1(t,x) = 4\left(C_1t + \frac{1}{2}C_2t^2 + \frac{1}{3}C_3t^3 + \frac{1}{4}C_4t^4\right)(1-x)e^{-\lambda x} \tag{4.476}$$

The first-order approximate solution of Eq. 4.470 can be written from Eqs. 4.473, 4.476 and 4.5 as

$$\begin{aligned}\overline{p}_{OHAM}(t,x) &= p_0(t,x) + p_1(t,x) = \\ &= 4\left(1 + C_1t + \frac{1}{2}C_2t^2 + \frac{1}{3}C_3t^3 + \frac{1}{4}C_4t^4\right)(1-x)e^{-\lambda x}\end{aligned} \tag{4.477}$$

From Eqs. 2.37 we obtain

$$\begin{aligned}C_1 &= 1.537867846, C_2 = 1.234185712, \\ C_3 &= -4.080804984, C_4 = 1.949527168\end{aligned} \tag{4.478}$$

and therefore, from Eqs. 4.477 and 4.478 we obtain

$$\begin{aligned}\overline{p}_{OHAM}(t,x) = 4(1 &+ 1.537867846t + 0.617092856t^2 - \\ &-1.360268328t^3 + 0.487381792t^4)(1-x)e^{-\lambda x}\end{aligned} \tag{4.479}$$

where $\lambda = 2.5569290855$.

If we consider the approximate solution using the homotopy analysis method, the first three terms are

$$\begin{aligned} p_0(t,x) &= 4(1-x)e^{-\lambda x} \\ p_1(t,x) &= 4h(1-\lambda)(1-x)\frac{1}{2}e^{-\lambda t} \\ p_2(t,x) &= 4\left[(h^2+h)t + \frac{1}{2}(2-\lambda)h^2t^2\right](1-\lambda)(1-x)e^{-\lambda x} \end{aligned} \tag{4.480}$$

The second-order approximate solution obtained by means of HAM can be written as

$$\begin{aligned} &\overline{p}_{HAM}(t,x) = p_0(t,x) + p_1(t,x) + p_2(t,x) = \\ &= 4\left[1 + (1-\lambda)(h^2+2h)t + \frac{1}{2}h^2(\lambda-1)(\lambda-2)t^2\right](1-x)e^{-\lambda x} \end{aligned} \tag{4.481}$$

where the parameter h can be determined using the square residual error

$$J(h) = \int_0^\infty \int_0^1 \left[\frac{\partial \overline{p}_{HAM}}{\partial t} + \frac{\partial \overline{p}_{HAM}}{\partial x} + \frac{\overline{p}_{HAM}}{1-x} + \overline{p}_{HAM}\int_0^1 \overline{p}_{HAM}(t,\eta)d\eta\right]^2 dtd\eta \tag{4.482}$$

or

$$\begin{aligned} &J(h) = 0.171363h^8 + 1.77053h^7 + 6.09553h^6 + 10.2483h^5 + \\ &+2.48229h^4 - 6.51398h^3 + 0.976435h^2 + 2.88614h + 1 \end{aligned} \tag{4.483}$$

From the condition 4.438 we obtain h = −0.486375. Hence, the second-order approximate solution obtained using HAM becomes

$$\overline{p}_{HAM}(t,x) = 4(1 + 1.146194626t + 0.102560758t^2)(1-x)e^{-\lambda x} \tag{4.484}$$

On the other hand, using four terms within homotopy analysis method we obtain

$$\begin{aligned} &p_3(x,t) = \\ &= 4(1-\lambda)(1-x)e^{-\lambda x}\left[h(h+1)^2t + h^2(h+1)(2-\lambda)t^2 + \frac{1}{6}h^3(\lambda^2-6\lambda+6)t^3\right] \end{aligned} \tag{4.485}$$

The third-order approximate solution obtained by means of HAM becomes

$$\begin{aligned} &\overline{p}^*_{HAM}(t,x) = p_0(t,x) + p_1(t,x) + p_2(t,x) + p_3(t,x) = \\ &= 4\left[1 + (h-\lambda)(h^3+3h^2+3h)t + \frac{1}{2}(2h^3+h^2)(1-\lambda)(2-\lambda)t^2 - \right. \\ &\left. -\frac{1}{6}(\lambda^3 - 7\lambda^2 + 12\lambda - 6)h^3t^3\right] \end{aligned} \tag{4.486}$$

Table 4.121 The solutions of Eq. 4.470 obtained by HAM (4.484), HAM (4.487), OHAM (4.479) and exact solution (4.488)

t	$\overline{p}_{HAM}(t)$ (4.484)	$\overline{p}_{HAM}(t)$ (4.487)	$\overline{p}_{OHAM}(t)$ (4.479)	$p_{ex}(t)$ (4.488)
0	1	1	1	1
0.2	1.233341356	1.291652556	1.322154948	1.322363682
0.4	1.474887572	1.571538629	1.639301796	1.639197014
0.6	1.724638648	1.831208528	1.914246777	1.914253096
0.8	1.982594686	2.062211562	2.128407903	2.128421663
1	2.24875384	2.25610104	2.282074166	2.281489071

Table 4.122 Comparison between the errors obtained by means of HAM (4.484), HAM (4.487), OHAM (4.479)

t	ε_{HAM} (4.484)	ε_{HAM} (4.487)	ε_{OHAM} (4.479)
0	0	0	0
0.2	8.9002326E-02	3.06911E-02	1.88734E-04
0.4	1.64309442E-01	6.7658385E-02	1.04782E-04
0.6	1.89614447E-01	8.3044568E-02	6.318E-06
0.8	1.45827077E-01	6.6209101E-02	1.376E-05
1	3.2733687E-02	2.538803E-02	5.85095E-04

From the condition 4.439 we obtain h = −0.6231378, such that the solution 4.486 becomes

$$\overline{p}^{*}_{HAM}(t,x) = 4\big(1 + 1.473596171t - 0.041459906t^2 - \\ -0.176035225t^3\big)(1-x)e^{-\lambda x} \tag{4.487}$$

The exact solution of Eq. 4.469 is [134, 138]:

$$p_{ex}(t,x) = 4(1-x)e^{-\lambda x}\frac{\lambda}{(\lambda-1)e^{-\lambda t}+1} \tag{4.488}$$

From Eqs. 4.479, 4.484, 4.487 and 4.488, we remark that

$$p(t,x) = 4p(t)(1-x)e^{-\lambda x} \tag{4.489}$$

In Table 4.121 we have tabulated some values of the approximate solutions obtained by means of OHAM, HAM, the exact solution and in Table 4.122 we present the corresponding relative error $\varepsilon = \left|\overline{p}_{app}(t) - p_{ex}(t)\right|$. From these tables one can see that all methods give accurate solutions, but the results obtained by OHAM after only one iteration are more accurate. The first-order approximate solution obtained by OHAM is more accurate than the second-order and third-order approximate solution obtained by HAM.

In Fig. 4.118 are graphically presented the corresponding approximate solutions obtained by HAM, OHAM and the exact solution:

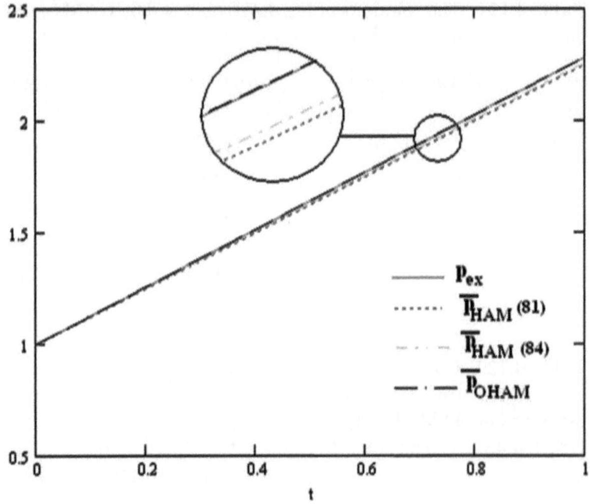

Fig. 4.118 Comparison between the present solution $\overline{p}(t)_{OHAM}$ given by Eq. 4.492, the HAM approximate solution given by Eq. 4.490 and Eq. 4.491 and the exact solution 4.493

$$\overline{p}_{HAM}(t) = 1 + 1.146194626t + 0.102560758t^2, from(4.484) \tag{4.490}$$

$$\begin{aligned}\overline{p}_{HAM}(t) &= 1 + 1.473576171t- \\ &-0.041459906t^2 - 0.176035225t^3, from(4.487)\end{aligned} \tag{4.491}$$

$$\begin{aligned}\overline{p}_{OHAM}(t) &= 1 + 1.537867846t + 0.617092856t^2- \\ &-1.360268328t^3 + 0.48738179t^4 from(4.479)\end{aligned} \tag{4.492}$$

$$p_{ex}(t) = \frac{\lambda}{(\lambda - 1)e^{-\lambda t} + 1}, \quad \text{from (4.488)} \tag{4.493}$$

Figure 4.118 shows that the solution obtained by OHAM is nearly identical with the exact solution and is better than HAM solutions.

4.10 Volterra's Population Model

The study of Volterra integral equations originated with the work of Volterra in population dynamics [139]. Scudo indicates that Volterra proposed a model for a population $\hat{u}(t)$ of identical individuals which can be written in the form:

$$\frac{d\hat{u}}{d\hat{t}} = a\hat{u} - b\hat{u}^2 - c\hat{u}\int_0^{\hat{t}} \hat{u}(\tau)d\tau \quad ; \quad \hat{u}(0) = \hat{u}_0 \tag{4.494}$$

where $\hat{u} = \hat{u}(t)$ and $a > 0$ is the birth rate coefficient, $b > 0$ is the crowding coefficient and $c > 0$ is the toxicity coefficient. The coefficient c indicates the essential behavior of the population evolution before its level falls to zero in the long run. The last term contains the integral that indicates the "total metabolism" or total amount of toxins produced since time zero.

Several scales and population scales may be employed [140]. However, we shall scale time and population by introducing the non-dimensional variables

$$t = \frac{c\hat{t}}{b}, u = \frac{b\hat{u}}{a}$$

producing the following differential equation:

$$\beta\dot{u}(t) = u(t) - u^2(t) - u(t)\int_0^t u(x)dx \quad , \quad \mathrm{u}(0) = \alpha \tag{4.495}$$

where $\beta = \frac{c}{ab}$, $\alpha = \frac{b\hat{u}_0(0)}{a}$ and dot denotes differentiation with respect to t.

In [140] Small showed that if β is large, where the populations are strongly sensitive to toxins, the solution is proportional to $\mathrm{sech}^2(t)$. In this case the solution $u(t)$ has a smaller amplitude. Furthermore, for small β, where populations are weakly sensitive to toxins, the author showed that a rapid rise occurs along the logistic curve (curve given by Eq. 4.494 with $c = 0$) that will rich a peak and then followed by a slow exponential decay.

For the nonlinear differential Eq. 4.495, the linear operator, the function g(t) and the nonlinear operator are

$$L[u(t)] = \dot{u}(t), \quad g(t) = \frac{2\alpha\lambda}{(\lambda t + 1)^3} \tag{4.496}$$

$$N[u(t)] = (\beta - 1)\dot{u}(t) - \frac{2\alpha\lambda}{(\lambda t + 1)^3} - u(t) + u^2(t) + u(t)\int_0^t u(x)dx \tag{4.497}$$

where λ is an unknown parameter.

The equation 4.4 is defined as

$$\dot{u}_0 + \frac{2\alpha\lambda}{(\lambda t + 1)^3} = 0, \quad u_0(0) = \alpha \tag{4.498}$$

and we find the following solution

$$u_0(t) = \frac{\alpha}{(\lambda t + 1)^2} \tag{4.499}$$

From Eqs. 4.499 and 4. 497 it follows that

$$N[u_0(t)] = \frac{\alpha^2 - \alpha\lambda}{\lambda(\lambda t + 1)^2} - \frac{\alpha^2 + 2\alpha\beta\lambda}{\lambda(\lambda t + 1)^3} + \frac{\alpha^2}{(\lambda t + 1)^4} \tag{4.500}$$

The equation 4.9 can be written in the form

$$\dot{u}_1 = H(t, C_i)\left[\frac{\alpha^2 - \alpha\lambda}{\lambda(\lambda t + 1)^2} - \frac{\alpha^2 + 2\alpha\beta\lambda}{\lambda(\lambda t + 1)^3} + \frac{\alpha^2}{(\lambda t + 1)^4}\right] \tag{4.501}$$

where we choose the optimal auxiliary function H(t,C_i) as follows:

$$H(t, C_i) = C_1 + \frac{C_2}{\lambda t + 1} + \frac{C_3}{(\lambda t + 1)^2} + \frac{C_4}{(\lambda t + 1)^3} + \frac{C_5}{(\lambda t + 1)^4} \tag{4.502}$$

Substituting Eq. 4.502 into Eq. 4.501 yields:

$$\begin{aligned} u_1(t, C_i) = {} & K - \frac{(\alpha^2 - \alpha\lambda)C_1}{\lambda^2(\lambda t + 1)} + \frac{(\alpha^2 + 2\alpha\beta\lambda)C_1 - (\alpha^2 - \alpha\lambda)C_2}{2\lambda^2(\lambda t + 1)^2} - \\ & - \frac{\alpha^2\lambda C_1 - (\alpha^2 + 2\alpha\beta\lambda)C_2 + (\alpha^2 - \alpha\lambda)C_3}{3\lambda^2(\lambda t + 1)^3} - \\ & - \frac{\alpha^2\lambda C_2 - (\alpha^2 + 2\alpha\beta\lambda)C_3 + (\alpha^2 - \alpha\lambda)C_4}{4\lambda^2(\lambda t + 1)^4} - \\ & - \frac{\alpha^2\lambda C_3 - (\alpha^2 + 2\alpha\beta\lambda)C_4 + (\alpha^2 - \alpha\lambda)C_5}{5\lambda^2(\lambda t + 1)^5} - \\ & - \frac{\alpha^2\lambda C_4 - (\alpha^2 + 2\alpha\beta\lambda)C_5}{6\lambda^2(\lambda t + 1)^6} - \frac{\alpha^2 C_5}{7\lambda(\lambda t + 1)^7} \end{aligned} \tag{4.503}$$

where

$$\begin{aligned} K = {} & \frac{3\alpha^2 + (2\alpha^2 - 6\alpha - 6\alpha\beta)\lambda}{6\lambda^2}C_1 + \frac{2\alpha^2 + (3\alpha^2 - 6\alpha - 8\alpha\beta)\lambda}{12\lambda^2}C_2 + \\ & + \frac{5\alpha^2 + (12\alpha^2 - 20\alpha - 30\alpha\beta)\lambda}{60\lambda^2}C_3 + \frac{3\alpha^2 + (10\alpha^2 - 15\alpha - 24\alpha\beta)\lambda}{60\lambda^2}C_4 + \\ & + \frac{7\alpha^2 + (30\alpha^2 - 42\alpha - 70\alpha\beta)\lambda}{210\lambda^2}C_5 \end{aligned}$$

From Eqs. 4.499 and 4.503 it is obtained the first-order approximate solution of Eq. 4.495

$$\overline{u}(t, C_i) = u_0(t) + u_1(t, C_i) \tag{4.504}$$

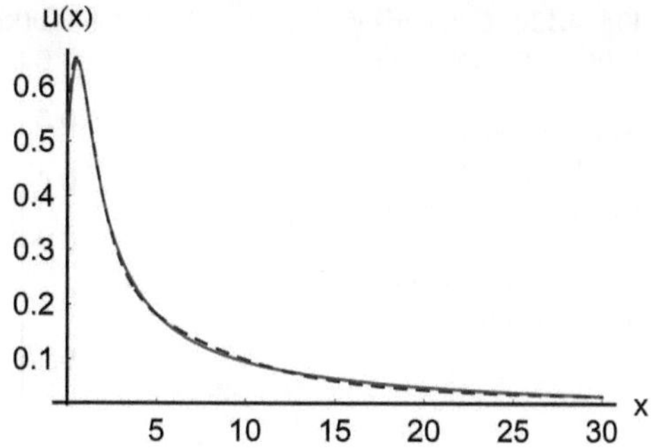

Fig. 4.119 Comparison between the approximate solution 4.504 and numerical solution for $\alpha = \beta = 0.5$: (*red solid line*) numerical results; (*blue dashed line*) approximate solution (colour figure online)

4.10.1 Numerical Examples

In order to show the efficiency of the Optimal Homotopy Asymptotic Method, we present four examples of obtaining analytical solutions.

4.10.1.1 In the first case, for $\alpha = \beta = 0.5$, one can get

$$\begin{aligned} C_1 &= \text{-}0.5504066789402998,\ C_2 = 7.594876945672639, \\ C_3 &= \text{-}35.025732472282854,\ C_4 = 61.959419517247554, \\ C_5 &= \text{-}35.232796139447984,\ \lambda = 0.16729798115829292 \end{aligned}$$

4.10.1.2 For $\alpha = 0.5$, $\beta = 0.9$, one obtains

$$\begin{aligned} C_1 &= -0.5156614123568719, C_2 = 5.7294407189238425, \\ C_3 &= -21.62951837310287, C_4 = 33.36266407955183, \\ C_5 &= -17.40746095238141, \lambda = 0.11701282409945546 \end{aligned}$$

4.10.1.3 If $\alpha = 0.6$, $\beta = 0.4$, the optimal value of the convergence-control parameters are

$$\begin{aligned} C_1 &= -0.19695462960950288, C_2 = 3.1372176079850536, \\ C_3 &= -19.455034735761895, C_4 = 43.79328348293068, \\ C_5 &= -28.841691738897314, \lambda = 0.22572737854735855 \end{aligned}$$

4.10.1.4 In the last case, we consider $\alpha = \beta = 0.8$, and therefore it holds that

$$\begin{aligned} C_1 &= -0.0592435030613214, C_2 = 0.9738196178724576, \\ C_3 &= -5.3949630880364925, C_4 = 12.395476399043675, \\ C_5 &= -8.33454578084606, \lambda = 0.1965298006638166 \end{aligned}$$

As it can be seen from Figs. 4.119, 4.120, 4.121, 4.122, a very good accuracy is achieved.

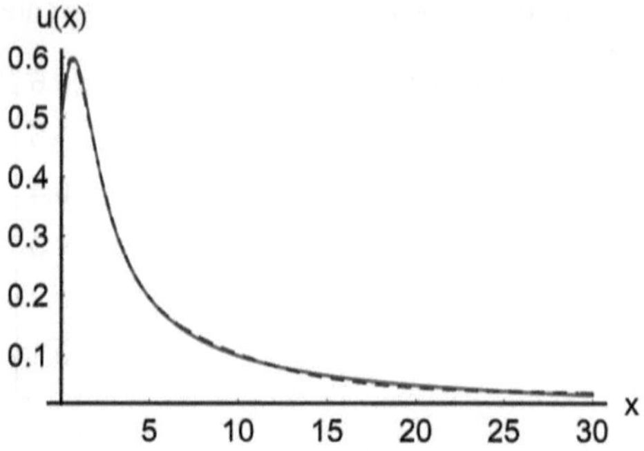

Fig. 4.120 Comparison between the approximate solution 4.504 and numerical solution for $\alpha = 0.5$, $\beta = 0.9$: (*red solid line*) numerical results; (*blue dashed line*) approximate solution (colour figure online)

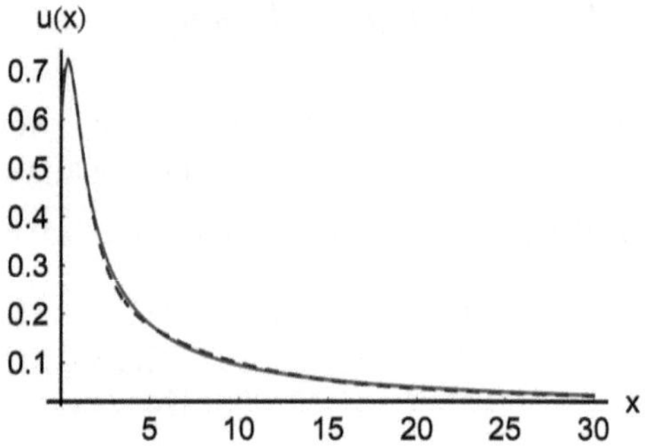

Fig. 4.121 Comparison between the approximate solution 4.504 and numerical solution for $\alpha = 0.6$, $\beta = 0.4$: (*red solid line*) numerical results; (*blue dashed line*) approximate solution (colour figure online)

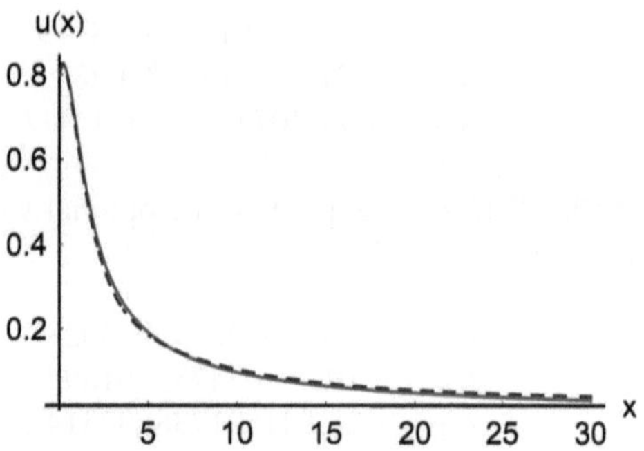

Fig. 4.122 Comparison between the approximate solution 4.504 and numerical solution for $\alpha = \beta = 0.8$: (*red solid line*) numerical results; (*blue dashed line*) approximate solution (colour figure online)

4.11 Lotka-Volterra Model with Three Species

Systems of nonlinear differential equations arise in many scientific models such as biological systems and are used in various fields as engineering, chemistry, ecology and so on. In 1925, A.J. Lotka [141] developed the notion of an evolutionary system based on two fundamental changes, those involving matter between components of a system and those involving exchanges of energy. In 1926, V. Volterra [142] developed the well-known mathematical models of multispecies interaction.

These models, the predator, prey and competition models are known today as Lotka-Volterra models. Lotka-Volterra equations describe variations of population densities of few species that compete for the same resources. However, the ecological system is often affected by environmental changes and other human activities. In many practical situations, it is often the case that one of the species maybe suffers a significant loss or increase in density for some reason at some transitory time slots. These models, for instance can describe the competing fish species which are exploited by human activities, can also describe the dynamics of normal and tumour cells in a changing environment under the effects of the chemotherapy. These models are also applicable in case we are interested in the existence and stability of tumour-free solution and how treatment affects the interaction of tumour and normal cells. Although simplistic, these few models are still used as the foundation for mathematical models in biology. These models can also describe the time history of a biological system and are used in various fields as engineering, chemistry, biology or mathematics. In fact the Lotka-Volterra model is one of the most popular ones to demonstrate a simple nonlinear control system. The accurate solutions of the Lotka-Volterra equations may become a difficult task either if the equations are stiff or when the number of species is large.

In what follows we consider a nonlinear differential equation described by the following time dependent Lotka-Volterra model with three species [143]:

$$\begin{aligned} \dot{x} &= x(1 - x - \alpha y - \beta z) \\ \dot{y} &= y(1 - \beta x - y - \alpha z) \\ \dot{z} &= z(1 - \alpha x - \beta y - z) \end{aligned} \tag{4.505}$$

were dot denotes derivative with respect to time t, α and β are known parameters not necessary small. Initial conditions are

$$x(0) = A, y(0) = B, z(0) = C \tag{4.506}$$

In accordance with Eq. 4.1 we have

$$L[x(t)] = \dot{x}(t) + x(t), \quad g_1(t) = 0 \tag{4.507}$$

$$L[y(t)] = \dot{y}(t) + y(t), \quad g_2(t) = 0 \tag{4.508}$$

$$L[z(t)] = \dot{z}(t) + z(t), \quad g_3(t) = 0 \tag{4.509}$$

$$N_1(x, y, z) = -2x + x^2 + \alpha xy + \beta xz \tag{4.510}$$

$$N_2(x, y, z) = -2y + \beta xy + y^2 + \alpha yz \tag{4.511}$$

$$N_3(x, y, z) = -2z + \alpha xz + \beta yz + z^2 \tag{4.512}$$

where the index i (i = 1,2,3) denotes the order number of the equation which appears in Eq.(4.505).

The initial approximations are obtained from the equations

$$L[x_0(t)] = 0, \quad x_0(0) = A \tag{4.513}$$

$$L[y_0(t)] = 0, \quad y_0(0) = B \tag{4.514}$$

$$L[z_0(t)] = 0, \quad z_0(0) = C \tag{4.515}$$

These equations have the solutions:

$$x_0(t) = Ae^{-t} \tag{4.516}$$

$$y_0(t) = Be^{-t} \tag{4.517}$$

$$z_0(t) = Ce^{-t} \tag{4.518}$$

For the initial approximations given by Eq. 4.516-4.518, the nonlinear operators 4.510-4.512 become, respectively

$$N_1[x_0(t), y_o(t), z_0(t)] = -2Ae^{-t} + (A^2 + \alpha AB + \beta AC)e^{-2t} \tag{4.519}$$

$$N_2[x_0(t), y_o(t), z_0(t)] = -2Be^{-t} + (B^2 + \alpha BC + \beta AB)e^{-2t} \tag{4.520}$$

$$N_3[x_0(t), y_o(t), z_0(t)] = -2Ce^{-t} + (C^2 + \alpha AC + \beta BC)e^{-2t} \tag{4.521}$$

The governing equation for the first approximations $x_1(t)$, $y_1(t)$, $z_1(t)$, are respectively:

$$\dot{x}_1(t) + x_1(t) = H_1(t, C_i)\left[-2Ae^{-t} + (A^2 + \alpha AB + \beta AC)e^{-2t}\right], x_1(0) = 0 \tag{4.522}$$

$$\dot{y}_1(t) + y_1(t) = H_2(t, C_j)\left[-2Be^{-t} + (B^2 + \alpha BC + \beta AB)e^{-2t}\right], y_1(0) = 0 \tag{4.523}$$

$$\dot{z}_1(t) + z_1(t) = H_3(t, C_k)\left[-2Ce^{-t} + (C^2 + \alpha AC + \beta BC)e^{-2t}\right], z_1(0) = 0 \tag{4.524}$$

It is natural to choose the optimal auxiliary functions H_1, H_2 and H_3 as follows:

$$H_1(t, C_i) = \frac{1}{A}\left[C_1e^t + C_2 + (C_3t + C_4)e^{-t} + (C_5t + C_6)e^{-2t}\right] \tag{4.525}$$

$$H_2(t, C_j) = \frac{1}{B}\left[C_7e^t + C_8 + (C_9t + C_{10})e^{-t} + (C_{10}t + C_{12})e^{-2t}\right] \tag{4.526}$$

$$H_3(t, C_k) = \frac{1}{C}\left[C_{13}e^t + C_{14} + (C_{15}t + C_{16})e^{-t} + (C_{17}t + C_{18})e^{-2t}\right] \tag{4.527}$$

Substituting Eqs. 4.525-4.527 into Eqs. 4.522-4.524 and then solving these equations we find the following solutions:

$$\begin{aligned} x_1(t) = & -2C_1 + (M_1t + M_2)e^{-t} + \\ & +(M_3t + M_4)e^{-2t} + (M_5t + M_6)e^{-3t} + (M_7t + M_8)e^{-4t} \end{aligned} \tag{4.528}$$

$$y_1(t) = -2C_7 + (N_1 t + N_2)e^{-t} + (N_3 t + N_4)e^{-2t} + \\ +(N_5 t + N_6)e^{-3t} + (N_7 t + N_8)e^{-4t} \tag{4.529}$$

$$z_1(t) = -2C_{13} + (P_1 t + P_2)e^{-t} + (P_3 t + P_4)e^{-2t} + \\ +(P_5 t + P_6)e^{-3t} + (P_7 t + P_8)e^{-4t} \tag{4.530}$$

where

$$M_1 = (A+\alpha B+\beta C)C_1 - 2C_2, M_2 = 2C_1 + (A+\alpha B+\beta C)C_2 + \\ +\left(\frac{A+\alpha B+\beta C}{4} - 2\right)C_3 + \left(\frac{A+\alpha B+\beta C}{2} - 2\right)C_4 + \left(\frac{A+\alpha B+\beta C}{9} - \frac{1}{2}\right)C_5 + \\ +\left(\frac{A+\alpha B+\beta C}{3} - 1\right)C_6, M_3 = 2C_3, M_4 = -(A+\alpha B+\beta C)C_2 + 2C_3 + 2C_4 \\ M_5 = C_5 - \frac{A+\alpha B+\beta C}{2}C_3, M_6 = -\frac{A+\alpha B+\beta C}{4}(C_3 + 2C_4) + \frac{C_5}{2} + C_6, \\ M_7 = -\frac{A+\alpha B+\beta C}{3}C_5, M_8 = -\frac{A+\alpha B+\beta C}{9}(C_5 + 3C_6), \\ N_1 = (B+\alpha C+\beta A)C_7 - 2C_8, N_2 = 2C_7 + (B+\alpha C+\beta A)C_8 + \\ +\left(\frac{B+\alpha C+\beta A}{4} - 2\right)C_9 + \left(\frac{B+\alpha C+\beta A}{2} - 2\right)C_{10} + \left(\frac{B+\alpha C+\beta A}{9} - \frac{1}{2}\right)C_{11} + \\ +\left(\frac{B+\alpha C+\beta A}{3} - 1\right)C_{12}, N_3 = 2C_9, N_4 = -(B+\alpha C+\beta A)C_8 + 2C_9 + 2C_{10}, \\ N_5 = C_{11} - \frac{B+\alpha C+\beta A}{2}C_9, N_6 = -\frac{B+\alpha C+\beta A}{4}(C_9 + 2C_{10}) + \frac{C_{11}}{2} + C_{12}, \\ N_7 = -\frac{B+\alpha C+\beta A}{3}C_{11}, N_8 = -\frac{B+\alpha C+\beta A}{9}(C_{11} + 3C_{12}), \\ P_1 = (C+\alpha A+\beta B)C_{13} - 2C_{14}, P_2 = 2C_{13} + (C+\alpha A+\beta B)C_{14} + \\ +\left(\frac{C+\alpha A+\beta B}{4} - 2\right)C_{15} + \left(\frac{C+\alpha A+\beta B}{2} - 2\right)C_{16} + \left(\frac{C+\alpha A+\beta B}{9} - \frac{1}{2}\right)C_{17} + \\ +\left(\frac{C+\alpha A+\beta B}{3} - 1\right)C_{18}, P_3 = 2C_{15}, P_4 = -(C+\alpha A+\beta B)C_{14} + 2C_{15} + 2C_{16}, \\ P_5 = C_{17} - \frac{C+\alpha A+\beta B}{2}C_{15}, P_6 = -\frac{C+\alpha A+\beta B}{4}(C_{15} + 2C_{16}) + \frac{C_{17}}{2} + C_{18}, \\ P_7 = -\frac{C+\alpha A+\beta B}{3}C_{17}, P_8 = -\frac{C+\alpha A+\beta B}{9}(C_{17} + 3C_{18})$$

From Eqs. 4.516-4.518, 4.528-4.530 and 4.5 one can get the first-order approximate solution of Eqs. 4.505 in the form

$$\bar{x}(t, C_i) = -2C_1 + (M_1 t + M_2 + A)e^{-t} + (M_3 t + M_4)e^{-2t} + \\ +(M_5 t + M_6)e^{-3t} + (M_7 t + M_8)e^{-4t} \tag{4.531}$$

$$\bar{y}(t, C_j) = -2C_7 + (N_1 t + N_2 + B)e^{-t} + (N_3 t + N_4)e^{-2t} + \\ +(N_5 t + N_6)e^{-3t} + (N_7 t + N_8)e^{-4t} \tag{4.532}$$

$$\bar{z}(t, C_k) = -2C_{13} + (P_1 t + P_2 + C)e^{-t} + (P_3 t + P_4)e^{-2t} + \\ +(P_5 t + P_6)e^{-3t} + (P_7 t + P_8)e^{-4t} \tag{4.533}$$

4.11.1 Numerical Examples

Substituting Eqs. 4.531-4.533 into Eq. 4.505 it results the residuals

$$R_1(t, C_i) = \dot{\overline{x}} - \overline{x}(1 - \overline{x} - \alpha\overline{y} - \beta\overline{z}) \quad (4.534)$$

$$R_2(t, C_j) = \dot{\overline{y}} - \overline{y}(1 - \beta\overline{x} - \overline{y} - \alpha\overline{z}) \quad (4.535)$$

$$R_3(t, C_k) = \dot{\overline{z}} - \overline{z}(1 - \alpha\overline{x} - \beta\overline{y} - \overline{z}) \quad (4.536)$$

Making collocation in six points for each considered equation,

$$R_1(t_i, C_i) = 0 \quad (4.537)$$

$$R_2(t_j, C_j) = 0 \quad (4.538)$$

$$R_3(t_k, C_k) = 0 \quad (4.539)$$

and solving the system of 18 nonlinear algebraic equations, we obtain the optimal values of the convergence-control parameters C_1, C_2, ..., C_{18}, and therefore the solution 4.531-4.533 in the first approximation is well determined.

We consider the following two cases:

4.11.1.1 In the first case, when A = 0.25, B = 0.4, C = 0.55, $\alpha = 0.1$ and $\beta = 0.2$, we obtain:

$$\begin{aligned}
&C_1 = -0.38377461048433403, C_2 = 0.43338085766203926\\
&C_3 = -1.2169469116473035, C_4 = 1.7447158938823284\\
&C_5 = -1.0843699449735535, C_6 = -2.041606321683714\\
&C_7 = -0.38533615203149424, C_8 = -0.35707541132021486\\
&C_9 = 1.4002088757456126, C_{10} = -1.7081166804867691\\
&C_{11} = 1.0084083746269483, C_{12} = 2.0484874474228385\\
&C_{13} = -0.3847629560631131, C_{14} = -0.3318898501704133\\
&C_{15} = 0.6299711061496708, C_{16} = -0.7532441171857128\\
&C_{17} = 0.4825220558272324, C_{18} = 0.918893316376898
\end{aligned} \quad (4.540)$$

By substituting 4.540 into Eqs. 4.531-4.533, we obtain:

$$\overline{x}(t) = 0.767549 + e^{-t}(1.0909 - 1.02027\, t) + e^{-2t}(0.882186 - 2.43389\, t) + \\ + e^{-3t}(-2.81104 - 0.840981\, t) + e^{-4t}(0.320408 + 0.144583\, t) \quad (4.541)$$

$$\overline{y}(t) = 0.770672 + e^{-t}(-2.34098 + 0.519556\, t) + e^{-2t}(-0.435493 + \\ + 2.80042\, t) + e^{-3t}(2.80721 + 0.654856\, t) + e^{-4t}(-0.401412 - 0.169749\, t) \quad (4.542)$$

$$\bar{z}(t) = 0.769526 + e^{-t}(-1.25831 + 0.41176\,t) + e^{-2t}(-0.0291582+ +1.25994\,t) + e^{-3t}(1.30368 + 0.276207\,t) + e^{-4t}(-0.235742-0.105351\,t) \quad (4.543)$$

4.11.1.2 In the last case we consider A = 0.2, B = 0.5, C = 0.7, α = 0.2, β = 0.1, such that the optimal values of the convergence-control parameters are:

$$\begin{aligned}
C_1 &= -0.3840028243134681, C_2 = 0.7648967565222846\\
C_3 &= -2.2161091884312287, C_4 = 3.026486301463203\\
C_5 &= -1.9203994515616976, C_6 = -3.6017582483461097\\
C_7 &= -0.38445734513009877, C_8 = -0.09672215553194186\\
C_9 &= -0.34452941469854825, C_{10} = 0.7362043420869233\\
C_{11} &= -0.4052608135482258, C_{12} = -0.753835433688793\\
C_{13} &= -0.3855706588868133, C_{14} = -0.9135043850109701\\
C_{15} &= 5.431045980408729, C_{16} = -10.038227135554257\\
C_{17} &= 6.305103284852999, C_{18} = 10.609192887787772
\end{aligned} \quad (4.544)$$

By means of Eq. 4.544 the first-order approximate solution becomes:

$$\bar{x}(t) = 0.768006 + e^{-t}(2.48795 -1.67187\,t) + e^{-2t}(1.33774 -4.43222\,t)+ +e^{-3t}(-4.91687-1.51042\,t) + e^{-4t}(0.523167 + 0.236849\,t) \quad (4.545)$$

$$\bar{y}(t) = 0.768915 + e^{-t}(-0.169098-0.0602975\,t) + e^{-2t}(0.847186 - -0.689059\,t) + e^{-3t}(-1.14257-0.291566\,t) + e^{-4t}(0.195563 + 0.089157\,t) \quad (4.546)$$

$$\bar{z}(t) = 0.771141 + e^{-t}(-4.88546 + 1.52241\,t) + e^{-2t}(-8.49269+ +10.8621\,t) + e^{-3t}(16.6542 + 4.15984\,t) + e^{-4t}(-3.3472-1.66034\,t) \quad (4.547)$$

It is easy to verify the accuracy of the obtained solutions 4.541-4.543 and 4.545-4.547 if we graphically compare the analytical solutions with the numerical one. Figs. 4.123, 4.124, 4.125 and 4.126, 4.127, 4.128 show the comparison between the

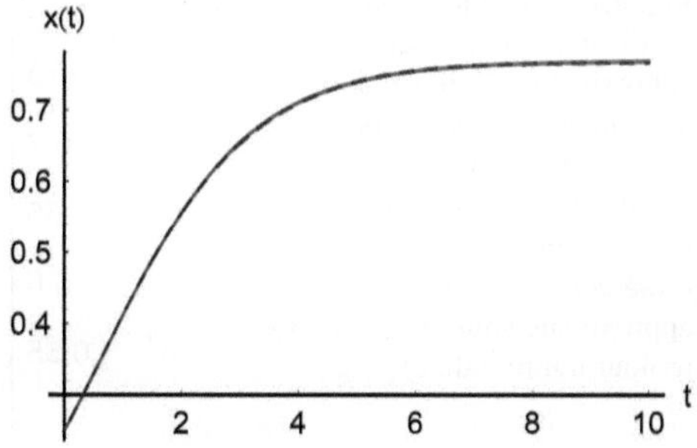

Fig. 4.123 Comparison between the first-order approximate solution 4.541 and numerical results for A = 0.25, B = 0.4, C = 0.55, α = 0.1, β = 0.2: (*red solid line*) numerical results; (*blue dashed line*) approximate solution (colour figure online)

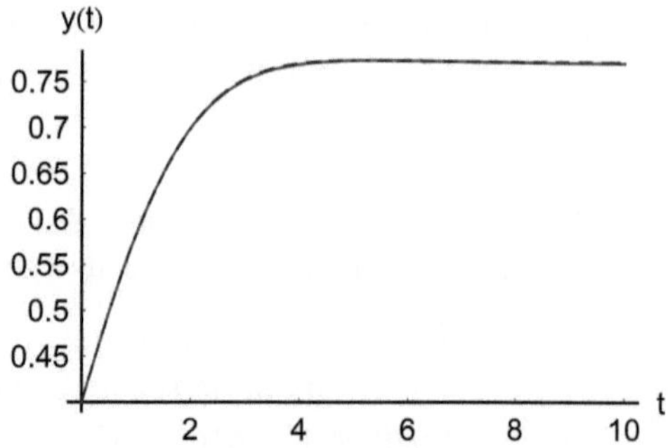

Fig. 4.124 Comparison between the first-order approximate solution 4.542 and numerical results for A = 0.25, B = 0.4, C = 0.55, α = 0.1, β = 0.2: (*red solid line*) numerical results; (*blue dashed line*) approximate solution (colour figure online)

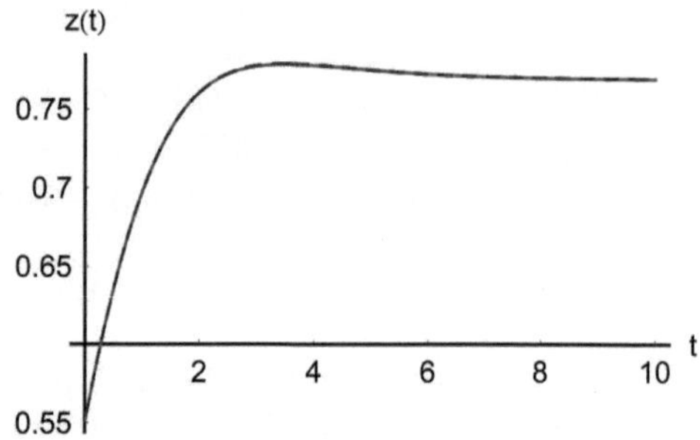

Fig. 4.125 Comparison between the first-order approximate solution 4.543 and numerical results for A = 0.25, B = 0.4, C = 0.55, α = 0.1, β = 0.2: (*red solid line*) numerical results; (*blue dashed line*) approximate solution (colour figure online)

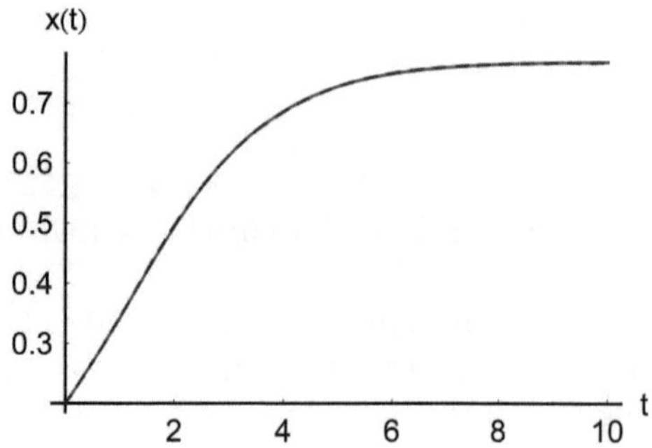

Fig. 4.126 Comparison between the first-order approximate solution 4.545 and numerical results for A = 0.2, B = 0.5, C = 0.7, α = 0.2, β = 0.1: (*red solid line*) numerical results; (*blue dashed line*) approximate solution (colour figure online)

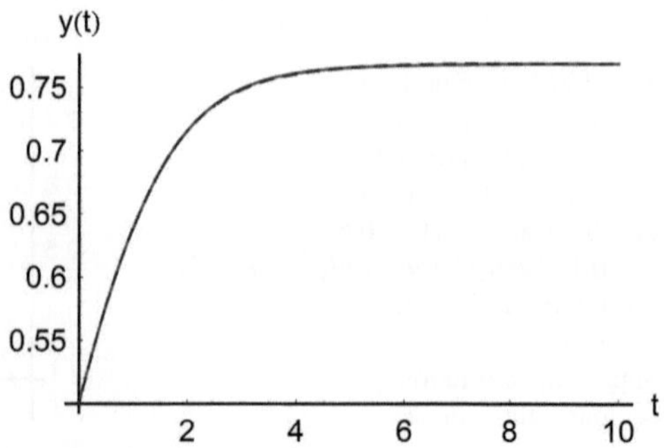

Fig. 4.127 Comparison between the first-order approximate solution 4.546 and numerical results for A = 0.2, B = 0.5, C = 0.7, α = 0.2, β = 0.1: (*red solid line*) numerical results; (*blue dashed line*) approximate solution (colour figure online)

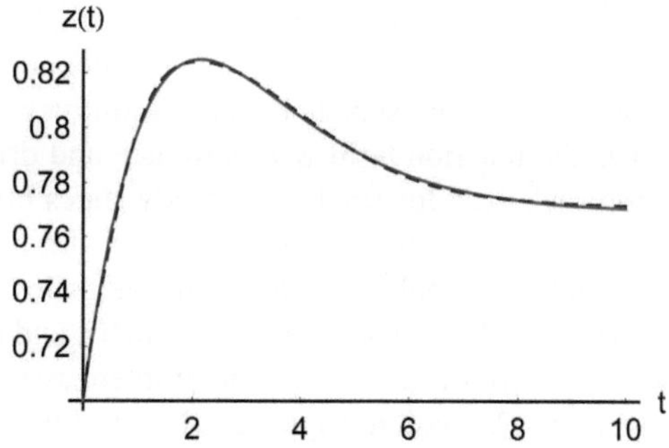

Fig. 4.128 Comparison between the first-order approximate solution 4.547 and numerical results for A = 0.2, B = 0.5, C = 0.7, α = 0.2, β = 0.1: (*red solid line*) numerical results; (*blue dashed line*) approximate solution (colour figure online)

present solutions and the numerical integration results obtained by a fourth-order Runge–Kutta method.

It can be seen from Figs. 4.123, 4.124, 4.125, 4.126, 4.127, 4.128 that the solutions obtained by OHAM are in very good agreement with numerical integration results. Our method gives analytic solutions valid globally in time unlike other known methods, for instance Adomian decomposition method, which unfortunately does not guarantee analytic solutions valid globally in time as proved by Repaci [92].

4.12 Bratu's Problem

The one-dimensional Bratu problem has a long history. Bratu's own article appeared in 1914 [144]; generalization are sometimes called the "Liouville-Gelfand" or "Liouville-Bratu-Gelfand" problem in honour of Gelfand and the nineteenth century work of the French mathematician J. Liouville [145, 146]. Bratu's problem is used in a large variety of applications such as the Chandrasekhar model of the expansion of the universe, questions in geometry and relativity concerning the Chandrasekhar model, fuel ignition model of the thermal combustion, the model of the thermal reaction process, chemical reaction theory, radiative heat transfer, nanotechnology and so on [147]. Moreover, Wan et al obtained a Bratu equation for electrospinning process [148].

The standard form of Bratu's boundary value problem is given by

$$u''(x) + \lambda e^{u(x)} = 0 \tag{4.548}$$

with the boundary conditions

$$u(0) = 0, u(1) = 0 \tag{4.549}$$

where λ is a positive parameter, known as the Frank-Kamenetskii parameter [148] and the prime denotes differentiation with respect to x.

If a nontrivial solution of Bratu's equation exists, then the competing forces of cooling (due to diffusion from the boundary) and heating (due to the positive reaction term) must be in balance. Intuitively, one might expect that for large values of λ, the reaction term will dominate and drive the temperature to infinity (explosion), whereas for smaller λ, steady states might be possible [149].

There are some analytic approaches for solving nonlinear Bratu's problem such as Chebyshev polynomials [150], the collocation in terms of Lambert W-function [151]. In [146] are determined existence and multiplicity results for radial solutions of the Liouville-Bratu-Gelfand problem associated with a class of quasilinear radial operators. Variational approach is used for mathematical models of electrospinning [152] and a numerical technique based on the variational method using the predictor-corrector technique is given in [153].

Smooth approximate solutions of the one-dimensional Bratu's problem are developed by using non-polynomial spline functions and are given in [154]. Adomian decomposition method is applied in [155] and a numerical technique based on the Laplace-Adomian decomposition method are used in [156] for one-dimensional Bratu problem and two variables are considered in [157]. The two-point boundary value problem (TPBVP) is applied in a wide variety of problems in engineering and science [158], homotopy analysis method is introduced to predict the multiplicity of the solutions of Bratu's equation [159] and different perturbation-iteration algorithms depending upon the number of Taylor expansion terms are proposed in [160].

4.12.1 The Exact Solution of Bratu's Problem 4.548

Multiplying Eq. 4.548 by $u'(x)$, and then integrating yields

$$u'^2(x) + 2\lambda e^{u(x)} = C \tag{4.550}$$

where C is the integration constant. Taking into account Eqs. 4.549, we obtain $C = u'^2(0) + 2\lambda$ and then Eq. 4.550 becomes

$$u'^2(x) + 2\lambda e^{u(x)} = u'^2(0) + 2\lambda \tag{4.551}$$

To solve Eq. 4.551 we set

$$t = \sqrt{u'^2(0) + 2\lambda - 2\lambda e^{u(x)}} \tag{4.552}$$

with the condition

$$e^{u(x)} \leq 1 + \frac{u'^2(0)}{2\lambda} \tag{4.553}$$

Separating variables, the Eq. 4.551 can be written as

$$\frac{2dt}{t^2 - u'^2(0) - 2\lambda} = dx$$

and then integrating, we obtain

$$\frac{1}{\theta}\ln\frac{t-\theta}{t+\theta} = x + \frac{1}{\theta}\ln\frac{u'(0)-\theta}{u'(0)+\theta} \tag{4.554}$$

where

$$\theta^2 = u'^2(0) + 2\lambda \tag{4.555}$$

Rearranging and substituting Eq. 4.552 into Eq. 4.554, one can get:

$$e^{u(x)} = \frac{4\theta^2 e^{\theta x}}{[u'(0) + \theta - (u'(0) - \theta)e^{\theta x}]^2} \tag{4.556}$$

We remark that the condition 4.553 is fulfilled in accordance with Eq. 4.550. From Eq. 4.556, we obtain after simple manipulations:

$$u(x) = -2\ln\left[\cosh\frac{\sqrt{u'^2(0) + 2\lambda}}{2}x - \frac{u'(0)}{\theta}\sinh\frac{\sqrt{u'^2(0) + 2\lambda}}{2}x\right] \tag{4.557}$$

where θ is given by Eq. 4.555

Now, using Eq. 4.549_2 into Eq. 4.557 and taking into account Eq. 4.555 we obtain the following relation between $u'(0)$ and λ:

$$\lambda e^{\frac{1}{2}\sqrt{u'^2(0) + 2\lambda}} - \lambda - u'^2(0)\sqrt{u'^2(0) + 2\lambda} = 0 \tag{4.558}$$

In conclusion, with $u'(0)$ obtained from Eq. 4.558 in respect to the parameter λ, the exact solution given by Eq. 4.557 is well-determined.

From Eq. 4.558 we deduce that for $\lambda < \lambda_{cr}$, the problem 4.548 and 4.549 has two solutions, for $\lambda = \lambda_{cr}$ admits only one solution and for $\lambda > \lambda_{cr}$ does not have a solution. For $u'(0) = 3.999754001$ into Eq. 4.558 we can obtain the value of λ_{cr}:

$$\lambda_{cr} = 3.51383071 \tag{4.559}$$

Remark 1 Using Eq. 4.555 and the notation

$$\sinh\frac{\theta}{4} = \frac{u'(0)}{\sqrt{2\lambda}}; \cosh\frac{\theta}{4} = \frac{\theta}{\sqrt{2\lambda}} \tag{4.560}$$

then the exact solution 4.557 can be written in the form

$$u(x) = -2\ln\frac{\cosh\frac{(2x-1)\theta}{4}}{\cosh\frac{\theta}{4}} \tag{4.561}$$

where θ is the solution of Eq. 4.560. This result is known [154–156].

Remark 2 The exact solution of Eqs. 4.548 and 4.549 has the property

$$u(x) = u(1-x), \forall x \in [0,1] \tag{4.562}$$

To prove this property, we remark that if x ∈ [0,1], then 1-x ∈ [0,1] and using the notation 1-x = y and the identity $d^2(1-x) = d^2y$, one can get:

$$u''(1-x) = \frac{d^2u(1-x)}{dx^2} = -\left(-\frac{d^2u(x)}{dx^2}\right) = u''(x) \tag{4.563}$$

On the other hand, from Eqs. 4.548 and 4.563 we obtain $e^{u(1-x)} = e^{u(x)}$. Because the exponential function is injective, the conclusion is immediate.

4.12.2 Solutions of the Bratu's Problem by Means of OHAM

The linear and nonlinear operators are respectively ($g(x) = 0$):

$$L(u(x)) = u'', N(u(x)) = \lambda e^{u(x)}, x \in [0,1] \tag{4.564}$$

and the initial conditions are given by Eq. 4.549.

The Equations 4.4 and 4.9 are respectively:

$$u_0''(x) = 0, u_0(0) = u_0(1) = 0 \tag{4.565}$$

$$u_1''(x) = H(x, C_i)\left[\lambda e^{u_0(x)}\right], u_1(0) = u_1(1) = 0 \tag{4.566}$$

The Eq. 4.565 has the solution $u_0 = 0$. If this result is substituted into Eq. 4.566 we obtain the following equation:

$$u_1''(x) = \lambda H(x, C_i), u_1(0) = u_1(1) = 0 \tag{4.567}$$

Taking into account the property 4.562 we choose the auxiliary function H in the form:

$$H(x, C_i) = \frac{1}{\lambda}\left[C_1x(1-x) + C_2x^2(1-x)^2 + C_3x^3(1-x)^3 + \ldots + C_nx^n(1-x)^n\right]'' \tag{4.568}$$

where C_i are unknown parameters at this moment and n is an arbitrary positive integer number. Alternatively, the function H could be chosen in other forms such as:

$$H(x, C_i) = \frac{1}{\lambda}\left[C_1'\ln(1+x(1-x)) + C_2'\left(1+x^2(1-x)^2\right) + \ldots + C_n'(1+x^n(1-x)^n)\right]'' \tag{4.569}$$

or

$$H(t, C_i) = \frac{1}{\lambda}\left[C_1''e^{x(1-x)} + C_2''e^{x^2(1-x)^2} + \ldots + C_n''e^{x^n(1-x)^n}\right]'' \tag{4.570}$$

or

$$H(t, C_i) = \frac{1}{\lambda}\left[C_1^*\ln\left(1+\sqrt{x(1-x)}\right) + C_2^*\ln\left(1+\sqrt[3]{x(1-x)}\right) + \ldots + C_n^*\left(1+\sqrt[n]{x(1-x)}\right)\right]'' \tag{4.571}$$

In what follows, we consider only Eq. 4.568 and n = 4. In this case it is clear that

$$u_1(x) = C_1x(1-x) + C_2x^2(1-x)^2 + C_3x^3(1-x)^3 + C_4x^4(1-x)^4 \tag{4.572}$$

The first-order approximate solution of Eqs. 4.548 and 4.549 is

$$\begin{aligned}\overline{u}(x) = u_0(x) + u_1(x) = C_1x(1-x) + C_2x^2(1-x)^2 + \\ + C_3x^3(1-x)^3 + C_4x^4(1-x)^4\end{aligned} \tag{4.573}$$

4.12.3 Numerical Examples

We will show that, using the algorithm described above, with the help of a computer program we can obtain the optimal values of the convergence-control parameters C_i, i = 1,2,3,4. These parameters are obtained using collocation method.

4.12.3.1 In the first case let be $\lambda = 1$, and then we obtain the following two sets of solutions for the optimal convergence-control parameters:

$$C_1 = 0.5493504955765137; C_2 = 0.04937291621622237;$$
$$C_3 = 0.007045986451157435; C_4 = 0.001418054070501909 4 \quad (4.574)$$

and

$$C_1^{'} = 9.752026223543547; C_2^{'} = 36.02774215930639$$
$$C_3^{'} = -176.11247507970032; C_4^{'} = 551.0772185049452 \quad (4.575)$$

The first-order approximate solutions 4.573 can be written in the forms:

$$\overline{u}_{1,1}(x) = 0.5493504955765137x(1-x) + 0.0493729162162237x^2(1-x^2)+$$
$$+0.007045986451157 35x^3(1-x)^3 + 0.001418054070501909 4x^4(1-x^4) \quad (4.576)$$

and respectively

$$\overline{u}_{1,2}(x) = 9.752026223543547x(1-x) + 36.02774215930639x^2(1-x)^2-$$
$$-176.1124750797003x^3(1-x)^3 + 551.0772185049452x^4(1-x)^4 \quad (4.577)$$

On the other hand, from Eq. 4.558 for $\lambda = 1$, we obtain two solutions for $u'(0)$:

$$u_1^{'}(0) = 0.549352729, u_2^{'}(0) = 10.846899019 \quad (4.578)$$

and therefore, from Eq. 4.557 we deduce that there exists two exact solutions.

We remark that in this first case and later also in the second case, we obtain by OHAM two approximate solutions given by Eqs. 4.576 and 4.577, independent of Eq. 4.557.

It is easy to verify the accuracy of the obtained solutions if we graphically compare these analytical solutions with the corresponding exact ones. Figs. 4.129 and 4.130 show the comparison between the present solutions given by Eqs. 4.576

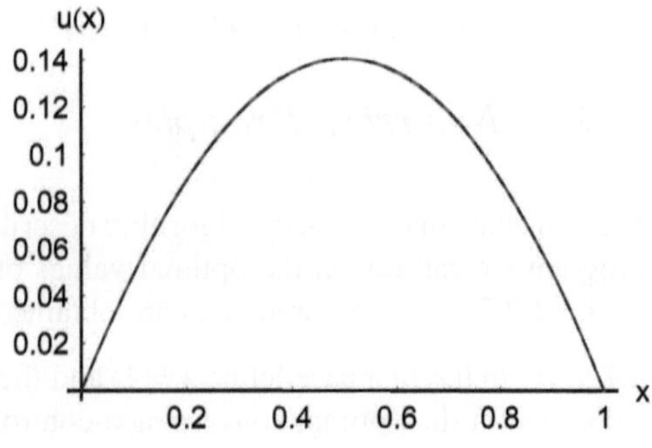

Fig. 4.129 Comparison between the approximate solution 4.576 and the exact solution 4.557 for $\lambda = 1$ and u'(0) = 0.549352729: (*red solid line*) exact solution, (*blue dashed line*) approximate solution (colour figure online)

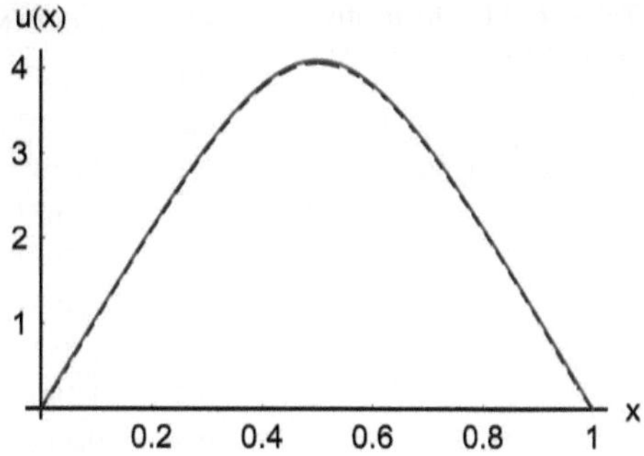

Fig. 4.130 Comparison between the approximate solution 4.577 and the exact solution 4.557 for λ = 1 and u'(0) = 10.846899019: (*red solid line*) exact solution, (*blue dashed line*) approximate solution (colour figure online)

Table 4.123 Comparison between the first approximate solution 4.576 given by OHAM, the solutions given in [156] and [160] and exact solution 4.557 for λ = 1 and u'(0) = 0.549352729

x	Approximate solution 4.576 (one iteration)	Approximate solution [156] (5 iterations)	Approximate solution [160] (2 iterations)	Exact solution 4.557
0.1	0.04984669478588	0.0498448112	0.04983	0.04984679126784
0.2	0.08918981564379	0.0891859956	0.08915	0.089189934673454
0.3	0.11760896040246	0.1176032408	0.11756	0.117609095834281
0.4	0.13479010739219	0.1347825488	0.13473	0.134790253971493
0.5	0.14053906396965	0.1405297477	0.14048	0.1405392145077411
0.6	0.13479010739219	0.1347791409	0.13473	0.1347902539714935
0.7	0.11760896040246	0.1175965240	0.11756	0.1176090958342818
0.8	0.08918981564379	0.0891899350	0.08915	0.0891899346734544
0.9	0.04984669478588	0.0498348222	0.04983	0.04984679126784

and 4.577 respectively and the exact solutions 4.557 in the case λ = 1 and for u'(0) given by Eq. 4.578.

Lin et al [158] applying a two-point boundary value problem algorithm, found the same values for u'(0) into Eqs. 4.578 using 15 iterations.

Within Tables 4.123, 4.124, 4.125, 4.126, 4.127, 4.128 we present a comparison between the results of the present method and exact or other known results [156, 160].

It can be seen from Tables 4.123 and 4.125 that the analytical solutions obtained by OHAM are extremely accurate.

4.12.3.2 In the second case we consider λ = 2. The first-order approximate solutions become:

$$\overline{u}_{2,1}(x) = 1.2481037441426102x(1-x) + 0.24860575243610394x^2(1-x)^2 + \\ +0.0768704100976 1027x^3(1-x)^3 + 0.0403035769940576x^4(1-x)^4 \tag{4.579}$$

Table 4.124 The relative error of the approximate solutions $\varepsilon = |\overline{u}_{approx} - u_{exact}|$ for different methods

x	ε OHAM	ε [156]	ε [160]
0.1	9.6478×10^{-8}	1.9787×10^{-6}	1.6791×10^{-5}
0.2	1.1902×10^{-8}	3.9394×10^{-6}	3.9934×10^{-5}
0.3	1.3543×10^{-7}	5.8548×10^{-6}	4.9095×10^{-5}
0.4	1.4657×10^{-7}	7.7038×10^{-6}	6.0251×10^{-5}
0.5	1.5053×10^{-7}	9.4665×10^{-6}	5.9214×10^{-5}
0.6	1.4657×10^{-7}	1.1111×10^{-6}	6.0251×10^{-5}
0.7	1.3543×10^{-7}	1.2571×10^{-5}	4.9095×10^{-5}
0.8	1.1902×10^{-8}	1.3475×10^{-5}	3.9934×10^{-5}
0.9	9.6478×10^{-8}	1.1967×10^{-5}	1.6791×10^{-5}

Table 4.125 Comparison between the second approximate solution 4.577 given by OHAM and the exact solution 4.557 for $\lambda = 1$ and u'(0) = 10.846899019

x	Approximate solution 4.577	Exact solution 4.557	ε OHAM
0.1	1.07727725358	1.077273316982309	3.9366×10^{-6}
0.2	2.12243166303	2.12239234143816	3.9321×10^{-5}
0.3	3.07751178977	3.0773951011770544	1.6668×10^{-4}
0.4	3.80944733899	3.8061519597709563	3.2953×10^{-3}
0.5	4.09062840250	4.091467246118674	8.3884×10^{-4}
0.6	3.80944733899	3.8061519597709563	3.2953×10^{-3}
0.7	3.07751178977	3.0773951011770544	1.6668×10^{-4}
0.8	2.12243166303	2.12239234143816	3.9321×10^{-5}
0.9	1.07727725358	1.077273316982309	3.9366×10^{-6}

Table 4.126 Comparison between the first solution 4.579 given by OHAM, solutions given in [156] and the exact solution 4.557 for $\lambda = 2$ and u'(0) = 1.248217519

x	Approximate solution 4.579 (one iteration)	Approximate solution [156] (5 iterations)	Exact solution 4.557	ε OHAM	ε [156]
0.1	0.1144017264162	0.1122817141	0.1144107433914860	9.01739×10^{-5}	2.12904×10^{-5}
0.2	0.2064021808771	0.2021094162	0.2064191167322770	1.69358×10^{-5}	4.20969×10^{-5}
0.3	0.2738555796198	0.2676925058	0.2738793121851143	2.37325×10^{-5}	6.18680×10^{-5}
0.4	0.3150609640793	0.3070874506	0.3150893646906812	2.84006×10^{-5}	8.00191×10^{-5}
0.5	0.3289223315683	0.3193532294	0.3289524218983754	3.28952×10^{-5}	9.59919×10^{-5}
0.6	0.3150609640793	0.3041598403	0.3150893646906812	2.84006×10^{-5}	1.09295×10^{-5}
0.7	0.2738555796198	0.2619458909	0.2738793121851143	2.37325×10^{-5}	1.19933×10^{-5}
0.8	0.1206402180877	0.1940413072	0.2064193167322770	1.69358×10^{-5}	1.23778×10^{-5}
0.9	0.1144017264142	0.1035373785	0.1144107433914861	9.01739×10^{-5}	1.08733×10^{-5}

Table 4.127 Comparison between the second solution 4.580 given by OHAM and the exact solution 4.557 for $\lambda = 2$ and $u'(0) = 8.268763182$

x	Approximate solution 4.580 given by OHAM (one iteration)	Exact solution 4.557	ε OHAM
0.1	0.813466080	0.8134746094509364	8.5294×10^{-6}
0.2	1.583099011	1.57961185308176	3.4871×10^{-3}
0.3	2.245005391	2.2452582168242805	2.5282×10^{-4}
0.4	2.719375273	2.719806581962672	4.3130×10^{-4}
0.5	2.895023453	2.895531265825199	5.0781×10^{-4}
0.6	2.719375273	2.71980658196267	4.3130×10^{-4}
0.7	2.245005391	2.2452582168242805	2.5282×10^{-4}
0.8	1.583099011	1.57961185308176	3.4871×10^{-3}
0.9	0.813466080	0.8134746094509364	8.5294×10^{-6}

Table 4.128 Comparison between the solution 4.581 given by OHAM and the exact solution 4.557 for $\lambda = 3.513830712$ and $u'(0) = 3.999754001$

x	Approximate solution 4.581 (one iteration)	Exact solution 4.557	ε OHAM
0.1	0.379623363641	0.3798805377205356	2.57174×10^{-4}
0.2	0.707892156645	0.7080655046375435	1.73347×10^{-4}
0.3	0.964731360565	0.9648288768794109	9.75163×10^{-4}
0.4	1.129688683896	1.1297364148964748	4.77310×10^{-4}
0.5	1.186712490182	1.1867567073280167	4.42171×10^{-4}
0.6	1.129688683896	1.1297364148964748	4.77310×10^{-4}
0.7	0.964731360565	0.9648288768794109	9.75163×10^{-4}
0.8	0.707892156645	0.7080655046375435	1.73347×10^{-4}
0.9	0.379623363641	0.3798805977205356	2.57174×10^{-4}

$$\begin{aligned}\overline{u}_{2,2}(x) &= 7.63720308326216x(1-x) + 20.995895806939686x^2(1-x)^2 - \\ &-82.4431106130786x^3(1-x)^3 + 246.57311626375208x^4(1-x)^4\end{aligned} \tag{4.580}$$

Figs. 4.131 and 4.132 show the comparison between the present solutions given by Eqs. 4.579 and 4.580 respectively and exact solutions 4.557 in the case $\lambda = 2$ and $u'(0)$ given by Eq. 4.558.

In Tables 4.126 and 4.127, the exact solutions for the case $\lambda = 2$ derived from Eq. 4.557 are compared with approximate solutions obtained by the OHAM and with the numerical solution obtained by the Laplace-Adomian decomposition method using five iterations [156].

4.12.3.3 In the last case, we consider $\lambda = \lambda_{cr} = 3.51383071$, when Eq. 4.558 has a unique solution given by $u'(0) = 3.999754001$. Using the collocation method we obtain also a unique solution, which lead to the approximate solution:

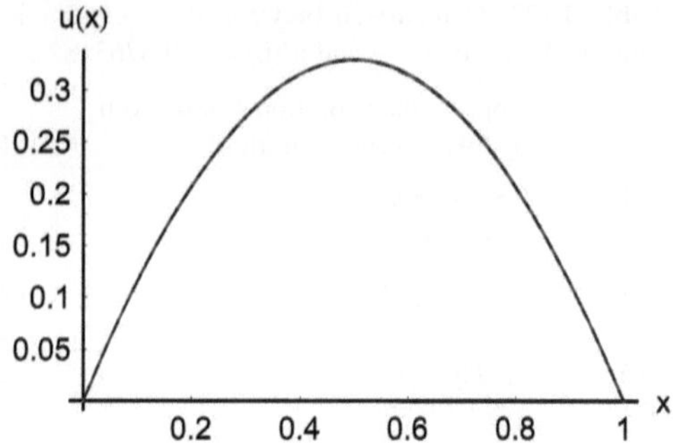

Fig. 4.131 Comparison between the approximate solution 4.579 and the exact solution 4.557 for $\lambda = 2$ and $u'(0) = 1.248217519$: (*red solid line*) exact solution, (*blue dashed line*) approximate solution (colour figure online)

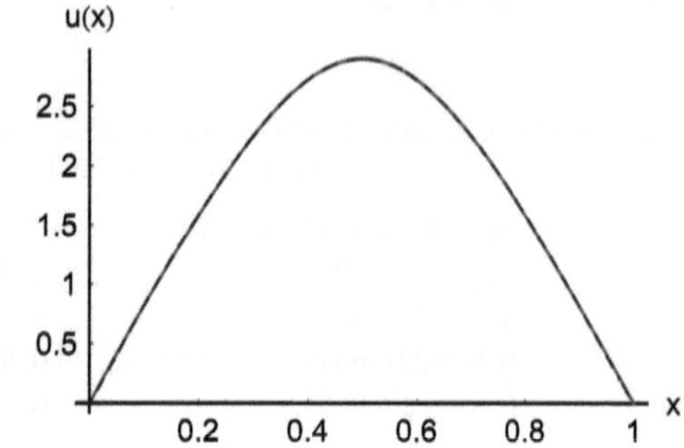

Fig. 4.132 Comparison between the approximate solution 4.580 and the exact solution 4.557 for $\lambda = 2$ and $u'(0) = 8.268763182$: (*red solid line*) exact solution, (*blue dashed line*) approximate solution (colour figure online)

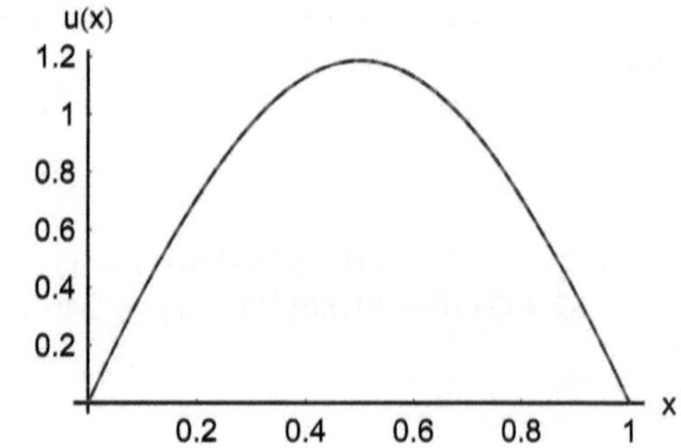

Fig. 4.133 Comparison between the approximate solution 4.581 and the exact solution 4.557 for $\lambda = 3.513830712$ and $u'(0) = 3.999754001$

$$\overline{u}(x) = 3.9862645043250926x(1-x) + 2.4615199977739213x^2(1-x)^2 + \\ +0.667701362683833x^3(1-x)^3 + 6.622343794799803x^4(1-x)^4 \tag{4.581}$$

Fig. 4.133 shows the comparison between the present solution given by Eq. 4.581 and the exact solution 4.557 in this last case.

In Table 4.128, the exact solution for the case $\lambda = 3.513830712$ and $u'(0) = 3.999754001$ derived from Eq. 4.557 is compared with the approximate solution obtained by OHAM after only one iteration.

From Figs. 4.129, 4.130, 4.131, 4.132, 4.133 and Tables 4.123, 4.124, 4.125, 4.126, 4.127, 4.128 it can be seen that the analytical solutions of the Bratu's problem obtained by OHAM are very accurate compared to the exact solutions or with solutions obtained by other procedures. We remark that our approach needs only one iteration for obtaining these accurate results, compared to 5 iterations needed in [156] and 2 iterations neded in [160]. Also, all branches of solutions for $\lambda < \lambda_{cr}$ can be obtained directly without the help of Eq. 4.558. As we know so far, it is for the first time in the literature when there are effectively and explicitely presented all the solutions in this way for the same value of λ, $\lambda < \lambda_{cr}$.

4.13 Lane-Emden Equation

Consider the following nonlinear singular initial value problem known as the Lane-Emden equation [161–163]:

$$\begin{aligned} &y''(x) + \frac{2}{x}y'(x) + f(y(x)) = 0, x > 0 \\ &y(0) = A, y'(0) = 0 \end{aligned} \tag{4.582}$$

where the primes denote differentiation with respect to the independent variable x, f(y(x)) is a nonlinear function of y and A is a constant.

The Eq. 4.582 can be written in the form

$$xy''(x) + 2y'(x) + xf(y(x)) = 0 \tag{4.583}$$

Using the method of variation of constants, the solution of Eq. 4.582 becomes [164]

$$y(x) = A + \int_0^x \left(\frac{t^2}{x} - t\right) f(y(t))dt, x > 0 \tag{4.584}$$

The Equation 4.584 is known as Volterra integral equation of the first kind. In this section we consider the case

$$f(y) = -8e^{A-y} - 4\sqrt{e}^{A-y} \tag{4.585}$$

The exact solution of Eq. 4.582 and taking into account 4.585 is

$$y(x) = A + 2\ln(1 + x^2) \tag{4.586}$$

If $f(y(x)) = y^n(x)$, then the Eq. 4.582 is the classical Lane-Emden equation that was used to model the thermal behavior of a spherical cloud of gas acting under the mutual attraction of its molecules [165] and subject to the classical laws of thermodynamics. The Lane-Emden equation was used to model several phenomena in mathematical physics and astrophysics such as the theory of stellar structure, the thermal behavior of a spherical cloud of gas, isothermal gas spheres and theory of thermionic currents [166]. This type of equation has received considerable attention in the last years. In 1983, Nobel physics laureate Subrahmanyan Chandrasekhar (1910–1995) introduced the function $f(y(x)) = (y^2 - c)^{3/2}$ with c a positive parameter, in his theory of white dwarf stars. For $f(y(x)) = e^{y(x)}$ into Eq. 4.583, Momoniat and Harley [167] obtained an approximate implicit solution by reducing it to first-order equation using Lie group analysis and determining a power series solution of the reduced equations. By means of a semi-inverse method, J.H.He [168] obtained a variational principle for the Lane-Emden equation. Adomain decomposition method was applied by Wazwaz [169] with an alternative framework designed to overcome the difficulty of the singular point. In [170], Yousefi used integral operator to convert Lane-Emden equation to integral equations and then applied Legendre wavelet approximations. The modified and the standard Adomian decomposition method are used by Hosseini and Nasabzadeh [171] to solve Lane-Emden problem and Marinca and Herisanu applied OHAM to solve this problem [172].

4.13.1 Solution of Eq. 4.584 by Means of OHAM

In the general case of Volterra integral equation of the first kind

$$y(x) - A - \int_0^x h(x,t) f(y,t)dt = 0, \quad x > 0 \tag{4.587}$$

where $h(x,t)$ is a known function of variables x and t, the family of equations 4.6 can be rewritten as

$$y_0(x) - A + p[y_1(t) - H(x, C_i)N(y_0(t))] = 0 \tag{4.588}$$

from which $y_0(t)$ and $y_1(t)$ can be obtained as

$$y_0(x) - A = 0, \quad y_0(0) = A \tag{4.589}$$

$$y_1(x) - H(t, C_i)N(y_0(t)) = 0, \quad y_1(0) = 0 \tag{4.590}$$

From Eq. 4.589 we have

$$y_0(x) = A \tag{4.591}$$

and therefore, the first approximation y_1 is obtained from

$$y_1(x) = -H(t, C_i)\int_0^x h(x,t)f(y_0(t))dt \tag{4.592}$$

where

$$N(y(x)) = -\int_0^x h(x,t)f(y(t))dt \tag{4.593}$$

and

$$h(x,t) = \frac{t^2}{x} - t \tag{4.594}$$

Using Eqs. 4.591, 4.594. 4.593 and 4.592 one can get

$$y_1(x, C_i) = -H(x, C_i)\left[\frac{x^2}{6}f(A)\right] \tag{4.595}$$

If we choose the auxiliary function H such as

$$H(x, C_i) = -\frac{6}{x^2 f(A)}\sum_{i=1}^{s} C_i \ln^2(1+ix) \tag{4.596}$$

where C_i, i = 1,2,. . .,s are unknown parameters, then the first approximation becomes

$$y_1(x, C_i) = \sum_{i=1}^{s} C_i \ln^2(1+ix) \tag{4.597}$$

But the possibility of choosing the function $H(x, C_i)$ is not unique. We can choose for example

$$H^*(x, C_i) = -\frac{6}{xf(A)}\sum_{i=1}^{s} C_i \ln(1+ix) \tag{4.598}$$

$$\widetilde{H}(x, C_i) = -\frac{6}{f(A)}\sum_{i=1}^{s} C_i \ln(1 + ix^2) \tag{4.599}$$

and so on.

Now, the first-order approximate solution of Eq. 4.582 or 4.584 becomes

$$\overline{y}(x, C_i) = y_0(x) + y_1(x, C_i) \tag{4.600}$$

or

$$\overline{y}(x, C_i) = A + \sum_{i=1}^{s} C_i \ln^2(1 + ix) \tag{4.601}$$

4.13.2 Numerical Examples

We will show that the error of the solution 4.601 decreases when the number of terms in the auxiliary function $H(x, C_i)$ increases. In Eqs. 4.582 or 4.584 we consider A = 1 and three cases for s.

Case 4.13.2.1 Following the procedure described above, for s = 5 it is obtained

$$C_1 = 9.750090326718208,\ C_2 = -101.81554047755736,$$
$$C_3 = 265.1437168943899,\ C_4 = -266.8566983525972,$$
$$C_5 = 93.82049950862766$$

The first-order approximate solution 4.601 becomes in this case

$$\begin{aligned}\overline{y}(x) = 1 + 9.75009\ln^2(1 + x) - 101.816\ln^2(1 + 2x) + \\ +265.144\ln^2(1 + 3x) - 266.857\ln^2(1 + 4x) + 93.8205\ln^2(1 + 5x)\end{aligned} \tag{4.602}$$

Case 4.13.2.2 For s = 6 it is obtained

$$C_1 = -37.069654314883756,\ C_2 = 791.8325447375765,$$
$$C_3 = -4186.767518124741,\ C_4 = 8703.957653001464,$$
$$C_5 = -7856.000786313658,\ C_6 = 2584.0873372088868$$

The first-order approximate solution 4.601 can be written in this case

$$\begin{aligned}\overline{y}(x) = 1 - 37.0697\ln^2(1 + x) + 791.833\ln^2(1 + 2x) - \\ -4186.77\ln^2(1 + 3x) + 8703.96\ln^2(1 + 4x) - 7856\ln^2(1 + 5x) + \\ +2584.09\ln^2(1 + 6x)\end{aligned} \tag{4.603}$$

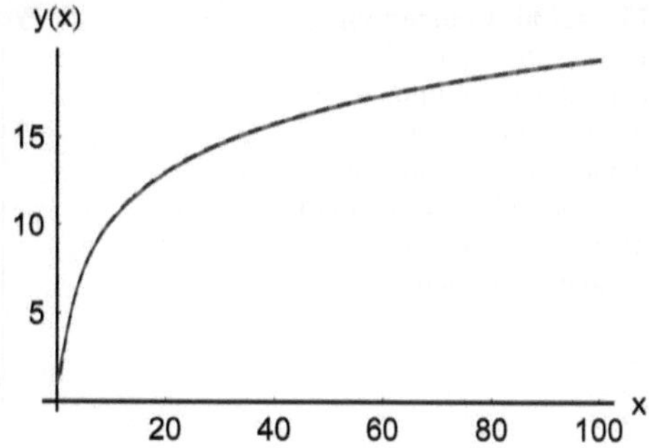

Fig. 4.134 Comparison between the first-order approximate solution 4.602 and the exact solution 4.586: (*red solid line*) exact solution, (*blue dashed line*) solution (colour figure online)

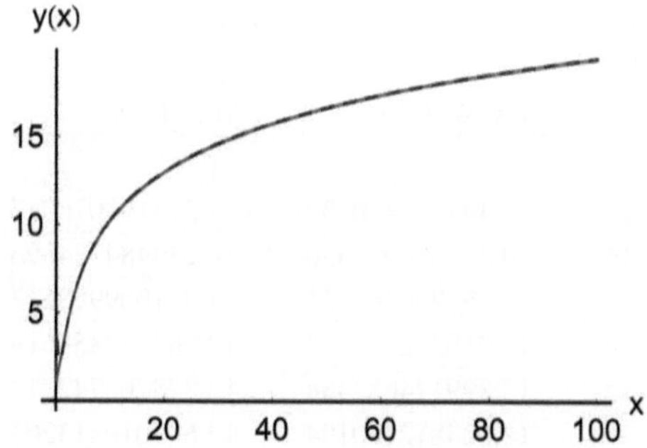

Fig. 4.135 Comparison between the first-order approximate solution 4.603 and the exact solution 4.586: (*red solid line*) exact solution, (*blue dashed line*) approximate solution (colour figure online)

Case 4.13.2.3 For $s = 7$ one can get

$$\begin{aligned}
&C_1 = \text{-}111.43562586830211, C_2 = 4812.874337962244,\\
&C_3 = \text{-}43058.39987057521, C_4 = 149147.01173457873,\\
&C_5 = \text{-}241191.88944445073, C_6 = 183963.9503862947,\\
&C_7 = \text{-}53562.10219670979
\end{aligned}$$

In this case the first-order approximate solution is obtained as:

$$\begin{aligned}
&\overline{y}(x) = 1 - 111.436\ln^2(1 + x) + 4812.874\ln^2(1 + 2x) -\\
&-43058.399\ln^2(1 + 3x) + 149147.011\ln^2(1 + 4x) -\\
&-241191.889\ln^2(1 + 5x) + 183963.95\ln^2(1 + 6x) \text{-} 53562.102\ln^2(1 + 7x)
\end{aligned} \tag{4.604}$$

Figs. 4.134, 4.135, 4.136 show the comparison between the present solutions 4.602-4.604 respectively and exact solution. It can be seen that the solutions obtained by OHAM are very accurate, being nearly identical with the exact solution. However, the analytical solutions obtained by our procedure prove to be more accurate along with an increased number of terms in the auxiliary function H (x,C_i). From Tables 4.129 and 4.130 we can conclude that the analytical solutions given by Eq. 4.602-4.604 are very accurate comparing to the exact solution.

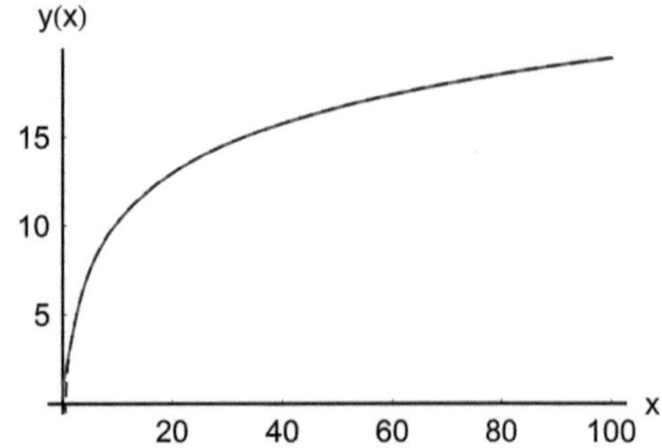

Fig. 4.136 Comparison between the first-order approximate solution 4.604 and the exact solution 4.586: (*red solid line*) exact solution, (*blue dashed line*) approximate solution (colour figure online)

Table 4.129 The results of the first-order approximate solution 4.602-4.604 and exact solution 4.586

x	$\bar{y}(x)$ 4.602	$\bar{y}(x)$ 4.603	$\bar{y}(x)$ 4.604	$y(x)$
0	1	1	1	1
5	7.494711460132	7.516193076047	7.516193076153	7.516193076042
10	10.243884623530	10.230484524596	10.230241033714	10.230241033682
15	11.862974364423	11.841069998547	11.841069998452	11.841069998544
20	13.010342792114	12.987922854619	12.987922854954	12.987922854613
25	13.899174682454	13.878700742214	13.878700742032	13.878700742200
30	14.624812150194	14.607010515203	14.607000233372	14.607010515216
35	15.238068629412	15.223056992996	15.222993817413	15.223024232992
40	15.769216694914	15.756889705415	15.756711103953	15.756767425993
45	16.237743325929	16.227918480639	16.227553626522	16.227637369617
50	16.656929491938	16.649405613832	16.648782585980	16.648891861755
55	17.036232647662	17.030813611141	17.029863301897	17.029993788696
60	17.382626127250	17.379129952518	17.377787979086	17.377933727297
65	17.701399287075	17.699660882644	17.697868346469	17.698022396352
70	17.996659491529	17.996530640986	17.994234252022	17.994389089819
75	18.271660011134	18.273008225514	18.270160109736	18.270307978099
80	18.529021580181	18.531728555302	18.528285900596	18.528419014284
85	18.770886477853	18.774846495027	18.770771120209	18.770881823415
90	18.999028377941	19.004146796723	18.999404727946	18.999485579660
95	19.214932377479	19.221124274263	19.215685315895	19.215729160773
100	19.419854419340	19.427043386240	19.420880733523	19.420880733953

Table 4.130 The absolute error $\varepsilon = |\overline{y}(x) - y(x)|$ between the first approximate solutions 4.602-4.604 and exact solution 4.586.

x	ε Eq.602	ε Eq. 4.603	ε Eq. 4.604
0	0	0	0
5	2.14×10^{-2}	4.82×10^{-12}	1.1×10^{-10}
10	1.36×10^{-2}	2.43×10^{-4}	3.1×10^{-11}
15	2.19×10^{-2}	2.45×10^{-12}	9.2×10^{-11}
20	2.24×10^{-2}	6.54×10^{-12}	3.4×10^{-10}
25	2.04×10^{-2}	1.40×10^{-11}	1.6×10^{-10}
30	1.78×10^{-2}	1.36×10^{-11}	1.0×10^{-5}
35	1.50×10^{-2}	3.27×10^{-5}	3.0×10^{-5}
40	1.24×10^{-2}	1.22×10^{-4}	5.6×10^{-5}
45	1.01×10^{-2}	2.81×10^{-4}	8.3×10^{-5}
50	8.03×10^{-3}	5.13×10^{-4}	1.0×10^{-4}
55	6.23×10^{-3}	8.19×10^{-4}	1.3×10^{-4}
60	4.69×10^{-3}	1.19×10^{-3}	1.4×10^{-4}
65	3.37×10^{-3}	1.63×10^{-3}	1.5×10^{-4}
70	2.27×10^{-3}	2.14×10^{-3}	1.5×10^{-4}
75	1.35×10^{-3}	2.70×10^{-3}	1.4×10^{-4}
80	6.02×10^{-4}	3.30×10^{-3}	1.3×10^{-4}
85	4.65×10^{-6}	3.96×10^{-3}	1.1×10^{-4}
90	4.57×10^{-4}	4.66×10^{-3}	8.0×10^{-5}
95	7.96×10^{-4}	5.39×10^{-3}	4.3×10^{-5}
100	1.02×10^{-3}	6.16×10^{-3}	4.2×10^{-10}

4.14 Oscillations of a Uniform Cantilever Beam Carrying an Intermediate Lumped Mass and Rotary Inertia

The problems related to large-amplitude oscillations of non-linear engineering structures have received considerable attention in the past years [173, 174]. Engineering structures undergoing large-amplitude oscillations often involve discretizing the structure, when free vibration analysis is performed, and results in a temporal problem having inertia and static non-linearities. Such problems are not amenable to exact treatment because of their complexity and approximate techniques must be resorted to [175].

The problem of large-amplitude vibration of a uniform cantilever beam approached in this section is of practical interest because many engineering structures can be modelled as a slender, flexible cantilever beam carrying a lumped mass with rotary inertia at an intermediate point along its span hence they exhibits large-amplitude vibration.

Often linearization techniques are employed in order to approximate such non-linear problems. One may question the accuracy of using such a linear mode method, which are a frequently used method in the analysis of non-linear continuous systems to approximate the large amplitude non-linear behavior [175].

In this section we consider a clamped beam at the base, free at the tip, which caries a lumped mass and rotary inertia at an arbitrary intermediate point along its span. The beam is considered to be uniform of constant length and mass per unit length and the thickness of this conservative beam is assumed to be small compared to the length so that the effects of rotary inertia and shearing deformation will be ignored. Moreover, the beam is assumed to be inextensible, which implies that the length of beam's neutral axis remains constant during the motion. These assumptions are the same as those used in references [174, 175]. In these conditions one can derive the discrete, single-mode, of order three nonlinearities, beam temporal problem [175, 176]:

$$\frac{d^2u}{dt^2} + u + \alpha u^2 \frac{d^2u}{dt^2} + \alpha u \left(\frac{du}{dt}\right)^2 + \beta u^3 = 0 \tag{4.605}$$

subject to the initial conditions

$$u(0) = A, \frac{du}{dt}(0) = 0 \tag{4.606}$$

This system, where α and β are modal constants which result from the discretization procedure [175], describes the large-amplitude free vibrations of the considered slender inextensible cantilever beam, which is assumed undergoing planar flexural vibrations. The third and fourth-terms in Eq. 4.605 represent inertia-type cubic non-linearity arising from the inextensibility assumption. The last term is a static-type cubic non-linearity associated with the potential energy stored in bending.

In order to analytically solve this problem, we introduce a new independent variable

$$\tau = \Omega t \tag{4.607}$$

then Eqs. 4.605 and 4.606 become respectively

$$u'' + \Omega^{-2}u + \alpha u^2 u'' + \alpha u u'^2 + \beta \Omega^{-2} u^3 = 0 \tag{4.608}$$

and

$$u(0) = A, u'(0) = 0 \tag{4.609}$$

where the prime denotes differentiation with respect to τ. The new independent variable τ is chosen in such a way that the solution of Eq. 4.608, which satisfies the assigned initial conditions in Eq. 4.609 is a periodic function of τ, of period 2π. The period of the corresponding non-linear oscillation is given by $T = 2\pi/\Omega$. Here, both periodic solution $u(\tau)$ and frequency Ω depend on A.

Under the transformation

$$u(\tau) = Ax(\tau) \tag{4.610}$$

the equation 4.608 becomes

$$x'' + \Omega^{-2}x + ax^2x'' + axx'^2 + b\Omega^{-2}x^3 = 0 \tag{4.611}$$

where $a = \alpha A^2$, $b = \beta A^2$ and the initial conditions become

$$x(0) = 1, x'(0) = 0 \tag{4.612}$$

Corresponding to nonlinear differential equation 4.611, the linear and the nonlinear operators are, respectively ($g(\tau) = 0$)

$$L[x(\tau)] = x'' + x \tag{4.613}$$

$$N[x(\tau), \Omega] = (\Omega^{-2} - 1)x + ax^2x'' + axx'^2 + b\Omega^{-2}x^3 \tag{4.614}$$

The equation 4.4 becomes

$$x_0'' + x_0 = 0, x_0(0) = 1, x_0'(0) = 0 \tag{4.615}$$

whose solution is

$$x_0(\tau) = \cos\tau \tag{4.616}$$

Substituting Eq. 4.616 into Eq. 4.614, we have

$$N[x_0(\tau), \Omega] = \left(\Omega^{-2} - 1 - \frac{\alpha}{2} + \frac{3}{4}\beta\Omega^{-2}\right)\cos\tau + \left(\frac{1}{4}b\Omega^{-2} - \frac{a}{2}\right)\cos 3\tau \tag{4.617}$$

The differential equation corresponding to the first approximation is obtained from Eq. 4.9:

$$x_1'' + x_1 = H(\tau, C_i)N[x_0(\tau), \Omega], x_1(0) = x_1'(0) = 0 \tag{4.618}$$

If we choose the auxiliary function as

$$H(\tau, C_i) = C_1 + 2C_2\cos 2\tau + 2C_3\cos 4\tau + 2C_4\cos 6\tau \tag{4.619}$$

then, substituting Eq. 4.619 into Eq. 4.618, it holds that

$$x_1'' + x_1 = \Big[\Big(\Omega^{-2} - 1 - \frac{a}{2} + \frac{3}{4}b\Omega^{-2}\Big)(C_1 + C_2) + \Big(\frac{1}{4}b\Omega^{-2} - \frac{a}{2}\Big)(C_2 + C_3)\Big]\cos\tau + [M(C_2 + C_3) + NC_2]\cos 5\tau + \big[MC_4 + NC_3\big]\cos 7\tau + NC_4\cos 9\tau \tag{4.620}$$

where

$$M = \Omega^{-2} - 1 - \frac{a}{2} + \frac{3}{4}b\Omega^{-2}, N = \frac{1}{4}b\Omega^{-2} - \frac{a}{2} \tag{4.621}$$

Now, avoiding the presence of a secular term into Eq. 4.620 needs

$$\Omega^2 = \frac{4(C_1 + C_2) + b(3C_1 + 4C_2 + C_3)}{4(C_1 + C_2) + 2a(C_1 + 2C_2 + C_3)} \tag{4.622}$$

From Eq. 4.620 it follows that

$$x_1(\tau, C_i) = \frac{1}{8}[M(C_2 + C_3) + N(C_1 + C_4)](\cos\tau - \cos 3\tau) + \frac{1}{24}[M(C_3 + C_4) + NC_2](\cos\tau - \cos 5\tau) + \frac{1}{48}[MC_4 + NC_3](\cos\tau - \cos 7\tau) + \frac{1}{80}NC_4(\cos\tau - \cos 9\tau) \tag{4.623}$$

Taking into consideration Eqs. 4.616, 4.623 and 4.5, the first-order approximate solution of Eqs. 4.611 and 4.612 is given by

$$\overline{x}(\tau, C_i) = \Big[1 + \frac{1}{8}M\Big(C_2 + \frac{4}{3}C_3 + \frac{1}{2}C_4\Big) + \frac{1}{8}N\Big(C_1 + \frac{1}{3}C_2 + \frac{1}{6}C_3 + \frac{11}{10}C_4\Big)\Big]\cos\tau - \frac{1}{8}[M(C_2 + C_3) + N(C_1 + C_4)]\cos 3\tau - \frac{1}{24}[M(C_3 + C_4) + NC_2]\cos 5\tau - \frac{1}{48}[MC_4 + NC_3]\cos 7\tau - \frac{1}{80}NC_4\cos 9\tau \tag{4.624}$$

From Eqs. 4.607, 4.610 and 4.624 one retrieves the first-order approximate solution of Eqs. 4.605 and 4.606:

$$\overline{u}(t) = A\overline{x}(\Omega t, C_i) \tag{4.625}$$

where Ω is obtained from Eq. 4.622 and $\overline{x}$ is given by Eq. 4.624.

4.14.1 Numerical Examples

In this analysis, periodic solutions are analyzed for the cantilever beam under study. Beside the role of the large amplitude A, a special role is played by the modal constants α and β, which depends on the inertia parameters of the attached inertia element with mass M and rotary inertia J. We do not approach the simplest cases when the modal constants α, β are small (0.1 or 0.2), because in these cases it is easy to achieve accurate periodic solutions even for large amplitudes using known procedures. Difficulties appear when these modal constants become larger [177, 178] and the oscillator experiences large amplitudes. Here, the meaning of "large" implies the fact that the peak amplitude reach a value where the non-linear terms are of an order comparable to that of the linear ones. More specific, the amplitude may be of the order of beam length.

We illustrate the accuracy of our procedure for large modal constants and large amplitudes comparing the obtained approximate analytical solutions with the numerical integration results obtained using a fourth-order Runge–Kutta method.

We further consider larger values for the modal constants α and β (1 or 2) and we also consider large values of the initial amplitude A (5 or 10). In order to prove the accuracy of the obtained results, two examples are analysed.

4.14.1.1 For the modal parameters $\alpha = 1$, $\beta = 1$ and the amplitude $A = 5$ we obtain the optimal values of the convergence-control parameters and the frequency as

$$C_1 = -7.3016943310218085,\ C_2 = 5.3979433890884065,$$
$$C_3 = -2.606980491307443,\ C_4 = 0.7290056441762671$$
$$\Omega = 1.3430597699023183$$

and consequently, the first-order approximate periodic solution becomes

$$\bar{u}(t) = 5.40272 \cos \Omega t - 0.560413 \cos 3\Omega t + 0.199132 \cos 5\Omega t - 0.0579088 \cos 7\Omega t + 0.0164666 \cos 9\Omega t \tag{4.626}$$

For the same modal parameters α and β, but the amplitude $A = 10$ we obtain the values

$$C_1 = -11.520140042593244,\ C_2 = 8.861684779641141,$$
$$C_3 = -4.604180230562646,\ C_4 = 1.3461139729697187$$
$$\Omega = 1.3892801150147203$$

and the first-order approximate solution in this case will be

$$\bar{u}(t) = 11.1181 \cos \Omega t - 1.55397 \cos 3\Omega t + 0.562487 \cos 5\Omega t - 0.188971 \cos 7\Omega t + 0.0623374 \cos 9\Omega t \tag{4.627}$$

4.14.1.2 For the modal parameters $\alpha = 2$, $\beta = 2$ and the amplitude $A = 5$ we obtain

$$C_1 = -3.613573923966412,\ C_2 = 3.2609390151395137,$$
$$C_3 = -2.9385052114968686,\ C_4 = 1.522818293800589$$
$$\Omega = 1.3717391311967528$$

The first-order approximate periodic solution 4.625 becomes

$$\overline{u}(t) = 5.57175\cos\Omega t - 0.824198\cos 3\Omega t + 0.300824\cos 5\Omega t - 0.118258\cos 7\Omega t + 0.0698858\cos 9\Omega t \quad (4.628)$$

In the last case, for the same modal parameters α and β, but $A = 10$, we have

$$C_1 = -3.7616027776924024,\ C_2 = 3.4276606494056003,$$
$$C_3 = -3.1027715866469263,\ C_4 = 1.604099478392615$$
$$\Omega = 1.4001459851261737$$

and the first-order approximate periodic solution in the form

$$\overline{u}(t) = 11.1933\cos\Omega t - 1.71471\cos 3\Omega t + 0.611362\cos 5\Omega t - 0.23934\cos 7\Omega t + 0.149372\cos 9\Omega t \quad (4.629)$$

Figures 4.137 and 4.138 present a comparison of the obtained analytical solutions 4.626 and 4.627 with numerical ones for the modal parameters $\alpha = \beta = 1$ and amplitude $A = 5$ and $A = 10$, respectively, for a period of motion. Similar comparison are presented in Figs. 4.139 and 4.140 for modal parameters $\alpha = \beta = 2$.

It can be seen from the above Figures that the solutions obtained by the proposed procedure are nearly identical with the solutions obtained by numerical integration. Additionally, we remark from Table 4.131 a very good agreement between the numerical and approximate values of the frequencies.

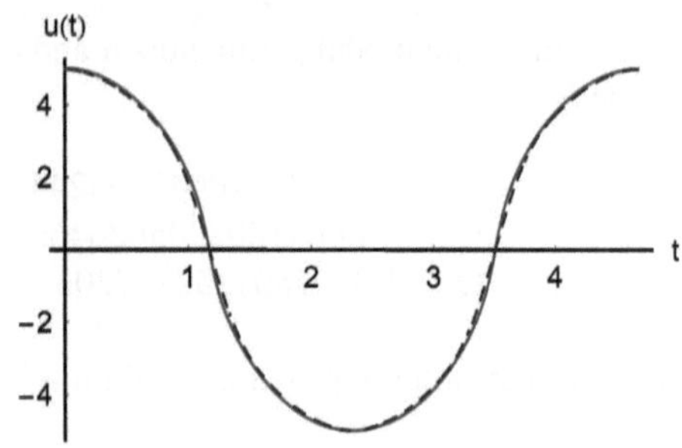

Fig. 4.137 Comparison of the results for $\alpha = \beta = 1$, $A = 5$: (*red solid line*) numerical solution; (*blue dashed line*) OHAM solution 4.626 (colour figure online)

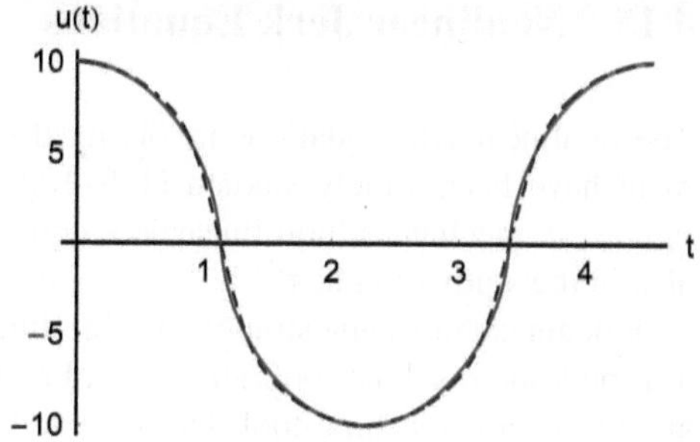

Fig. 4.138 Comparison of the results for $\alpha = \beta = 1$, $A = 10$: (*red solid line*) numerical solution; (*blue dashed line*) OHAM solution 4.627 (colour figure online)

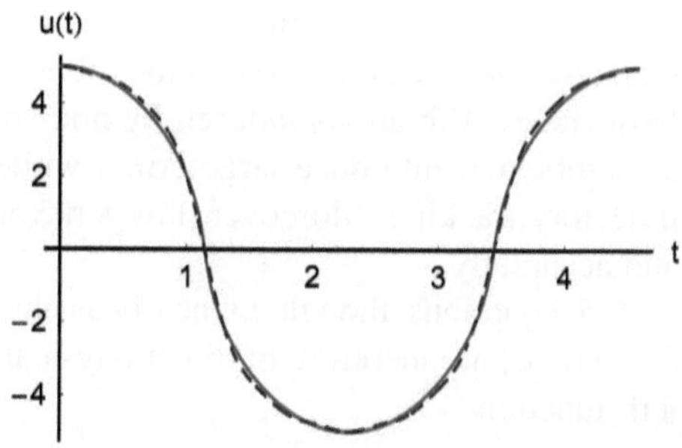

Fig. 4.139 Comparison of the results for $\alpha = \beta = 2$, $A = 5$: (*red solid line*) numerical solution; (*blue dashed line*) OHAM solution 4.628 (colour figure online)

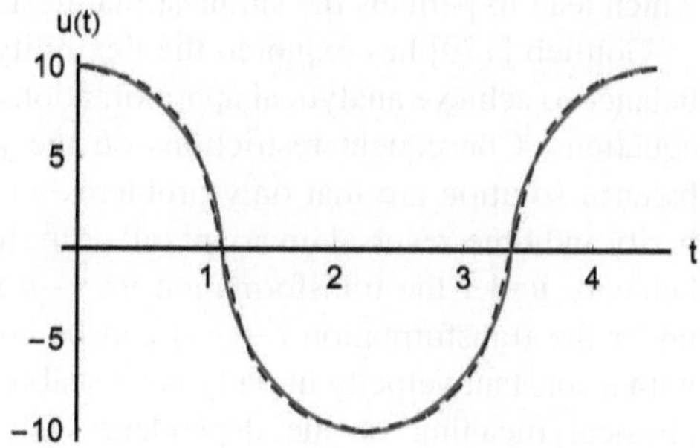

Fig. 4.140 Comparison of the results for $\alpha = \beta = 2$, $A = 10$: (*red solid line*) numerical solution; (*blue dashed line*) OHAM solution 4.629 (colour figure online)

Table 4.131 Comparison between the numerical and approximate frequencies

Case	$\Omega_{numerical}$	$\Omega_{approximate}$
4.14.1.1.a	1.3428753142728111	1.3430597699023183
4.14.1.1.b	1.3892796992331535	1.3892801150147203
4.14.1.2.a	1.3713182459632058	1.3717391311967528
4.14.1.2.b	1.4000618102244748	1.4001459851261737

4.15 Nonlinear Jerk Equations

The nonlinear jerk equations involving the third temporal derivative of displacement have been widely studied [179–182]. As well as originally being of some interest in mechanics, non-linear jerk equations are finding increasing importance also in the study of chaos.

Jerk appears in some structures exhibiting rotating and translating motions, such as robots and machine-tools structures. From a practical perspective, excessive jerk arising at some machine-tools leads to excitation of vibrations in components in the machine assembly, accelerated wear in the transmission and bearing elements, noisy operations and large contouring errors at discontinuities (such as corners) in the machining path. Also in the case of robots, limiting jerk (defined as the time derivative of the acceleration of the manipulator joints) is very important because high jerk values can wear out of the robot structure, and heavily excite its resonance frequencies. Vibrations induced by non-smooth trajectories can damage the robot actuators, and introduce large errors while the robot is performing tasks such as trajectory tracking. Moreover, low-jerk trajectories can be executed more rapidly and accurately.

Jerk equations, though not nearly as common as acceleration (or force) equation $\ddot{u} = f(u, \dot{u})$ are therefore of direct physical interest. Moreover, simple forms of the jerk function

$$\dddot{u} + f(u, \dot{u}, \ddot{u}) = 0 \tag{4.630}$$

which lead to perhaps the simplest manifestation of chaos have found in [182].

Gottlieb [179] has explored the flexibility of applying the method of harmonic balance to achieve analytical approximations of periodic solutions to nonlinear jerk equations. Consequent restrictions on the jerk equations amenable to harmonic balance solution are that only problems which have zero initial acceleration and parity and time-reversal invariant (all terms have the same space-parity of reflective behavior under the transformation $u \to -u$ and time-parity of reflective behavior under the transformation $t \to -t$) can be considered. This situation of taking off with a constant velocity initially is a feasible condition and it depends on the actual physical meaning of the dependent variable u and the interpretation of the governing equation. Wu et al [180] proposed an improved harmonic balance method for determining the periodic solutions of nonlinear jerk equations. Ma et al. [181] applied homotopy perturbation method to the jerk equations.

Following Gottlieb [179], the most general jerk function with invariance of the time-reversal and space-reversal and which has only cubic non-linearities may be written as

$$\dddot{u} + \alpha\dot{u}^3 + \beta u^2\dot{u} + \gamma u\dot{u}\ddot{u} + \delta\dot{u}\ddot{u}^2 + \lambda\dot{u} = 0 \tag{4.631}$$

where the parameters α, β, γ, δ and λ are constants and the dot denotes derivative with respect to time. The corresponding initial conditions are

$$u(0) = 0, \dot{u}(0) = A, \ddot{u}(0) = 0 \tag{4.632}$$

Here, at least one of α, β, γ and δ should be non-zero. In addition, if $\delta = 0$, we require $\gamma \neq -2\alpha$ such that the jerk equation is simply not the time-derivative of an acceleration equation.

A new independent variable $\tau = \omega t$ and a new dependent variable $u = \frac{A}{\omega}x$ are introduced. Thus, Eqs. 4.631 and 4.632 can be written as

$$\Omega^2 \left(x''' + \delta A^2 x' x''^2\right) + \Omega\left(\alpha A^2 x'^3 + \gamma A^2 x x' x'' + \lambda x'\right) + \beta A^2 x^2 x' = 0 \tag{4.633}$$

$$x(0) = 0, x'(0) = 1, x''(0) = 0 \tag{4.634}$$

where prime denotes differentiation with respect to τ and $\Omega = \omega^2$. The new independent variable is chosen such that the solution of Eqs. 4.633 and 4.634 is a periodic function of τ of period 2π. The corresponding period of the non-linear jerk equation is given by $T = 2\pi/\sqrt{\Omega}$.

The linear operator is defined by

$$L[x(\tau)] = \Omega^2(x''' + x) \tag{4.635}$$

and $g(\tau) = 0$, while the non-linear operator is defined by

$$N[x(\tau), \Omega] = \Omega^2\left(-x + \delta A^2 x' x''^2\right) + \Omega\left(\alpha A^2 x'^3 + \gamma A^2 x x' x'' + \lambda x'\right) + \beta A^2 x^2 x' \tag{4.636}$$

The equation 4.4 becomes

$$x_0''' + x_0 = 0, \quad x_0(0) = 0, x_0'(0) = 1, x_0''(0) = 0 \tag{4.637}$$

The above equation has the solution

$$x_0(\tau) = \sin\tau \tag{4.638}$$

From Eqs. 4.638 and 4.636 we can deduce that

$$N[x_0(\tau), \Omega] = M\cos\tau + N\cos 3\tau \tag{4.639}$$

where

$$M = \left(\frac{1}{4}\delta A^2 - 1\right)\Omega^2 + \left(\lambda + \frac{3\alpha - \gamma}{4}A^2\right)\Omega + \frac{1}{4}\beta A^2$$
$$N = -\frac{1}{4}\delta A^2\Omega^2 + \frac{1}{4}\Omega(\alpha + \gamma)A^2 - \frac{1}{4}\beta A^2 \tag{4.640}$$

The first approximation is obtained from Eq. 4.9:

$$x_1''' + x_1 = H(\tau, C_i)(M\cos\tau + N\cos 3\tau) \tag{4.641}$$

where the auxiliary function H obviously should be chosen as

$$H(\tau, C_i) = 2C_1\sin 2\tau + 2C_2\sin 4\tau + 2C_3\sin 6\tau \tag{4.642}$$

Substituting Eq. 4.642 into Eq. 4.641, after simple manipulations we obtain

$$\begin{aligned} &x_1''' + x_1 = [MC_1 + N(C_2 - C_1)]\sin\tau + [M(C_1 + C_2) + NC_3]\sin 3\tau + \\ &+[M(C_2 + C_3) + NC_1]\sin 5\tau + (MC_3 + NC_2)\sin 7\tau + NC\sin 9\tau, \\ &x_1(0) = x_1'(0) = x_1''(0) = 0 \end{aligned} \tag{4.643}$$

Avoiding the presence of a secular term in Eq. 4.643 needs

$$MC_1 + N(C_2 - C_1) = 0 \tag{4.644}$$

From Eqs. 4.640 and 4.644 we can find the expression from which we can obtain the frequency:

$$\begin{aligned} &[(\delta A^2 - 4)C_1 - \delta A^2(C_2 - C_1)]\Omega^2 + [(4\lambda + 3\alpha A^2 - \gamma A^2)C_1 + \\ &+(\alpha + \gamma)A^2(C_2 - C_1)]\Omega + \beta A^2(2C_1 - C_2) = 0 \end{aligned} \tag{4.645}$$

The solution of Eq. 4.643 is given by

$$\begin{aligned} &x_1(\tau, C_i) = -\frac{1}{26}[M(C_1 + C_2) + NC_3](\sin 3\tau - 3\sin\tau) - \frac{1}{124}[M(C_2 + \\ &+C_3) + NC_1](\sin 5\tau - 5\sin\tau) - \frac{1}{342}(MC_3 + NC_2)(\sin 7\tau - 7\sin\tau) - \\ &-\frac{1}{728}NC_3(\sin 9\tau - 9\sin\tau) \end{aligned} \tag{4.646}$$

From Eqs. 4.638, 4.646, 4.5 and taking into account that $\overline{u}(t) = \frac{A}{\omega}\overline{x}(\tau)$ one can get the first-order approximate solution

$$\begin{aligned}\overline{u}(t, C_i) = \frac{A}{\sqrt{\Omega}}\Big[\sin\sqrt{\Omega t} - \frac{M(C_1 + C_2) + NC_3}{26}\left(\sin 3\sqrt{\Omega t} - 3\sin\sqrt{\Omega t}\right) - \\ - \frac{M(C_2 + C_3) + NC_1}{124}\left(\sin 5\sqrt{\Omega t} - 5\sin\sqrt{\Omega t}\right) - \\ - \frac{MC_3 + NC_2}{342}\left(\sin 7\sqrt{\Omega t} - 7\sin\sqrt{\Omega t}\right) - \frac{NC_3}{728}\left(\sin 9\sqrt{\Omega t} - 9\sin\sqrt{\Omega t}\right)\Big]\end{aligned} \tag{4.647}$$

4.15.1 Numerical Examples

By setting $\alpha = \beta = 1$, $\delta = \gamma = \lambda = 0$ (jerk function containing velocity-cubed and velocity times displacements squared) in Eq. 4.631, the governing equation will be

$$\dddot{u} + \dot{u}\left(\dot{u}^2 + u^2\right) = 0 \tag{4.648}$$

subject to the initial conditions

$$u(0) = 0, \quad \dot{u}(0) = A, \quad \ddot{u}(0) = 0 \tag{4.649}$$

In what follows we analyze three distinct cases.

4.15.1.1 The case A = 0.5

The optimal convergence-control parameters and the frequency are:

$$C_1 = 3.802562708,\ C_2 = 2.838580474,\ C_3 = -11.0262723, \Omega = 0.378591766$$

and therefore the first-order approximate periodic solution given by Eq. 4.647 becomes

$$\begin{aligned}\overline{u}(t) = 0.848712787\sin\omega t - 0.011340699\sin 3\omega t + \\ +4.39531627\cdot 10^{-4}\sin 5\omega t + 3.998190208\cdot 10^{-6}\sin 7\omega t - \\ -4.78011974\cdot 10^{-4}\sin 9\omega t\end{aligned} \tag{4.650}$$

where $\omega = \sqrt{\Omega} = 0.615298112$.

4.15.1.2 The case A = 5

One can get

$$\begin{aligned}C_1 = 8.960443247\cdot 10^{-4},\ C_2 = 7.760907805\cdot 10^{-4}, \\ C_3 = -3.456380568\cdot 10^{-3},\ \Omega = 18.30055092\end{aligned}$$

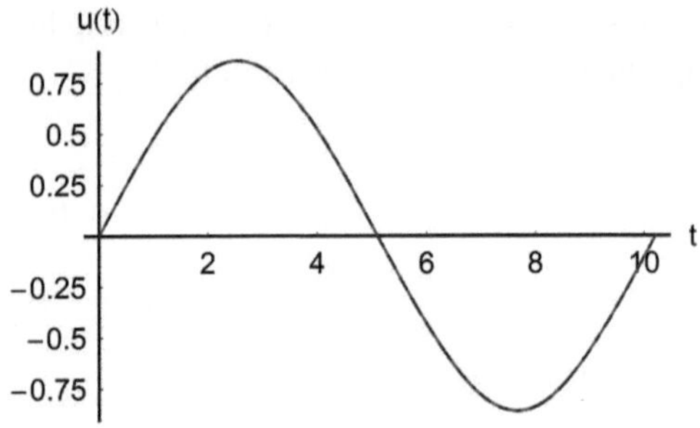

Fig. 4.141 Comparison between the approximate solution 4.650 and numerical results of Eq.4.648 in the case A = 0.5: (*red solid line*) numerical solution; (*blue dashed line*) approximate solution (colour figure online)

The first-order approximate periodic solution 4.647 can be written in the form

$$\overline{u}(t) = 1.119804295 \sin \omega t + 0.015712573 \sin 3\omega t - 5.475444777 \cdot 10^{-4} \sin 5\omega t - \\ -1.158056227 \cdot 10^{-4} \sin 7\omega t + 6.000232077 \cdot 10^{-4} \sin 9\omega t \tag{4.651}$$

where $\omega = \sqrt{\Omega} = 4.277914319$.

4.15.1.3 For the last case, A = 10 it holds:

$$C_1 = 5.658220098 \cdot 10^{-5},\ C_2 = 4.920659599 \cdot 10^{-5},\\ C_3 = -2.204987242 \cdot 10^{-4},\ \Omega = 72.132959048$$

The first-order approximate periodic solution 4.647 is

$$\overline{u}(t) = 1.1255457003 \sin \omega t + 0.016646767 \sin 3\omega t - 0.000578405 \sin 5\omega t - \\ -0.000125288 \sin 7\omega t + 0.000634189 \sin 9\omega t \tag{4.652}$$

where $\omega = \sqrt{\Omega} = 8.493112449$.

It is easy to verify the accuracy of the obtained solution if we graphically compare the analytical results with the numerical simulation results. Figs. 4.141, 4.142, 4.143 show the comparison between the present solutions and the numerical integration results obtained by a fourth-order Runge–Kutta method.

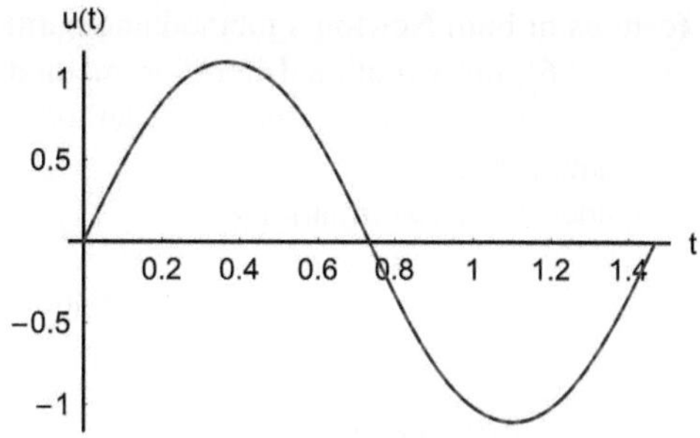

Fig. 4.142 Comparison between the approximate solution 4.651 and numerical results of Eq.4.648 in the case A = 5: (*red solid line*) numerical solution; (*blue dashed line*) approximate solution (colour figure online)

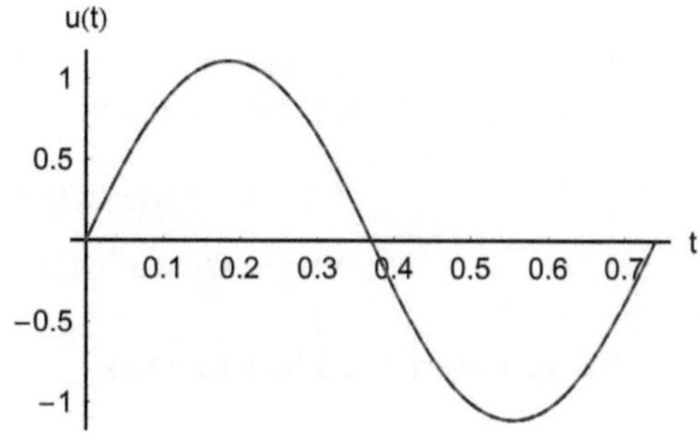

Fig. 4.143 Comparison between the approximate solution 4.652 and numerical results of Eq.4.648 in the case A = 10: (*red solid line*) numerical solution; (*blue dashed line*) approximate solution (colour figure online)

4.16 Nonlinear Oscillator with Discontinuities

Consider the following antisymetric constant force oscillator with discontinuities

$$\ddot{u} + sign(u) = 0 \tag{4.653}$$

with the initial conditions:

$$u(0) = A \quad , \quad \dot{u}(0) = 0 \tag{4.654}$$

This equation was studied first by Awrejcewicz and Andrianov [183]. The function *sign(u)* is defined as

$$sign(u) = \begin{cases} 1 & \text{if } u > 0 \\ -1 & \text{if } u \leq 0 \end{cases}$$

There exists no small parameter in the Eq. 4.653 and therefore the traditional perturbation methods cannot be applied directly. For such problems concerning nonlinear oscillators with discontinuities, the homotopy perturbation method is employed by Belendez et al [184]. A modified Lindstedt-Poincare method is used by Liu [185], an analytical approximate technique which incorporates silent features of both Newton's method and harmonic balance method is applied by Wu et al [186], the variational iteration method is employed by Rafei et al [187], the Optimal Homotopy Asymptotic Method is successfully used by Marinca and Herisanu [188].

Under the transformations

$$\begin{aligned} \tau &= \omega t, \\ u(t) &= Ax(\tau) \end{aligned} \tag{4.655}$$

the original Eq. 4.653 becomes

$$A\Omega^2 x''(\tau) + signx(\tau) = 0 \tag{4.656}$$

where the prime denotes the derivative with respect to τ.

For the Eq. 4.656, the linear and nonlinear operators ($g(\eta) = 0$) are, respectively

$$L[x(\tau), \Omega] = x'' + x \tag{4.657}$$

$$N[x(\tau), \Omega] = x''(A\Omega^2 - 1) - x + signx \tag{4.658}$$

The equation 4.4 can be written as:

$$x_0'' + x_0 = 0 \ , \ \ x_0(0) = 1 \ , \ \ x_0'(0) = 0 \tag{4.659}$$

and has the solution

$$x_0(\tau) = \cos\tau \tag{4.660}$$

If we note $f(x) = sign(x)$, where $x = x_0 + \alpha$, $\alpha \in R$, then

$$f(x) = f(x_0) + f'(x_0)\alpha + \frac{1}{2}f''(x_0)\alpha^2 + \ldots$$

But $f'(x_0) = f''(x_0) = \ldots = 0$ and therefore we obtain:

$$sign(x) = sign(x_0) = sign(\cos\tau)$$

From Eqs. 4.658 and 4.660 it is obtained

$$N[x_0(\tau), \Omega] = -A\Omega^2\cos\tau + sign(\cos\tau) \tag{4.661}$$

Taking into account the identity

$$sign(\cos\tau) = \frac{4}{\pi}\left(\cos\tau - \frac{1}{3}\cos 3\tau + \frac{1}{5}\cos 5\tau - \frac{1}{7}\cos 7\tau + \ldots\right) \tag{4.662}$$

The equation 4.661 becomes:

$$N[x_0(\tau),\Omega] = \left(\frac{4}{\pi} - A\Omega^2\right)\cos\tau - \frac{4}{3\pi}\cos 3\tau + \frac{4}{5\pi}\cos 5\tau - \\ -\frac{4}{7\pi}\cos 7\tau + \frac{4}{9\pi}\cos 9\tau - \frac{4}{11\pi}\cos 11\tau \tag{4.663}$$

The first approximation $x_1(\tau)$ is obtained from Eq. 4.9

$$x_1'' + x_1 = H(\tau, C_i)\left[\left(\frac{4}{\pi} - A\Omega^2\right)\cos\tau - \frac{4}{3\pi}\cos 3\tau + \frac{4}{5\pi}\cos 5\tau - \right. \\ \left. -\frac{4}{7\pi}\cos 7\tau + \frac{4}{9\pi}\cos 9\tau - \frac{4}{11\pi}\cos 11\tau + \dots\right] \tag{4.664}$$

where the optimal auxiliary function H can be chosen as follows:

$$H(\tau, C_i) = C_1 + 2C_2\cos 2\tau + 2C_3\cos 4\tau + 2C_4\cos 6\tau + \\ +2C_5\cos 8\tau + 2C_6\cos 10\tau \tag{4.665}$$

Substituting Eq. 4.665 into Eq. 4.664 we have

$$x_1' + x_1 = \left[\left(\frac{4}{\pi} - A\Omega^2\right)(C_1 + C_2) - \frac{4(C_1 + C_3)}{3\pi} + \frac{4(C_3 + C_4)}{5\pi} + \right. \\ \left. + \frac{4(C_4 + C_5)}{7\pi} + \frac{4(C_5 + C_6)}{9\pi} + \frac{4C_6}{11\pi}\right]\cos\tau + M_3\cos 3\tau + M_5\cos 5\tau + \\ +M_7\cos 7\tau + M_9\cos 9\tau + M_3\cos 11\tau + \dots \tag{4.666}$$

where

$$M_3 = \left(\frac{4}{\pi} - A\Omega^2\right)(C_2 + C_3) + \frac{4}{\pi}\left(-\frac{C_1}{5} + \frac{C_2}{5} - \frac{C_3}{7} - \frac{2C_4}{9} + \frac{6C_5}{55} - \frac{C_6}{7}\right)$$

$$M_5 = \left(\frac{4}{\pi} - A\Omega^2\right)(C_3 + C_4) + \frac{4}{\pi}\left(\frac{C_1}{5} - \frac{10C_2}{21} + \frac{C_3}{9} - \frac{C_4}{11} + \frac{C_5}{3} + \frac{C_6}{5}\right)$$

$$M_7 = \left(\frac{4}{\pi} - A\Omega^2\right)(C_4 + C_5) + \frac{4}{\pi}\left(-\frac{C_1}{7} + \frac{14C_2}{45} - \frac{14C_3}{33} - \frac{C_6}{3}\right)$$

$$M_9 = \left(\frac{4}{\pi} - A\Omega^2\right)(C_5 + C_6) + \frac{4}{\pi}\left(-\frac{C_1}{9} + \frac{18C_2}{77} - \frac{C_3}{5} - \frac{C_4}{3}\right)$$

$$M_{11} = \left(\frac{4}{\pi} - A\Omega^2\right)C_6 + \frac{4}{\pi}\left(-\frac{C_1}{11} + \frac{C_2}{9} - \frac{C_3}{7} + \frac{C_4}{5} - \frac{C_5}{3}\right)$$

Of course, we impose the avoidance of secular term into Eq. 4.666, i.e.

$$\Omega^2 = \frac{4}{\pi A(C_1 + C_2)}\left(C_1 + \frac{2}{3}C_2 - \frac{2}{15}C_3 + \frac{2}{35}C_4 - \frac{2}{63}C_5 + \frac{2}{99}C_6\right) \tag{4.667}$$

From Eqs. 4.5, 4.660, 4.655 and solving Eq. 4.666 we obtain the first-order approximate solution of Eqs. 4.653 and 4.654:

$$\begin{aligned}&\overline{u}(t, C_i) = A\cos\Omega t + \frac{AM_3}{8}(\cos\Omega t - \cos 3\Omega t) + \frac{AM_5}{24}(\cos\Omega t - \cos 5\Omega t) + \\ &\frac{AM_7}{48}(\cos\Omega t - \cos 7\Omega t) + \frac{AM_9}{80}(\cos\Omega t - \cos 9\Omega t) + \\ &+ \frac{AM_{11}}{120}(\cos\Omega t - \cos 11\Omega t)\end{aligned} \tag{4.668}$$

where Ω is given by Eq. 4.667.

4.16.1 Numerical Examples

The substitution of Eq. 4.668 into Eq. 4.653 results in a residual which reads

$$R(t, C_i) = \ddot{\overline{u}}(t, C_i) + sign\overline{u}(t, C_i), \quad i = 1, 2, \ldots, 6 \tag{4.669}$$

The optimal convergence-control parameters are determined from Eqs. 4.668 and 4.669 using collocation method.

As numerical examples we consider two cases.

4.16.1.1 In the first case A = 1. It holds that

$$\begin{aligned}\overline{u}(t) &= 1.030870213\cos\Omega t - 0.035427958\cos 3\Omega t + \\ &+ 0.006680062\cos 5\Omega t - 0.002554785\cos 7\Omega t + \\ &+ 0.000689697\cos 9\Omega t - 0.000257229\cos 11\Omega t\end{aligned} \tag{4.670}$$

where we find $\Omega = 1.115347472$

4.16.1.2 In the last case A = 10 for which we have

$$\begin{aligned}\overline{u}(t) &= 10.254180017\cos\Omega t - 0.288893863\cos 3\Omega t + \\ &+ 0.042787575\cos 5\Omega t - 0.009387227\cos 7\Omega t + \\ &+ 0.000897934\cos 9\Omega t + 0.000415563\cos 11\Omega t\end{aligned} \tag{4.671}$$

with $\Omega = 0.351220283$.

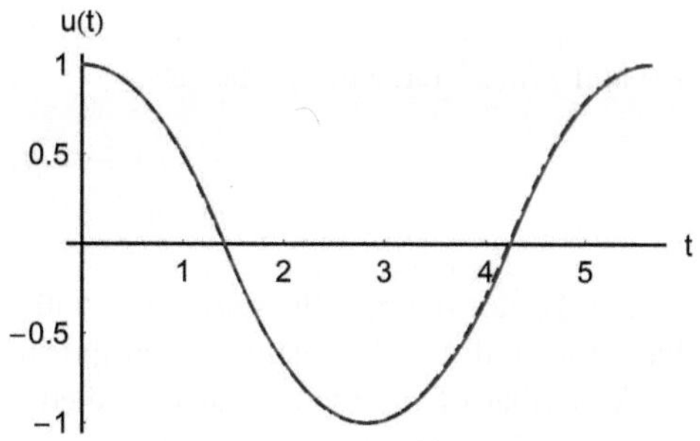

Fig. 4.144 Comparison between the approximate solution 4.670 and numerical solution of Eq. 4.653 for A = 1: (*red solid line*) numerical solution; (*blue dashed line*) approximate solution (colour figure online)

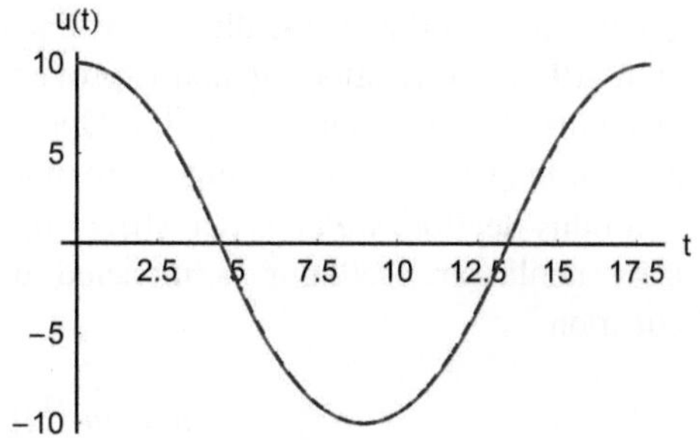

Fig. 4.145 Comparison between the approximate solution 4.671 and numerical solution of Eq. 4.653 for A = 10: (*red solid line*) numerical solution; (*blue dashed line*) approximate solution (colour figure online)

Figures 4.144 and 4.145 show the comparison between the present solutions and the numerical integration results obtained by a fourth-order Runge–Kutta method.

It is clear that the solutions obtained by OHAM are nearly identical with the solutions given by the numerical method. Additionally, we remark that the exact values of the frequencies are $\Omega_{ex} = 1.110720735$ for the case $A = 1$ and $\Omega_{ex} = 0.351240736$ in case $A = 10$, which means that very good approximations are found also for the frequencies.

4.17 Truly Nonlinear Oscillators

Following Mickens and Oyedeji [189] or Mickens [190, 191], the most general form of truly nonlinear oscillator is given by the following differential equation

$$\ddot{u} + g(u) = \varepsilon f(u, \dot{u}), \tag{4.672}$$

subject to the initial conditions $u(0) = A, \dot{u}(0) = 0$.

In Eq. 4.672, ε is a positive parameter, not necessarily small, and the functions g (u) and $f(u, \dot{u})$ have the properties:

$$g(-u) = -g(u)$$
$$f(-u, -\dot{u}) = -f(u, \dot{u})$$

where the dot denotes the derivative with respect to variable t and g(u) does not have for small u a dominant term proportional to u.

A number of researchers have studied this problem. For example, Senator and Bapat [192] used a perturbation technique even the nonlinearity is not small and Das and Chaterjee [193] applied multiple scales via Galerkin projections in the study of nonlinear oscillations. Harmonic balance based averaging is proposed by Chatterjeee [194] and so-called quasi-linearization method is developed by Krivec et al [195]. An iteration method is proved by Mickens [190, 191] which was then generalized by Lim and Wu [196]. Optimal variational method is introduced and applied in [197] for calculating approximate periodic solutions for these systems.

In this Section, we consider $g(u) = au^{1/3} + bu^3$, $f(u, \dot{u}) = 0$ and therefore the truly nonlinear oscillator is modeled by the following nonlinear differential equation

$$\ddot{u} + au^{1/3} + bu^3 = 0 \tag{4.673}$$

with the initial conditions $u(0) = A, \quad \dot{u}(0) = 0$, where a, b and A > 0 are known parameters. We remark that in Eq. 4.673 there exists no small or large parameter.

4.17.1 Solution of Eqs. 4.673 Using OHAM

Introducing a new independent variable τ and a new unknown function u as

$$\tau = \Omega t, \quad u(t) = Ax(\tau) \tag{4.674}$$

where Ω is the frequency of the system 4.673, then this becomes

$$\Omega^2 x'' + aA^{-2/3}x^{1/3} + bA^2x^3 = 0 \tag{4.675}$$

where prime denotes differentiation with respect to the new variable τ.

The initial conditions become $x(0) = 1, \quad x'(0) = 0$.

The linear and the nonlinear operator corresponding to Eq. 4.675 are, respectively ($g(\tau) = 0$):

$$\begin{aligned} L[x(\tau)] &= \Omega^2(x'' + x) \\ N[x(\tau), \Omega] &= -\Omega^2 x + aA^{-2/3}x^{1/3} + bA^2x^3 \end{aligned} \tag{4.676}$$

The initial approximation x_0 can be found from Eq. 4.4:

$$\Omega^2\left(x_0''+x_0\right)=0,\quad x_0(0)=1,\quad x'(0)=0 \tag{4.677}$$

which has the solution

$$x_0(\tau)=\cos\tau \tag{4.678}$$

For the initial approximation given by Eq. 4.678, the nonlinear operator given by Eq. 4.676 is written as follows [197]:

$$N[x_0(\tau),\Omega]=-\Omega^2\cos\tau+\alpha a A^{-2/3}\left(\cos\tau-\frac{1}{5}\cos 3\tau+\frac{1}{10}\cos 5\tau-\frac{7}{110}\cos 7\tau+\right.$$
$$\left.+\frac{1}{22}\cos 9\tau+\frac{13}{374}\cos 11\tau\right)+\frac{bA^2}{4}(3\cos\tau+\cos 3\tau)$$

or after some manipulations:

$$\begin{aligned}N[x_0(\tau),\Omega]&=M_1\cos\tau+M_2\cos 3\tau+M_3\cos 5\tau+M_4\cos 7\tau+\\&+M_5\cos 9\tau+M_6\cos 11\tau\end{aligned} \tag{4.679}$$

where

$$M_1=-\Omega^2+\alpha aA^{-2/3}+\frac{3}{4}bA^2, M_2=\frac{bA^2}{4}-\frac{\alpha aA^{-2/3}}{5}, M_3=\frac{\alpha aA^{-2/3}}{10},$$
$$M_4=-\frac{7\alpha aA^{-2/3}}{110}, M_5=\frac{\alpha aA^{-2/3}}{22}, M_6=-\frac{3\alpha aA^{-2/3}}{34}, \alpha=1.15959526696393$$

The first approximation x_1 can be obtained from Eq. 4.9

$$\Omega^2\left(x_1''+x_1\right)=H(\tau,C_i)N[x_0(\tau),\Omega],\quad x_1(0)=x_1'(0)=0 \tag{4.680}$$

where the auxiliary function $H(t,C_i)$ has the form

$$H(\tau,C_i)=C_1+2C_2\cos 2\tau+2C_3\cos 4\tau+2C_4\cos 8\tau \tag{4.681}$$

Substituting Eq. 4.681 into Eq. 4.680 we obtain

$$\begin{aligned}\Omega^2\left(x_1''+x_1\right)&=\left[M_1(C_1+C_2)+M_2(C_2+C_3)+M_3C_3+(M_4+\right.\\&\left.+M_5)C_4\right]\cos\tau+\left[M_1(C_2+C_3)+M_2C_1+M_3(C_2+C_4)+M_4C_3+\right.\\&\left.+M_6C_4\right]\cos 3\tau+\left[M_1C_3+M_2(C_2+C_4)+M_3C_1+M_5C_3\right]\cos 5\tau+\\&+\left[M_2C_3+M_3C_2+M_4C_1+M_5C_2+M_6C_3\right]\cos 7\tau+\left[M_1C_4+\right.\\&\left.+M_3C_3+M_4C_2+M_5C_1\right]\cos 9\tau+\ldots\end{aligned} \tag{4.682}$$

No secular term in Eq. 4.682 requires that

$$\Omega^2=\frac{\alpha a A^{-2/3}}{C_1+C_2}\left(C_1+\frac{4}{5}C_2-\frac{C_3}{10}-\frac{C_4}{55}\right)+\frac{bA^2}{4(C_1+C_2)}(3C_1+4C_2+C_3)$$

Now, solving Eq. 4.682 and taking into consideration Eqs. 4.678, 4.674 and 4.5, we obtain the first-order approximate periodic solution of Eq. 4.673:

$$\begin{aligned}&\overline{u}(t,C_i)=(A-D_3-D_5-D_7-D_9)\cos\Omega t+D_3\cos 3\Omega t+D_5\cos 5\Omega t+\\&+D_7\cos 7\Omega t+D_9\cos 9\Omega t\end{aligned}\qquad(4.683)$$

where

$$\begin{aligned}D_3&=-\frac{A}{8\Omega^2}[M_2C_1+(M_1+M_3)C_2+(M_1+M_4)C_3+(M_3+M_6)C_4]\\D_5&=-\frac{A}{24\Omega^2}[M_3C_1+(M_2+M_4)C_2+(M_1+M_5)C_3+M_2C_4]\\D_7&=-\frac{A}{48\Omega^2}[M_4C_1+(M_3+M_5)C_2+(M_2+M_6)C_3+M_1C_4]\\D_9&=-\frac{A}{80\Omega^2}[M_5C_1+(M_4+M_6)C_2+M_3C_3+M_1C_4]\end{aligned}$$

4.17.2 Numerical Examples

We consider the following six special cases:

4.17.2.1 In the first case we consider a = b = 1 and A = 5 and we obtain:

$$\begin{aligned}&C_1=-1.003889993; C_2=0.094754331; C_3=0.016952992;\\&C_4=-0.217591016; \Omega=4.287879\end{aligned}$$

The first-order approximate periodic solution 4.683 of Eq. 4.673 becomes

$$\begin{aligned}&\overline{u}(t)=4.782655816\cos\Omega t+0.207611523\cos 3\Omega t+0.0089174779\cos 5\Omega t+\\&+0.0000173909\cos 7\Omega t+0.000641271\cos 9\Omega t\end{aligned}\qquad(4.684)$$

The value of Ω obtained by numerical integration in this case is $\Omega=4.281323$.

4.17.2.2 In the second case, for a = b = 1 and A = 10, we have

$$\begin{aligned}&C_1=-1.002306678; C_2=0.086310365; C_3=0.030561786;\\&C_4=-0.21510348; \Omega=8.489084\end{aligned}$$

and therefore the approximate solution 4.683 is

$$\overline{u}(t) = 9.553601665\cos\Omega t + 0.427295364\cos 3\Omega t + 0.018167221\cos 5\Omega t - \\ -0.000276806\cos 7\Omega t + 0.001212555\cos 9\Omega t \tag{4.685}$$

In this case the value of Ω obtained by numerical integration is $\Omega = 8.485057$.

4.17.2.3 In the third case, we consider a = b = 2 and A = 5 and thus

$$C_1 = \text{-}1.000970955; C_2 = 0.143610482; C_3 = \text{-}0.03478336; \\ C_4 = \text{-}0.194790838; \Omega = 6.060244999$$

The approximate solution of Eq. 4.673 can be written as

$$\overline{u}(t) = 4.785977624\cos\Omega t + 0.207130994\cos 3\Omega t + 0.004390675\cos 5\Omega t + \\ +0.001885855\cos 7\Omega t + 0.000614849\cos 9\Omega t \tag{4.686}$$

The result of numerical integration for the frequency Ω in this case is $\Omega = 6.059008$.

4.17.2.4 In the case a = b = 2 and A = 10 we get

$$C_1 = -1.004636276; C_2 = 0.089356985; C_3 = 0.027586708; \\ C_4 = -0.220395287; \Omega = 12.005012$$

and the approximate solution of Eq. 4.673 will be

$$\overline{u}(t) = 9.551905331\cos\Omega t + 0.428314915\cos 3\Omega t + 0.018547139\cos 5\Omega t - \\ -0.000011309\cos 7\Omega t + 0.001243922\cos 9\Omega t \tag{4.687}$$

By numerical integration we get in this case $\Omega = 12.003983$.

4.17.2.5 for a = b = 3 and A = 5 we get

$$C_1 = -1.195723159; C_2 = 0.090980365; C_3 = 0.035819648; \\ C_4 = -0.355909562; \Omega = 7.4366745$$

and the first-order approximate solution of Eq. 4.673 becomes

$$\overline{u}(t) = 4.733264666\cos\Omega t + 0.247058495\cos 3\Omega t + 0.018741431\cos 5\Omega t - \\ -0.00001041\cos 7\Omega t + 0.000945816\cos 9\Omega t \tag{4.688}$$

In this case, the numerical integration result for the frequency is $\Omega = 7.434185$.

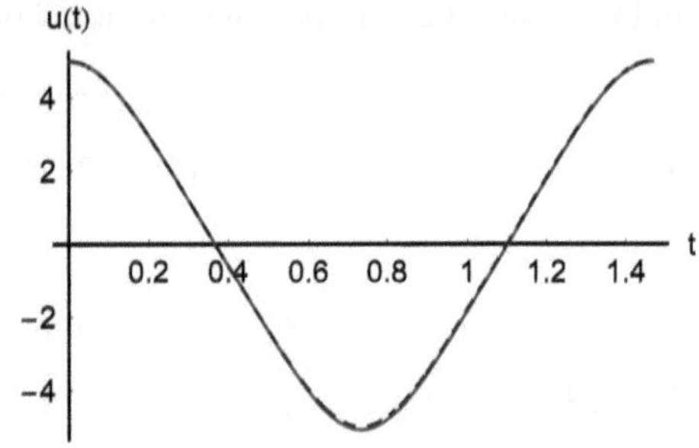

Fig. 4.146 Comparison between the approximate solution 4.684 and numerical solution of Eq. 4.673 for a = b = 1, A = 5: (*red solid line*) numerical; (*blue dashed line*) analytical solution (colour figure online)

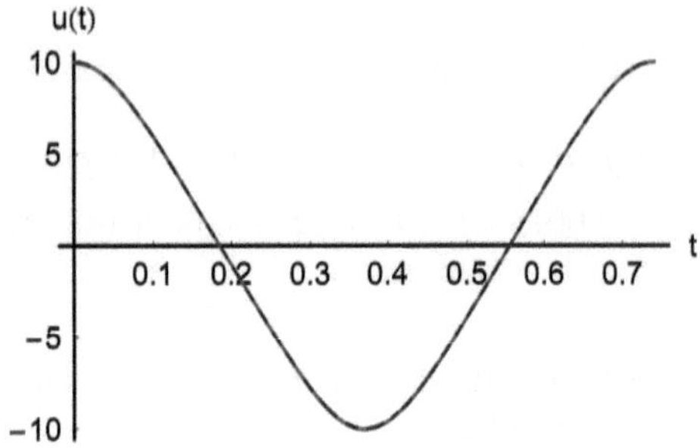

Fig. 4.147 Comparison between the approximate solution 4.685 and numerical solution of Eq. 4.673 for a = b = 1, A = 10: (*red solid line*) numerical; (*blue dashed line*) analytical solution (colour figure online)

4.17.2.6 In the last case, we consider a = b = 3 and A = 10 and therefore

$$C_1 = -0.984080628; C_2 = -0.126549201; C_3 = 0.267533946;$$
$$C_4 = -0.032155728; \Omega = 14.704743$$

and the approximate solution of Eq. 4.673 can be written in the form

$$\bar{u}(t) = 9.582557961 \cos \Omega t + 0.418183568 \cos 3\Omega t + 0.018095665 \cos 5\Omega t - \\ -0.019013623 \cos 7\Omega t + 0.000176428 \cos 9\Omega t \tag{4.689}$$

Figures 4.146, 4.147, 4.148, 4.149, 4.150, 4.151 present a comparison between the solutions 4.684-4.689 obtained by OHAM and the numerical integration results for Eq. 4.673. Thus, it is easier to emphasize the accuracy of the obtained results, since within these graphical representations the analytical results are nearly identical with numerical ones. Also, the approximate frequencies obtained by OHAM are in very good agreement with those obtained by numerical integration, which also proves the validity of the approximate results.

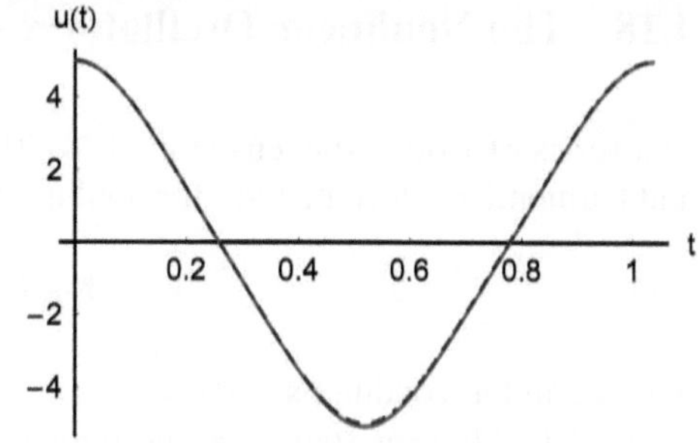

Fig. 4.148 Comparison between the approximate solution 4.686 and numerical solution of Eq. 4.673 for a = b = 2, A = 5: (*red solid line*) numerical; (*blue dashed line*) analytical solution (colour figure online)

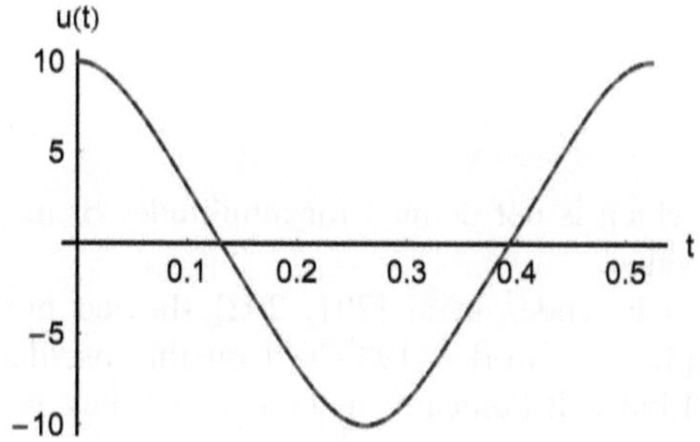

Fig. 4.149 Comparison between the approximate solution 4.687 and numerical solution of Eq. 4.673 for a = b = 2, A = 10: (*red solid line*) numerical; (*blue dashed line*) analytical solution (colour figure online)

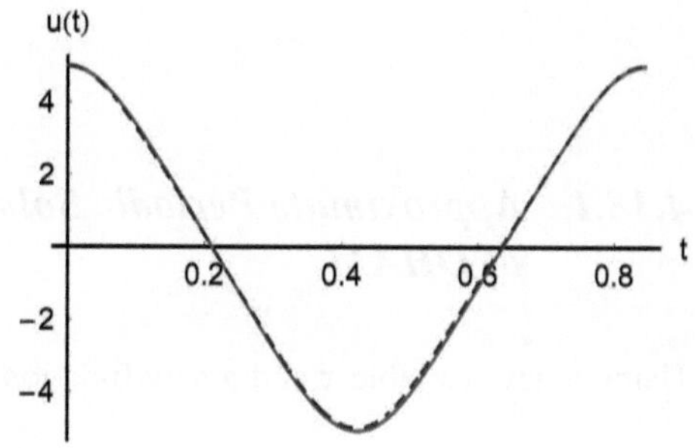

Fig. 4.150 Comparison between the approximate solution 4.688 and numerical solution of Eq. 4.673 for a = b = 3, A = 5: (*red solid line*) numerical; (*blue dashed line*) analytical solution (colour figure online)

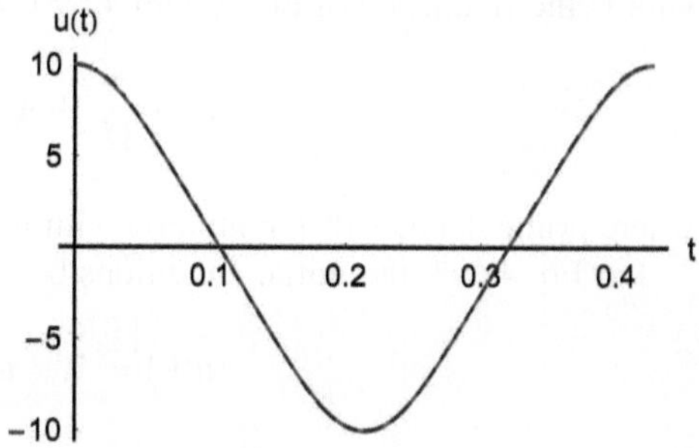

Fig. 4.151 Comparison between the approximate solution 4.689 and numerical solution of Eq. 4.673 for a = b = 3, A = 10: (*red solid line*) numerical; (*blue dashed line*) analytical solution (colour figure online)

4.18 The Nonlinear Oscillator $\ddot{x} + (1 + \dot{x}^2)x = 0$

In a series of works Mickens et al [198–200] used the Lindstedt-Poincare method and harmonic balance method for nonlinear oscillator

$$\ddot{x} + \left(1 + \dot{x}^2\right)x = 0 \tag{4.690}$$

with the initial conditions $x(0) = A, \quad \dot{x}(0) = 0$.

In [200] Mickens states that the frequency of this oscillator is singular and not defined at finite values of A when classical perturbation procedures are used. By means of first-order harmonic balance method [200] the frequency is

$$\omega(A) = \frac{2}{\sqrt{4 - A^2}} \tag{4.691}$$

which is not defined for amplitudes of magnitude equal to or larger than two in value.

Belendez et al [201, 202] showed how the difficulties arising in Mickens' previous works [198–200] on this oscillator can be resolved using a modified Lindstedt-Poincare method. Also, they determined [201] the period as $T(A) = 4\sqrt{\frac{2}{2+A^2}}K(q)$ in terms of a complete elliptic integral of the first kind K(q) where $q = \frac{A}{\sqrt{2+A^2}}$.

4.18.1 *Approximate Periodic Solution of Eq. 4.690 by Means of OHAM*

Using a new variable τ and a new function u such that

$$\tau = \Omega t, \quad x(t) = Au(\tau) \tag{4.692}$$

with Ω the frequency of the system 4.690, we obtain

$$u'' + \frac{u}{\Omega^2} + A^2 u u'^2 = 0 \tag{4.693}$$

where prime denotes differentiation with respect to τ.

For Eq. 4.693, the initial conditions become

$$u(0) = 1, \quad u'(0) = 0 \tag{4.694}$$

The linear and nonlinear operators are, respectively ($g(\tau) = 0$)

$$L[u(\tau)] = u'' + u \tag{4.695}$$

$$N[u(\tau), \Omega] = u\left(\frac{1}{\Omega^2} - 1\right) + A^2 u u'^2 \tag{4.696}$$

From the equation

$$u_0'' + u_0 = 0, \quad u_0(0) = 1, u_0'(0) = 0 \tag{4.697}$$

it follows that

$$u_0(\tau) = \cos \tau \tag{4.698}$$

The nonlinear operator 4.696 for Eq. 4.698 becomes

$$N[u_0(\tau), \Omega] = \left(1 + \frac{A^2}{4} + \frac{1}{\Omega^2}\right) \cos \tau - \frac{A^2}{4} \cos 3\tau \tag{4.699}$$

The linear differential equation for the first approximation is

$$u_1'' + u_1 = H(\tau, C_i)\left[\left(1 + \frac{A^2}{4} + \frac{1}{\Omega^2}\right) \cos \tau - \frac{A^2}{4} \cos 3\tau\right], \quad u_1(0) = u_1'(0) = 0 \tag{4.700}$$

We have the freedom to choose the auxiliary function in the form

$$H(\tau, C_i) = C_1 + 2C_2 \cos 2\tau + 2C_3 \cos 4\tau + 2C_4 \cos 6\tau \tag{4.701}$$

By means of this last expression, Eq. 4.700 can be put as:

$$u_1'' + u_1 = \left[\left(1 + \frac{A^2}{4} + \frac{1}{\Omega^2}\right)(C_1 + C_2) - \frac{A^2}{4}(C_2 + C_3)\right] \cos \tau + \left[\left(1 + \frac{A^2}{4} + \right.\right.$$
$$\left.\left. + \frac{1}{\Omega^2}\right)(C_2 + C_3) - \frac{A^2}{4}(C_1 + C_4)\right] \cos 3\tau + \left[\left(1 + \frac{A^2}{4} + \frac{1}{\Omega^2}\right)(C_3 + C_4) - \right.$$
$$\left. - \frac{A^2}{4} C_2\right] \cos 5\tau + \left[\left(1 + \frac{A^2}{4} + \frac{1}{\Omega^2}\right) C_4 - \frac{A^2}{4} C_3\right] \cos 7\tau - \frac{A^2}{4} C_4 \cos 9\tau \tag{4.702}$$

Avoiding the secular term into Eq. 4.702, we obtain

$$\Omega^2 = \frac{4(C_1 + C_2)}{A^2(C_3 - C_1) - 4(C_1 + C_2)}$$

Taking into account the solution of Eq. 4.702 and Eqs. 4.692 and 4.5, one can obtain the first-order approximate periodic solution of Eq. 4.690

$$\begin{aligned} x(t) &= A\cos\Omega t + A\left[\frac{A^2}{32}(C_1 + C_4) - \frac{1}{8}\left(1 + \frac{A^2}{4} + \frac{1}{\Omega^2}\right)(C_2 + C_3)\right](\cos 3\Omega t - \\ &- \cos\Omega t) + A\left[\frac{A^2}{96}C_2 - \frac{1}{24}\left(1 + \frac{A^2}{4} + \frac{1}{\Omega^2}\right)(C_3 + C_4)\right](\cos 5\Omega t - \cos\Omega t) + \\ &+ A\left[\frac{A^2}{192}C_3 - \frac{1}{48}\left(1 + \frac{A^2}{4} + \frac{1}{\Omega^2}\right)C_4\right](\cos 7\Omega t - \cos\Omega t) + \\ &+ \frac{A^3}{320}C_4(\cos 9\Omega t - \cos\Omega t) \end{aligned} \quad (4.703)$$

4.18.2 Numerical Examples

In what follows we consider four cases in order to prove the accuracy of the OHAM.

4.18.2.1 For $A = 0.5$, the convergence-control parameters and the frequency are:
$C_1 = -0.031087449$, $C_2 = 0.032063813$,
$C_3 = -0.000814183$, $C_4 = 0.0000684702$, $\Omega = 1.032584242$
The first-order approximate solution 4.732 becomes

$$\begin{aligned} x(t) &= 0.503957224\cos\Omega t - 0.004028121\cos 3\Omega t + \\ &+ 0.000072827\cos 5\Omega t - 1.9568047\cdot 10^{-6}\cos 7\Omega t + 2.6746195\cdot 10^{-8}\cos 9\Omega t \end{aligned} \quad (4.704)$$

4.18.2.2 In the case $A = 1$ we obtain
$C_1 = -0.127127459$, $C_2 = 0.142521092$,
$C_3 = -0.017827985$, $C_4 = 0.005222895$, $\Omega = 1.135867917$
and therefore the approximate solution can be written as

$$\begin{aligned} x(t) &= 1.033122338\cos\Omega t - 0.035373646\cos 3\Omega t + \\ &+ 0.002548189\cos 5\Omega t - 0.000313203\cos 7\Omega t + 0.000016321\cos 9\Omega t \end{aligned} \quad (4.705)$$

4.18.2.3 If $A = 1.5$, then
$C_1 = -0.302003499$, $C_2 = 0.371534993$,

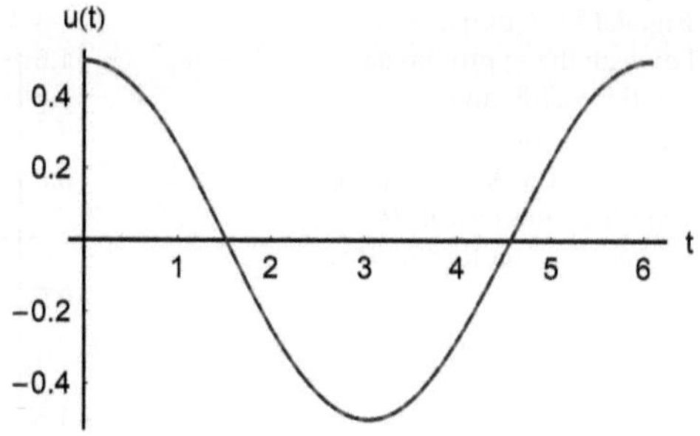

Fig. 4.152 Comparison between the approximate solution 4.704 and numerical solution of Eq. 4.690 for A = 0.5: (*red solid line*) numerical; (*blue dashed line*) analytical solution (colour figure online)

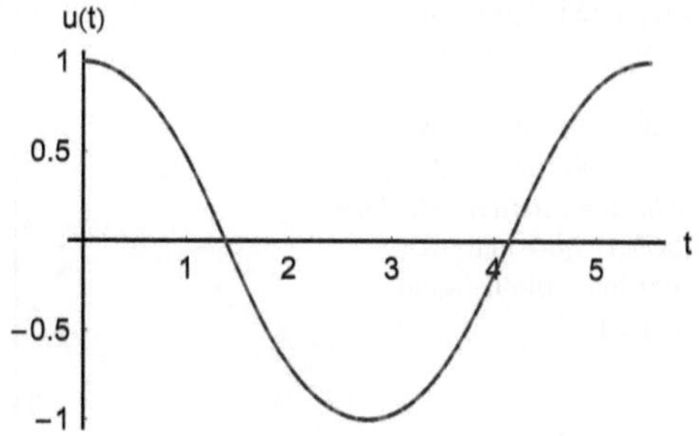

Fig. 4.153 Comparison between the approximate solution 4.705 and numerical solution of Eq. 4.690 for A = 1: (*red solid line*) numerical; (*blue dashed line*) analytical solution (colour figure online)

$C_3 = -0.108406832$, $C_4 = 0.041475872$, $\Omega = 1.329005134$
The approximate solution of Eq. 4.690 gives

$$x(t) = 1.614759468 \cos \Omega t - 0.132498703 \cos 3\Omega t + \\ +0.021966396 \cos 5\Omega t - 0.004664602 \cos 7\Omega t + 0.0004374408 \cos 9\Omega t \tag{4.706}$$

4.18.2.4 For A = 2, we obtain
$C_1 = -0.321403419$, $C_2 = 0.462361215$,
$C_3 = -0.129905494$, $C_4 = 0.042591199$, $\Omega = 1.670038093$
On the other hand, the approximate solution has the form

$$x(t) = 2.218573404 \cos \Omega t - 0.265731244 \cos 3\Omega t + \\ +0.055691347 \cos 5\Omega t - 0.009598286 \cos 7\Omega t + 0.001064779 \cos 9\Omega t \tag{4.707}$$

Figures 4.152, 4.153, 4.154, 4.155 show the comparison between the approximate solutions 4.704-4.707 of Eq. 4.690, and numerical solutions. It can be seen the accuracy of the analytical results in comparison with numerical ones. Also, from

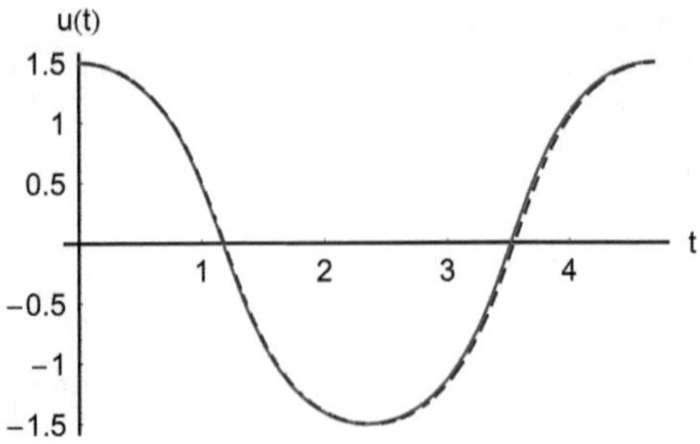

Fig. 4.154 Comparison between the approximate solution 4.706 and numerical solution of Eq. 4.690 for A = 1.5: (*red solid line*) numerical; (*blue dashed line*) (colour figure online)

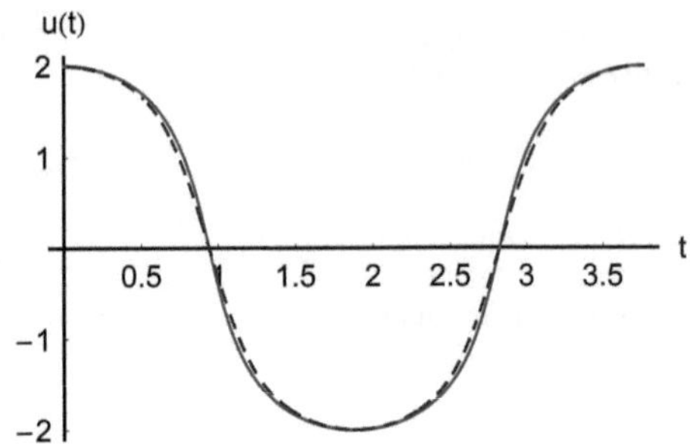

Fig. 4.155 Comparison between the approximate solution 4.707 and numerical solution of Eq. 4.690 for A = 2: (*red solid line*) numerical; (*blue dashed line*) analytical solution (colour figure online)

Table 4.132 Comparison between the numerical and approximate values of the frequency for Eq. 4.690

A	Ω_{num}	Ω_{app}
0.5	1.03198	1.03258
1	1.13678	1.13586
1.5	1.33963	1.32900
2	1.67046	1.67003

Table 4.132, a high accuracy is observed also for the approximate frequencies when comparing to numerical ones.

4.19 Nonlinear Oscillators with Quadratic and Cubic Nonlinearities

Nonlinear oscillators with quadratic and cubic nonlinearities often occur in mechanics. Some examples are asymmetric vibrations of pendulums, shells and curved beams [203]. The method of harmonic balance is used in [204] for solving nonlinear oscillatory problems. It had been found to work well when all terms in the dependent variable have odd parity. Nayfeh and Mook [204] have cautioned against use of the method when terms of mixed parity are involved, pointing out that for full consistency a second harmonic term (as well as a constant) must be taken into account in the solution expression. But, in contrast, Gottlieb [205] showed that a second harmonic in term is not needed for large amplitude, to obtain accurate expressions of the frequency. Belhaq and Lakrad [206] applied the harmonic balance method involving the Jacobian elliptic functions to show the approximation of the period obtained by Gottlieb using the trigonometric harmonic balance method can be significantly improved. The method of harmonic balance is also used by Hu [207] to calculate first-order approximations to the periodic solutions of a mixed parity nonlinear oscillator. These solutions are more accurate than the second approximate solution obtained by the Lindstedt-Poincare method for large amplitudes. Marinca and Herisanu solved this problem in [208] by means of the variational iteration method.

We consider the free oscillation of a nonlinear oscillator with quadratic and cubic nonlinearities [208]:

$$\ddot{u} + \omega^2 u + au^2 + bu^3 = 0 \tag{4.708}$$

where a and b are known constants. The initial conditions are given by

$$u(0) = A, \quad \dot{u}(0) = 0 \tag{4.709}$$

Introducing the dependent variable

$$u = v - \frac{a}{3b} \tag{4.710}$$

The quadratic nonlinearity in Eq. 4.708 is eliminated to obtain

$$\ddot{v} + \left(\omega^2 - \frac{a^2}{3b}\right)v + bv^3 - k = 0 \tag{4.711}$$

where

$$k = \frac{\omega^2 a}{3b} - \frac{2a^3}{22b^2} \tag{4.712}$$

The initial conditions for Eq. 4.711 are

$$v(0) = A + \frac{a}{3b}, \quad \dot{v}(0) = 0 \tag{4.713}$$

Introducing a new variable and a new function

$$\tau = \Omega t, \quad v(t) = \left(A + \frac{a}{3b}\right)x(\tau) \tag{4.714}$$

where Ω is the frequency of the system 4.708, then this equation can be written as

$$x'' + \frac{\alpha}{\Omega^2}x + \frac{\beta}{\Omega^2}x^3 + \frac{\gamma}{\Omega^2} = 0 \tag{4.715}$$

where prime denotes differentiation with respect to τ and

$$\alpha = \omega^2 - \frac{a^2}{3b},\ \beta = b\left(A + \frac{a}{3b}\right)^2,\ \gamma = -\frac{k}{A + \dfrac{a}{3b}} = \frac{2a^3 - 9ab\omega^2}{9b(a + 3bA)} \tag{4.716}$$

The initial conditions become $x(0) = 1, \quad \dot{x}(0) = 0$.

4.19.1 Solutions of Eq. 4.715 by Means of OHAM

For the nonlinear differential equation with constant coefficients 4.715 we should choose the linear and nonlinear operator in the form $(g(\tau) = 0)$:

$$\begin{aligned} L[x(\tau)] &= x'' + x \\ N[x(\tau), \Omega] &= \left(\frac{\alpha}{\Omega^2} - 1\right)x + \frac{\beta}{\Omega^2}x^3 + \frac{\gamma}{\Omega^2} \end{aligned} \tag{4.717}$$

The initial approximation x_0 is obtained from the linear equation

$$x_0'' + x_0 = 0 \quad x_0(0) = 1, \dot{x}_0(0) = 0 \tag{4.718}$$

whose solution is

$$x_0(\tau) = \cos\tau \tag{4.719}$$

From Eqs. 4.719 and 4.717 one can get

$$N[x_0(\tau), \Omega] = \frac{\gamma}{\Omega^2} + \left(\frac{4\alpha + 3\beta}{4\Omega^2} - 1\right)\cos\tau + \frac{\beta}{4\Omega^2}\cos 3\tau \tag{4.720}$$

The equation 4.9 for the first approximation $x_1(\tau)$ is

$$\begin{aligned} &x_1'' + x_1 = H(\tau, C_i)\left[\frac{\gamma}{\Omega^2} + \left(\frac{4\alpha + 3\beta}{4\Omega^2} - 1\right)\cos\tau + \frac{\beta}{4\Omega^2}\cos 3\tau\right], \\ &x_1(0) = \dot{x}_1(0) = 0 \end{aligned} \tag{4.721}$$

where the auxiliary function H can be chosen as follows:

$$H(\tau, C_i) = C_1 + 2C_2 \cos 2\tau + 2C_3 \cos 4\tau \tag{4.722}$$

Substituting Eq. 4.722 into Eq. 4.721 it follows that

$$\begin{aligned} x_1'' + x_1 &= \frac{\gamma C_1}{\Omega^2} + \left[\left(\frac{4\alpha + 3\beta}{4\Omega^2} - 1\right)(C_1 + C_2) + \frac{\beta}{4\Omega^2}(C_2 + C_3)\right] \cos\tau + \\ &+ \frac{2\gamma C_2}{\Omega^2} \cos 2\tau + \left[\left(\frac{4\alpha + 3\beta}{4\Omega^2} - 1\right)(C_2 + C_3) + \frac{\beta C_1}{4\Omega^2}\right] \cos 3\tau + \\ &+ \frac{2\gamma C_3}{\Omega^2} \cos 4\tau + \left[\left(\frac{4\alpha + 3\beta}{4\Omega^2} - 1\right) C_3 + \frac{\beta C_2}{4\Omega^2}\right] \cos 5\tau + \frac{\beta C_3}{4\Omega^2} \cos 7\tau \end{aligned} \tag{4.723}$$

Avoiding the presence of a secular term into Eq. 4.723 needs

$$\Omega^2 = \alpha + \frac{3}{4}\beta + \frac{\beta}{4}\frac{C_2 + C_3}{C_1 + C_2}$$

The solution of Eq. 4.723 becomes

$$\begin{aligned} x_1(\tau) &= \frac{\gamma C_1}{\Omega^2}(1 - \cos\tau) + \frac{2\gamma C_2}{3\Omega^2}(\cos\tau - \cos 2\tau) + \left[\left(\frac{4\alpha + 3\beta}{4\Omega^2} - 1\right)\frac{C_2 + C_3}{8} + \right. \\ &\left. + \frac{\beta C_1}{32\Omega^2}\right](\cos\tau - \cos 3\tau) + \frac{2\gamma C_3}{15\Omega^2}(\cos\tau - \cos 4\tau) + \left[\frac{\beta C_2}{96\Omega^2} + \right. \\ &\left. + \left(\frac{4\alpha + 3\beta}{4\Omega^2} - 1\right)\frac{C_3}{24}\right](\cos\tau - \cos 5\tau) + \frac{\beta C_3}{192\Omega^2}(\cos\tau - \cos 7\tau) \end{aligned} \tag{4.724}$$

From Eqs. 4.719, 4.724, 4.5, 4.714 and 4.710, the first-order approximate periodic solution of Eqs. 4.708 and 4.709 can be written as

$$\begin{aligned} \overline{u}(t, C_i) &= \left(A + \frac{a}{3b}\right)\overline{x}(\Omega t, C_i) - \frac{a}{3b} = \left(A + \frac{a}{3b}\right)\left[\frac{\gamma C_1}{\Omega^2} + \left(1 - \frac{\gamma C_1}{\Omega^2}\right)\cos\Omega t + \right. \\ &+ \frac{2\gamma C_2}{3\Omega^2}(\cos\Omega t - \cos 2\Omega t) + \frac{(4\alpha + 3\beta - 4\Omega^2)(C_2 + C_3) + \beta C_1}{32\Omega^2}(\cos\Omega t - \\ &- \cos 3\Omega t) + \frac{2\gamma C_3}{15\Omega^2}(\cos\Omega t - \cos 4\Omega t) + \frac{\beta C_2 + (4\alpha + 3\beta - 4\Omega^2)C_3}{96\Omega^2}(\cos\Omega t - \\ &\left. - \cos 5\Omega t) + \frac{\beta C_3}{192\Omega^2}(\cos\Omega t - \cos 7\Omega t)\right] - \frac{a}{3b} \end{aligned} \tag{4.725}$$

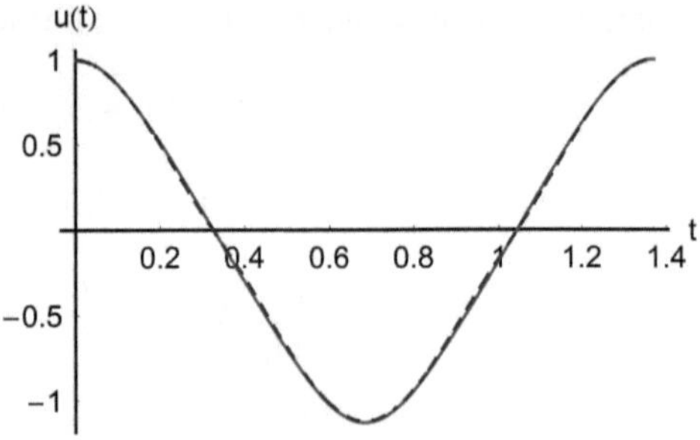

Fig. 4.156 Comparison between the approximate solution 4.725 and numerical results of Eq. 4.708 in the case a = 5, b = 25, A = 1: (*red solid line*) numerical; (*blue dashed line*) analytical solution (colour figure online)

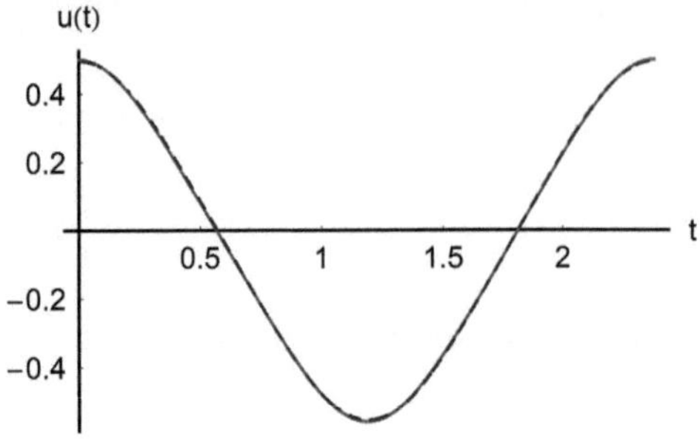

Fig. 4.157 Comparison between the approximate solution 4.725 and numerical results of Eq. 4.708 in the case a = 3, b = 30, A = 0.5: (*red solid line*) numerical; (*blue dashed line*) analytical solution (colour figure online)

4.19.2 Numerical Examples

We illustrate the efficiency of our procedure by comparing the above presented periodic solutions with numerical ones obtained by means of a fourth-order Runge-Kuta method.

4.19.2.1 In the first case we consider a = 5, b = 25, ω = 1 and A = 1. Using the collocation method, the optimal values of the convergence-control parameters C_i and the frequency Ω are obtained as:

$$C_1 = -1.232774407151, C_2 = 0.378928865166,$$
$$C_3 = -0.2647580043003, \Omega = 4.587934928$$

4.19.2.2 In the last case, for a = 3, b = 30, ω = 1 and A = 0.5, we obtain

$$C_1 = \text{-}0.808395094369, C_2 = 0.061628250168,$$
$$C_3 = 0.053994675033, \Omega = 2.640017632$$

Figures 4.156 and 4.157 show a comparison between the present analytical solutions and the numerical integration results obtained using a fourth-order Runge–Kutta method. It can be observed that the first-order approximate periodic solutions obtained through OHAM are almost identical with the numerical simulation results in the considered cases for various values of the parameters a, b and A.

4.20 Damped Oscillator with Fractional-Order Restoring Force

In this section we consider the nonlinear oscillator with damping and fractional-order restoring force [92]

$$\ddot{x} + 2k\ddot{x} + ax + bx|x|^{n-1} = 0 \tag{4.726}$$

where k, a and b are constants and $n > 0$.

The initial conditions are given by

$$x(0) = A, \quad \dot{x}(0) = 0 \tag{4.727}$$

where the dot denotes derivative with respect to time t.

Under the transformations

$$\tau = kt, \;\; x = Ae^{-\tau}y(\tau) \tag{4.728}$$

the original Eqs. 4.726 and 4.727 become

$$y'' + (\alpha - 1)y + \beta e^{(1-n)\tau}y|y|^{n-1} = 0, \quad y(0) = y'(0) = 1 \tag{4.729}$$

where prime denotes the derivative with respect to τ and

$$\alpha = \frac{a}{k^2}, \;\; \beta = \frac{b}{k^2}A^{n-1} \tag{4.730}$$

The linear operator, the function g(τ) and the nonlinear operator, corresponding to Eq. 4.729 are respectively

$$\begin{aligned} &L[y(\tau)] = y'' + \gamma^2 y \\ &g(\tau) = -e^{-\lambda\tau}\left(\lambda^2 \cos\gamma\tau + 2\lambda\gamma \sin\gamma\tau\right) \\ &N[y(\tau)] = (\alpha - 1 - \gamma^2)y(\tau) + \beta e^{(1-n)\tau}y(\tau)|y(\tau)|^{n-1} + \\ &+e^{-\lambda\tau}\left(\lambda^2 \cos\gamma\tau + 2\lambda\gamma \sin\gamma\tau\right) \end{aligned} \tag{4.731}$$

where γ is an arbitrary constant unknown at this moment.

To determine the initial approximation, Eq. 4.4 can be written as

$$y_0'' + \gamma^2 y_0 - e^{-\lambda t}\left(\lambda^2 \cos\gamma\tau + 2\lambda\gamma \sin\gamma\tau\right) = 0 \tag{4.732}$$

and therefore

$$y_0(\tau) = e^{-\lambda t} \cos \gamma\tau \tag{4.733}$$

We point out that at this moment the initial conditions are not completely fulfilled because $y_0'(0) = -\lambda$. Substituting Eq. 4.733 into Eq. 4.732 we find the following expression

$$\begin{aligned} N[y_0(\tau)] &= \left[(\alpha - 1 + \lambda^2 - \gamma^2)\cos\gamma\tau + 2\lambda\gamma\sin\gamma\tau\right]e^{-\lambda t} + \\ &+ \beta e^{-(\lambda n + n - 1)\tau}\cos\gamma\tau|\cos\gamma\tau|^{n-1} \end{aligned} \tag{4.734}$$

Now, taking into account that

$$\cos\gamma\tau|\cos\gamma\tau|^{n-1} = |\cos\gamma\tau|^n sign(\cos\gamma\tau) \tag{4.735}$$

and that the Fourier series expansion of this last expression is known in the form

$$\begin{aligned} |\cos\gamma\tau|^n sign(\cos\gamma\tau) &= a_{1n}\cos\gamma\tau + a_{3n}\cos 3\gamma\tau + \\ &+ a_{5n}\cos 5\gamma\tau + a_{7n}\cos 7\gamma\tau + \ldots \end{aligned} \tag{4.736}$$

where

$$a_{2j+1,n} = \frac{4}{\pi}\int_0^{\pi/2} (\cos u)^n \cos(2j+1)u\,du, \quad j = 0, 1, 2\ldots \tag{4.737}$$

then Eq. 4.734 becomes

$$\begin{aligned} N[y_0(\tau)] &= \left[(\alpha - 1 + \lambda^2 - \gamma^2)\cos\gamma\tau + 2\lambda\gamma\sin\gamma\tau\right]e^{-\lambda t} + \\ &+ \beta e^{-(\lambda n + n - 1)\tau}(a_{1n}\cos\gamma\tau + a_{3n}\cos 3\gamma\tau + a_{5n}\cos 5\gamma\tau\ldots) \end{aligned} \tag{4.738}$$

We assumed that the auxiliary function $H(\tau, C_i)$ which appears in Eq. 4.9 of the first approximation y_1 is given by

$$H(\tau, C_i) = C_1 + 2C_2\cos 2\gamma\tau + 2C_3\sin 2\gamma\tau \tag{4.739}$$

Substituting this equation and Eq. 4.738 into Eq. 4.9, the equation in $y_1(\tau)$ is derived as

$$
\begin{aligned}
&y_1'' + \gamma^2 y_1 = e^{-\lambda\tau}\{\left[(C_1+C_2)(\lambda^2-\gamma^2+\alpha-1)+2\lambda\gamma C_3\right]\cos\gamma\tau+\\
&+\left[2\lambda\gamma(C_1-C_2)+(\lambda^2-\gamma^2+\alpha-1)C_3\right]\sin\gamma\tau+\left[(\lambda^2-\gamma^2+\alpha-1)C_2-\right.\\
&\left.-2\lambda\gamma C_3\right]\cos 3\gamma\tau+\left[2\lambda\gamma C_2+(\lambda^2-\gamma^2+\alpha-1)C_3\right]\sin 3\gamma\tau\}+\\
&+e^{-(\lambda n+n-1)\tau}\left[(a_{1n}C_1+a_{1n}C_2+a_{3n}C_2)\cos\gamma\tau+(a_{1n}C_3-\right.\\
&-a_{3n}C_3)\sin\gamma\tau+(a_{3n}C_1+a_{1n}C_2+a_{5n}C_2)\cos 3\gamma\tau+(a_{1n}C_3-\\
&-a_{5n}C_3)\sin 3\gamma\tau+(a_{5n}C_1+a_{3n}C_2+a_{7n}C_2)\cos 5\gamma\tau+(a_{3n}C_3-\\
&-a_{7n}C_3)\sin 5\gamma\tau+(a_{7n}C_1+a_{5n}C_2)\cos 7\gamma\tau+\\
&\left.+a_{5n}C_3\sin 7\gamma\tau+a_{7n}C_2\cos 9\gamma\tau+a_{7n}C_3\sin 9\gamma\tau+\ldots\right]\\
&y_1(0)=0,\ y_1'(0)=\lambda+1
\end{aligned}
\tag{4.740}
$$

After we will solve this equation, and taking into consideration Eqs. 4.733, 4.5, 4.728 and finally Eq. 4.727, we find the first-order approximation:

$$
\begin{aligned}
&\bar{x}(t,C_i)=Ae^{-kt}\{-(M_1+M_3+P_1+P_3+P_5+P_7+P_9)\cos\gamma kt+\left[\frac{\lambda+1}{\gamma}-\right.\\
&-N_1-3N_3-Q_1-3Q_3-5Q_5-7Q_7-9Q_9+\frac{\lambda n+n-1}{\gamma}(P_1+P_3+P_5+\\
&\left.+P_7+P_9)\right]\sin\gamma kt\}+Ae^{-(\lambda+1)kt}\left[(1+M_1)\cos\gamma kt+M_3\cos 3\gamma kt+\right.\\
&\left.+N_1\sin\gamma kt+N_3\sin 3\gamma kt\right]+Ae^{-(\lambda+1)nkt}(P_1\cos k\gamma t+P_3\cos 3\gamma kt+\\
&+P_5\cos 5\gamma kt+P_7\cos 7\gamma kt+P_9\cos 9\gamma kt+Q_1\sin\gamma kt+Q_3\sin 3\gamma kt+\\
&+Q_5\sin 5\gamma kt+Q_7\sin 7\gamma kt+Q_9\sin 9\gamma kt)
\end{aligned}
\tag{4.741}
$$

where

$$M_1=\frac{\lambda(3\gamma^2+\lambda^2+\alpha-1)C_1+\lambda(\lambda^2-5\gamma^2+\alpha-1)C_2+2\gamma(2\lambda^2-\gamma^2+\alpha-1)C_3}{\lambda(\lambda^2+4\gamma^2)}$$

$$N_1=\frac{2\gamma(\gamma^2-\alpha+1)C_1-2\gamma(2\lambda^2-\gamma^2+\alpha-1)C_2+\lambda(\lambda^2-5\gamma^2+\alpha-1)C_3}{\lambda(\lambda^2+4\gamma^2)}$$

$$M_3=\frac{\left[\lambda^4+3\lambda^2\gamma^2+8\gamma^4+(\alpha-1)(\lambda^2-8\gamma^2)\right]C_2+2\lambda\gamma(2\lambda^2+5\gamma^2+3\alpha-3)C_3}{\lambda^4+64\gamma^4+20\lambda^2\gamma^2}$$

$$N_3=\frac{-2\lambda\gamma(2\lambda^2+5\gamma^2+3\alpha-3)C_2+\left[(\lambda^4+3\lambda^2\gamma^2+8\gamma^4+(\alpha-1)(\lambda^2-8\gamma^2)\right]C_3}{\lambda^4+64\gamma^4+20\lambda^2\gamma^2}$$

$$P_1=\beta\frac{a_{1n}(n\lambda+n-1)C_1+(a_{1n}+a_{3n})(n\lambda+n-1)C_2+2\gamma(a_{1n}-a_{3n})C_3}{(n\lambda+n-1)\left[4\gamma^2+(n\lambda+n-1)^2\right]}$$

$$Q_1=\beta\frac{-2a_{1n}\gamma C_1-2\gamma(a_{1n}+a_{3n})C_2+(a_{1n}-a_{3n})(n\lambda+n-1)C_3}{(n\lambda+n-1)\left[4\gamma^2+(n\lambda+n-1)^2\right]}$$

$$
\begin{aligned}
P_3 &= \beta\frac{\left[(n\lambda+n-1)^2-8\gamma^2\right][a_{3n}C_1+(a_{1n}+a_{5n})C_2]+6(a_{1n}-a_{5n})(n\lambda+n-1)\gamma C_3}{(n\lambda+n-1)^4+64\gamma^4+20\gamma^2(n\lambda+n-1)^2}\\
Q_3 &= \beta\frac{-(n\lambda+n-1)[6\gamma a_{3n}C_1+6\gamma(a_{1n}+a_{5n})C_2]+(a_{1n}-a_{5n})\left[(n\lambda+n-1)^2-8\gamma^2\right]C_3}{(n\lambda+n-1)^4+64\gamma^4+20\gamma^2(n\lambda+n-1)^2}\\
P_5 &= \beta\frac{\left[(n\lambda+n-1)^2-24\gamma^2\right][a_{5n}C_1+(a_{3n}+a_{7n})C_2]+10(a_{3n}-a_{7n})(n\lambda+n-1)\gamma C_3}{(n\lambda+n-1)^4+576\gamma^4+52\gamma^2(n\lambda+n-1)^2}\\
Q_5 &= \beta\frac{-(n\lambda+n-1)[10a_{5n}\gamma C_1+(a_{3n}+a_{7n})\gamma C_2]+(a_{3n}-a_{7n})\left[(n\lambda+n-1)^2-24\gamma^2\right]C_3}{(n\lambda+n-1)^4+576\gamma^4+52\gamma^2(n\lambda+n-1)^2}\\
P_7 &= \beta\frac{\left[(n\lambda+n-1)^2-48\gamma^2\right][a_{7n}C_1+a_{5n}C_2]+14a_{5n}(n\lambda+n-1)\gamma C_3}{(n\lambda+n-1)^4+2304\gamma^4+100\gamma^2(n\lambda+n-1)^2}\\
Q_7 &= \beta\frac{-(n\lambda+n-1)(14a_{7n}\gamma C_1+14a_{5n}\gamma C_2)+a_{5n}\left[(n\lambda+n-1)^2-48\gamma^2\right]C_3}{(n\lambda+n-1)^4+2304\gamma^4+100\gamma^2(n\lambda+n-1)^2}\\
P_9 &= \beta\frac{a_{7n}\left[(n\lambda+n-1)^2-80\gamma^2\right]C_2+18a_{7n}\gamma(n\lambda+n-1)C_3}{(n\lambda+n-1)^4+6400\gamma^4+162\gamma^2(n\lambda+n-1)^2}\\
Q_9 &= \beta\frac{-18a_{7n}(n\lambda+n-1)C_2+a_{7n}\left[(n\lambda+n-1)^2-80\gamma^2\right]C_3}{(n\lambda+n-1)^4+6400\gamma^4+162\gamma^2(n\lambda+n-1)^2}
\end{aligned}
\tag{4.742}
$$

4.20.1 Numerical Examples

We illustrate the accuracy of this approach by comparing the above approximate solutions with numerical integration results.

4.20.1.1 In the first considered case, it is assumed that a = 1, b = 1, k = 0.1, n = 3. A = 0.5. This corresponds to the oscillator with a linear plus cubic restoring force. It should be noted that the values of the parameters of nonlinearity are not small.

From Eq. 4.737 one finds

$$a_{1,3} = 0.75, \quad a_{3,3} = 0.25, \quad a_{5,3} = a_{7,3} = 0$$

By means of a collocation approach, one obtains

$$
\begin{aligned}
&C_1 = -1.383098578,\ C_2 = 0.099734261,\ C_3 = 0.046154175\\
&\lambda = -0.018123796,\ \gamma = 10.2735580407
\end{aligned}
$$

The first-order approximate solution 4.741 becomes

$$\bar{x}(t) = 0.5e^{-0.1t}(0.6838704\cos\gamma kt - 23.006699\sin\gamma kt)+$$
$$+0.5e^{-0.098187t}\big(0.348293\cos\gamma kt + 0.000753\cos 3\gamma kt + 22.512585\sin\gamma kt+$$
$$+0.0004007\sin 3\gamma kt\big) + 0.5e^{-0.294562t}\big(- 0.040723\cos\gamma kt + 0.008042\cos 3\gamma kt-$$
$$-0.000235\cos 5\gamma kt + 0.582473\sin\gamma kt + 0.000117\sin 3\gamma kt - 0.000115\sin 5\gamma kt\big) \tag{4.743}$$

4.20.1.2 For the second case, we consider a = 0, b = 1, k = 0.1, n = 3, A = 0.5 and we obtain

$$C_1 = 0.1824268303, C_2 = -0.756523239, C_3 = 0.226871951,$$
$$\lambda = 0.243375367, \gamma = 1.686844496$$

and the first-order approximate solution is

$$\bar{x}(t) = 0.5e^{-0.1t}(0.281902\cos\gamma kt + 1.031895\sin\gamma kt)+$$
$$+0.5e^{-0.124337t}\big(1.099599\cos\gamma kt - 0.123572\cos 3\gamma kt - 2.867017\sin\gamma kt+$$
$$+0.051785\sin 3\gamma kt\big) + 0.5e^{-0.373012t}\big(- 0.636471\cos\gamma kt + 0.317918\cos 3\gamma kt+$$
$$+0.060624\cos 5\gamma kt + 1.166979\sin\gamma kt + 0.295938\sin 3\gamma kt - 0.011076\sin 5\gamma kt\big) \tag{4.744}$$

4.20.1.3 In the last case, if a = b = 1, k = 0.1, n = 3/2, A = 0.5, then

$$a_{1,3/2} = 0.915311716,\ a_{3,3/2} = 0.101701301,\ a_{5,3/2} = -0.023469531,$$
$$a_{7,3} = 0.009663924$$

and

$$C_1 = 0.113055436, C_2 = 0.333431285, C_3 = -1.156410646,$$
$$\lambda = 2.623175804, \gamma = 11.3$$

The approximate solution becomes

$$\bar{x}(t) = 0.5e^{-0.1t}(0.622368\cos\gamma kt - 0.867651\sin\gamma kt)+$$
$$+0.5e^{-0.362317t}\big(1.0510606\cos\gamma kt - 0.051059\cos 3\gamma kt + 1.326588\sin\gamma kt-$$
$$-0.053288\sin 3\gamma kt\big) + 0.5e^{-0.543476t}\big(- 0.580183\cos\gamma kt - 0.041094\cos 3\gamma kt-$$
$$-0.001175\cos 5\gamma kt + 0.000112\cos 7\gamma kt - 0.000028\cos 9\gamma kt - 0.426098\sin\gamma kt+$$
$$+0.064289\sin 3\gamma kt + 0.002402\sin 5\gamma kt - 0.000309\sin 7\gamma kt + 0.000076\sin 9\gamma kt\big) \tag{4.745}$$

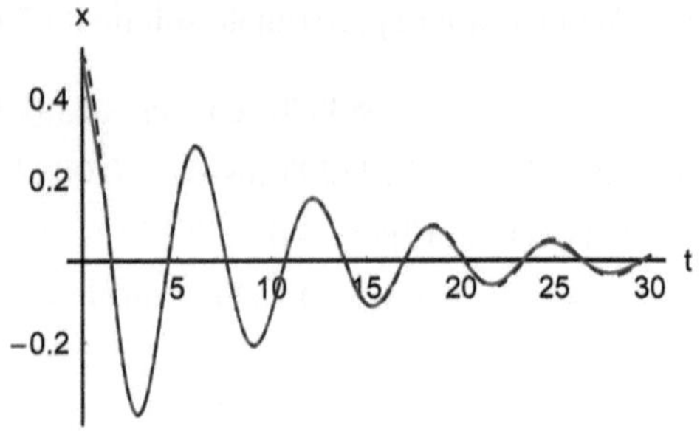

Fig. 4.158 Comparison between the approximate result 4.743 (*blue dashed line*) and numerical result (*red solid line*) for a = b = 1, A = 0.5, k = 0.1, n = 3 (colour figure online)

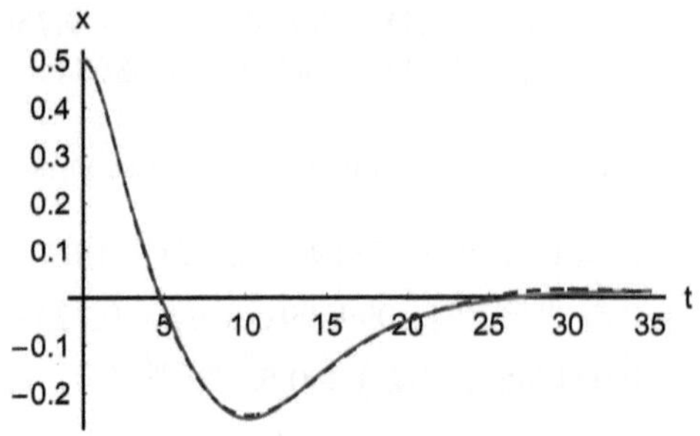

Fig. 4.159 Comparison between the approximate result 4.744 (*blue dashed line*) and numerical result (*red solid line*) for a = 0, b = 1, A = 0.5, k = 0.1, n = 3 (colour figure online)

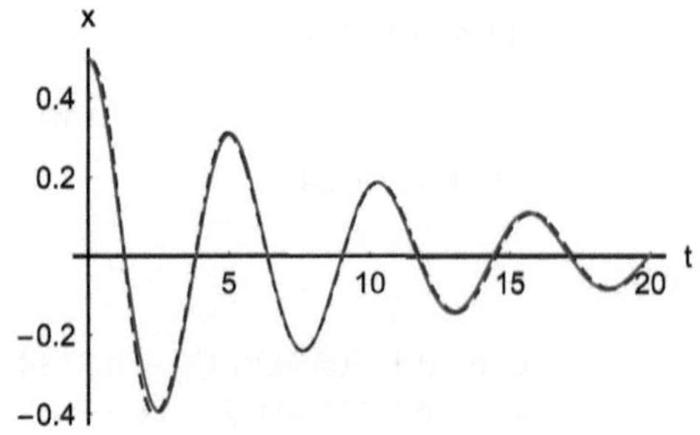

Fig. 4.160 Comparison between the approximate result 4.745 (*blue dashed line*) and numerical result (*red solid line*) for a = b = 1, A = 0.5, k = 0.1, n = 3/2 (colour figure online)

From Figs. 4.158, 4.159, 4.160 we can observe that the first-order approximate solutions obtained through OHAM are highly accurate.

As an interesting particularity of this application, we underline the involvement here of some distinct features such as the non-zero function $g(\tau)$ given by Eq. 4.731, the initial approximation given by Eq. 4.733, which depends upon two unknown parameters, and the initial conditions which are partly fulfilled by the initial approximation and partly fulfilled by the first approximation and finally completely fulfilled by the first-order approximate solutions.

These approximate solutions are plotted in Figs. 4.158, 4.159, 4.160 along with the numerical solutions and a very good agreement between them was found. These solutions represent in the cases 4.20.1.1 and 4.20.1.3 oscillations with decreasing amplitude approaching to zero equilibrium point, which corresponds to stable focus in the phase plane. In the case 4.20.1.2 plotted in Fig. 4.159, there is no oscillation.

Figures 4.158, 4.159, 4.160 show a comparison between the first-order approximate analytical solutions of Eq. 4.726 and the numerical integration results. One can observe that the first-order approximate analytical results obtained through OHAM are nearly identical with the numerical simulation results for various values of the parameters.

4.21 The Oscillator with Cubic and Harmonic Restoring Force

Consider the nonlinear differential oscillator governed by the equation [92, 209]

$$\ddot{u} + u + au^3 + b\sin u = 0 \tag{4.746}$$

subject to the initial conditions

$$u(0) = A, \quad \dot{u}(0) = 0 \tag{4.747}$$

If Ω is the frequency of the system described by 4.746 and introducing a new independent variable and a new function

$$\tau = \Omega t, \quad u(t) = Ax(\tau) \tag{4.748}$$

then Eq. 4.746 becomes

$$x'' + \frac{1}{\Omega^2}x + \frac{aA^2}{\Omega^2}x^3 + \frac{b}{A\Omega^2}\sin Ax = 0, \quad x(0) = 1, \quad x'(0) = 0 \tag{4.749}$$

where the prime denotes the derivative with respect to the variable τ. From Eqs. 4.749 and 4.1 we can choose $g(\tau) = 0$ and the operators

$$L[x(\tau)] = x'' + x, \quad N[x(\tau)] = \left(\frac{1}{\Omega^2} - 1\right)x + \frac{aA^2}{\Omega^2}x^3 + \frac{b}{\Omega^2}\sin Ax \tag{4.750}$$

From Eq. 4.4:

$$x_0'' + x_0 = 0, \quad x_0(0) = 1, \quad x_0'(0) = 0 \tag{4.751}$$

one gets

$$x_0(\tau) = \cos\tau \tag{4.752}$$

From Eqs. 4.752 and 4.750 we find

$$N[x_0(\tau), \Omega] = \left(\frac{1}{\Omega^2} - 1\right)\cos\tau + \frac{aA^2}{\Omega^2}\cos^3\tau + \frac{b}{\Omega^2}\sin(\cos\tau) \tag{4.753}$$

In Eq. 4.753 the terms $\cos^3\tau$ and $\sin(\cos\tau)$ can be expanded in the power series

$$\cos^3\tau = \frac{\cos 3\tau + 3\cos\tau}{4} \tag{4.754}$$

$$\sin(\cos\tau) = \cos\tau - \frac{\cos^3\tau}{3!} + \frac{\cos^5\tau}{5!} - \frac{\cos^7\tau}{7!} + \frac{\cos^9\tau}{9!} + \ldots \tag{4.755}$$

$$\cos^{2n+1}\tau = \frac{1}{4^n}\sum_{k=0}^{n}\binom{2n+1}{n-k}\cos(2k+1)\tau \tag{4.756}$$

By using Eqs. 4.754, 4.755 and 4.756, by simple manipulations, Eq. 4.753 can be expressed in the form

$$\begin{aligned} N(x_0, \Omega) = {}& \left[\frac{1}{\Omega^2} - 1 + \frac{3aA^2}{4\Omega^2} + \frac{b}{\Omega^2}\left(1 - \frac{A^2}{8} + \frac{A^4}{192} - \frac{A^6}{9216} + \frac{A^8}{737280}\right)\right]\cos\tau + \\ & + \left[\frac{aA^2}{4\Omega^2} - \frac{bA^2}{\Omega^2}\left(\frac{1}{24} - \frac{A^2}{384} + \frac{A^4}{15360} - \frac{A^6}{1105920}\right)\right]\cos 3\tau + \frac{bA^4}{\Omega^2}\left(\frac{1}{1920} - \frac{A^2}{46080} + \right. \\ & \left. + \frac{A^4}{2580480}\right)\cos 5\tau - \frac{bA^6}{\Omega^2}\left(\frac{1}{322560} - \frac{A^2}{10321920}\right)\cos 7\tau + \frac{bA^8}{92897280}\cos 9\tau + .. \end{aligned} \tag{4.757}$$

From Eq. 4.757 we remark that $N(x_0)$ is a combination of the functions $\cos\tau$, $\cos 3\tau$, $\cos 5\tau$, ... and therefore we choose the auxiliary function $H(\tau, C_i)$ in the form

$$H(\tau, C_i) = C_1 + 2C_2\cos 2\tau + 2C_3\cos 4\tau + 2C_4\cos 6\tau \tag{4.758}$$

In this way, Eq. 4.9 can be written in the form:

$$\begin{aligned} & x_1'' + x_1 = \left[M(C_1 + C_2) + N(C_2 + C_3) + P(C_3 + C_4 + QC_4\right]\cos\tau + H.O.T., \\ & x_1(0) = x_1'(0) = 0 \end{aligned} \tag{4.759}$$

The solution of 4.759 is chosen that it contains no secular terms, which requires that coefficient of $\cos\tau$ disappear, i.e.

$$\begin{aligned}\Omega^2 &= 1+\frac{3aA^2}{4}+b\left(1-\frac{A^2}{8}+\frac{A^4}{192}-\frac{A^6}{9216}+\frac{A^8}{737280}+\ldots\right)+\\&+\frac{C_2+C_3}{C_1+C_2}\left[\frac{aA^2}{4}+bA^2\left(-\frac{1}{24}+\frac{A^2}{384}-\frac{A^4}{15360}+\frac{A^6}{1105920}\right)\right]+\\&+\frac{C_3+C_4}{C_1+C_2}bA^4\left(\frac{1}{1920}-\frac{A^2}{46080}+\frac{A^4}{2580480}\right)+\\&+\frac{C_4}{C_1+C_2}bA^6\left(-\frac{1}{322560}+\frac{A^2}{10321920}\right)\end{aligned}\tag{4.760}$$

The equation 4.759 becomes

$$\begin{aligned}x_1''+x_1 &= [N(C_1+C_4)+PC_2+M(C_2+C_3)+QC_3+RC_4]\cos 3\tau+\\&+[PC_1+NC_2+QC_2+M(C_3+C_4)+RC_3]\cos 5\tau+(QC_1+PC_2+\\&+RC_2+NC_3+MC_4)\cos 7\tau+(RC_1+QC_2+PC_3+NC_4)\cos 9\tau+\\&+[RC_2+QC_3+PC_4)\cos 11\tau+(RC_3+QC_4)\cos 13\tau+RC_4\cos 15\tau\end{aligned}\tag{4.761}$$

where

$$\begin{aligned}M &= \frac{1}{\Omega^2}-1+\frac{3aA^2}{4\Omega^2}+\frac{b}{\Omega^2}\left(1-\frac{A^2}{8}+\frac{A^4}{192}-\frac{A^6}{9216}+\frac{A^8}{737280}\right)\\N &= \frac{aA^2}{4\Omega^2}+\frac{bA^2}{\Omega^2}\left(-\frac{1}{24}+\frac{A^2}{384}-\frac{A^4}{15360}+\frac{A^6}{1105920}\right)\\P &= \frac{bA^4}{\Omega^2}\left(\frac{1}{1920}-\frac{A^2}{46080}+\frac{A^4}{2580480}\right)\\Q &= \frac{bA^6}{\Omega^2}\left(-\frac{1}{322560}+\frac{A^2}{10321920}\right), R=\frac{bA^8}{92897280\Omega^2}\end{aligned}\tag{4.762}$$

From Eqs. 4.752, 4.761, 4.5 and 4.748 we obtain the first-order approximate periodic solution of Eqs. 4.746 and 4.747:

$$\begin{aligned}\bar{u}(t) &= A\cos\Omega t + \frac{A}{8}[N(C_1+C_4)+(P+M)C_2+(M+Q)C_3+RC_4)(\cos\Omega t-\\ &-\cos 3\Omega t)+\frac{A}{24}[PC_1+(N+Q)C_2+(M+R)C_3+MC_4](\cos\Omega t-\\ &-\cos 5\Omega t)+\frac{A}{48}[QC_1+(P+R)C_2+NC_3+MC_4](\cos\Omega t-\cos 7\Omega t)+\\ &+\frac{A}{80}[RC_1+QC_2+PC_3+NC_4](\cos\Omega t-\cos 9\Omega t)+\frac{A}{120}(RC_1+\\ &+QC_3+PC_4)(\cos\Omega t-\cos 11\Omega t)+\frac{A}{168}(RC_3+QC_4)(\cos\Omega t-\\ &-\cos 13\Omega t)+\frac{A}{224}RC_4(\cos\Omega t-\cos 15\Omega t)\end{aligned} \tag{4.763}$$

4.21.1 Numerical Examples

We will illustrate the applicability, accuracy and effectiveness of the proposed approach by comparing the analytical approximate periodic solution given by Eq. 4.763 with numerical integration results. These comparisons are presented in the Figs. 4.161, 4.162, 4.163, 4.164 for several cases, where the optimal convergence-control parameters were determined from the residual of the initial equation

$$R(t,C_i)=\ddot{\bar{u}}(t,C_i)+\bar{u}(t,C_i)+a\bar{u}^3(t,C_i)+b\sin\bar{u}(t,C_i) \tag{4.764}$$

by means of a collocation approach:

$$R(t_1,C_i)=R(t_2,C_i)=R(t_3,C_i)=R(t_4,C_i)=0,\quad t_i\in(0,2\pi/\Omega),\quad i=1,2,3,4$$

where $\bar{u}(t,C_i)$ is given by Eq. 4.763.

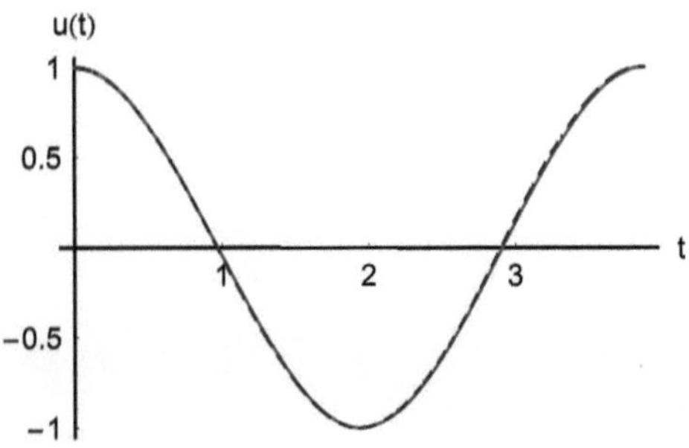

Fig. 4.161 Comparison between the approximate solution 4.763 (*blue dashed line*) and numerical solution (*red solid line*) for a = b = A = 1 (colour figure online)

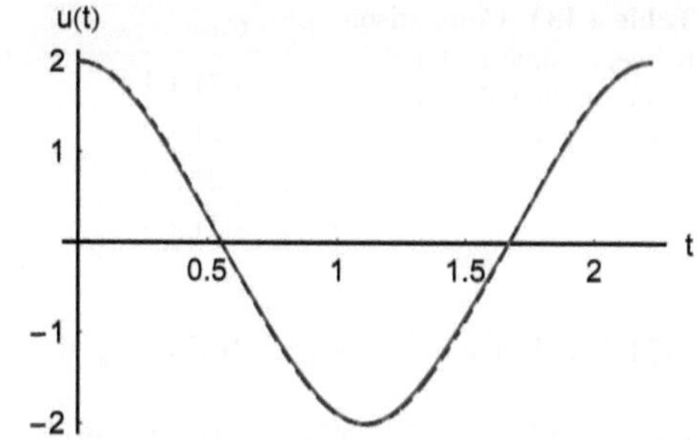

Fig. 4.162 Comparison between the approximate solution 4.763 (*blue dashed line*) and numerical solution (*red solid line*) for a = b = A = 2 (colour figure online)

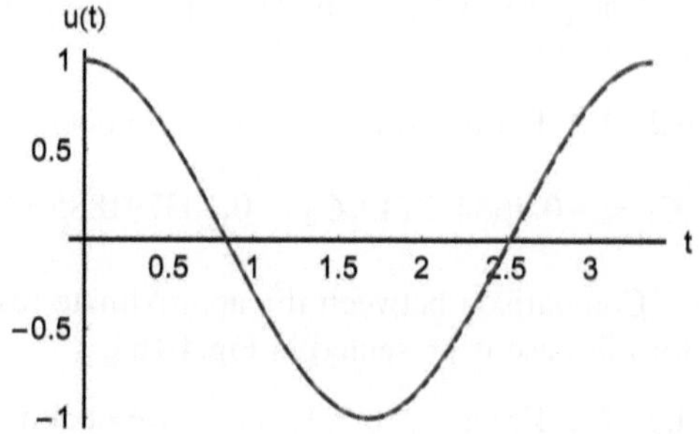

Fig. 4.163 Comparison between the approximate solution 4.763 (*blue dashed line*) and numerical solution (*red solid line*) for a = 1, b = 2, A = 1 (colour figure online)

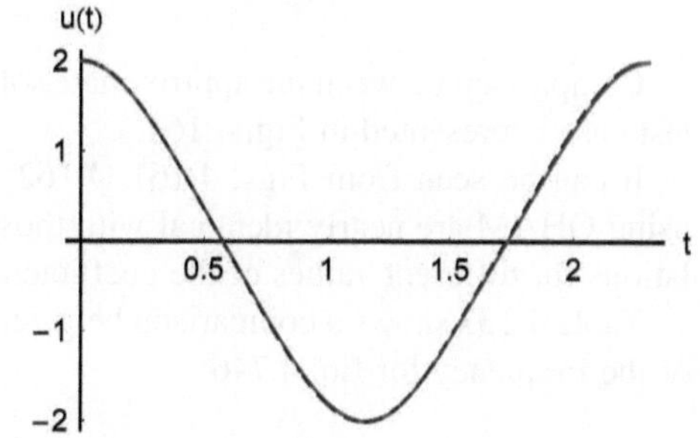

Fig. 4.164 Comparison between the approximate solution 4.763 (*blue dashed line*) and numerical solution (*red solid line*) for a = 2, b = 1, A = 2 (colour figure online)

4.21.1.1 For a = b = A = 1. Following the procedure described above, we obtain the optimal values of the unknown parameters:

$$C_1 = -0.723489144, C_2 = -0.241343924, C_3 = 0.18451862, C_4 = -0.103830855$$

Taking into account these optimal convergence-control parameters, in Fig. 4.161 is presented a comparison between the first-order approximate periodic solution and the solution obtained through numerical simulations.

Table 4.133 Comparison between numerical and approximate values of frequency for Eq. 4.746

Case	Ω_{num}	Ω_{aprox}
4.21.1.1	1.6198	1.62557
4.21.1.2	2.83047	2.82742
4.21.1.3	1.87271	1.86617
4.21.1.4	2.72034	2.72906

4.21.1.2 For a = b = A = 2, following the same procedure we obtain

$$C_1 = -0.738994741, C_2 = 0.032145602, C_3 = 0.031369091, C_4 = 0.1042003005$$

Comparison between the approximate and numerical results is presented in Fig. 4.162.

4.21.1.3 For a = 1, b = 2, A = 1, we obtain

$$C_1 = -0.968473749, C_2 = 0.103791885, C_3 = 0.034718048, C_4 = 0.049397703$$

Comparison between the approximate results and numerical simulation results for this case is presented in Fig.4.163.

4.21.1.4 For a = 2, b = 1, A = 2, we obtain

$$C_1 = -0.986445726, C_2 = -0.0124462601, C_3 = 0.080849813, C_4 = 0.035542424$$

Comparison between the approximate solution and numerical simulations for the last case is presented in Fig. 4.164.

It can be seen from Figs. 4.161, 4.162, 4.163, 4.164 that the results obtained using OHAM are nearly identical with those determined through numerical simulations for different values of the coefficients and amplitudes.

Table 4.133 shows a comparison between the numerical and approximate values of the frequency for Eq. 4.746.

4.22 Duffing Harmonic Oscillator

In [210–212] are presented the conservative nonlinear oscillatory systems which can often be modeled by potentials having a rational form. Such models also lead to differential equations for which the usual expansion in a small parameter, perturbation procedures do not apply [212]. An example is the Duffing-harmonic oscillator [212].

$$\frac{d^2x}{dt^2} + \frac{ax^3}{b + cx^2} = 0 \tag{4.765}$$

where x is the displacement, t is the time, a,b and c are non-negative parameters.

Defining

$$y = \sqrt{\frac{b}{c}}x, \quad \bar{t} = t\sqrt{\frac{c}{a}} \tag{4.766}$$

and dropping the bar on $\bar{t}$, gives the following non-dimensional equation

$$\ddot{y} + \frac{y^3}{1 + y^2} = 0 \tag{4.767}$$

where dot denotes derivative with respect to time t. For small and large y, respectively, the Eq. 4.767 can be written as

$$\ddot{y} + y^3 \approx 0 \text{ for small y} \tag{4.768}$$

$$\ddot{y} + y \approx 0 \text{ for large y} \tag{4.769}$$

We remark that for small y, the equation of motion 4.768 is that of a Duffing-type nonlinear oscillator, while for large y, the equation of motion approximates that of a linear harmonic oscillator. Hence, Eq. 4.767 is referred to as the Duffing-harmonic oscillator [212]. The restoring force in Eq. 4.767 is the same for both negative and positive amplitudes. Mickens [212] applied two non-standard finite difference schemes to numerically integrate the equation of motion 4.767 with the initial conditions

$$y(0) = A, \quad \dot{y}(0) = 0 \tag{4.770}$$

Tiwari et al [213] presented an approximate frequency-amplitude relation close to the exact one assuming a single-term solution and following the Ritz procedure. Also, they applied a rational harmonic balance approximation to the Eq. 4.767. Hu and Tang [214] used the first-order harmonic balance method via first Fourier coefficient to construct an approximate frequency-amplitude relation for a Duffing-harmonic oscillator. Herisanu and Marinca [210] proposed an optimization procedure, namely optimal variational iteration method for analytically solving the Eq. 4.767. This method provides a good approximation for both frequency and periodic solution. Ozis and Yildirim [215] applying the energy balance method obtained the frequency in the form

$$\omega^2 = 1 - \frac{2}{A^2}\ln\frac{2(1 + A^2)}{2 + A^2} \tag{4.771}$$

4.22.1 *Solutions of Duffing-Harmonic Oscillator Using OHAM*

By means of a new independent variable and a new function

$$\tau = \Omega t, \quad y(t) = Au(\tau) \tag{4.772}$$

the equation 4.767 can be written as

$$u'' + A^2u^2u'' + \frac{A^2}{\Omega^2}u^3 = 0 \tag{4.773}$$

with the initial conditions

$$u(0) = 1, \quad u'(0) = 0 \tag{4.774}$$

where prime denotes the derivative with respect to the variable τ.

The linear and nonlinear operator corresponding to Eq. 4.773 are respectively

$$L[u(\tau)] = u'' + u \tag{4.775}$$

$$N[u(\tau), \Omega] = -u + A^2u^2u'' + \frac{A^2}{\Omega^2}u^3 \tag{4.776}$$

The initial approximation is obtained from Eq. 4.4:

$$u_0'' + u_0 = 0, \quad u_0(0) = 1, \quad u_0'(0) = 0 \tag{4.777}$$

leading to the solution

$$u_0(\tau) = \cos\tau \tag{4.778}$$

Introducing Eq. 4.778 into Eq. 4.776, it follows that

$$N[u_0(\tau), \Omega] = \left[\frac{3}{4}A^2\left(\frac{1}{\Omega^2} - 1\right) - 1\right]\cos\tau + \frac{1}{4}A^2\left(\frac{1}{\Omega^2} - 1\right)\cos 3\tau \tag{4.779}$$

Choosing the auxiliary function such that

$$H(\tau, C_i) = C_1 + 2C_2\cos 2\tau + 2C_3\cos 4\tau + 2C_4\cos 6\tau \tag{4.780}$$

the equation in the first approximation 4.9 can be written as

$$u_1'' + u_1 = \left[\left(\frac{3}{4}A^2\left(\frac{1}{\Omega^2}-1\right)-1\right)(C_1+C_2)+\frac{1}{4}A^2\left(\frac{1}{\Omega^2}-1\right)(C_2+ +C_3)\right]\cos\tau+\left[\left(\frac{3}{4}A^2\left(\frac{1}{\Omega^2}-1\right)-1\right)(C_2+C_3)+\frac{1}{4}A^2\left(\frac{1}{\Omega^2}-1\right)(C_1+ +C_4)\right]\cos 3\tau+\left[\left(\frac{3}{4}A^2\left(\frac{1}{\Omega^2}-1\right)-1\right)(C_3+C_4)+ +\frac{1}{4}A^2\left(\frac{1}{\Omega^2}-1\right)C_2\right]\cos 5\tau+\left[\left(\frac{3}{4}A^2\left(\frac{1}{\Omega^2}-1\right)-1\right)C_4+ \frac{1}{4}A^2\left(\frac{1}{\Omega^2}-1\right)C_3\right]\cos 7\tau+\frac{1}{4}A^2\left(\frac{1}{\Omega^2}-1\right)C_4\cos 9\tau \tag{4.781}$$

No secular terms in Eq. 4.781 requires that

$$\Omega^2 = 1-\frac{4(C_1+C_2)}{4(C_1+C_2)+A^2(3C_1+4C_2+C_3)} \tag{4.782}$$

Taking into account Eqs. 4.778, 4.5, 4.772 and the solution $u_1(\tau)$ obtained from Eq. 4.781, we obtain the first-order approximate periodic solution of Eqs. 4.767 and 4.770:

$$\bar{y}(t, C_i) = A\cos\Omega t+\frac{A}{8}\left[\left(\frac{3}{4}A^2\left(\frac{1}{\Omega^2}-1\right)-1\right)(C_2+C_3)+ +\frac{1}{4}A^2\left(\frac{1}{\Omega^2}-1\right)(C_1+C_4)\right](\cos\Omega t-\cos 3\Omega t)+ +\frac{A}{24}\left[\left(\frac{3}{4}A^2\left(\frac{1}{\Omega^2}-1\right)-1\right)(C_3+C_4)+\frac{1}{4}A^2\left(\frac{1}{\Omega^2}-1\right)C_2\right](\cos\Omega t- -\cos 5\Omega t)+\frac{A}{48}\left[\left(\frac{3}{4}A^2\left(\frac{1}{\Omega^2}-1\right)-1\right)C_4+\frac{1}{4}A^2\left(\frac{1}{\Omega^2}-1\right)C_3\right](\cos\Omega t- -\cos 7\Omega t)+\frac{A^3}{320}\left(\frac{1}{\Omega^2}-1\right)C_4(\cos\Omega t-\cos 9\Omega t) \tag{4.783}$$

4.22.2 Numerical Examples

Applying the collocation approach, using the residual of the equation obtained from the initial Eq. 4.767

$$R(t, C_i) = \ddot{\bar{y}}+\bar{y}^2\ddot{\bar{y}}+\bar{y}^3$$

we present the following three cases:

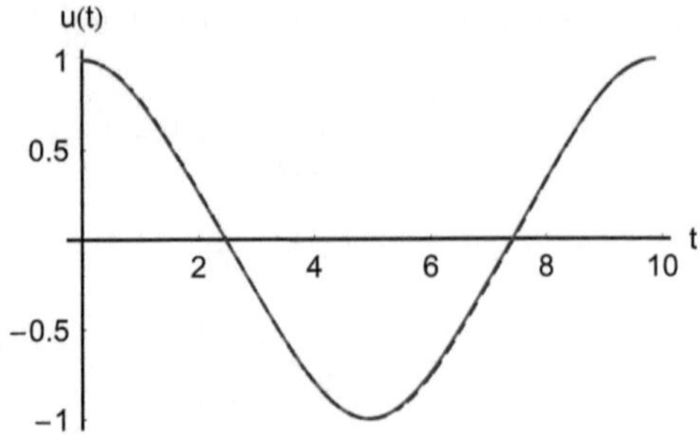

Fig. 4.165 The behavior of the present solution 4.784 and the numerical integration results of Eq. 4.767 for A = 1: numerical (*red solid line*), approximate (*blue dashed line*) (colour figure online)

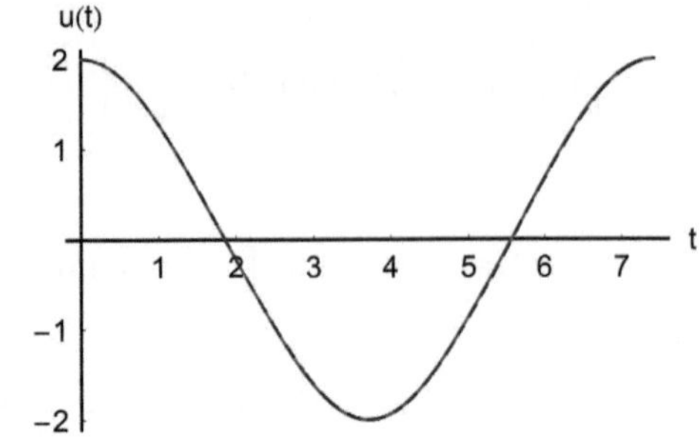

Fig. 4.166 The behavior of the present solution 4.785 and the numerical integration results of Eq. 4.767 for A = 2: numerical (*red solid line*), approximate (*blue dashed line*) (colour figure online)

4.22.2.1 For A = 1 one can get

$$C_1 = -0.654863918, C_2 = 0.128960075, C_3 = 0.0256826306, C_4 = 0.008951265$$
$$\bar{y}(t) = 0.974664888\cos\Omega t + 0.027737026\cos 3\Omega t - 0.002142566\cos 5\Omega t -$$
$$-0.000217997\cos 7\Omega t - 0.000041349\cos 9\Omega t \tag{4.784}$$

In Fig. 4.165 we compare the present solution 4.784 and numerical integration results.

4.22.2.2 For A = 2, the optimal values of the convergence-control parameters and the approximate frequency are:

$$C_1 = -0.369597738, C_2 = 2.809150613 \cdot 10^{-4}, C_3 = 0.17536749,$$
$$C_4 = -0.09212599$$

and the first-order approximate periodic solution given by Eq. 4.783 becomes:

$$\bar{y}(t) = 1.965122188\cos\Omega t + 0.037452984\cos 3\Omega t - 0.001316184\cos 5\Omega t -$$
$$-0.002171344\cos 7\Omega t + 0.000912356\cos 9\Omega t \tag{4.785}$$

In Fig. 4.166 we compare the approximate solution 4.785 with numerical results in case A = 2.

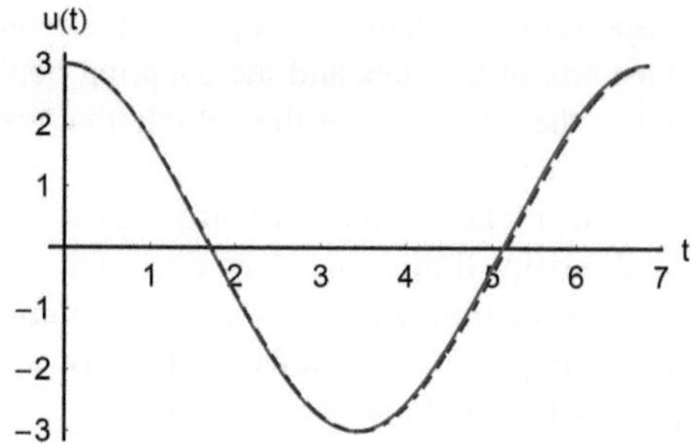

Fig. 4.167 The behavior of the present solution 4.786 and the numerical integration results of Eq. 4.767 for A = 3: numerical (*red solid line*), approximate (*blue dashed line*) (colour figure online)

Table 4.134 Comparison between the exact and approximate frequencies for different values of A

A	Ω_{ex}	Ω_{OHAM}
1	0.63678	0.635228
2	0.847626	0.846324
3	0.9196	0.912075

4.22.2.3 In the last case we consider A = 3. We have:

$$\begin{aligned}&C_1 = -0.230838188, C_2 = 2.316705959 \cdot 10^{-3}, C_3 = 0.180688596,\\&C_4 = -0.101983117\\&\bar{y}(t) = 2.973027451 \cos \Omega t + 0.031761841 \cos 3\Omega t -\\&-0.003714223 \cos 5\Omega t - 0.002814068 \cos 7\Omega t + 0.001738999 \cos 9\Omega t\end{aligned} \tag{4.786}$$

In Fig. 4.167 the present solution 4.786 is compared with the numerical integration results.

For the Duffing-harmonic oscillator, the exact frequency is

$$\Omega_{ex}(A) = \frac{\pi}{2}\left(\int_0^{\pi/2} A \cos \tau \left[A^2 \cos^2\tau + \ln\left(1 - A^2 \cos^2\tau \left(1 + A^2\right)^{-1}\right]^{-\frac{1}{2}} d\tau\right) \tag{4.787}$$

Table 4.134 presents a comparison between the exact and approximate values of the frequencies obtained through the proposed procedure.

4.23 The Oscillator with Linear and Cubic Elastic Restoring Force and Quadratic Damping

The oscillator with linear and cubic elastic restoring force and quadratic damping used as a test example in this section approximately describes those real systems for which the damping of the oscillations is produced by a turbulent liquid flow inside the damper. A typical example of an oscillator with quadratic damping is the suspension of a vehicle equipped with hydraulic shock absorbers. By neglecting the elasticity of the tires and the coupling between the vibrations of the front and rear axles, the vibrations of the vehicle that are symmetrical with respect to the longitudinal axis may be studied on the single-degree-of-freedom. Suspensions with nonlinear elastic and damping characteristics are frequently utilized, because nonlinearity limits displacements and velocities, reduces the extreme values of the acceleration, and leads to a more uniform dynamic loading of the suspension. The damping nonlinearity is usually achieved by using hysteretic or hydro-pneumatic shock absorbers [216].

We consider an oscillator with linear and cubic elastic restoring force and quadratic damping of the form

$$\ddot{u} + u + \alpha\dot{u}^2 + \beta u^3 = 0, \quad u(0) = A, \quad u'(0) = 0 \tag{4.788}$$

where dot denotes derivative with respect to time, and α, β, A are known constants, $A > 0$.

Making the transformations

$$\tau = \Omega t, \quad u(t) = Ax(\tau) \tag{4.789}$$

where Ω is the frequency of the system, Eq. 4.788 can be written as

$$x'' + \Omega^{-2}x + \alpha A x'^2 + \beta A^2\Omega^{-2}x^3 = 0 \tag{4.790}$$

with the initial conditions

$$x(0) = 1, \quad x'(0) = 0 \tag{4.791}$$

where prime denotes derivative with respect to the new variable τ.

In accordance with Eq. 4.1, the linear and nonlinear operators are given by

$$L[x(\tau)] = x'' + x \tag{4.792}$$

$$N[x(\tau), \Omega] = \left(\Omega^{-2} - 1\right)x + \alpha A x'^2 + \beta A^2\Omega^{-2}x^3 \tag{4.793}$$

and $g(\tau) = 0$.

The initial approximation x_0 will be determined from the equation

$$x_0'' + x_0 = 0, \quad x_0(0) = 1, \quad x_0'(0) = 0 \tag{4.794}$$

whose solution is

$$x_0(\tau) = \cos\tau \tag{4.795}$$

The nonlinear operator 4.793 for the initial approximation given by Eq. 4.795 can be expressed as

$$N[x_0(\tau),\Omega]=\left(\frac{3\beta A^2+4}{4\Omega^2}-1\right)\cos\tau-\frac{1}{2}\alpha A\cos 2\tau+\frac{\beta A^2}{4\Omega^2}\cos 3\tau+\frac{1}{2}\alpha A \quad (4.796)$$

If we choose the auxiliary function as

$$H(\tau,C_i)=C_1+2C_2\cos\tau+2C_3\cos 2\tau+2C_4\cos 3\tau+2C_5\cos 4\tau \quad (4.797)$$

the equation 4.9 becomes

$$x_1''+x_1=\left[\left(\frac{3\beta A^2+4}{4\Omega^2}-1\right)C_1+\frac{1}{2}\alpha AC_2+\left(\frac{\beta A^2+1}{4\Omega^2}-1\right)C_3-\frac{1}{2}\alpha AC_4+\right.$$
$$\left.+\frac{\beta A^2}{4\Omega^2}C_5\right]\cos\tau+B_0-3B_1\cos 2\tau-8B_2\cos 3\tau-15B_3\cos 4\tau-$$
$$-24B_4\cos 5\tau-35B_5\cos 6\tau-48B_6\cos 7\tau,\qquad x_1(0)=x_1'(0)=0$$
$$(4.798)$$

where

$$B_0=\frac{1}{2}\alpha AC_1+\left(\frac{3\beta A^2+4}{4\Omega^2}-1\right)C_2-\frac{1}{2}\alpha AC_3+\frac{\beta A^2}{4\Omega^2}C_4$$
$$B_1=\frac{1}{6}\alpha AC_1+\left(\frac{1}{3}-\frac{\beta A^2+1}{3\Omega^2}\right)C_2-\frac{1}{3}\alpha AC_3+\left(\frac{1}{3}-\frac{3\beta A^2+4}{12\Omega^2}\right)C_4+$$
$$+\frac{1}{6}\alpha AC_5$$
$$B_2=-\frac{\beta A^2}{32\Omega^2}C_1+\frac{1}{16}\alpha AC_2+\left(\frac{1}{8}-\frac{3\beta A^2+4}{32\Omega^2}\right)C_3-\frac{1}{8}\alpha AC_4+$$
$$+\left(\frac{1}{8}-\frac{3\beta A^2+4}{32\Omega^2}\right)C_5$$
$$B_3=-\frac{\beta A^2}{60\Omega^2}C_2+\frac{1}{30}\alpha AC_3+\left(\frac{1}{15}-\frac{3\beta A^2+4}{60\Omega^2}\right)C_4-\frac{1}{15}\alpha AC_5$$
$$B_4=-\frac{\beta A^2}{96\Omega^2}C_3+\frac{1}{48}\alpha AC_4+\left(\frac{1}{24}-\frac{3\beta A^2+4}{96\Omega^2}\right)C_5 \quad (4.799)$$
$$B_5=-\frac{\beta A^2}{140\Omega^2}C_4+\frac{1}{70}\alpha AC_5;\, B_6=-\frac{\beta A^2}{192\Omega^2}C_5$$

The condition required to eliminate the secular term in Eq. 4.798 is

$$\Omega^2=\frac{(3\beta A^2+4)C_1+4(\beta A^2+1)C_3+\beta A^2C_5}{2(2C_1-\alpha AC_2+2C_3+\alpha AC_4)} \quad (4.800)$$

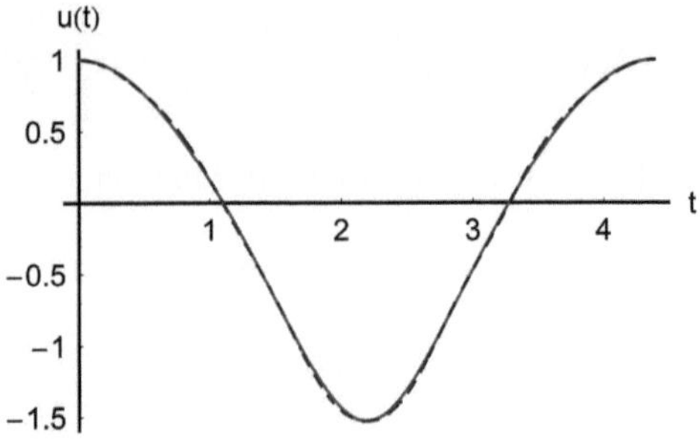

Fig. 4.168 Comparison between the approximate solution 4.802 and numerical solution of Eqs. 4.788 and 4.789 for $\alpha = 1/3$, $\beta = 1$, $A = 1$: numerical (*red solid line*), approximate (*blue dashed line*) (colour figure online)

The first-order approximate solution of Eq. 4.788 can be obtained solving Eq. 4.798 and by means of Eqs. 4.795, 4.5 and 4.789:

$$\begin{aligned}\bar{u}(t) &= AB_0 + A(1 - B_0 - B_1 - B_2 - B_3 - B_4 - B_5 - B_6)\cos\Omega t + \\ &+ AB_1 \cos 2\Omega t + AB_2 \cos 3\Omega t + AB_3 \cos 4\Omega t + AB_4 \cos 5\Omega t + \\ &+ AB_5 \cos 6\Omega t + AB_6 \cos 7\Omega t\end{aligned} \tag{4.801}$$

4.23.1 Numerical Examples

In order to show the validity and accuracy of the OHAM, we consider the following cases:

4.23.1.1 If $\alpha = 1/3$, $\beta = 1$, $A = 1$, the convergence-control parameters and the frequency Ω are determined by a collocation approach as:

$$\begin{aligned}&C_1 = -1.11297600564616;\ C_2 = -2.331751818465062;\\ &C_3 = 1.708450419812395;\ C_4 = 0.04529558689726963;\\ &C_5 = 2.281140631488463;\ \Omega = 1.4341036\end{aligned}$$

The first-order approximate solution given by Eq. 4.801 will be

$$\begin{aligned}\bar{u}(t) &= -0.117061 + 1.22195\cos\Omega t - 0.144089\cos 2\Omega t + 0.0408032\cos 3\Omega t - \\ &-0.012363\cos 4\Omega t + 0.00583336\cos 5\Omega t + \\ &+0.0107053\cos 6\Omega t - 0.00577683\cos 7\Omega t\end{aligned} \tag{4.802}$$

The approximate solution 4.802 and the numerical solution of Eq. 4.788 are presented in Fig. 4.168.

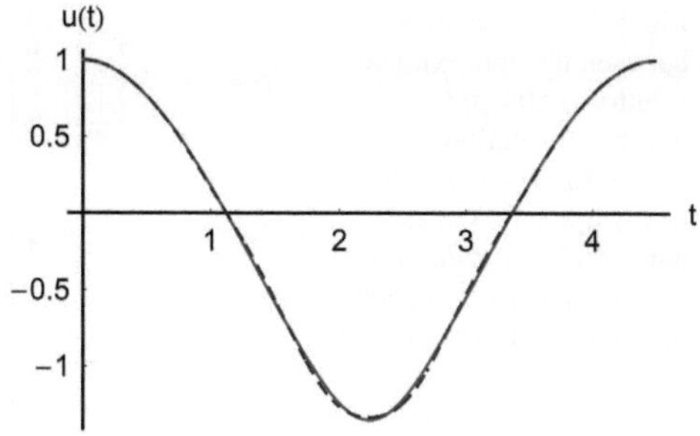

Fig. 4.169 Comparison between the approximate solution 4.803 and numerical solution of Eqs. 4.788 and 4.789 for $\alpha=1/4$, $\beta=1$, $A=1$: numerical (*red solid line*), approximate (*blue dashed line*) (colour figure online)

4.23.1.2 In this case $\alpha=1/4$, $\beta=1$ and $A=1$ and we have

$$\begin{aligned}&C_1=1.3830976579366836;\ C_2=-1.4348624216700008;\\&C_3=1.56785715145014;\ C_4=-1.6885127635352009;\\&C_5=0.710985521468802;\ \Omega=1.4014862\end{aligned}$$

The first-order approximate solution becomes

$$\begin{aligned}\overline{u}(t)&=-0.081559+1.14417\cos\Omega t-0.0960442\cos 2\Omega t+0.0394004\cos 3\Omega t+\\&+0.00111721\cos 4\Omega t-0.0138791\cos 5\Omega t+\\&+0.00867966\cos 6\Omega t-0.00188531\cos 7\Omega t\end{aligned} \tag{4.803}$$

In Fig. 4.169 we present this approximate solution in comparison with corresponding numerical solution.

4.23.1.3 If $\alpha=1/4$, $\beta=1$ and $A=2$ we obtain

$$\begin{aligned}&C_1=0.3376263819490409;\ C_2=-1.1378967342745472;\\&C_3=1.1278896877749445;\ C_4=-1.4228624298867956;\\&C_5=0.8638879365788908;\ \Omega=2.373378\end{aligned}$$

The first-order approximate solution becomes

$$\begin{aligned}\overline{u}(t)&=-0.240599+2.59356\cos\Omega t-0.53593\cos 2\Omega t+0.236103\cos 3\Omega t-\\&-0.0480581\cos 4\Omega t-0.0254596\cos 5\Omega t+\\&+0.0267754\cos 6\Omega t-0.00639017\cos 7\Omega t\end{aligned} \tag{4.804}$$

In Fig. 4.170 we present the approximate solution given by Eq. 4.804 and the numerical solution.

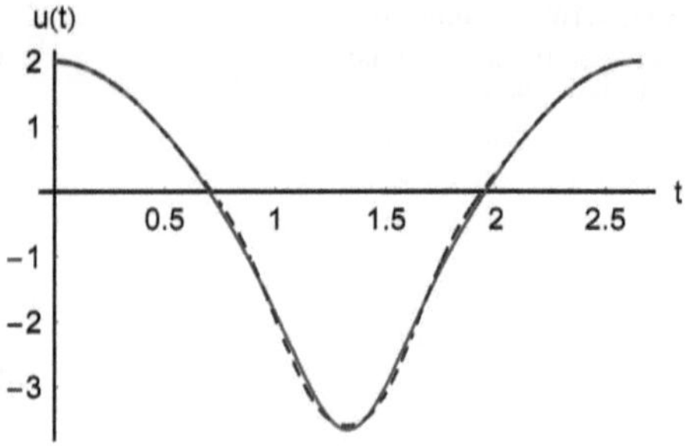

Fig. 4.170 Comparison between the approximate solution 4.804 and numerical solution of Eqs. 4.788 and 4.789 for $\alpha = 1/4$, $\beta = 1$, $A = 2$: numerical (*red solid line*), approximate (*blue dashed line*) (colour figure online)

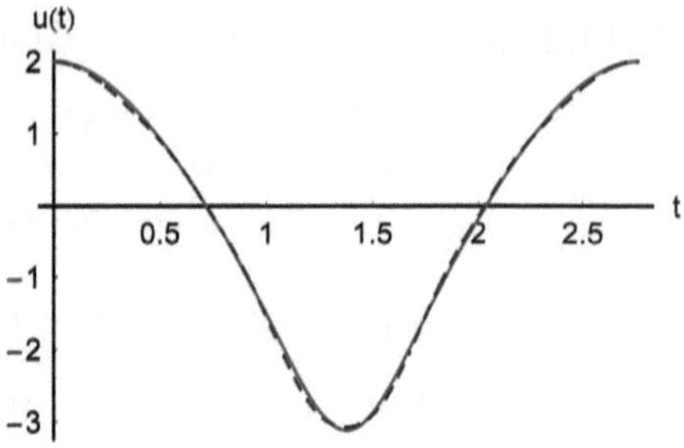

Fig. 4.171 Comparison between the approximate solution 4.805 and numerical solution of Eqs. 4.788 and 4.789 for $\alpha = 1/5$, $\beta = 1$, $A = 2$: numerical (*red solid line*), approximate (*blue dashed line*) (colour figure online)

4.23.1.4 In the last case, we consider $\alpha = 1/5$, $\beta = 1$ and $A = 2$, such that

$$\begin{aligned} &C_1 = -0.031354847858872485;\ C_2 = -1.4172470375817974;\\ &C_3 = 1.935582555818684;\ C_4 = -0.17494255572266407;\\ &C_5 = 1.6944002658680275;\ \Omega = 2.2857339 \end{aligned}$$

$$\begin{aligned} \overline{u}(t) &= -0.189372 + 2.39188\cos\Omega t - 0.362366\cos 2\Omega t +\\ &+0.160839\cos 3\Omega t - 0.00805104\cos 4\Omega t - 0.000693107\cos 5\Omega t +\\ &+0.021278\cos 6\Omega t - 0.0135131\cos 7\Omega t \end{aligned} \tag{4.805}$$

In Fig. 4.171 are shown the approximate solution and corresponding numerical one.

It is clear that the approximate results match very well with the numerical ones.

4.24 Generalized Duffing Equation

In 1918, Georg Wilhelm Christian Caspar Duffing (1861–1944), a German engineer published the monograph titled "Forced oscillations with variable natural frequency and their technical significance" [217]. This is the work for which he is famous. The book comprises seven chapters and five Appendices and contains results on the response of both linear and nonlinear oscillatory systems obtained analytically, graphically, numerically and experimentally.

The systems with nonlinear restoring force, a method of successive approximations are subsequently named after Duffing. Duffing was the author of nine publications including books, book chapters and journal articles. Although most physical systems cannot be described accurately by so called "Duffing equation", in many cases it is the first step in moving from a linear to a nonlinear system. We emphasize several systems which can be represented by various forms of the Duffing equation: the pendulum (where the series of the trigonometric function sinus is truncated to third order), geometrical nonlinearities (for example systems consisting of two linear springs and a mass), large deflection of a pinned-pinned beam with nonlinear stiffness, nonlinear isolators (used in engineering systems to isolate vibrating source from their surroundings), nonlinear cable vibrations, beam with nonlinear stiffness due to inplane tension, nonlinear electrical circuits. Besides those considered above, there are plates, shells, woofers, optical fibers, micromechanical structures, nanomechanical resonators, rotors, prisms in fluid flow, ultrasonic cutting systems. For more details see [218].

Researches recognize the Duffing equation as a model for different systems and many journal papers related to Duffing equation have been published. A survey has been carried out via SCOPUS to trace the journal papers that used the word "Duffing" in the title, abstract or keywords. Until the 30 March 2010, the number of such published papers was at least 2700 [218]. It is remarkable the diversity of disciplines in which Duffing's equation appears: 40% belong to engineering, 25% to physics and astronomy, 19% to mathematics, 6% to computer science and so on. The Duffing equation(s) formed the basis of many mathematical models of different systems.

There exists a wide body of literature dealing with the problem of approximate solutions of Duffing equation(s) with various methodologies. Many different approaches have been proposed such as: modified Lindstedt-Poincare method [219], variational iteration method [208], perturbation procedure with a linear transformation of time [220], modified iteration perturbation method [221], homotopy analysis method [222], the method of energy balance [223], and so on [224–227].

We consider the following generalized Duffing equation

$$\ddot{u} + \alpha u + \beta u^{2n+1} = 0 \tag{4.806}$$

where α and β are known parameters and n is an integer positive number. For $n = 1$ it is obtained the classical Duffing oscillator. The initial conditions to Eq. 4.806 are known:

$$u(0) = A, \quad u'(0) = 0$$

By means of the transformations

$$\tau = \Omega t, \quad u(t) = Ax(\tau) \tag{4.807}$$

where Ω is unknown frequency of the system, one can get

$$x'' + \frac{\alpha}{\Omega^2}x + \frac{\beta A^{2n}}{\Omega^2}x^{2n+1} = 0 \tag{4.808}$$

and

$$x(0) = 1, \quad x'(0) = 0 \tag{4.809}$$

The prime denotes derivative with respect to τ. The linear and nonlinear operators, corresponding to Eq. 4.808 are, respectively

$$L[x(\tau)] = x'' + x \tag{4.810}$$

$$N[x(\tau), \Omega] = \left(\frac{\alpha}{\Omega^2} - 1\right)x + \frac{\beta A^{2n}}{\Omega^2}x^{2n+1} \tag{4.811}$$

The initial approximation is determined from the equation

$$x_0'' + x_0 = 0, \quad x_0(0) = 1, \quad x_0'(0) = 0 \tag{4.812}$$

and therefore

$$x_0(\tau) = \cos\tau \tag{4.813}$$

From Eqs. 4.811 and 4.813 and making use of the identity

$$\cos^{2n+1}\tau = \frac{1}{4^n}\sum_{k=0}^{n}\binom{2n+1}{n-k}\cos(2k+1)\tau \tag{4.814}$$

where $\binom{n}{k} = \frac{n!}{k!(n-k)!}, 0! = 1, k! = 1 \cdot 2 \cdot \ldots k$, we get

$$N[x_0(\tau), \Omega] = \left[\frac{\alpha}{\Omega^2} - 1 + \frac{\beta}{\Omega^2}\left(\frac{A}{2}\right)^{2n}\right]\cos\tau + \\ + \frac{\beta}{\Omega^2}\left(\frac{A}{2}\right)^{2n}\sum_{k=1}^{n}\binom{2n+1}{n-k}\cos(2k+1)\tau \tag{4.815}$$

Taking into account the expression of the last equation, we choose the auxiliary function as

$$H(\tau, C_i) = C_1 + 2C_2\cos 2\tau + 2C_3\cos 4\tau \tag{4.816}$$

The first approximation is given by the linear equation

$$x_1'' + x_1 = H(\tau, C_i)N[x_0(\tau)], x_1(0) = 0, x_1'(0) = 0 \tag{4.817}$$

which can be rewritten as

$$\begin{aligned} x_1'' + x_1 = &\left[\left(\frac{\alpha}{\Omega^2} - 1 + \frac{\beta}{\Omega^2}\left(\frac{A}{2}\right)^{2n}\binom{2n+1}{n}\right)(C_1 + C_2) + \right. \\ &+ \frac{\beta}{\Omega^2}\left(\frac{A}{2}\right)^{2n}\binom{2n+1}{n-1}(C_2 + C_3) + \\ &\left. + \frac{\beta}{\Omega^2}\left(\frac{A}{2}\right)^{2n}\binom{2n+1}{n-1}\delta(n-2)C_3\right]\cos\tau + \sum_{k\geq 1} M_{2k+1}\cos(2k+1)\tau \end{aligned} \tag{4.818}$$

where

$$\begin{aligned} M_3 = &\left[\frac{\alpha}{\Omega^2} - 1 + \frac{\beta}{\Omega^2}\left(\frac{A}{2}\right)^{2n}\binom{2n+1}{n}\right](C_2 + C_3) + \frac{\beta}{\Omega^2}\left(\frac{A}{2}\right)^{2n}\left[\binom{2n+1}{n-1}C_1 + \right. \\ &\left. + \binom{2n+1}{n-2}C_2\delta(n-2) + \binom{2n+1}{n-3}C_3\delta(n-3)\right] \\ M_5 = &\left[\frac{\alpha}{\Omega^2} - 1 + \frac{\beta}{\Omega^2}\left(\frac{A}{2}\right)^{2n}\binom{2n+1}{n}\right]C_3 + \frac{\beta}{\Omega^2}\left(\frac{A}{2}\right)^{2n}\left[\binom{2n+1}{n-1}C_2 + \right. \\ &\left. + \binom{2n+1}{n-2}C_1\delta(n-2) + \binom{2n+1}{n-3}C_2\delta(n-3) + \binom{2n+1}{n-4}C_3\delta(n-4)\right] \\ M_7 = &\frac{\beta}{\Omega^2}\left(\frac{A}{2}\right)^{2n}\left[\binom{2n+1}{n-2}\delta(n-2)C_2 + \binom{2n+1}{n-3}C_1\delta(n-3) + \right. \\ &\left. + \binom{2n+1}{n-4}C_2\delta(n-4) + \binom{2n+1}{n-5}C_3\delta(n-5) + \binom{2n+1}{n-1}C_3\right] \\ M_9 = &\frac{\beta}{\Omega^2}\left(\frac{A}{2}\right)^{2n}\left[\binom{2n+1}{n-2}\delta(n-2)C_3 + \binom{2n+1}{n-3}C_2\delta(n-3) + \right. \\ &\left. + \binom{2n+1}{n-4}C_1\delta(n-4) + \binom{2n+1}{n-5}C_2\delta(n-5) + \binom{2n+1}{n-6}C_3\delta(n-6)\right] \\ M_{2n+1} = &\frac{\beta A^2}{\Omega^2}\left(\frac{A}{2}\right)^{2n}\left[\sum_{k=5}^{n}\binom{2n+1}{n-k}\delta(n-5)C_1 + \sum_{k=5}^{n+1}\binom{2n+1}{n+1-k}C_2\delta(n-4) + \right. \\ &\left. + \sum_{k=5}^{n-1}\binom{2n+1}{n-1-k}C_2\delta(n-6) + \sum_{k=5}^{n+2}\binom{2n+1}{n+2-k}C_3\delta(n-5) + \sum_{k=5}^{n-2}\binom{2n+1}{n-2-k}C_3\delta(n-7)\right] \\ &\text{if} \quad n \geq 5 \end{aligned} \tag{4.819}$$

and

$$\delta(k) = \begin{cases} 1 & k \geq 0 \\ 0 & k < 0 \end{cases}$$

Avoiding secular term in Eq. 4.818 leads to the equation for the frequency:

$$\begin{aligned} \Omega^2 &= \alpha + \beta\left(\tfrac{A}{2}\right)^{2n}\binom{2n+1}{n} + \beta\left(\tfrac{A}{2}\right)^{2n}\binom{2n+1}{n-1}\frac{C_2 + C_3}{C_1 + C_2} + \\ &+\beta\left(\tfrac{A}{2}\right)^{2n}\binom{2n+1}{n-2}\frac{C_3}{C_1 + C_2}\delta(n-2) \end{aligned} \tag{4.820}$$

The solution of Eq. 4.818 becomes

$$\begin{aligned} x_1(\tau) &= \frac{M_3}{8}(\cos\tau - \cos 3\tau) + \frac{M_5}{24}(\cos\tau - \cos 5\tau) + \\ &+\frac{M_7}{48}(\cos\tau - \cos 7\tau) + \sum_{k\geq 4}\frac{M_{2k+1}}{4k(k+1)}(\cos\tau - \cos(2k+1)\tau) \end{aligned} \tag{4.821}$$

To obtain the first-order approximate periodic solution 4.5, we use Eqs. 4.813, 4.821 and 4.807:

$$\begin{aligned} \overline{u}(t) &= A\cos\Omega t + \frac{AM_3}{8}(\cos\Omega t - \cos 3\Omega t) + \frac{AM_5}{24}(\cos\Omega t - \cos 5\Omega t) + \\ &+\frac{AM_7}{48}(\cos\Omega t - \cos 7\Omega t) + \sum_{k\geq 4}\frac{AM_{2k+1}}{4k(k+1)}(\cos\Omega t - \cos(2k+1)\Omega t) \end{aligned} \tag{4.822}$$

4.24.1 Numerical Examples

We consider $\alpha = A = 1$ and three cases for the parameter n.

4.24.1.1 In the first case, for $n = 1$, we have the subcases:

a) If $\beta = 1$, then one can get

$$\begin{aligned} &C_1 = -1.232242239, C_2 = 0.570023572, C_3 = -0.531705696 \\ &\overline{u}(t) = 0.979493163\cos\Omega t + 0.022147809\cos 3\Omega t - \\ &-0.0032366206\cos 5\Omega t + 0.001595647\cos 7\Omega t \end{aligned} \tag{4.823}$$

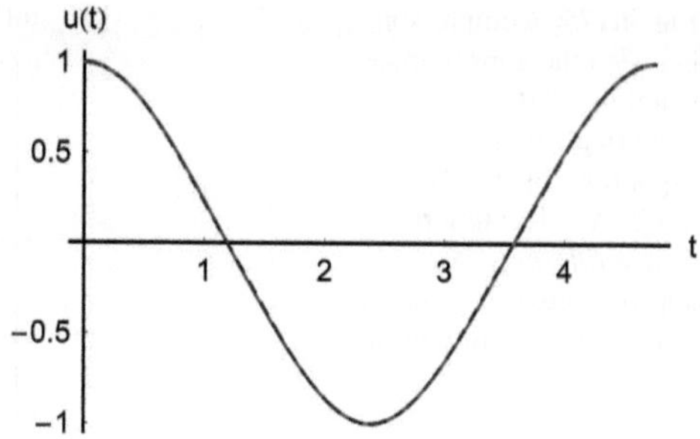

Fig. 4.172 Comparison between the approximate solution 4.823 and numerical solution of Eq. 4.806 for $\alpha=\beta=A=n=1$: numerical (*red solid line*), approximate (*blue dashed line*) (colour figure online)

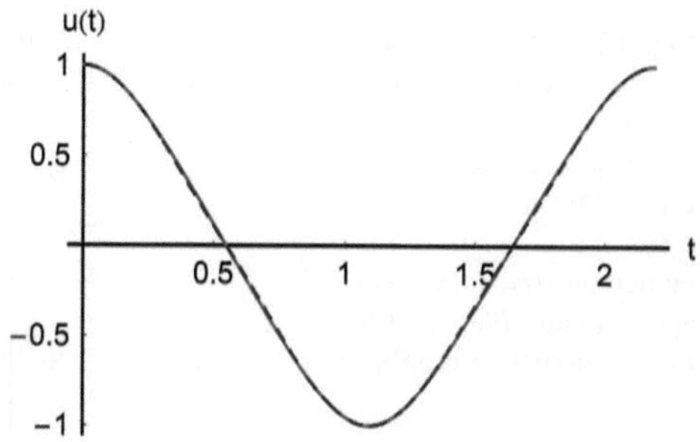

Fig. 4.173 Comparison between the approximate solution 4.824 and numerical solution of Eq. 4.806 for $\alpha=A=n=1$, $\beta=10$: numerical (*red solid line*), approximate (*blue dashed line*) (colour figure online)

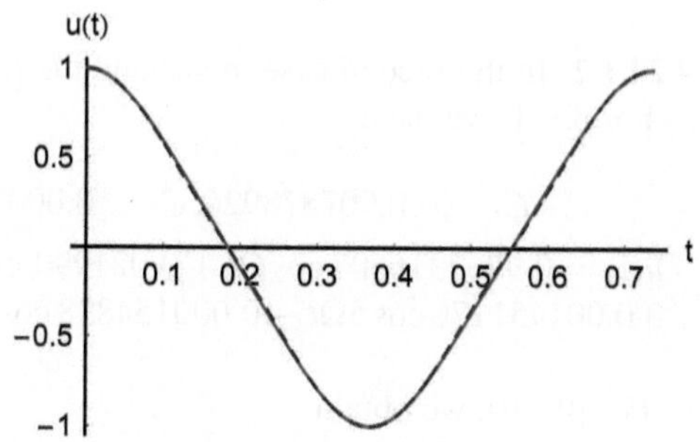

Fig. 4.174 Comparison between the approximate solution 4.825 and numerical solution of Eq. 4.806 for $\alpha=A=n=1$, $\beta=100$: numerical (*red solid line*), approximate (*blue dashed line*) (colour figure online)

b) If $\beta=10$, we obtain

$$C_1 = -1.209920877, C_2 = 0.310633723, C_3 = -0.208710491$$

$$\begin{aligned}\overline{u}(t) &= 0.956738231 \cos \Omega t + 0.0455769809 \cos 3\Omega t- \\ &-0.003638176 \cos 5\Omega t + 0.001322963 \cos 7\Omega t\end{aligned} \tag{4.824}$$

c) For $\beta=100$, it holds that

$$C_1 = -1.240228205, C_2 = 0.297203489, C_3 = -0.176931632$$

$$\begin{aligned}\overline{u}(t) &= 0.950092408 \cos \Omega t + 0.052571033 \cos 3\Omega t- \\ &-0.003929063 \cos 5\Omega t + 0.001265622 \cos 7\Omega t\end{aligned} \tag{4.825}$$

In Figs. 4.172, 4.173, 4.174 the present solutions 4.823-4.825 are compared with numerical integration results.

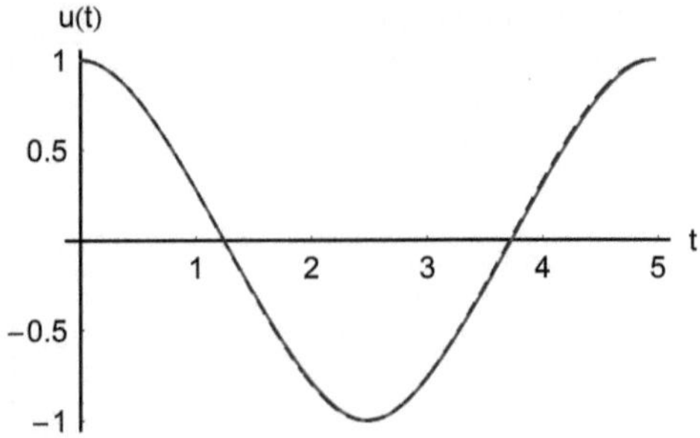

Fig. 4.175 Comparison between the approximate solution 4.826 and numerical solution of Eq. 4.806 for α = β = 1, n = 2, A = 1,: numerical (*red solid line*), approximate (*blue dashed line*) (colour figure online)

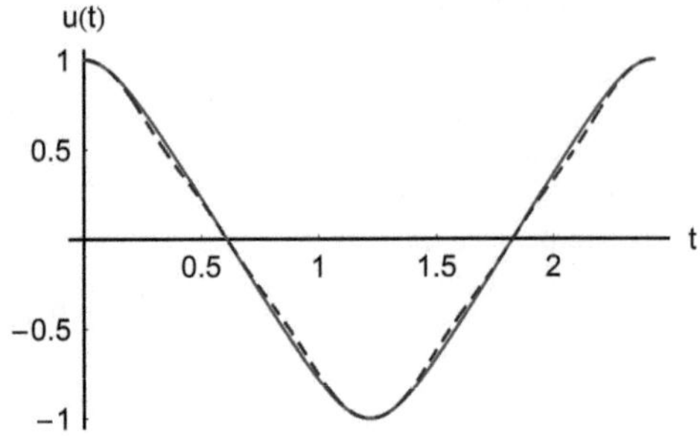

Fig. 4.176 Comparison between the approximate solution 4.827 and numerical solution of Eq. 4.806 for n = 2, α = A = 1, β = 10: numerical (*red solid line*), approximate (*blue dashed line*) (colour figure online)

4.24.1.2 In the second case, n = 2 and the following subcases are considered:
a) For β = 1, we have

$$C_1 = -0.907878926, C_2 = 0.000297002, C_3 = 0.038210401$$

$$\begin{aligned}\bar{u}(t) &= 0.9767316607\cos\Omega t + 0.021990553\cos 3\Omega t + \\ &+ 0.001451176\cos 5\Omega t - 0.000154838\cos 7\Omega t - 0.000018551\cos 9\Omega t\end{aligned} \tag{4.826}$$

b) For β = 10, we obtain

$$C_1 = -1.290297341, C_2 = -0.583086598, C_3 = 0.776894838$$

$$\begin{aligned}\bar{u}(t) &= 0.913449443\cos\Omega t + 0.080309455\cos 3\Omega t + \\ &+ 0.013598605\cos 5\Omega t - 0.006447198\cos 7\Omega t - 0.000910307\cos 9\Omega t\end{aligned} \tag{4.827}$$

c) For β = 100, one can get

$$C_1 = -0.954847192, C_2 = -0.349233289, C_3 = 0.520580985$$

$$\begin{aligned}\bar{u}(t) &= 0.928163629\cos\Omega t + 0.067862835\cos 3\Omega t + \\ &+ 0.009845645\cos 5\Omega t - 0.005157329\cos 7\Omega t - 0.000714782\cos 9\Omega t\end{aligned} \tag{4.828}$$

Figures 4.175, 4.176, 4.177 show a comparison between the approximate solution 4.826-4.828 and numerical solutions.

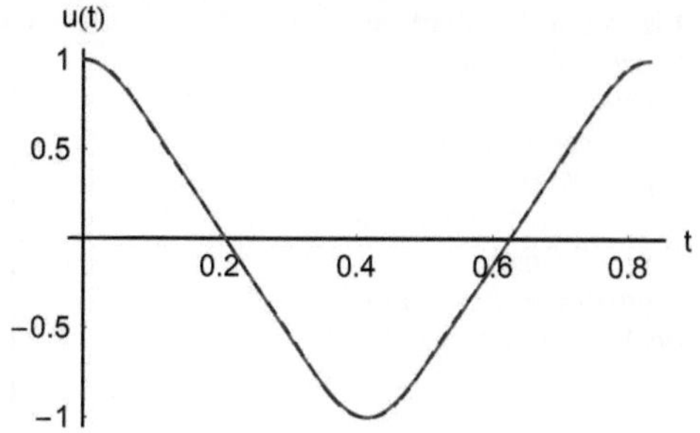

Fig. 4.177 Comparison between the approximate solution 4.828 and numerical solution of Eq. 4.806 for n = 2, α = A = 1, β = 100: numerical (*red solid line*), approximate (*blue dashed line*) (colour figure online)

4.24.1.3 In the last case, n = 3 and also 3 subcases are considered:
a) For β = 1 we obtain

$$\begin{aligned}&C_1 = -0.844024656, C_2 = 0.008318142, C_3 = 0.0302490908\\ &\overline{u}(t) = 0.9750888108\cos\Omega t + 0.022485865\cos 3\Omega t + 0.00242395\cos 5\Omega t +\\ &+0.000032081\cos 7\Omega t - 0.000028132\cos 9\Omega t - 2.578054\cdot 10^{-6}\cos 11\Omega t\end{aligned} \tag{4.829}$$

b) For β = 10 one can get

$$\begin{aligned}&C_1 = -0.833068275, C_2 = -0.6108809, C_3 = 0.710613181\\ &\overline{u}(t) = 0.918789456\cos\Omega t + 0.070442304\cos 3\Omega t + 0.01802527\cos 5\Omega t -\\ &-0.005600684\cos 7\Omega t - 0.001494129\cos 9\Omega t - 0.000162219\cos 11\Omega t\end{aligned} \tag{4.830}$$

c) For β = 100 we have

$$\begin{aligned}&C_1 = -0.716815935, C_2 = -0.738974792, C_3 = 0.852541076\\ &\overline{u}(t) = 0.909702039\cos\Omega t + 0.078262096\cos 3\Omega t + 0.02282931\cos 5\Omega t -\\ &-0.008370081\cos 7\Omega t - 0.002185779\cos 9\Omega t - 0.000237589\cos 11\Omega t\end{aligned} \tag{4.831}$$

In Figs. 4.178, 4.179, 4.180 the present solutions 4.829-4.831 are compared with the numerical integration results. From Figs. 4.172, 4.173, 4.174, 4.175, 4.176, 4.177, 4.178, 4.179, 4.180 one can see that the errors between the approximate analytical solutions and numerical integration results are very good.

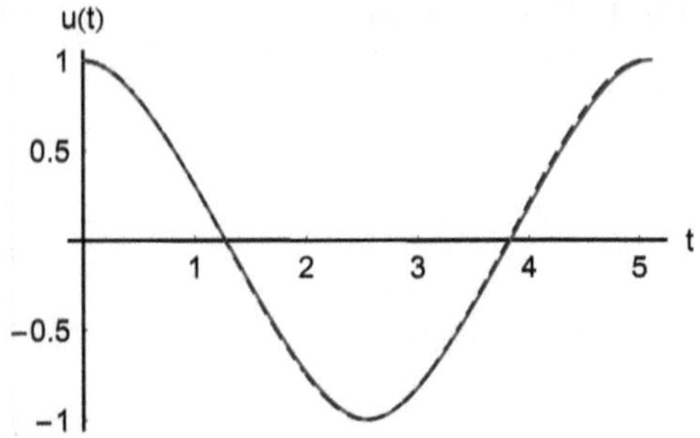

Fig. 4.178 Comparison between the approximate solution 4.829 and numerical solution of Eq. 4.806 for n = 3, α = β = A = 1: numerical (*red solid line*), approximate (*blue dashed line*) (colour figure online)

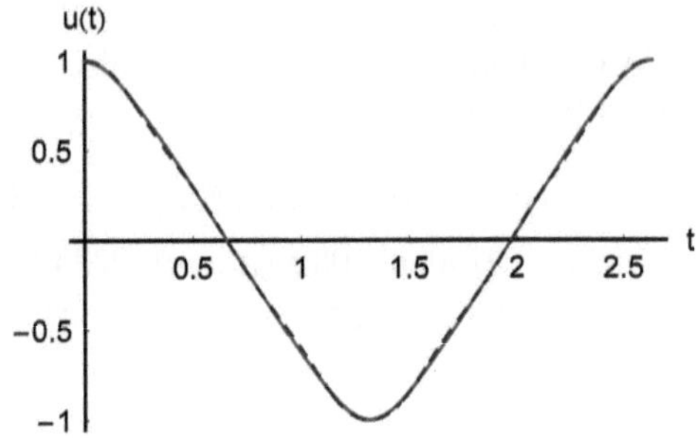

Fig. 4.179 Comparison between the approximate solution 4.830 and numerical solution of Eq. 4.806 for n = 3, α = A = 1, β = 10: numerical (*red solid line*), approximate (*blue dashed line*) (colour figure online)

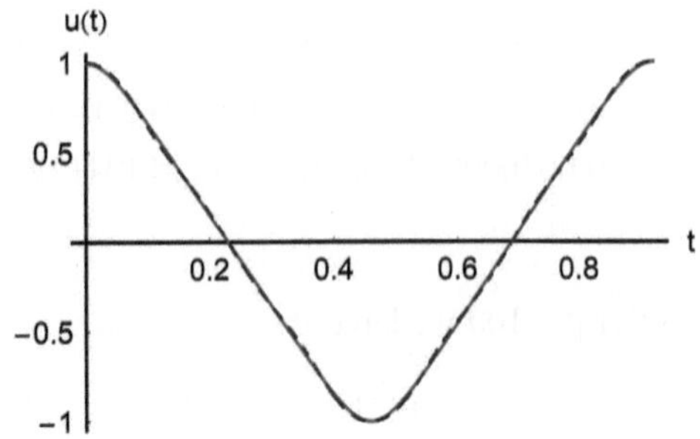

Fig. 4.180 Comparison between the approximate solution 4.831 and numerical solution of Eq. 4.806 for n = 3, α = A = 1, β = 100: numerical (*red solid line*), approximate (*blue dashed line*) (colour figure online)

4.25 Generalized Van der Pol Equation

The Van der Pol oscillator was originally proposed by the Dutch electrical engineer and physicist Balthasar Van der Pol (1889–1959) while he was working at Philips. At that time radio and vacuum tube technology were developed. Diode, triode or tetras were involved into electrical circuits forming an oscillatory system. Van der Pol considered that oscillations of the electrical circuit and published the experimental results in 1926 [228]. The mathematical model of an electrical ciruit with input–output relation of the tunnelled diode as a cubic function can be written in the so-called "the standard Van der Pol equation"

$$\ddot{u} + \omega^2 u + \varepsilon(u^2 - 1)\dot{u} = 0 \tag{4.832}$$

where ω and ε > 0 are known parameters. The mathematical model 4.832 corresponds also to a significant number of systems in biology, chemistry, economics, mechanics and so on. In dynamical systems, the Van der Pol oscillator is a nonconservative system with non-negative damping. It is worthy to mention some interesting regimes:

- For ε = 0, i.e. there is no damping function and Eq. 4.832 becomes the equation of a simple harmonic oscillator
- If u(0) and u are small, the system described by Eq. 4.832 becomes a linear differential equation with negative damping.
- If u is large, the quadrate term is dominant and the damping becomes positive. Fore more details see [229]

Van der Pol and Van der Mark [230] were the first to model the electric activity of the heart and of the human heartbeat based on the relaxing oscillations.

The generalized Van der Pol Eq. 4.832 has been considered by many researchers. Thus, Lienard [231] made the widest generalization of the Van der Pol equation in the form

$$\ddot{u} + f(u)\dot{u} + g(u) = 0 \tag{4.833}$$

where f and g are continuously differentiable functions with g an odd function and f an even function. The other generalization given by Lienard is the mixed Rayleigh-Lienard oscillator

$$\ddot{u} + f(u, \dot{u})\dot{u} + g(u) = 0 \tag{4.834}$$

The Eq. 4.833 is taken as the typical example of nonlinear self-excited vibration. Of course, different expressions for the functions f and g can be considered [232–235].

In this section we will consider nonlinear oscillations governed by a generalized Van der Pol equation of the form

$$\ddot{u} + u + \varepsilon(u^{2n} - 1)\dot{u} = 0 \tag{4.835}$$

where n is any positive integer. For n = 1, Eq. 4.835 reduces to the standard Van der Pol equation. This type of equation has been studied extensively by many authors using various methods. For instance Davies [165] obtained the steady-state solutions for various values of ε using numerical methods. Another kind of methods as described by Nayfeh [204] are perturbation methods including the method of multiple time scales [236] and the Lindstedt-Poincare method. Burton [237] presented a time transformation method for limit cycle analysis. For small ε to moderate ε (e.g. ε = 2) the results obtained by this method are more accurate than those obtained by classical perturbation approaches. Minorsky [238] concluded that for $n > 1$ into Eq. 4.835, the stationary amplitude decreases from 2 and when $0 < n < 1$ it increases from 2. Homotopy analysis method is applied by Liao [239] and Chen and Liu [240]. Herisanu and Marinca [241] determined periodic solution of Van der Pol equation for $\varepsilon = 0.6$ and $\varepsilon = 0.8$ using an iteration procedure.

Several other generalizations of the Van der Pol equation have been considered in the literature. Nguyen [242] has considered a generalized Van der Pol equation of the form

$$\ddot{u} + \omega^2 u = \varepsilon\left[1 + (u + q\cos\nu t)^2\right]\dot{u} \tag{4.836}$$

where ω, q and ν are constants independent of ε. Nguyen considered the special cases $\nu = 3\omega$ and $\nu = \omega$ and found that there are stable stationary oscillations with amplitudes greater and less than2. Obi [243] has analysed mathematically a generalized Van der Pol equation of the form

$$\ddot{u} + \omega^2 u = \varepsilon\left(1 - u^{2n+2}\right)\dot{u} + \varepsilon a\cos\omega t \tag{4.837}$$

where n is any positive integer, and a and ω are constants independent of ε.

Moremedi et al [244] used the derivative expansion procedure in the method of multiple scales to derive a uniformly valid perturbation solution of Eq. 4.835 which is valid for all values of the positive integer n. Kovacic and Mickens [245] considered a generalized Van der Pol type oscillator with power-form nonlinearities in the restoring and damping force. Amplitude of the limit cycle is obtained by the generalized Krylov-Bogoliubov method.

4.25.1 Solutions of the Van der Pol Equation 4.835 by Means of OHAM

The initial conditions corresponding to Eq. 4.835 are

$$u(0) = A, \quad u'(0) = 0 \tag{4.838}$$

but A is an unknown parameter.

Using the transformations

$$\tau = \Omega t, \quad u(t) = Ax(\tau) \tag{4.839}$$

the equation 4.835 can be written as follows

$$x'' + \Omega^{-2}x - \varepsilon\Omega^{-1}\left(1 - A^{2n}x^{2n}\right)x' = 0 \tag{4.840}$$

with the initial conditions

$$x(0) = 1, \quad x'(0) = 0 \tag{4.841}$$

where prime denotes derivative with respect to τ.

For the differential equation 4.840 we choose $g(\tau) = 0$ and the following operators:

$$L[x(\tau)] = x'' + x$$
$$N[x(\tau), \Omega] = (\Omega^{-2} - 1)x - \varepsilon\Omega^{-1}(1 - A^{2n}x^{2n})x' \tag{4.842}$$

Making use of the identity

$$\sin\tau\cos^{2n}\tau = \frac{1}{4^n}\sum_{k=0}^{n}\binom{2n}{n-k}\frac{2k+1}{n+k+1}\sin(2k+1)\tau \tag{4.843}$$

where $\binom{n}{0} = 1$, $\binom{n}{k} = \frac{n!}{k!(n-k)!}$, $k! = 1 \cdot 2 \cdot \ldots \cdot k$

The nonlinear operator N calculated for the initial approximation $x_0(\tau) = \cos\tau$ which is given by

$$x_0'' + x_0 = 0, \quad x_0 = 1, \quad x_0'(0) = 0, \tag{4.844}$$

becomes

$$N[x_0(\tau), \Omega] = \left(\frac{1}{\Omega^2} - 1\right)\cos\tau + \frac{\varepsilon}{\Omega}\sin\tau - \\ -\frac{\varepsilon}{\Omega}\left(\frac{A}{2}\right)^{2n}\sum_{k=1}^{n}\binom{2n}{n-k}\frac{2k+1}{n+k+1}\sin(2k+1)\tau \tag{4.845}$$

Having in view the expression of $N[x_0(\tau), \Omega]$ from the above equation, we can choose the auxiliary function as

$$H(\tau, C_i) = C_1 + 2C_2\cos 2\tau + 2C_3\sin 2\tau \tag{4.846}$$

The first approximation $x_1(\tau)$ is determined from Eq. 4.9 which is written in the form

$$x_1'' + x_1 = \left[\left(\frac{1}{\Omega^2} - 1\right)(C_1 + C_2) + \frac{\varepsilon C_3}{\Omega}\left(1 - \left(\frac{A}{2}\right)^{2n}\binom{2n}{n}\frac{1}{n+1} - \right.\right. \\ \left.\left. - \left(\frac{A}{2}\right)^{2n}\binom{2n}{n-1}\frac{3}{n+2}\right)\right]\cos\tau + \left[\left(\frac{1}{\Omega^2} - 1\right)C_3 + \right. \\ \left. + \frac{\varepsilon}{\Omega}\left(1 - \left(\frac{A}{2}\right)^{2n}\binom{2n}{n}\frac{C_1 - C_2}{n+1} - \left(\frac{A}{2}\right)^{2n}\binom{2n}{n-2}\frac{3C_2}{n+2}\right)\right]\sin\tau + H.O.T \tag{4.847}$$

By eliminating the secular terms in Eq. 4.847, we obtain

$$A = 2\sqrt[2n]{\frac{n(n+2)\left(C_1^2 - C_2^2 - C_3^2\right)}{\binom{2n}{n-1}\left[n\left(C_1^2 + 2C_2^2 - 4C_3^2 + 3C_1C_2\right) + 2\left(C_1^2 - C_2^2 - C_3^2\right)\right]}} \tag{4.848}$$

$$\begin{aligned}\Omega = &-\frac{3\varepsilon C_3}{2(C_1+C_2)}\frac{n\left(C_1^2 - C_1C_2 - 2C_2^2\right)}{n\left(C_1^2 + 2C_2^2 - 4C_3^2 + 3C_1C_2\right) + 2\left(C_1^2 - C_2^2 - C_3^2\right)} + \\ &+\sqrt{1 + \frac{9\varepsilon^2C_3^2}{4(C_1+C_2)}\frac{n^2\left(C_1^2 - C_1C_2 - 2C_2^2\right)^2}{\left[n\left(C_1^2 + 2C_2^2 - 4C_3^2 + 3C_1C_2\right) + 2\left(C_1^2 - C_2^2 - C_3^2\right)\right]}}\end{aligned} \tag{4.849}$$

The equation 4.847 can be rewritten as

$$\begin{aligned}x_1'' + x_1 = &\left[\left(\frac{1}{\Omega^2} - 1\right)C_2 - \frac{\varepsilon C_3}{\Omega}\left(1 - \left(\frac{A}{2}\right)^{2n}\left(\binom{2n}{n}\frac{1}{n+1} -\right.\right.\right. \\ &\left.\left.\left.-\binom{2n}{n-2}\frac{5}{n+3}\delta(n-2)\right)\right)\right]\cos 3\tau + \left[\left(\frac{1}{\Omega^2} - 1\right)C_3 + \right. \\ &+\frac{\varepsilon C_2}{\Omega}\left(1 - \left(\frac{A}{2}\right)^{2n}\left(\binom{2n}{n}\frac{1}{n+1} + \binom{2n}{n-2}\frac{5}{n+3}\delta(n-2)\right)\right) - \\ &\left.-\frac{\varepsilon C_3}{\Omega}\left(\frac{A}{2}\right)^{2n}\binom{2n}{n-1}\frac{3}{n+2}\right]\sin 3\tau + \frac{\varepsilon C_3}{\Omega}\left(\frac{A}{2}\right)^{2n}\left[\sum_{k=1}^{n}\binom{2n}{n-k}\frac{2k+1}{n+k+1} - \right. \\ &\left.-\sum_{k=1}^{n-2}\binom{2n}{n-k-2}\frac{2k+5}{n+k+3}\delta(n-2)\right]\cos(2k+3)\tau - \\ &-\frac{\varepsilon}{\Omega}\left(\frac{A}{2}\right)^{2n}\left[\sum_{k=1}^{n-1}\binom{2n}{n-k-1}\frac{2k+3}{n+k+2}\delta(n-2)C_1 + \sum_{k=1}^{n}\binom{2n}{n-k}\frac{2k+1}{n+k+1}C_2 + \right. \\ &\left.+\sum_{k=1}^{n-2}\binom{2n}{n-k-2}\frac{2k+5}{n+k+3}\delta(n-3)C_2\right]\sin(2k+3)\tau, \\ x_1(0) = &\, x_1'(0) = 0\end{aligned} \tag{4.850}$$

where $\delta(k) = \begin{cases} 1 & if \quad k \geq 0 \\ 0 & if \quad k < 0 \end{cases}$.

Now, solving Eq. 4.850 and taking into consideration Eqs. 4.5 and 4.839 we obtain the first-order approximate periodic solution of Eq. 4.836 after some manipulations:

$$\begin{aligned}
\overline{u}(t) &= A\cos\Omega t + A\left[\frac{1}{8}\left(1-\frac{1}{\Omega^2}\right)C_2+\right.\\
&+\frac{\varepsilon C_3}{8\Omega}\left(1+2\left(\frac{A}{2}\right)^{2n}\frac{n-3}{n^2+5n+6}\binom{2n+1}{n}\delta(n-3)\right](\cos 3\Omega t-\cos\Omega t)+\\
&+\left[\frac{1}{8}\left(1-\frac{1}{\Omega^2}\right)C_3-\frac{\varepsilon C_2}{8\Omega}\left(1-6\left(\frac{A}{2}\right)^{2n}\frac{n^2+1}{n^3+6n^2+11n+6}\binom{2n}{n}\right)+\right.\\
&\left.+\frac{\varepsilon C_1}{8\Omega}\left(\frac{A}{2}\right)^{2n}\frac{3}{n+2}\binom{2n}{n-1}\right)\Bigg](\sin 3\Omega t-3\sin\Omega t)+\\
&+\frac{\varepsilon C_3}{4\Omega}\left(\frac{A}{2}\right)^{2n}\left[\sum_{k=1}^{n-2}\frac{2k+5}{(k^2+3k+2)(n+k+3)}\binom{2n}{n-k-2}\delta(n-3)-\right.\\
&\left.-\sum_{k=1}^{n}\frac{2k+1}{(k^2+3k+2)(n+k+1)}\binom{2n}{n-k}\right](\cos(2k+3)\Omega t-\cos\Omega t)+ \\
&+\frac{\varepsilon}{4\Omega}\left(\frac{A}{2}\right)^{2n}\left[\sum_{k=1}^{n-1}\binom{2n}{n-k-1}\frac{2k+3}{(k^2+3k+2)(n+k+2)}\delta(n-2)C_1+\right.\\
&+\sum_{k=1}^{n}\binom{2n}{n-k}\frac{2k+1}{(k^2+3k+2)(n+k+1)}C_2+\\
&\left.+\sum_{k=1}^{n-2}\binom{2n}{n-k-2}\frac{2k+5}{(k^2+3k+2)(n+k+3)}\delta(n-3)C_2\right](\sin(2k+\\
&+3)\Omega t-(2k+3)\sin\Omega t)
\end{aligned} \tag{4.851}$$

where A and Ω are given by Eqs. 4.848 and 4.849, respectively.

4.25.2 Numerical Examples

In what follows we determine the first-order approximate periodic solution in three cases.

4.25.2.1 In the first case, for n = 1 (standard Van der Pol oscillator) we consider two sub-cases for the damping parameter ε.

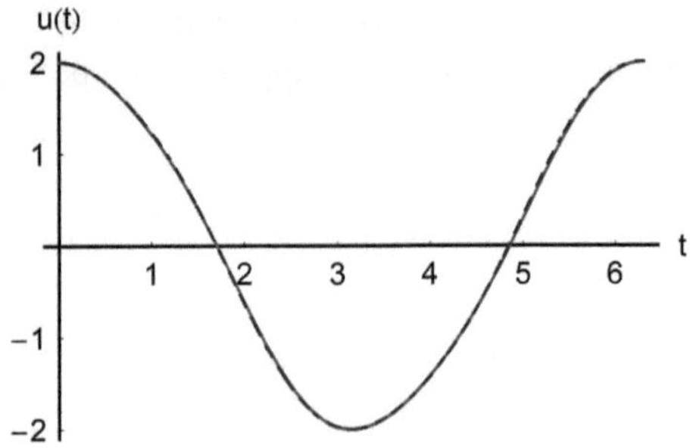

Fig. 4.181 Comparison between the approximate solution 4.852 and numerical solution for ε = 0.25 and n = 1: numerical (*red solid line*), approximate (*blue dashed line*) (colour figure online)

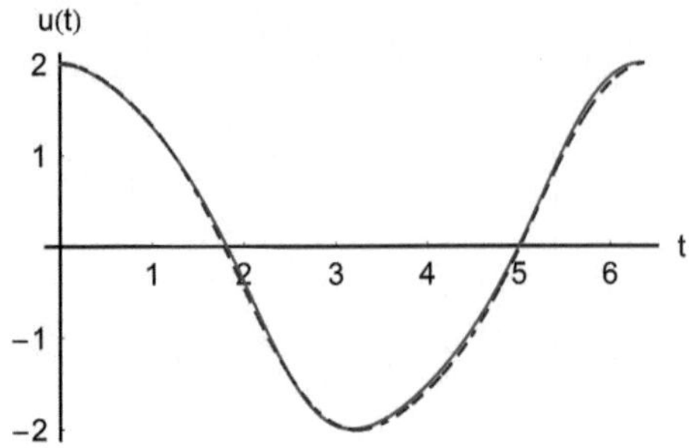

Fig. 4.182 Comparison between the approximate solution 4.853 and numerical solution for ε = 0.45 and n = 1: numerical (*red solid line*), approximate (*blue dashed line*) (colour figure online)

a) For ε = 0.25, we have

$$A = 2.000466865, C_1 = -1.022105281, C_2 = 0.000437566, C_3 = -0.006363676$$

such that the first-order approximate solution 4.851 becomes

$$\begin{aligned} u(t) &= 2.000334076 \cos \Omega t + 1.5221 \cdot 10^{-8} \cos 3\Omega t + 0.000132772 \cos 5\Omega t + \\ &+ 0.191875362 \sin \Omega t - 0.063973669 \sin 3\Omega t + 9.12946375 \cdot 10^{-6} \sin 5\Omega t \end{aligned} \tag{4.852}$$

b) For ε = 0.45 it is obtained

$$A = 2.0087240805, C_1 = -0.743880555, C_2 = 0.003802715, C_3 = -0.044324692$$

$$\begin{aligned} u(t) &= 2.006998962 \cos \Omega t + 0.00001784 \cos 3\Omega t + 0.0017072708 \cos 5\Omega t + \\ &+ 0.256198929 \sin \Omega t - 0.0856437607 \sin 3\Omega t + 0.0001464706 \sin 5\Omega t \end{aligned} \tag{4.853}$$

Figures 4.181, 4.182, 4.183, 4.184, 4.185, 4.186 illustrate that the displacements obtained by OHAM are very accurate comparing to numerical results.

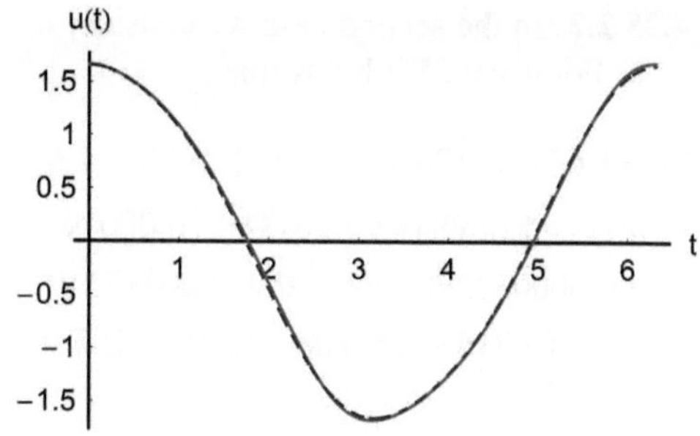

Fig. 4.183 Comparison between the approximate solution 4.854 and numerical solution for ε = 0.25 and n = 2: numerical (*red solid line*), approximate (*blue dashed line*) (colour figure online)

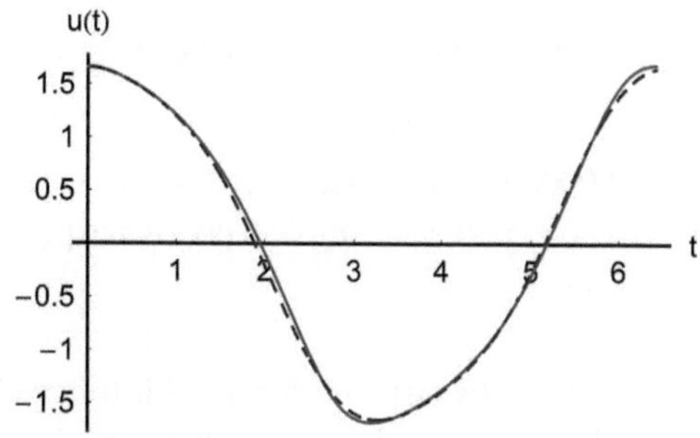

Fig. 4.184 Comparison between the approximate solution 4.855 and numerical solution for ε = 0.45 and n = 2: numerical (*red solid line*), approximate (*blue dashed line*) (colour figure online)

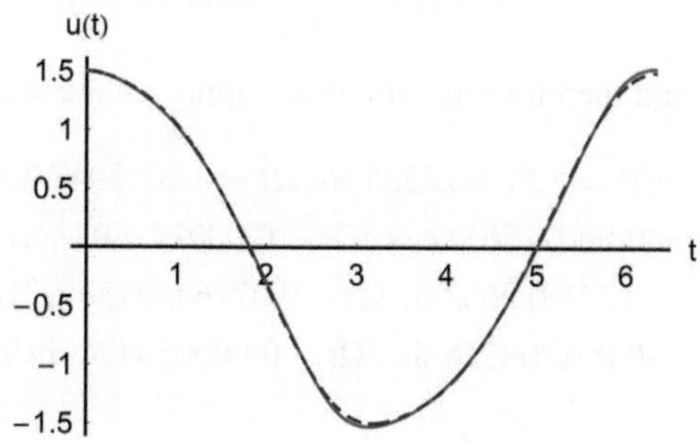

Fig. 4.185 Comparison between the approximate solution 4.856 and numerical solution for ε = 0.25 and n = 3: numerical (*red solid line*), approximate (*blue dashed line*) (colour figure online)

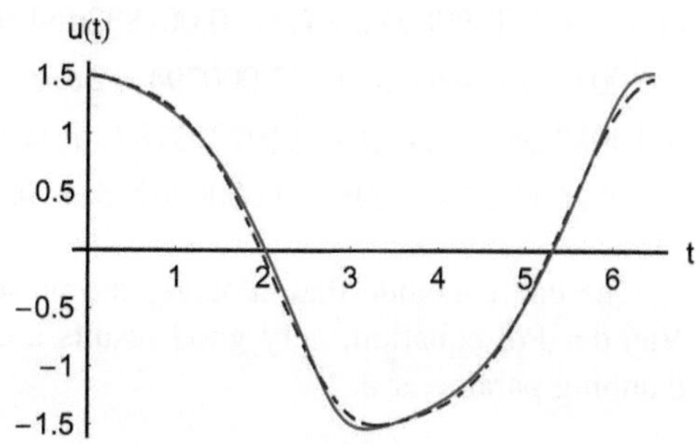

Fig. 4.186 Comparison between the approximate solution 4.857 and numerical solution for ε = 0.45 and n = 3: numerical (*red solid line*), approximate (*blue dashed line*) (colour figure online)

4.25.2.2 In the second case we consider n = 2 and two subcases
a) For ε = 0.25 it holds that

$$A = 1.659872174, C_1 = -0.215699654, C_2 = -0.007735692, C_3 = -0.007844913$$

$$\begin{aligned} u(t) &= 1.659841285 \cos \Omega t - 0.000195707 \cos 3\Omega t + 0.000194225 \cos 5\Omega t + \\ &+ 0.0000323709 \cos 7\Omega t + 0.203821191 \sin \Omega t - 0.064579863 \sin 3\Omega t - \\ &- 0.001971631 \sin 5\Omega t - 0.00003192 \sin 7\Omega t \end{aligned} \tag{4.854}$$

b) If ε = 0.45 we obtain

$$A = 1.659949448, C_1 = -0.191852066, C_2 = -0.007417731, C_3 = -0.012993328$$

$$\begin{aligned} u(t) &= 1.660047169 \cos \Omega t - 0.0005872304 \cos 3\Omega t + 0.000587411 \cos 5\Omega t - \\ &- 0.000097901 \cos 7\Omega t + 0.3311419 \sin \Omega t - 0.104872765 \sin 3\Omega t - \\ &- 0.003226473 \sin 5\Omega t - 0.000055891 \sin 7\Omega t \end{aligned} \tag{4.855}$$

4.25.2.3 In the last case for n = 3 it follows that
a) For ε = 0.25, the coefficients are

$$A = 1.500011237, C_1 = -0.920517426, C_2 = -0.060124862, C_3 = -0.032547639$$

and therefore the first-order approximate solutions becomes

$$\begin{aligned} u(t) &= 1.500422325 \cos \Omega t - 0.001394740 \cos 3\Omega t + \\ &+ 0.000728631 \cos 5\Omega t + 0.000227697 \cos 7\Omega t + 0.00002732 \cos 9\Omega t + \\ &+ 0.271919663 \sin \Omega t - 0.071685095 \sin 3\Omega t - 0.008890014 \sin 5\Omega t - \\ &- 0.001708576 \sin 7\Omega t - 0.000050474 \sin 9\Omega t \end{aligned} \tag{4.856}$$

b) For ε = 0.45 we obtain

$$A = 1.500097858, C_1 = -0.7527885409, C_2 = -0.052396118, C_3 = -0.061829219$$

$$\begin{aligned} \overline{u}(t) &= 1.501499893 \cos \Omega t - 0.004833464 \cos 3\Omega + \\ &+ 0.002541799 \cos 5\Omega t + 0.000794312 \cos 7\Omega t + 0.000095317 \cos 9\Omega t + \\ &+ 0.408359021 \sin \Omega t - 0.107373787 \sin 3\Omega t - 0.013451884 \sin 5\Omega t - \\ &- 0.002607322 \sin 7\Omega t - 0.000080775 \sin 9\Omega t \end{aligned} \tag{4.857}$$

One can conclude that adopting the present procedure for the solutions of the Van der Pol equation, very good results are obtained for moderate values of the damping parameter ε.

4.26 Oscillations of an Electrical Machine

Electrical machines are widely used in engineering applications and industry due to their reliability. They are dynamical systems encountering dynamical phenomena which can be detrimental to the system. From engineering point of view it is very important to predict the nonlinear dynamic behavior of complex dynamical systems, such as the electrical machines. This is a significant stage in the design process, before the machine is exploited in real conditions, avoiding in this way undesired dynamical phenomena which could damage the system. Basically, the electrical machines share the same dynamical problems with classical rotor systems, having specific sources of excitation, which lead to nonlinear vibration occurrence.

The main sources of dynamic problems are the unbalanced forces of the rotor [246], bad bearings or nonlinear bearings [247], mechanical looseness, misalignments, other electrical and mechanical faults which generate nonlinear vibration in the system. These problems are usually solved by numerical simulations, experimental investigations or by analytical developments [248, 249].

In what follows, the investigated electrical machine is considered to be supported by nonlinear bearings and the assumption made in development of the mathematical model is that these bearings are characterised by nonlinear stiffness of Duffing type. In the same time, the entire dynamical system is subjected to a parametric excitation caused by an axial thrust and a forcing excitation caused by an unbalanced force of the rotor, which is obviously harmonically shaped. In these conditions, the dynamical behavior of the investigated electrical machine will be governed by the following second-order strongly nonlinear differential equation:

$$m\ddot{u} + k_1(1 - q\sin\omega_2 t)u + k_2u^3 = f\sin\omega_1 t \tag{4.858}$$

with the initial conditions

$$u(0) = A, \dot{u}(0) = 0 \tag{4.859}$$

which can be written in the more convenient form:

$$\ddot{x} + \omega^2 x - \alpha x\sin\omega_2 t + \beta x^3 - \gamma\sin\omega_1 t = 0 \tag{4.860}$$

where $\omega^2 = \frac{k_1}{m}, \alpha = \frac{k_1 q}{m}, \beta = \frac{k_2}{m}, \gamma = \frac{f}{m}$, the dot denotes derivative with respect to time and A is the amplitude of the oscillations. Note that it is unnecessary to assume the existence of any small or large parameter in Eq. 4.860.

The equation 4.860 describes a system oscillating with an unknown period T. We switch to a scalar time $\tau = 2\pi t/T = \Omega t$. Under the transformations

$$\tau = \Omega t, u(t) = Ax(\tau) \tag{4.861}$$

the original Eq. 4.860 becomes

$$\Omega^2 x'' + \omega^2 x - \alpha x \sin\frac{\omega_2}{\Omega}\tau + \beta A^2 x^3 - \frac{\gamma}{A}\sin\frac{\omega_1}{\Omega}\tau = 0 \tag{4.862}$$

with the initial conditions

$$x(0) = 1, \quad x'(0) = 0 \tag{4.863}$$

where the prime denotes the derivative with respect to τ and Ω is the frequency.

The linear and nonlinear operators obtained from Eq. 4.862 are respectively

$$L[x(\tau)] = (x'' + x)\Omega^2 \tag{4.864}$$

$$N[x(\tau), \Omega] = (\omega^2 - \Omega^2)x - \alpha x \sin\frac{\omega_2}{\Omega}\tau + \beta A^2 x^3 - \frac{\gamma}{A}\sin\frac{\omega_1}{\Omega}\tau = 0 \tag{4.865}$$

The initial approximation can be obtained from the equation

$$x_0'' + x_0 = 0, \quad x(0) = 1, \quad x'(0) = 0 \tag{4.866}$$

is

$$x_0(\tau) = \cos\tau \tag{4.867}$$

Combining Eqs. 4.867 and 4.865 it holds that

$$\begin{aligned} N[x_0(\tau), \Omega] &= \left(\omega^2 - \Omega^2 + \frac{3}{4}\beta A^2\right)\cos\tau + \frac{1}{4}\beta A^2 \cos 3\tau - \\ &-\frac{1}{2}\alpha\left[\sin\left(\frac{\omega_2}{\Omega} + 1\right)\tau + \sin\left(\frac{\omega_2}{\Omega} - 1\right)\tau\right] + \frac{\gamma}{A}\sin\frac{\omega_1}{\Omega}\tau = 0 \end{aligned} \tag{4.868}$$

It is clear that the auxiliary function can be chosen as

$$H(\tau, C_i) = C_1 + 2C_2\cos 2\tau + 2C_3\cos 4\tau \tag{4.869}$$

The linear equation for the determination of the first-order approximation is

$$\Omega^2(x_1'' + x_1) = H(\tau, C_i)N[x_0(\tau), \Omega], \quad x_1(0) = x_1'(0) = 0 \tag{4.870}$$

or using Eqs. 4.869, 4.868:

$$\begin{aligned} \Omega^2(x_1'' + x_1) = &\left[\left(\omega^2 - \Omega^2 + \frac{3}{4}\beta A^2\right)(C_1 + C_2) + \frac{1}{4}\beta A^2 C_2\right]\cos\tau \\ &+ H.O.T. \end{aligned} \tag{4.871}$$

No secular term in Eq. 4.871 requires that

$$\Omega^2 = \omega^2 + \frac{1}{4}\beta A^2 \frac{3C_1 + 4C_2}{C_1 + C_2} \tag{4.872}$$

Developing Eq. 4.871 and making use of the Eq. 4.872 we have

$$\begin{aligned}
\Omega^2\left(x_1'' + x_1\right) &= \frac{1}{4}\beta A^2 \frac{C_1^2 + C_1C_2 - C_2^2}{4(C_1 + C_2)} \cos 3\tau + \frac{1}{4}\beta A^2 C_2 \cos 5\tau - \\
&- \frac{\alpha C_1}{2}\left[\sin\left(\frac{\omega_2}{\Omega} + 1\right)\tau + \sin\left(\frac{\omega_2}{\Omega} - 1\right)\tau\right] - \frac{1}{2}\alpha C_2\left[\sin\left(\frac{\omega_2}{\Omega} + 3\right)\tau + \sin\left(\frac{\omega_2}{\Omega} - 3\right)\tau + \right.\\
&\left. + \sin\left(\frac{\omega_2}{\Omega} + 1\right)\tau + \sin\left(\frac{\omega_2}{\Omega} - 1\right)\tau\right] - \frac{\gamma}{A} C_1 \sin\frac{\omega_1}{\Omega}\tau - \frac{\gamma}{A} C_2\left[\sin\left(\frac{\omega_1}{\Omega} + 2\right)\tau + \right.\\
&\left. + \sin\left(\frac{\omega_1}{\Omega} - 2\right)\tau\right] - \frac{\gamma}{A} C_3\left[\sin\left(\frac{\omega_1}{\Omega} + 4\right)\tau + \sin\left(\frac{\omega_2}{\Omega} - 4\right)\tau\right]
\end{aligned} \tag{4.873}$$

From Eqs. 4.867, 4.873 and 4.861, we find the following first-order approximate solution:

$$\begin{aligned}
\overline{u}(t) &= A\cos\Omega t + \left(\frac{\beta A^3\left(C_1^2 + C_1C_2 - C_2^2\right)}{32(C_1 + C_2)\Omega^2}\right)(\cos\Omega t - \cos 3\Omega t) + \\
&+ \left(\frac{\beta A^3 C_2}{96\Omega^2}\right)(\cos\Omega t - \cos 5\Omega t) + \frac{\alpha A\Omega^2 C_1}{2\omega_2(\omega_2 + 2\Omega)}\left[\sin(\omega_2 + \Omega)t - \right.\\
&\left. - \frac{(\omega_2 + \Omega)}{\Omega}\sin\Omega t\right] + \frac{\alpha A\Omega^2 C_1}{2\omega_2(\omega_2 - 2\Omega)}\left[\sin(\omega_2 - \Omega)t - \frac{\omega_2 - \Omega}{\Omega}\sin\Omega t\right] + \\
&+ \frac{\alpha A\Omega^2 C_2}{2\omega_2(\omega_2 + 2\Omega)}\left[\sin(\omega_2 + \Omega)t - \frac{\omega_2 + \Omega}{\Omega}\sin\Omega t\right] - \\
&- \frac{\alpha A\Omega^2 C_2}{2\omega_2(2\Omega - \omega_2)}\left[\sin(\omega_2 - \Omega)t - \frac{\omega_2 - \Omega}{\Omega}\sin\Omega t\right] - \frac{\gamma\Omega^2 C_1}{\Omega^2 - \omega_1^2}\left[\sin\omega_1 t - \right.\\
&\left. - \frac{\omega_1}{\Omega}\sin\Omega t\right] + \frac{\gamma\Omega^2 C_2}{3\Omega^2 + 4\Omega\omega_1 + \omega_1^2}\left[\sin(\omega_1 + 2\Omega)t - \frac{\omega_1 + 2\Omega}{\Omega}\sin\Omega t\right] +
\end{aligned} \tag{4.874}$$

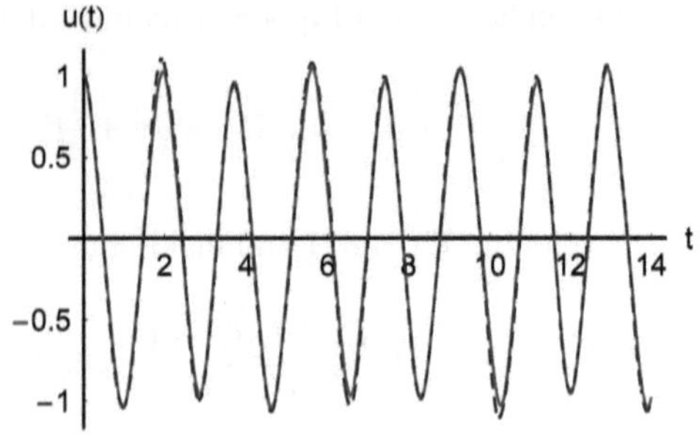

Fig. 4.187 Comparison of the results for Eq. 4.860: numerical (*red solid line*), approximate solution 4.874 (*blue dashed line*) (colour figure online)

$$+\frac{\gamma\Omega^2 C_2}{3\Omega^2-4\Omega\omega_1+\omega_1^2}\left[\sin\left(\omega_1-2\Omega\right)t-\frac{\omega_1-2\Omega}{\Omega}\sin\Omega t\right]+$$
$$+\frac{\gamma\Omega^2 C_3}{15\Omega^2+8\Omega\omega_1+\omega_1^2}\left[\sin\left(\omega_1+4\Omega\right)t-\frac{\omega_1+4\Omega}{\Omega}\sin\Omega t\right]+$$
$$+\frac{\gamma\Omega^2 C_3}{15\Omega^2-8\Omega\omega_1+\omega_1^2}\left[\sin\left(\omega_1-4\Omega\right)t-\frac{\omega_1-4\Omega}{\Omega}\sin\Omega t\right]+$$
$$+\frac{\alpha A\Omega^2 C_2}{2\left(8\Omega^2+6\Omega\omega_2+\omega_2^2\right)}\left[\sin\left(\omega_2+3\Omega\right)t-\frac{\omega_2+3\Omega}{\Omega}\sin\Omega t\right]+$$
$$+\frac{\alpha A\Omega^2 C_2}{2\left(8\Omega^2-6\Omega\omega_2+\omega_2^2\right)}\left[\sin\left(\omega_2-3\Omega\right)t-\frac{\omega_2-3\Omega}{\Omega}\sin\Omega t\right]$$

4.26.1 Numerical Example

In the case when A=1, $\omega_1=1.1$, $\omega_2=1.5$, $\omega=1.581$, $\alpha=2.75$, $\beta=12.5$, $\gamma=0.2$, we obtain:

$$C_1 = \text{-}0.080681272297,\ C_2 = 0.006225379112,\ C_3 = 1.085587531696$$

Figure 4.187 shows the comparison between the approximate solution of Eq. 4.860 and the numerical integration results.

It can be seen that the solution obtained by our procedure is nearly identical with that given by the numerical method.

4.27 Dynamic Analysis of a Rotating Electric Machine

Rotating electric machines are elastic systems exhibiting nonlinear vibrations in their working regime. The main causes that determine the occurrences of undesirable vibration in these dynamic systems are the nonlinear bearings, which support the rotating machine, the unbalanced forces, the mechanical looseness or misalignments. All these factors can affect the integrity of the system, therefore being highly detrimental [92].

In this Section, a rotating electrical machine supported by nonlinear bearings is investigated. The main sources of undesirable nonlinear vibration for this dynamic system are the nonlinear bearings, which support the rotating machine, the unbalanced forces and misalignments. All these factors can affect the integrity of the system, being therefore highly detrimental. The rotating machine under study is modelled as a 2-DOF system, assuming that the rotor mass is lumped at the midpoint, the rotor speed is constant, the axial and torsional vibrations and the mass of the shaft are negligible, as well as the damping. In these simplifying hypotheses, the equations of motion describing the dynamic behavior of the system are [92]:

$$\begin{aligned} &\ddot{x} + \left(\frac{1}{s_1^2} + \frac{1}{2cs^2}\right)x + \frac{\alpha}{s^2}x^3 = 0 \\ &\ddot{y} + \left(\frac{1}{s_1^2} + \frac{1}{2cs^2}\right)y + \frac{\alpha}{s^2}y^3 + \frac{f}{s^2} = 0 \end{aligned} \tag{4.875}$$

where x is the horizontal and y the vertical displacement and s_1, s, c, α and f are known parameters.

The initial conditions of the system 4.875 are

$$x(0) = A_1, \quad \dot{x}(0) = 0, \quad y(0) = A_2, \quad \dot{y}(0) = 0 \tag{4.876}$$

Using the notations

$$B_1 = \frac{1}{s_1^2} + \frac{1}{2cs^2}, \quad B_2 = \frac{\alpha}{s^2}, B_3 = \frac{f}{s^2} \tag{4.877}$$

and transformations

$$\tau_1 = \Omega_1 t, \quad \tau_2 = \Omega_2 t, \quad x(t) = A_1 u(\tau_1), \quad y(t) = A_2 v(\tau_2) \tag{4.878}$$

the system 4.875 becomes

$$u'' + \frac{B_1}{\Omega_1^2}u + \frac{A_1^2 B_2}{\Omega_1^2}u^3 = 0 \tag{4.879}$$

$$v'' + \frac{B_1}{\Omega_2^2}v + \frac{A_2^2 B_2}{\Omega_2^2}v^3 + \frac{B_3}{A_2\Omega_2^2} = 0 \tag{4.880}$$

with the initial conditions

$$u(0) = 1, \quad u'(0) = 0, \quad v(0) = 1, \quad v'(0) = 0 \tag{4.881}$$

For each Eq. 4.879 and 4.880, the linear and nonlinear operators are respectively

$$L[u(\tau_1)] = u'' + u \tag{4.882}$$

$$L[v(\tau_2)] = v'' + v \tag{4.883}$$

$$N[u(\tau_1), \Omega_1] = \left(\frac{B_1}{\Omega_1^2} - 1\right)u + \frac{A_1^2 B_2}{\Omega_1^2}u^3 \tag{4.884}$$

$$N[v(\tau_2), \Omega_2] = \left(\frac{B_1}{\Omega_2^2} - 1\right)v + \frac{A_2^2 B_2}{\Omega_2^2}v^3 + \frac{B_3}{A_2\Omega_2^2} \tag{4.885}$$

The initial approximations are obtained from

$$u_0'' + u_0 = 0, \quad u_0(0) = 1, \quad u'(0) = 0 \tag{4.886}$$

$$v_0'' + v_0 = 0, \quad v_0(0) = 1, \quad v'(0) = 0 \tag{4.887}$$

The solution of Eqs. 4.886 and 4.887 are respectively

$$u_0(\tau_1) = \cos\tau_1 \tag{4.888}$$

$$v_0(\tau_2) = \cos\tau_2 \tag{4.889}$$

From Eqs. 4.888 and 4.884 one gets

$$N[u_0(\tau_1), \Omega_1] = \left(\frac{B_1}{\Omega_1^2} - 1 + \frac{3A_1^2 B_2}{4\Omega_1^2}\right)\cos\tau_1 + \frac{A_1^2 B_2}{4\Omega_1^2}\cos 3\tau_1 \tag{4.890}$$

and from Eqs. 4.889 and 4.885 it follows that

$$N[v_0(\tau_2), \Omega_2] = \left(\frac{B_1}{\Omega_2^2} - 1 + \frac{3A_2^2 B_2}{4\Omega_2^2}\right)\cos\tau_2 + \frac{A_2^2 B_2}{4\Omega_2^2}\cos 3\tau_2 + \frac{B_3}{A_2\Omega_2^2} \tag{4.891}$$

The first approximations u_1 and v_1 are given by the equations

$$u_1'' + u_1 = H_1(\tau_1, C_i)N[u_0(\tau_1), \Omega_1], \quad u_1(0) = u_1'(0) = 0 \tag{4.892}$$

$$v_1'' + v_1 = H_2(\tau_2, C_j)N[v_0(\tau_2), \Omega_2], \quad v_1(0) = v_1'(0) = 0 \tag{4.893}$$

For the auxiliary functions, we have the freedom to choose

$$H_1(\tau_1, C_i) = C_1 + 2C_2\cos 2\tau_1 + 2C_3\cos 4\tau_1 \tag{4.894}$$

$$H_2(\tau_2, C_j) = C_4 + 2C_5\cos 2\tau_2 + 2C_6\cos 4\tau_2 \tag{4.895}$$

Substituting Eqs. 4.894 and 4.890 into Eq. 4.892 we obtain the following equation

$$u_1'' + u_1 = \left[\left(\frac{B_1}{\Omega_1^2} - 1 + \frac{3A_1^2B_2}{4\Omega_1^2}\right)(C_1 + C_2) + \frac{A_1^2B_2}{4\Omega_1^2}(C_2 + C_3)\right]\cos\tau_1 +$$
$$+\left[\left(\frac{B_1}{\Omega_1^2} - 1 + \frac{3A_1^2B_2}{4\Omega_1^2}\right)(C_2 + C_3) + \frac{A_1^2B_2}{4\Omega_1^2}C_1\right]\cos 3\tau_1 + \tag{4.896}$$
$$+\left[\left(\frac{B_1}{\Omega_1^2} + \frac{3B_2A_1^2}{4\Omega_1^2} - 1\right)C_3 + \frac{A_1^2B_2}{4\Omega_1^2}C_2\right]\cos 5\tau_1 + \frac{A_1^2B_2}{4\Omega_1^2}C_3\cos 7\tau_1,$$

The elimination of secular term in the last equation requires

$$\Omega_1^2 = B_1 + A_1^2B_2 + \frac{A_1^2B_2(C_3 - C_1)}{4(C_1 + C_2)} \tag{4.897}$$

The solution of Eq. 4.896 is given by

$$u_1(\tau_1) = \left[\left(\frac{1}{8} - \frac{4B_1 + 3A_1^2B_2}{32\Omega_1^2}\right)(C_2 + C_3) - \frac{A_1^2B_2}{32\Omega_1^2}C_1\right](\cos 3\tau_1 - \cos\tau_1) +$$
$$+\left[\left(\frac{1}{24} - \frac{4B_1 + 3A_1^2B_2}{96\Omega_1^2}\right)C_3 - \frac{A_1^2B_2}{96\Omega_1^2}C_2\right](\cos 5\tau_1 - \cos\tau_1) -$$
$$-\frac{A_1^2B_2C_3}{192\Omega_1^2}(\cos 7\tau_1 - \cos\tau_1) \tag{4.898}$$

From Eqs. 4.888, 4.898 and 4.5 we obtain the first-order approximate solution as

$$\overline{x}(t) = A_1\cos\Omega_1 t + \left[\left(\frac{A_1}{8} - \frac{4A_1B_1 + 3A_1^3B_2}{32\Omega_1^2}\right)(C_2 + C_3) - \frac{A_1^3B_2}{32\Omega_1^2}C_1\right](\cos 3\Omega_1 t -$$
$$-\cos\Omega_1 t) + \left[\left(\frac{A_1}{24} - \frac{4A_1B_1 + 3A_1^3B_2}{96\Omega_1^2}\right)C_3 - \frac{A_1^3B_2}{96\Omega_1^2}C_2\right](\cos 5\Omega_1 t - \cos\Omega_1 t) -$$
$$-\frac{A_1^3B_2C_3}{192\Omega_1^2}(\cos 7\Omega_1 t - \cos\Omega_1 t) \tag{4.899}$$

Now, substituting Eqs. 4.895 and 4.891 into Eq. 4.893 we obtain

$$\begin{aligned}
v_1'' + v_1 = &\left[\left(\frac{B_1}{\Omega_2^2} - 1 + \frac{3A_2^2B_2}{4\Omega_2^2}\right)(C_4 + C_5) + \frac{A_2^2B_2}{4\Omega_2^2}(C_5 + C_6)\right]\cos\tau_2 + \\
&+\left[\left(\frac{B_1}{\Omega_2^2} - 1 + \frac{3A_2^2B_2}{4\Omega_2^2}\right)(C_5 + C_6) + \frac{A_2^2B_2}{4\Omega_2^2}C_4\right]\cos 3\tau_2 + \\
&+\left[\left(\frac{B_1}{\Omega_2^2} - 1 + \frac{3B_2A_2^2}{4\Omega_2^2}\right)C_6 + \frac{A_2^2B_2}{4\Omega_2^2}C_5\right]\cos 5\tau_2 + \frac{A_2^2B_2}{4\Omega_2^2}C_6\cos 7\tau_2 + \\
&+\frac{2B_2C_5}{A_2\Omega_2^2}\cos 2\tau_2 + \frac{2B_3C_6}{A_2\Omega_2^2}\cos 4\tau_2 + \frac{B_3C_4}{A_2\Omega_2^2}
\end{aligned} \tag{4.900}$$

No secular terms in Eq. 4.900 requires

$$\Omega_2^2 = B_1 + A_2^2B_2 + \frac{A_2^2B_2(C_6 - C_4)}{4(C_4 + C_5)} \tag{4.901}$$

The equation 4.900 has the solution

$$\begin{aligned}
v_1(\tau_2) = &\left[\left(\frac{1}{8} - \frac{4B_1 + 3A_2^2B_2}{32\Omega_2^2}\right)(C_5 + C_6) - \frac{A_2^2B_2}{32\Omega_2^2}C_4\right](\cos 3\tau_2 - \cos\tau_2) + \\
&+\left[\left(\frac{1}{24} - \frac{4B_1 + 3A_2^2B_2}{96\Omega_2^2}\right)C_6 - \frac{A_2^2B_2}{96\Omega_2^2}C_5\right](\cos 5\tau_2 - \cos\tau_2) - \\
&-\frac{A_2^2B_2C_6}{192\Omega_2^2}(\cos 7\tau_2 - \cos\tau_2) + \frac{B_3C_4}{A_2\Omega_2^2}(1 - \cos\tau_2) - \frac{2B_3C_5}{3A_2\Omega_2^2}(\cos 2\tau_2 - \\
&-\cos\tau_2) - \frac{2B_3C_6}{15A_2\Omega_2^2}(\cos 4\tau_2 - \cos\tau_2)
\end{aligned} \tag{4.902}$$

In this way, the first-order approximate solution is given by

$$\begin{aligned}
\bar{y}(t) = &A_2\cos\Omega_2 t + \left[\left(\frac{A_2}{8} - \frac{4A_2B_1 + 3A_2^3B_2}{32\Omega_2^2}\right)(C_5 + C_6) - \frac{A_2^3B_2}{32\Omega_2^2}C_4\right](\cos 3\Omega_2 t - \\
&-\cos\Omega_2 t) + \left[\left(\frac{A_2}{24} - \frac{4A_2B_1 + 3A_2^3B_2}{96\Omega_2^2}\right)C_6 - \frac{A_2^3B_2C_5}{96\Omega_2^2}\right](\cos 5\Omega_2 t - \cos\Omega_2 t) - \\
&-\frac{A_2^3B_2C_6}{192\Omega_2^2}(\cos 7\Omega_2 t - \cos\Omega_2 t) + \frac{B_3C_4}{\Omega_2^2}(1 - \cos\Omega_2 t) - \frac{2B_3C_5}{3\Omega_2^2}(\cos 2\Omega_2 t - \\
&-\cos\Omega_2 t) - \frac{2B_3C_6}{15\Omega_2^2}(\cos 4\Omega_2 t - \cos\Omega_2 t)
\end{aligned} \tag{4.903}$$

4.27.1 Numerical Examples

We illustrate the accuracy of the OHAM by comparing obtained approximate solutions with the numerical integration results.

4.27.1.1 In the first case, we consider $s_1 = 0.75$, $s = 2$, $\alpha = 6.9$, $c = 0.5$, $f = 0.013$, such that $B_1 = 2.02778$, $B_2 = 1.725$, $B_3 = 0.00325$, $A_1 = 0.7$, $A_2 = 0.5$. The convergence-control parameters are obtained using the collocation method:

$$C_1 = -0.8841345865635, C_2 = -0.08737840736561, C_3 = 0.12501951586607$$
$$C_4 = -0.9527802249412, C_5 = -0.025599944500578, C_6 = 0.0630880156178$$

The first-order approximate solution of the system 4.875 can be written as

$$\begin{aligned}\bar{x}(t) &= 0.693803 \cos \Omega_1 t + 0.00615051 \cos 3\Omega_1 t + \\ &+ 0.000191701 \cos 5\Omega_1 t - 0.00014519 \cos 7\Omega_1 t\end{aligned} \tag{4.904}$$

$$\begin{aligned}&\bar{y}(t) = -0.00131931 + 0.498584 \cos \Omega_2 t + 0.0000236321 \cos 2\Omega_2 t + \\ &+ 0.00273123 \cos 3\Omega_2 t - 0.0000116477 \cos 4\Omega_2 t + \\ &+ 0.0000221851 \cos 5\Omega_2 t - 0.0000301867 \cos 7\Omega_2 t\end{aligned} \tag{4.905}$$

4.27.1.2 In the last case, we consider $s_1 = 0.9$, $s = 2$, $\alpha = 6.5$, $c = 0.4$, $f = 0.012$, $B_1 = 1.54707$, $B_2 = 1.625$, $B_3 = 0.003$, $A_1 = 0.2$, $A_2 = 0.8$ and therefore the results are

$$C_1 = -0.9909197339333, C_2 = 0.006064212647292, C_3 = 0.007023356693426$$
$$C_4 = -0.9059640183354, C_5 = -0.05141851688738, C_6 = 0.1314172156624$$

$$\begin{aligned}\bar{x}(t) &= 0.1997485706 \cos \Omega_1 t + 0.000252249 \cos 3\Omega_1 t - \\ &- 5.225828315 \cdot 10^{-7} \cos 5\Omega_1 t - 2.980315822 \cdot 10^{-7} \cos 7\Omega_1 t\end{aligned} \tag{4.906}$$

$$\begin{aligned}&\bar{y}(t) = -0.00117895 + 0.79111 \cos \Omega_2 t + 0.0000446081 \cos 2\Omega_2 t + \\ &+ 0.0101422 \cos 3\Omega_2 t - 0.0000228022 \cos 4\Omega_2 t + \\ &+ 0.000152019 \cos 5\Omega_2 t - 0.000247024 \cos 7\Omega_2 t\end{aligned} \tag{4.907}$$

Figures 4.188, 4.189, 4.190, 4.191 show a comparison between the present analytical solutions and the numerical integration results. One can observe that the first-order approximate analytical results obtained through OHAM are almost identical with the numerical simulation results in all considered cases, for different values of the parameters.

Fig. 4.188 Comparison between the approximate solution 4.904 and numerical solution in the case 4.27.1.1: numerical (*red solid line*), approximate solution (*blue dashed line*) (colour figure online)

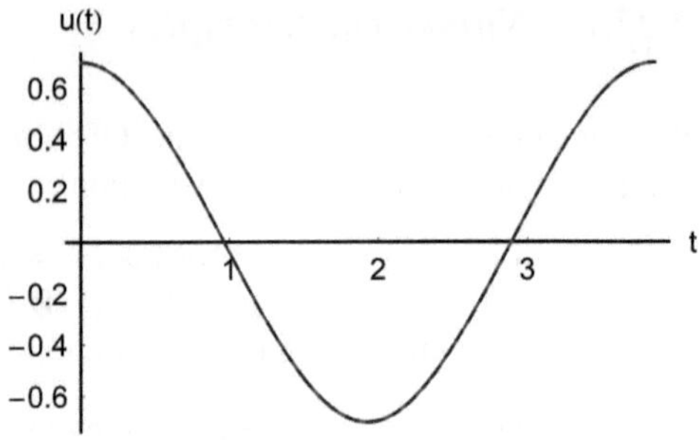

Fig. 4.189 Comparison between the approximate solution 4.905 and numerical solution in the case 4.27.1.1: numerical (*red solid line*), approximate solution (*blue dashed line*) (colour figure online)

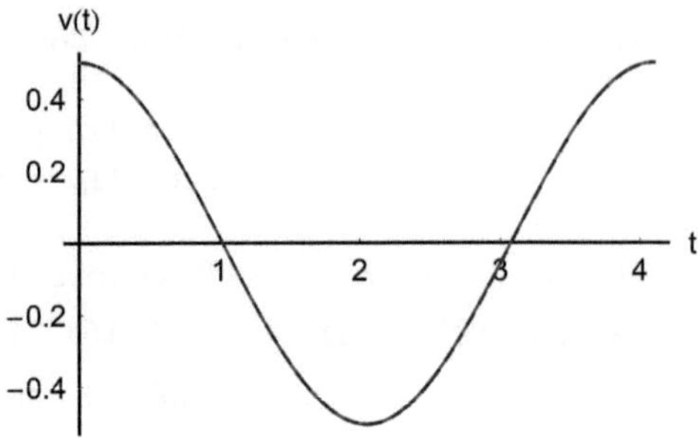

Fig. 4.190 Comparison between the approximate solution 4.906 and numerical solution in the case 4.27.1.2: numerical (*red solid line*), approximate solution (*blue dashed line*) (colour figure online)

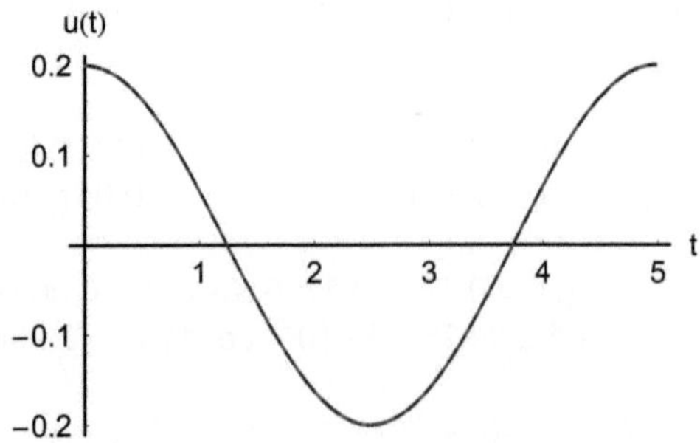

Fig. 4.191 Comparison between the approximate solution 4.907 and numerical solution in the case 4.27.1.2: numerical (*red solid line*), approximate solution (*blue dashed line*) (colour figure online)

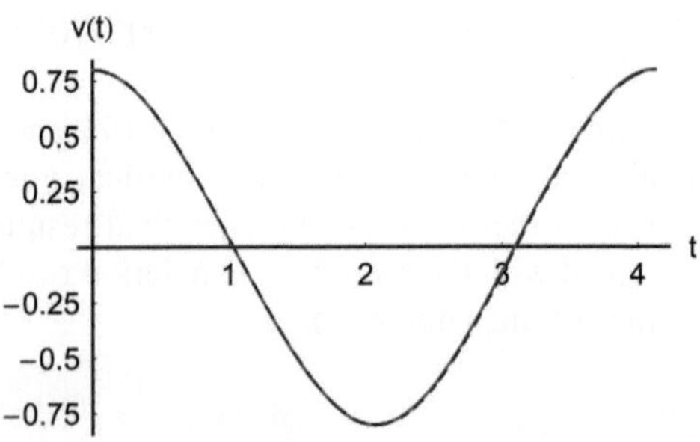

4.28 A Non-conservative Oscillatory System of a Rotating Electrical Machine

Rotating electrical machines are complex engineering systems which combine electrical and mechanical concepts. From mechanical point of view, these systems inherently involve the presence of a rotor, which often raises some dynamical problems, typical for rotor dynamics.

In what follows, we consider that the rotating electrical machine is modeled as a non-conservative oscillatory system subject to a parametric excitation caused by an axial thrust and a forcing excitation caused by an unbalanced force of the rotor. This system can be described by the second-order differential equation with variable coefficients [250]:

$$m\ddot{u} + c\dot{u} + k(1 - \lambda \sin(\omega_2 t))u = f \sin(\omega_1 t) \tag{4.908}$$

subject to the initial conditions

$$u(0) = A, \dot{u}(0) = 0 \tag{4.909}$$

where m is the lumped mass of the rotor, c is the linear damping coefficient, k is the shaft stiffness, ω_1 and ω_2 are the forcing and the parametric frequencies, respectively, and f and $k\lambda$ are the amplitudes of forcing and parametric excitations, respectively, and A is a known parameter.

Eq. 4.908 can be written in the dimensionless form

$$\ddot{u} + 2\mu\dot{u} + \omega^2 u - \alpha u \sin(\omega_2 t) - \beta \sin(\omega_1 t) = 0 \tag{4.910}$$

where

$$\mu = \frac{c}{2m}, \quad \omega^2 = \frac{k}{m}, \quad \alpha = \frac{k\lambda}{m}, \quad \beta = \frac{f}{m} \tag{4.911}$$

Introducing a new function

$$u(t) = e^{-\mu t} v(t) \tag{4.912}$$

the equation 4.910 can be written as

$$\ddot{v} + \left(\omega^2 - \mu^2\right)v - \alpha v \sin \omega_2 t - \beta e^{\mu t} \sin \omega_1 t = 0 \tag{4.913}$$

By means of a new variable

$$\tau = \Omega t, \quad \Omega = \sqrt{\omega^2 - \mu^2} \tag{4.914}$$

and of a new parameters

$$\alpha^* = \frac{\alpha}{\Omega^2}, \quad \beta^* = \frac{\beta}{\Omega^2}, \quad \mu^* = \frac{\mu}{\Omega}, \quad \omega_1^* = \frac{\omega_1}{\Omega}, \quad \omega_2^* = \frac{\omega_2}{\Omega}, \tag{4.915}$$

The equation 4.914 becomes

$$v'' + v - \alpha v \sin \omega_2 \tau - \beta e^{\mu\tau} \sin \omega_1 \tau = 0 \tag{4.916}$$

where the prime denotes derivative with respect to τ. In Eq. 4.916, the star is omitted.

The initial conditions 4.909 are

$$v(0) = A, \quad v'(0) = \mu A \tag{4.917}$$

The linear and nonlinear operators corresponding to Eq. 4.916 are respectively

$$L[v(\tau)] = v'' + v \tag{4.918}$$

$$N[v(\tau)] = -\alpha v \sin \omega_2 \tau - \beta e^{\mu\tau} \sin \omega_1 \tau \tag{4.919}$$

The initial approximation is given by the equation

$$v_0'' + v_0 = 0, \quad v_0(0) = A, \quad v_0'(0) = \mu A \tag{4.920}$$

and therefore

$$v_0(\tau) = A(\cos \tau + \mu \sin \tau) \tag{4.921}$$

From Eqs. 4.921 and 4.919 one can get

$$\begin{aligned} N[v_0(\tau)] = -\frac{1}{2}\alpha A[\sin(\omega_2+1)\tau + \sin(\omega_2-1)\tau] - \\ -\frac{1}{2}\mu\alpha A[\cos(\omega_2-1)\tau - \cos(\omega_2+1)\tau] - \beta e^{\mu\tau}\sin\omega_1\tau \end{aligned} \tag{4.922}$$

The auxiliary function is chosen in the form

$$H(\tau, C_i) = (2C_1 + 4C_2 \cos \tau + 4C_3 \sin \tau)e^{-\mu\tau} \tag{4.923}$$

The equation which gives the first approximation is

$$v_1'' + v_1 = H(\tau, C_i)N[v_0(\tau)], \quad v_1(0) = v_1'(0) = 0 \tag{4.924}$$

or using Eqs. 4.923 and 4. 922

$$\begin{aligned} v_1'' + v_1 &= -2\beta C_1 \sin\omega_1\tau - 2\beta C_2 \sin(\omega_1 + 1)\tau - 2\beta C_2 \sin(\omega_2 - 1)\tau + \\ &+ 2\beta C_3 \cos(\omega_1 + 1)\tau - 2\beta C_3 \cos(\omega_1 - 1)\tau + e^{-\mu\tau}\big[-\alpha A C_1 \sin(\omega_2 + 1)\tau - \\ &- A C_1 \sin(\omega_2 - 1)\tau - \mu\alpha A C_1 \cos(\omega_2 - 1)\tau + \mu\alpha A C_1 \cos(\omega_2 + 1)\tau + \\ &+ (\mu\alpha A C_3 - \alpha A C_2)\sin(\omega_2 - 2)\tau + (\mu\alpha A C_3 - \alpha A C_2)\sin(\omega_2 + 2)\tau + \\ &+ 2(\mu\alpha A C_3 - \alpha A C_2)\sin\omega_2\tau + (\mu\alpha A C_2 - \alpha A C_3)\cos(\omega_2 + 2)\tau - \\ &- (\mu\alpha A C_2 + \alpha A C_3)\cos(\omega_2 - 2)\tau\big] \end{aligned} \tag{4.925}$$

From the above equation, we find the first approximation $v_1(\tau)$.

The first-order approximate solution is obtained from Eqs. 4.912, 4.5, 4.921, 4.925, 4.914 and 4.915:

$$\begin{aligned} \bar{u}(t) &= e^{-\mu t}\Big[(A - M_1 - M_2 - M_3 - M_4 - M_5)\cos\Omega t + \Big(\frac{\mu}{\Omega}(A + M_1 + \\ &+ M_2 + M_3 + M_4 + M_5) - \Big(1 + \frac{\omega_2}{\Omega}\Big)N_1 + \Big(1 - \frac{\omega_2}{\Omega}\Big)N_2 - \Big(2 + \frac{\omega_2}{\Omega}\Big)N_3 + \\ &+ \Big(2 - \frac{\omega_2}{\Omega}\Big)N_4 - \frac{\omega_2}{\Omega}N_5\Big)\sin\Omega t - \frac{2\beta\Omega^2 C_1}{\Omega^2 - \omega_1^2}\Big(\sin\omega_1 t - \frac{\omega_1}{\Omega}\sin\Omega t\Big) + \\ &+ \frac{2\beta\Omega^2 C_2}{\omega_1(\omega_1 + 2\Omega)}\Big(\sin(\omega_1 + \Omega)t - \Big(1 + \frac{\omega_1}{\Omega}\Big)\sin\Omega t\Big) + \\ &+ \frac{2\beta\Omega^2 C_2}{\omega_2(\omega_2 - 2\Omega)}\Big(\sin(\omega_1 - \Omega)t + \Big(1 - \frac{\omega_1}{\Omega}\Big)\sin\Omega t\Big) - \\ &- \frac{2\beta\Omega^2 C_3}{\omega_1(\omega_1 + 2\Omega)}(\cos(\omega_1 + \Omega)t - \cos\Omega t) + \\ &+ \frac{2\beta\Omega^2 C_3}{\omega_1(\omega_1 - 2\Omega)}(\cos(\omega_1 - \Omega)t - \cos\Omega t)\Big] + e^{-2\mu t}\big[M_1\cos(\omega_2 + \Omega)t + \\ &+ N_1\sin(\omega_2 + \Omega)t + M_2\cos(\omega_2 - \Omega)t + N_2\sin(\omega_2 - \Omega)t + \\ &+ M_3\cos(\omega_2 + 2\Omega)t + N_3\sin(\omega_2 + 2\Omega)t + M_4\cos(\omega_2 - 2\Omega)t + \\ &+ N_4\sin(\omega_2 - 2\Omega)t + M_5\cos\omega_2 t + N_5\sin\omega_2 t\big] \end{aligned} \tag{4.926}$$

where

$$M_1 = \frac{(\mu^2 - \omega_2^2 - 4\omega_2\Omega - 2\Omega^2)\frac{\mu}{\Omega}\alpha A C_1}{(\mu^2 - \omega_2^2 - 2\omega_2\Omega)^2 + 4\mu^2(\omega_2 + \Omega)^2}$$

$$M_2 = \frac{(\omega_2^2 - \mu^2 - 4\mu\omega_2 + 2\mu\Omega)\frac{\alpha A\mu}{\Omega}C_1}{(\mu^2 - \omega_2^2 + 2\omega_2\Omega)^2 + 4\mu^2(\omega_2 - \Omega)^2}$$

$$N_1 = \frac{(\omega_2^2\Omega + 2\omega_2\Omega^2 - 3\mu^2\Omega - 2\mu^2\omega_2)\frac{\alpha A}{\Omega}C_1}{(\mu^2 - \omega_2^2 - 2\omega_2\Omega)^2 + 4\mu^2(\omega_2 + \Omega)^2}$$

$$N_2 = \frac{(\Omega\omega_2^2 - 3\mu^2\Omega - 2\omega_2\Omega^2 + 2\mu^2\omega_2)\frac{\alpha A}{\Omega}C_1}{(\mu^2 - \omega_2^2 + 2\omega_2\Omega)^2 + 4\mu^2(\omega_2 - \Omega)^2}$$

$$M_3 = \frac{\left(\mu^3 - \omega_2^2\mu - 6\omega_2\mu\Omega - 7\mu\Omega^2\right)\frac{\alpha A}{\Omega}C_2 + \left(2\mu^2\omega_2 + 3\mu^2\Omega + \omega_2^2\Omega + 4\omega_2\Omega^2 + 3\Omega^3\right)\frac{\alpha A}{\Omega}C_3}{\left(\mu^2 - \omega_2^2 - 4\omega_2\Omega - 3\Omega^2\right)^2 + 4\mu^2(\omega_2 + 2\Omega)^2}$$

$$N_3 = \frac{\left(\omega_2^2\Omega + 4\omega_2\Omega^2 + 3\Omega^3 - 2\mu^2\omega_2 - 5\mu^2\Omega\right)\frac{\alpha A}{\Omega}C_2 + \left(\mu^3 - \omega_2^2\mu - 2\omega_2\mu\Omega + \mu\Omega^2\right)\frac{\alpha A}{\Omega}C_3}{\left(\mu^2 - \omega_2^2 - 4\omega_2\Omega - 3\Omega^2\right)^2 + 4\mu^2(\omega_2 + 2\Omega)^2}$$

$$M_4 = \frac{\left(\omega_2^2\mu - \mu^3 - 6\omega_2\mu\Omega + 7\mu\Omega^2\right)\frac{\alpha A}{\Omega}C_2 + \left(\omega_2^2\Omega - 5\mu^2\Omega - 4\omega_2\Omega^2 + 3\Omega^3 + 2\mu^2\omega_2\right)\frac{\alpha A}{\Omega}C_3}{\left(\mu^2 - \omega_2^2 + 4\omega_2\Omega - 3\Omega^2\right)^2 + 4\mu^2(\omega_2 - 2\Omega)^2}$$

$$N_4 = \frac{\left(2\mu^2\omega_2 - 5\mu^2\Omega + \omega_2^2\Omega - 4\omega_2\Omega^2 + 3\Omega^3\right)\frac{\alpha A}{\Omega}C_2 + \left(6\mu\omega_2\Omega - 7\mu\Omega^2 + \mu^3 - \omega_2^2\mu\right)\frac{\alpha A}{\Omega}C_3}{\left(\mu^2 - \omega_2^2 + 4\omega_2\Omega - 3\Omega^2\right)^2 + 4\mu^2(\omega_2 - 2\Omega)^2}$$

$$M_5 = \frac{4\mu\omega_2(\mu C_3 - \Omega C_2)\frac{\alpha A}{\Omega}}{\left(\mu^2 - \omega_2^2 + \Omega^2\right)^2 + 4\mu^2\omega_2^2}, N_5 = \frac{\left(\mu^2 - \omega_2^2 + \Omega^2\right)(\mu C_3 - \Omega C_2)\frac{2\alpha A}{\Omega}}{\left(\mu^2 - \omega_2^2 + \Omega^2\right)^2 + 4\mu^2\omega_2^2}$$

To prove the efficiency and accuracy of the OHAM, we have considered the special case of a certain working regime characterized by

$$\mu = 0.2, \omega = 5\sqrt{2}, \alpha = 1.5, \beta = 1, \omega_1 = 1, \omega_2 = 3 \tag{4.927}$$

In these conditions, the optimal values of the convergence-control parameters are

$$C_1 = -0.006563018652, C_2 = -0.00236445114, C_3 = -0.002264487794$$

The first-order approximate solution of Eqs. 4.910 and 4.909 becomes

$$\begin{aligned}
\overline{u}(t) &= e^{-0.2t}(0.017224\cos 6.0682t + 0.967867\cos 7.0682t + \\
&+ 0.01494849\cos 8.0682t + 0.0133941\sin t - 0.00707153\sin 6.0682t + \\
&+ 0.050622\sin 7.0682t - 0.015608\sin 8.0682t) + e^{-0.4t}(4.922\cdot 10^{-6}\cos 3t - \\
&- 2.331578\cdot 10^{-6}\cos 4.0682t + 0.0000203\cos 10.0682t - \\
&- 0.0000498\cos 11.1364t - 0.0000131\cos 17.1364t + 0.00016817\sin 3t - \\
&- 0.00029402\sin 4.0682t - 0.00019004\sin 10.0682t + \\
&+ 0.0000436148\sin 11.1364t - 0.00001453\sin 17.13647t)
\end{aligned} \tag{4.928}$$

Figure 4.192 shows the comparison between the numerical solution of Eqs. 4.910 and the approximate solution given by Eq. 4.928. One can observe that the results obtained for this non-conservative oscillatory system through OHAM is quasi-identical to that obtained via numerical integration, which proves the validity of the proposed procedure for non-conservative oscillators.

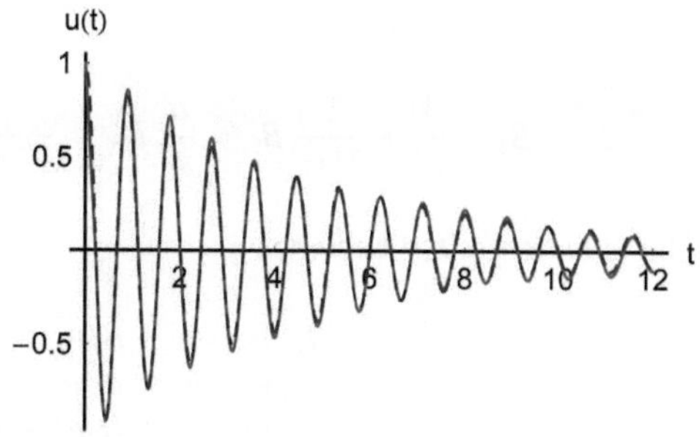

Fig. 4.192 Comparison between the approximate solution 4.928 and numerical solution in the case : numerical (*red solid line*), approximate solution (*blue dashed line*) (colour figure online)

4.29 Nonlinear Dynamics of an Electrical Machine Rotor-Bearing System

The rotating electrical machine considered in this Section is modelled as a flexible rotor supported by two journal bearings with nonlinear suspension [251]. In order to simplify the mathematical modelling of such a system, one can assume that the rotor mass and the bearing mass are lumped at the midpoint, the rotor speed is constant, the axial and torsional vibrations and the mass of the shaft are negligible, as well as the damping. In these hypotheses the motion of the system is governed by the following non-linear system of four coupled non-linear differential equations [252]:

$$\begin{aligned}
&\ddot{u}_1 + \frac{1}{s_1^2}u_1 + \frac{\alpha}{s^2}u_1^3 - \frac{1}{2cs^2}(u_2 - u_1) = 0\\
&\ddot{u}_2 + \frac{1}{s^2}(u_2 - u_1) = 0\\
&\ddot{u}_3 + \frac{1}{s_1^2}u_3 + \frac{\alpha}{s^2}u_3^3 - \frac{1}{2cs^2}(u_4 - u_3) + \frac{f}{s^2} = 0\\
&\ddot{u}_4 + \frac{1}{s^2}(u_4 - u_3) + \frac{f}{s^2} = 0
\end{aligned} \tag{4.929}$$

where u_1, u_2 and u_3, u_4 are the horizontal and vertical displacements of the bearing centre and rotor centre, respectively. The rest of the parameters are considered constants for a given working regime:

$$\begin{aligned}
&s_1^2 = c_{om}c_p s^2, c_{om} = \frac{m_0}{m}, c_p = \frac{k_p}{k_1},\\
&s^2 = \frac{\omega_1}{\omega_n}, \alpha = \frac{k_2 c^2}{k_p c_{om}}, f = \frac{mg}{ck_p}
\end{aligned} \tag{4.930}$$

where ω_1 is the rotational speed of the shaft, k_p is the stiffness of the shaft, k_1 and k_2 are the stiffness of the linear and nonlinear springs which support the bearing housing, m and m_0 are the masses lumped at the rotor mid-point and at the bearing mid-point and c is the clearance.

Using the notations

$$B_1 = \frac{1}{s^2} + \frac{1}{2cs^2}, B_2 = \frac{\alpha}{s^2}, B_3 = -\frac{1}{2cs^2}, B_4 = \frac{f}{s^2}, \omega^2 = \frac{1}{s^2} \tag{4.931}$$

the equation 4.929 becomes

$$\begin{aligned}
&\ddot{u}_1 + B_1 u_1 + B_2 u_1^3 + B_2 u_2 = 0 \\
&\ddot{u}_2 + \omega^2 u_2 - \omega^2 u_1 = 0 \\
&\ddot{u}_3 + B_1 u_3 + B_2 u_3^3 + B_3 u_4 + B_4 = 0 \\
&\ddot{u}_4 + \omega^2 u_4 - \omega^2 u_3 + B_4 = 0
\end{aligned} \tag{4.932}$$

The initial conditions are

$$u_i(0) = A_i, \quad \dot{u}_i(0) = 0, \quad i = 1, 2, 3, 4 \tag{4.933}$$

Making the transformations

$$\tau_i = \Omega_i t, \quad u_i = A_i x_i(\tau_i) \tag{4.934}$$

where Ω_i are the frequencies of u_i, respectively, the equation 4.932 can be rewritten in the form

$$\begin{aligned}
&x_1'' + x_1 + \left(\frac{B_1}{\Omega_1^2} - 1\right)x_1 + \frac{B_2}{\Omega_1^2}A_1^2 x_1^3 + \frac{B_3 A_2}{\Omega_1^2 A_1}x_2 = 0 \\
&x_2'' + x_2 + \left(\frac{\omega^2}{\Omega_2^2} - 1\right)x_2 - \frac{\omega^2 A_1}{\Omega_2^2 A_2}x_1 = 0 \\
&x_3'' + x_3 + \left(\frac{\omega^2}{\Omega_3^2} - 1\right)x_3 + \frac{B_2 A_3^2}{\Omega_3^2}x_3^3 + \frac{B_3 A_4}{\Omega_3^2 A_3}x_4 + \frac{B_4}{\Omega_3^2 A_3} = 0 \\
&x_4'' + x_4 + \left(\frac{\omega^2}{\Omega_4^2} - 1\right)x_4 - \frac{\omega^2 A_3}{\Omega_4^2 A_4}x_3 + \frac{B_4}{\Omega_4^2 A_4} = 0
\end{aligned} \tag{4.935}$$

with the initial conditions

$$x_i(0) = 1, \quad x_i'(0) = 0, i = 1, 2, 3, 4 \tag{4.936}$$

The linear and nonlinear operators for Eqs. 4.935 are respectively

$$L[x_i(\tau)] = x_i'' + x_i, \quad i = 1, 2, 3, 4 \tag{4.937}$$

$$N[x_1(\tau), \Omega_1] = \left(\frac{B_1}{\Omega_1^2} - 1\right)x_1 + \frac{B_2}{\Omega_1^2}A_1^2 x_1^3 + \frac{B_3 A_2}{\Omega_1^2 A_1}x_2 \tag{4.938}$$

$$N[x_2(\tau),\Omega_2]=\left(\frac{\omega^2}{\Omega_2^2}-1\right)x_2-\frac{\omega^2A_1}{\Omega_2^2A_2}x_1 \tag{4.939}$$

$$N[x_3(\tau),\Omega_3]=\left(\frac{\omega^2}{\Omega_3^2}-1\right)x_3+\frac{B_2A_3^2}{\Omega_3^2}x_3^3+\frac{B_3A_4}{\Omega_3^2A_3}x_4+\frac{B_4}{\Omega_3^2A_3} \tag{4.940}$$

$$N[x_4(\tau),\Omega_4]=\left(\frac{\omega^2}{\Omega_4^2}-1\right)x_4-\frac{\omega^2A_3}{\Omega_4^2A_4}x_3+\frac{B_4}{\Omega_4^2A_4} \tag{4.941}$$

The initial approximations are obtained from the equations

$$x_{i0}''+x_{i0},\quad x_{i0}(0)=1,\quad x_{i0}'(0)=0,\quad i=1,2,3,4 \tag{4.942}$$

such that

$$x_{i0}(\tau)=\cos\tau_i,\quad i=1,2,3,4 \tag{4.943}$$

From Eqs. 4.943 and 4.938, we have for i = 1:

$$N[x_{10}(\tau_1),\Omega_1]=\left(\frac{B_1}{\Omega_1^2}-1+\frac{3B_2A_1^2}{4\Omega_1^2}\right)\cos\tau_1+\frac{B_2A_1^2}{4\Omega_1^2}\cos3\tau_1+\frac{B_3A_2}{\Omega_1^2A_1}\cos\tau_2 \tag{4.944}$$

The first approximation of the first variable will be obtained from the equation

$$x_{11}''+x_{11}=H_1(\tau_1,C_i)N[x_{10}(\tau_1),\Omega_1] \tag{4.945}$$

The auxiliary function H_1 can be put in the form

$$H_1(\tau_1,C_i)=C_1+2C_2\cos2\tau_1+2C_3\cos4\tau_1 \tag{4.946}$$

By means of Eqs. 4.944 and 4.946, the Eq. 4.945 becomes

$$\begin{aligned}x_{11}''+x_{11}=&\left[\left(\frac{4B_1+3B_2A_1^2}{4\Omega_1^2}-1\right)(C_1+C_2)+\frac{B_2A_1^2}{4\Omega_1^2}(C_2+C_3)\right]\cos\tau_1+\\&+\left[\frac{B_2A_1^2}{4\Omega_1^2}C_1+\left(\frac{4B_1+3B_2A_1^2}{4\Omega_1^2}-1\right)(C_2+C_3)\right]\cos3\tau_1+\left[\frac{B_2A_1^2}{4\Omega_1^2}C_2+\right.\\&\left.+\left(\frac{4B_1+3B_2A_1^2}{4\Omega_1^2}-1\right)C_3\right]\cos5\tau_1+\frac{B_2A_1^2}{4\Omega_1^2}C_3\cos7\tau_1+\frac{B_3A_2C_1}{\Omega_1^2A_1}\cos\tau_2+\\&+\frac{B_3A_2C_2}{\Omega_1^2A_1}[\cos(2\tau_1+\tau_2)+\cos(2\tau_1-\tau_2)]+\\&+\frac{B_3A_2C_3}{\Omega_1^2A_1}[\cos(4\tau_1+\tau_2)+\cos(4\tau_1-\tau_2)]\end{aligned} \tag{4.947}$$

Avoiding the presence of secular term in Eq. 4.947, we obtain the frequency

$$\Omega_1^2 = B_1 + \frac{B_2 A_1^2 (3C_1 + 4C_2 + C_3)}{4(C_1 + C_2)} \tag{4.948}$$

Using the solution of Eq. 4.947 and Eqs. 4.943, 4.934 and 4.5, we obtain the first-order approximate solution

$$\begin{aligned}
\overline{u}_1(t) = {} & A_1 \cos\Omega_1 t + \left[\left(\frac{A_1}{8} - \frac{4A_1B_1 + 3A_1^3B_2}{32\Omega_1^2}\right)(C_2 + C_3) - \right. \\
& \left. - \frac{A_1^3 B_2 C_1}{32\Omega_1^2}\right](\cos 3\Omega_1 t - \cos\Omega_1 t) + \left[\left(\frac{A_1}{24} - \frac{4A_1B_1 + 3A_1^3B_2}{96\Omega_1^2}\right) C_3 - \right. \\
& \left. - \frac{A_1^3 B_2 C_2}{96\Omega_1^2}\right](\cos 5\Omega_1 t - \cos\Omega_1 t) - \frac{A_1^3 B_2 C_3}{192\Omega_1^2}(\cos 7\Omega_1 t - \cos\Omega_1 t) + \\
& + \frac{B_3 A_2 C_1}{\Omega_1^2 - \Omega_2^2}(\cos\Omega_2 t - \cos\Omega_1 t) - \\
& - \frac{B_3 A_2 C_2}{3\Omega_1^2 + 4\Omega_1\Omega_2 + \Omega_2^2}(\cos(2\Omega_1 + \Omega_2)t - \cos\Omega_1 t) - \\
& - \frac{B_3 A_2 C_2}{3\Omega_1^2 - 4\Omega_1\Omega_2 + \Omega_2^2}(\cos(2\Omega_1 - \Omega_2)t - \cos\Omega_1 t) - \\
& - \frac{B_3 A_2 C_3}{15\Omega_1^2 + 8\Omega_1\Omega_2 + \Omega_2^2}(\cos(4\Omega_1 + \Omega_2)t - \cos\Omega_1 t) - \\
& - \frac{B_3 A_2 C_2}{15\Omega_1^2 - 8\Omega_1\Omega_2 + \Omega_2^2}(\cos(4\Omega_1 - \Omega_2)t - \cos\Omega_1 t)
\end{aligned} \tag{4.949}$$

From Eqs. 4.939 and 4.943 for i = 2, we have

$$N[x_{20}(\tau_2), \Omega_2] = \left(\frac{\omega^2}{\Omega_2^2} - 1\right)\cos\tau_2 - \frac{\omega^2 A_1}{\Omega_2^2 A_2}\cos\tau_1 \tag{4.950}$$

For the second variable, the first approximation is given by the equation

$$x_{21}'' + x_{21} = H_2(\tau_2, C_i) N[x_{20}(\tau_2), \Omega_2] \tag{4.951}$$

where the auxiliary function is chosen as

$$H_2(\tau_2, C_i) = C_4 + 2C_5 \cos(\tau_1 + \tau_2) + 2C_6 \cos(\tau_1 - \tau_2) \tag{4.952}$$

Taking into account Eqs. 4.952 and 4.950, the Eq. 4.951 will be

$$x_{21}'' + x_{21} = \left[\left(\frac{\omega^2}{\Omega_2^2} - 1\right)C_4 - \frac{\omega^2 A_1}{\Omega_2^2 A_2}(C_5 + C_6)\right]\cos\tau_2 + \left[\left(\frac{\omega^2}{\Omega_2^2} - 1\right)(C_5 + \right.$$
$$\left. + C_6) - \frac{\omega^2 A_1}{\Omega_2^2 A_2}C_4\right]\cos\tau_1 + \left(\frac{\omega^2}{\Omega_2^2} - 1\right)C_5\cos(\tau_1 + 2\tau_2) - \frac{\omega^2 A_1}{\Omega_2^2 A_2}C_5\cos(2\tau_1 +$$
$$+\tau_2) + \left(\frac{\omega^2}{\Omega_2^2} - 1\right)C_6\cos(2\tau_2 - \tau_1) - \frac{\omega^2 A_1}{\Omega_2^2 A_2}C_6\cos(2\tau_1 - \tau_2) \tag{4.953}$$

Eliminating the secular term in Eq. 4.953 we obtain

$$\Omega_2^2 = \omega^2 - \frac{\omega^2 A_1(C_5 + C_6)}{A_2 C_4} \tag{4.954}$$

The first-order approximate solution is obtained from the solution of Eq. 4.953 and from Eqs. 4.5, 4.943 for n = 2 and 4.934:

$$\overline{u}_2(t) = A_2\cos\Omega_2 t + \frac{A_2(\omega^2 - \Omega_2^2)(C_5 + C_6) - \omega^2 A_1 C_4}{\Omega_2^2 - \Omega_1^2}(\cos\Omega_1 t -$$
$$- \cos\Omega_2 t) + \frac{A_2(\Omega_2^2 - \omega^2)C_5}{3\Omega_2^2 + 4\Omega_1\Omega_2 + \Omega_1^2}[(\cos(\Omega_1 + 2\Omega_2)t - \cos\Omega_2 t] +$$
$$+\frac{\omega^2 A_1 C_5}{4\Omega_1(\Omega_1 + \Omega_2)}[(\cos(2\Omega_1 + \Omega_2)t - \cos\Omega_2 t)] +$$
$$+\frac{A_2(\Omega_2^2 - \omega^2)C_6}{3\Omega_2^2 - 4\Omega_1\Omega_2 + \Omega_1^2}[(\cos(\Omega_1 - 2\Omega_2)t - \cos\Omega_2 t] +$$
$$+\frac{\omega^2 A_1 C_6}{4\Omega_1(\Omega_1 - \Omega_2)}[(\cos(2\Omega_1 - \Omega_2)t - \cos\Omega_2 t)] \tag{4.955}$$

From Eqs. 4.940 and 4.943 for n = 3, it is obtained

$$N[x_{30}(\tau_3), \Omega_3] = \left(\frac{B_1}{\Omega_3^2} - 1 + \frac{3B_2A_3^2}{4\Omega_3^2}\right)\cos\tau_3 + \frac{B_2A_3^2}{4\Omega_3^2}\cos 3\tau_3 +$$
$$+\frac{B_3A_4}{\Omega_2^2 A_3}\cos 3\tau_4 + \frac{B_4}{\Omega_3^2 A_3} \tag{4.956}$$

The first approximation for the third variable is obtained from the equation

$$x_{31}'' + x_{31} = H_3(\tau_3, C_i)N[x_{30}(\tau_3), \Omega_3] \tag{4.957}$$

The auxiliary function is given by

$$H_3(\tau_3, C_i) = C_7 + 2C_8 \cos 2\tau_3 + 2C_9 \cos 4\tau_3 \tag{4.958}$$

such that, from Eqs. 4.958, 4.956 and 4.957 we find the following equation

$$\begin{aligned}
x_{31}'' + x_{31} = &\left[\left(\frac{4B_1 + 3B_2A_3^2}{4\Omega_3^2} - 1\right)(C_7 + C_8) + \frac{B_2A_3^2}{4\Omega_3^2}(C_8 + C_9)\right] \cos\tau_3 + \\
&+\left[\left(\frac{4B_1 + 3B_2A_3^2}{4\Omega_3^2} - 1\right)(C_8 + C_9) + \frac{B_2A_3^2}{4\Omega_3^2}C_7\right] \cos 3\tau_3 + \\
&+\left[\left(\frac{4B_1 + 3B_2A_3^2}{4\Omega_3^2} - 1\right)C_9 + \frac{B_2A_3^2}{4\Omega_3^2}C_8\right] \cos 5\tau_3 + \frac{B_2A_3^2C_9}{4\Omega_3^2} \cos 7\tau_3 + \\
&+\frac{2B_4C_8}{\Omega_3^2A_3} \cos 2\tau_3 + \frac{2B_4C_9}{\Omega_3^2A_3} \cos 4\tau_3 + \frac{B_4C_7}{\Omega_3^2A_3} + \frac{B_3A_4C_8}{\Omega_3^2A_3}[\cos(2\tau_3 + \tau_4) + \\
&+\cos(2\tau_3 - \tau_4)] + \frac{B_3A_4C_9}{\Omega_3^2A_3}[\cos(4\tau_3 + \tau_4) + \cos(4\tau_3 - \tau_4)] + \\
&+\frac{B_3A_4C_7}{\Omega_3^2A_3} \cos\tau_4
\end{aligned} \tag{4.959}$$

No secular term in Eq. 4.959 lead to the determination of the frequency

$$\Omega_3^2 = B_1 + \frac{3}{4}B_2A_3^2 + B_2A_3^2\frac{C_8 + C_9}{4(C_7 + C_8)} \tag{4.960}$$

The first-order approximate solution can be written from Eqs. 4.959, 4.943 for n = 3, 4.934 and 4.5:

$$\begin{aligned}
\bar{u}_3(t) = &A_3 \cos\Omega_3 t + \left[\left(\frac{A_3}{8} - \frac{4A_3B_1 + 3A_3^3B_2}{32\Omega_3^2}\right)(C_8 + C_9) - \right. \\
&\left. -\frac{A_3^3B_2C_7}{32\Omega_3^2}\right](\cos 3\Omega_3 t - \cos\Omega_3 t) + \left[\left(\frac{A_3}{24} - \frac{4A_3B_1 + 3A_3^3B_2}{96\Omega_3^2}\right)C_9 - \right. \\
&\left. -\frac{A_3^3B_2C_8}{96\Omega_3^2}\right](\cos 5\Omega_3 t - \cos\Omega_3 t) - \frac{A_3^3B_2C_9}{192\Omega_3^2}(\cos 7\Omega_3 t - \cos\Omega_3 t) + \\
&+\frac{2B_4C_8}{3\Omega_3^2}(\cos 2\Omega_3 t - \cos\Omega_3 t) - \frac{2B_4C_9}{15\Omega_3^2}(\cos 4\Omega_3 t - \cos\Omega_3 t) + \\
&+\frac{B_4C_7}{\Omega_3^2}(1 - \cos\Omega_3 t) - \\
&-\frac{B_3A_4C_8}{3\Omega_3^2 + 4\Omega_3\Omega_4 + \Omega_4^2}(\cos(2\Omega_3 + \Omega_4)t - \cos\Omega_3 t) - \\
&-\frac{B_3A_4C_8}{3\Omega_3^2 - 4\Omega_3\Omega_4 + \Omega_4^2}(\cos(2\Omega_3 - \Omega_4)t - \cos\Omega_3 t) -
\end{aligned}$$

$$-\frac{B_3A_4C_9}{15\Omega_3^2+8\Omega_3\Omega_4+\Omega_4^2}(\cos(4\Omega_3+\Omega_4)t-\cos\Omega_3 t)- \\ -\frac{B_3A_4C_9}{15\Omega_3^2-8\Omega_3\Omega_4+\Omega_4^2}(\cos(4\Omega_3-\Omega_4)t-\cos\Omega_3 t)+ \\ +\frac{B_3A_4C_7}{\Omega_3^2-\Omega_4^2}(\cos\Omega_4 t-\cos\Omega_3 t) \tag{4.961}$$

Now, from Eqs. 4.941 and 4.943 for i = 4 we have

$$N[x_{40}(\tau_4),\Omega_4]=\left(\frac{\omega^2}{\Omega_4^2}-1\right)\cos\tau_4-\frac{\omega^2A_3}{\Omega_4^2A_4}\cos\tau_3+\frac{B_4}{\Omega_4^2A_4} \tag{4.962}$$

The first approximation for the fourth variable is obtained by solving the equation

$$x_{41}''+x_{41}=H_4(\tau_4,C_i)N[x_{40}(\tau_4),\Omega_4] \tag{4.963}$$

where the auxiliary function is

$$H_4(\tau_4,C_i)=C_{10}+2C_{11}\cos(\tau_3+\tau_4)+2C_{12}\cos(\tau_3-\tau_4) \tag{4.964}$$

Substituting Eqs. 4.964 and 4.962 into Eq. 4.963, we obtain

$$x_{41}''+x_{41}=\left[\left(\frac{\omega^2}{\Omega_4^2}-1\right)C_{10}-\frac{\omega^2A_3}{\Omega_4^2A_4}(C_{11}+C_{12})\right]\cos\tau_4+ \\ +\left[\left(\frac{\omega^2}{\Omega_4^2}-1\right)(C_{11}+C_{12})-\frac{\omega^2A_3}{\Omega_4^2A_4}C_{10}\right]\cos\tau_3+\frac{2B_4C_{11}}{\Omega_4^2A_4}\cos(\tau_3+\tau_4)+ \\ +\frac{2B_4C_{12}}{\Omega_4^2A_4}\cos(\tau_3-\tau_4)+\left(\frac{\omega^2}{\Omega_4^2}-1\right)C_{11}\cos(\tau_3+2\tau_4)+ \\ +\left(\frac{\omega^2}{\Omega_4^2}-1\right)C_{12}\cos(\tau_3-2\tau_4)-\frac{\omega^2A_3C_{11}}{\Omega_4^2A_4}\cos(2\tau_3+\tau_4)- \\ -\frac{\omega^2A_3C_{12}}{\Omega_4^2A_4}\cos(2\tau_3-\tau_4)+\frac{B_4C_{10}}{\Omega_4^2A_4} \tag{4.965}$$

Avoiding the secular term in Eq. 4.965 we can obtain the frequency

$$\Omega_4^2=\omega^2-\frac{\omega^2A_3}{A_4}\frac{C_{11}+C_{12}}{C_{10}} \tag{4.966}$$

The first-order approximate solution can be determined from Eqs. 4.965, 4.943 for i = 4, 4.5 and 4.934:

$$\begin{aligned}
\overline{u}_4(t) &= A_4 \cos \Omega_4 t + \frac{A_4(\omega^2 - \Omega_4^2)(C_{11} + C_{12}) - \omega^2 A_3 C_{10}}{\Omega_4^2 - \Omega_3^2}\big(\cos \Omega_3 t - \\
&- \cos \Omega_4 t\big) - \frac{2B_4C_{11}}{\Omega_3(\Omega_3 + 2\Omega_4)}\big[\big(\cos(\Omega_3 + \Omega_4)t - \cos \Omega_4 t\big] - \\
&- \frac{2B_4C_{12}}{\Omega_3(\Omega_3 - 2\Omega_4)}\big[\big(\cos(\Omega_3 - \Omega_4)t - \cos \Omega_4 t\big] - \\
&- \frac{A_4(\omega^2 - \Omega_4^2)C_{11}}{3\Omega_4^2 + 4\Omega_3\Omega_4 + \Omega_3^2}\big[\big(\cos(\Omega_3 + 2\Omega_4)t - \cos \Omega_4 t\big] - \\
&- \frac{A_4(\omega^2 - \Omega_4^2)C_{12}}{3\Omega_4^2 - 4\Omega_3\Omega_4 + \Omega_3^2}\big[\big(\cos(\Omega_3 - 2\Omega_4)t - \cos \Omega_4 t\big] + \\
&+ \frac{\omega^2 A_3 C_{11}}{4\Omega_3(\Omega_3 + \Omega_4)}[(\cos(2\Omega_3 + \Omega_4)t - \cos \Omega_4 t)] + \\
&+ \frac{\omega^2 A_3 C_{12}}{4\Omega_3(\Omega_3 - \Omega_4)}[(\cos(2\Omega_3 - \Omega_4)t - \cos \Omega_4 t)] + \\
&+ \frac{B_4 C_{10}}{\Omega_4^2}(1 - \cos \Omega_4 t)
\end{aligned} \tag{4.967}$$

4.29.1 Numerical Examples

For $B_1 = 1.37827$, $B_2 = 0.133455$, $B_3 = -0.0372504$, $B_4 = 0.00124245$, using the collocation method we obtain the frequencies:

$$\Omega_1 = 1.18231846250, \Omega_2 = 0.2363612189683, \Omega_3 = 1.183115695,$$

$$\Omega_4 = 0.236972271$$

and the values of the convergence-control parameters

$$\begin{aligned}
&C_1 = 0.254339389, C_2 = 0.679531519, C_3 = -0.164675359, C_4 = -1.327612893, \\
&C_5 = -0.043060136, C_6 = -0.006009376, C_7 = 0.1462105902, C_8 = 0.666667057, \\
&C_9 = 0.168113325, C_{10} = -1.174943839, C_{11} = 0.074249382, C_{12} = -0.110297346
\end{aligned}$$

The first-order approximate solution of the system 4.929 becomes

$$\begin{aligned}
&\overline{u}_1(t) = 0.398398485 \cos \Omega_1 t + 5.636 \cdot 10^{-6} \cos 3\Omega_1 t - 0.000049051 \cos 5\Omega_1 t + \\
&+ 5.242959367 \cdot 10^{-6} \cos 7\Omega_1 t + 0.002359031 \cos(2\Omega_1 - \Omega_2) + \\
&+ 0.000585138 \cos(4\Omega_1 - \Omega_2) - 0.003531592 \cos \Omega_2 + \\
&+ 0.002359031 \cos(2\Omega_1 + \Omega_2) - 0.000131922 \cos(4\Omega_1 + \Omega_2)
\end{aligned} \tag{4.968}$$

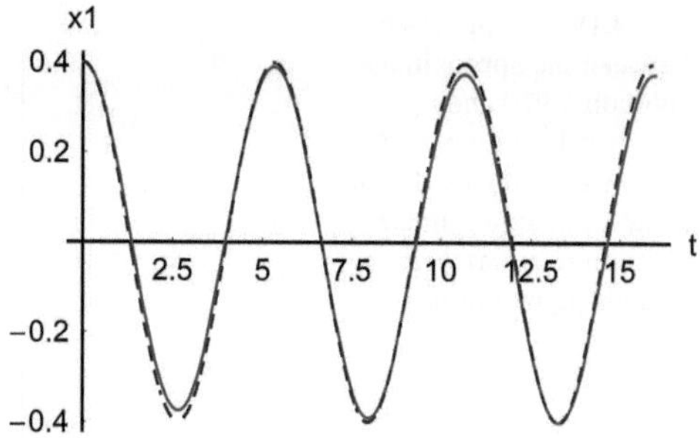

Fig. 4.193 Comparison between the approximate solution 4.968 and numerical solution of the Eq. 4.929: numerical (*red solid line*), approximate solution (*blue dashed line*) (colour figure online)

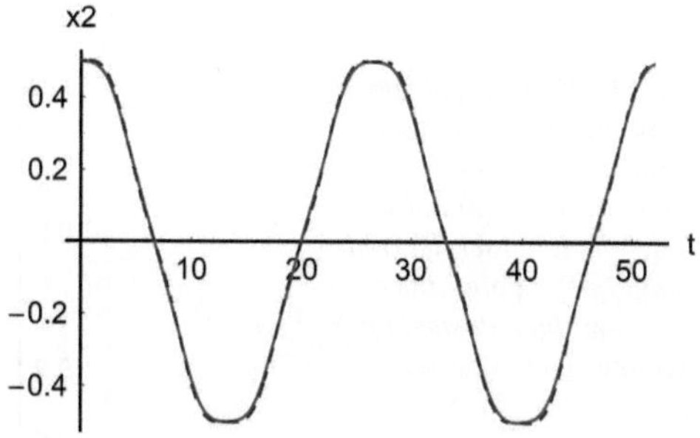

Fig. 4.194 Comparison between the approximate solution 4.969 and numerical solution of the Eq. 4.929: numerical (*red solid line*), approximate solution (*blue dashed line*) (colour figure online)

$$\begin{aligned}\overline{u}_2(t) = &-0.0227605\cos\Omega_1 + 0.0000114354\cos(\Omega_1 - 2\Omega)- \\ &-0.0000309485\cos(\Omega_1 - \Omega_2) + 0.522914\cos\Omega_2- \\ &-0.000147853\cos(2\Omega_1 + \Omega_2) + 0.0000136629\cos(\Omega_1 + 2\Omega_2)\end{aligned} \tag{4.969}$$

$$\begin{aligned}\overline{u}_3(t) = &\,0.000129778 + 0.395640399\cos\Omega_3 - 0.000394495\cos 2\Omega_3+ \\ &+0.000135587\cos 3\Omega_3 - 0.000019895\cos 4\Omega_3 - 0.0000314005\cos 5\Omega_3- \\ &-5.342709195\cdot 10^{-6}\cos 7\Omega_3 + 0.003961994\cos(2\Omega_3 - \Omega_4)+ \\ &+0.000166464\cos(4\Omega_3 - \Omega_4) - 0.002026786\cos\Omega_4+ \\ &+0.002309286\cos(2\Omega_3 + \Omega_4) + 0.000134409\cos(4\Omega_3 + \Omega_4)\end{aligned} \tag{4.970}$$

$$\begin{aligned}\overline{u}_4(t) = &-0.0259957 - 0.0201179\cos\Omega_3 + 0.000174419\cos(\Omega_3 - 2\Omega_4)+ \\ &+0.00032666\cos(\Omega_3 - \Omega_4) - 0.000567242\cos(2\Omega_3 - \Omega_4) + 0.546\cos\Omega_4- \\ &\,0.0000941102\cos(\Omega_3 + \Omega_4) + 0.000254412\cos(2\Omega_3 + \Omega_4)+ \\ &+0.000019503\cos(\Omega_3 + 2\Omega_4)\end{aligned} \tag{4.971}$$

It is easy to verify the accuracy of the obtained results if we compare this analytical solution with the numerical ones. It can be seen from Figs.4.193, 4.194, 4.195, 4.196 that the analytical solution obtained by OHAM is very accurate.

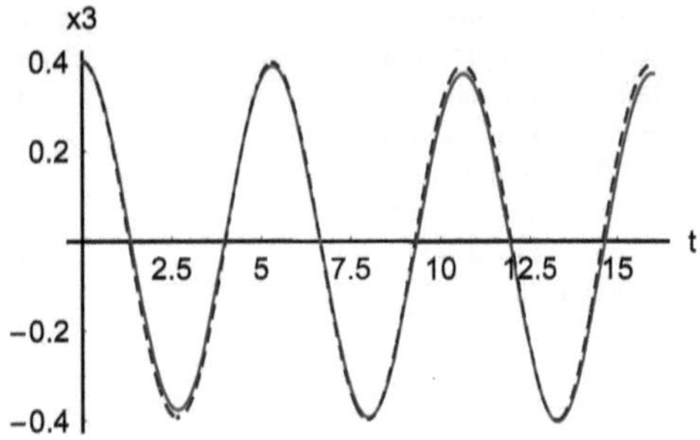

Fig. 4.195 Comparison between the approximate solution 4.970 and numerical solution of the Eq. 4.929: numerical (*red solid line*), approximate solution (*blue dashed line*) (colour figure online)

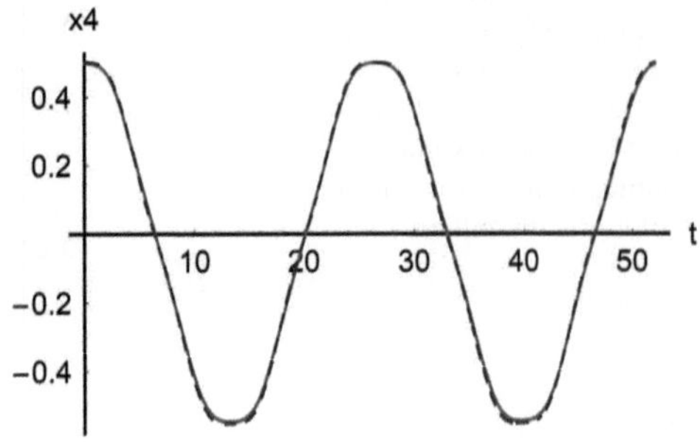

Fig. 4.196 Comparison between the approximate solution 4.971 and numerical solution of the Eq. 4.929: numerical (*red solid line*), approximate solution (*blue dashed line*) (colour figure online)

References

1. K.C. Chen, M.R. Char, J.W. Cleaver, Temperature field in non-Newtonian flow over a stretching plate. J. Math. Anal. Appl. **151**, 301–307 (1990)
2. S. Bariş, Steady three-dimensional flow of a Walters' B' fluid in a vertical channel. Turkish J. Eng. Env. Sci. **26**, 385–394 (2002)
3. P.D. Ariel, On the flow of an elastico-viscous fluid near a rotating disk. J. Comput. Appl. Math. **154**, 1–25 (2003)
4. R. Mahapatra, S. Dholey, A.S. Gupta, Oblique stagnation-point flow of an incompressible visco-elastic fluid towards a stretching surface. Int. J. Non-Linear Mech. **42**, 484–499 (2007)
5. A. Postelnicu, Heat and mass transfer by natural convection at a stagnation point in a porous medium considering Soret and Dufour effects. Heat Mass Transf. **46**, 831–840 (2010)
6. T.B. Chang, A. Mehmood, O.A. Bég, M. Narahari, M.N. Islam, F. Ameen, Numerical study of transient free convective mass transfer in Walters-B viscoelsctic flow with wall suction. Commun. Nonlinear Sci. Numer. Simul. **16**, 216–225 (2011)
7. O.U. Mehmood, S. Shafie, N. Mustapha, Peristaltic transport of Walters' B fluid in an asymmetric channel. Int. J. Appl. Math. Mech. **7**, 1–19 (2011)
8. S.A.M. Tonekaboni, R. Abkar, R. Khoeliar, On the study of viscoelastic Walters' B fluid in boundary layer flows, Math. Probl. Eng. Art. ID 861508 (2012)
9. H. Amatousse, H.A. Abderrahmane, N. Mehidi, Traveling waves on a falling weakly viscoelastic fluid film. Int. J. Eng. Sci. **54**, 27–41 (2012)
10. M.C. Lin, C.K. Chen, Finite amplitude long-wave instability of a viscoelastic fluid during spin coating. Appl. Math. Model. **36**, 2536–2549 (2012)

11. B. Rushi Kumar, R. Sivaraj, MHD viscoelastic fluid non-Darcy flow over a vertical cone and a flat plate. Int. Commun. Heart Mass Transf. **40**, 1–6 (2013)
12. D.W. Beard, K. Walters, Elastico-viscous boundary layer flows. Part I: Two-dimensional flow near a stagnation point. Proc. Camb. Philos. Soc. **60**, 667–674 (1964)
13. A.A. Joneidi, G. Domairry, M. Babaelahi, Homotopy analysis method to Walter's B fluid in a vertical channel with porous wall. Meccanica **45**, 887–868 (2010)
14. K. Batchelor, *An introduction to fluid dynamics, Cambridge University Press*, 1967
15. J.G. Oldroyd, On the formulation of rheological equations of state. Proc. R. Soc. Lond. A **200**, 523–541 (1950)
16. K. Strauss, Model der Ströming die sich am Rakelnesser einer Beschichtungs-anlage einstelt. Rheol. Acta **14**, 1058–1065 (1975)
17. C. Hancock, E. Lewis, N.K. Moffat, Effects of inertia in forced corner flows. J. Fluid Mech. **112**, 315–327 (1981)
18. K.R. Rajagopal, R.K. Bhatnagar, Exact solutions for some simple flows of an Oldroyd-B fluid. Acta Mech. **113**, 233–239 (1995)
19. S. Bariş, Flow of an Oldroyd 6-constant fluid between intersecting planes, one of which is moving. Acta Mech. **147**, 125–135 (2001)
20. T. Hayat, M. Khan, S. Asghar, Homotopy analysis of MHD flows of an Oldroyd 8-constant fluid. Acta Mech. **168**, 213–232 (2004)
21. T. Hayat, M. Khan, M. Ayub, Couette and Poiseuille flows of an Oldroyd 6-constant fluid with magnetic field. J. Math. Anal. Appl. **298**, 225–244 (2004)
22. Y. Wang, T. Hayat, K. Hutter, On non-linear magnetohydrodynamic problems of an Oldroyd 6-constant fluid. Int. J. Non-Linear Mech. **40**, 49–58 (2005)
23. M. Sajid, A.M. Siddiqui, T. Hayat, Wire coating analysis using MHD Oldroyd 8-constant fluid. Int. J. Eng. Sci. **45**, 381–392 (2007)
24. M. Khan, T. Hayat, M. Ayub, Numerical study of partial slip of the MHD flow of an Oldroyd 8-constant fluid. Comput. Math. Appl. **53**, 1088–1097 (2007)
25. T. Hayat, M. Khan, M. Ayub, The effect of the slip condition on flows of an Oldroyd 6-constant fluid. J. Comput. Appl. Math. **202**, 402–413 (2007)
26. R. Ellahi, T. Hayat, T. Javed, S. Asghar, On the analytic solution of nonlinear flow problem involving Oldroyd 8-constant fluid. Math. Comput. Modell. **48**, 1191–1200 (2008)
27. M. Sajid, T. Hayat, Thin film flow of an Oldroyd 8-constant fluid: an exact solution. Phys. Lett. A **372**, 1827–1830 (2008)
28. D. Vieru, C. Fetecău, C. Fetecău, Flow of a generalized Oldroyd-B fluid due to a constantly accelerating plate. Appl. Math. Comput. **201**, 834–842 (2008)
29. T. Hayat, R. Ellahi, F.M. Mahomed, Exact solution of a thin film flow of an Oldroyd 6-constant fluid over a moving belt. Commun. Nonlinear. Sci. Numer. Simul. **14**, 133–139 (2009)
30. R. Ellahi, T. Hayat, F.M. Mahomed, A. Zeeshan, Exact solutions for flow of an Oldroyd 8-constant fluid with nonlinear slip conditions. Commun. Nonlinear. Sci. Numer. Simul. **15**, 322–330 (2010)
31. C. Fetecău, T. Hayat, J. Zierep, M. Sajid, Energetic balance for the Rayleigh-Stokes problem of an Oldroyd-B fluid. Nonlinear Anal. Real World Appl. **12**, 1–13 (2011)
32. L. Zheng, Z. Guo, X. Zhang, 3D flow of a generalized Oldroyd-B fluid induced by a constant pressure gradient between two side walls perpendicular to a plate. Nonlinear Anal. Real World Appl. **12**, 3499–3508 (2011)
33. R.A. Shah, S. Islam, A.M. Siddiqui, T. Haroon, Wire coating analysis with Oldroyd 8-constant fluid by optimal homotopy asymptotic method. Comput. Math. Appl. **63**, 695–707 (2012)
34. T. Hayat, S. Zaib, S. Asghar, A.A. Hendi, Exact solutions in generalized Oldroyd-B fluid. Appl. Math. Mech. Eng. Ed. **34**(4), 411–426 (2012)
35. V.M. Falkner, S.W. Skan, Some approximate solutions of the boundary layer equations. Philos. Mag. **12**, 865–816 (1931)

36. D.R. Hartree, On the equation occurring in Falkner and Skan's approximate treatment of the equations of the boundary layer. Proc. Camb. Philos. Soc. **33**, 223–239 (1937)
37. K. Stewartson, Further solutions of the Falkner-Skan equation. Proc. Camb. Philos. Soc. **50**, 454–465 (1954)
38. S.P. Hastings, Reversed flow solutions of the Falkner-Skan equation, SIAM. J. Appl. Math. **22**(2), 329–334 (1972)
39. E.F.F. Botta, F.J. Hut, A.E.P. Veldman, The role of periodic solution in the Falknar-Skan problem for $\lambda \geq 0$. J. Eng. Math. **20**(1), 81–93 (1986)
40. K.Y. Yih, Uniform suction/blowing effect on forced convection about a wedge uniform heat flow. Acta Mech. **128**(3–4), 173–181 (1998)
41. A. Asaithambi, A finite difference method for the Falkner-Skan equation. Appl. Math. Comput. **92**(23), 135–141 (1998)
42. M.B. Zaturska, W.N. Banks, A new method of the Falkner-Skan equation. Acta Mech. **152**, 197–201 (2001)
43. B.L. Kuo, Application of the differential transformation method to the solutions of Falkner-Skan flow. Acta Mech. **164**, 161–174 (2003)
44. N.S. Elgazery, Numerical solution for the Falkner-Skan equation. Chaos Solitons Fractals **35**, 738–746 (2008)
45. E. Alizadeh, M. Farhadi, K. Sedighi, N.R.E. Kebria, A. Ghafourian, Solution of the Falkner-Skan equation for wedge by Adomian decomposition method. Commun. Nonlinear Sci. Numer. Simul. **14**(724–733) (2009)
46. S. Abbasbandy, T. Hayat, Solution of the MHD Falkner-Skan flow by Hankel-Padé method. Phys. Lett. A **373**(3), 731–734 (2009)
47. S. Bernhard, S. Mohlenkamp, A. Tilgner, Transient integral boundary layer method to calculate the translesional pressure drop and the fractional flow reserve in myocardical bridges, BioMed Eng Online. **5**, 42. doi: 101186.1478-925x-5-42 (2006)
48. A. Pirkhedri, H.H.S. Javadi, K. Parand, N. Fatahi, S. Lotfi, Solving MHD Falkner-Skan boundary-layer equation using collocation method based on rational Legendre function with transformed Hermite-Gauss node. World Appl. Sci. J. **13**(10), 2224–2230 (2011)
49. K. Ahmad, R. Nazar, I. Pop, Falkner-Skan for gravity-driven film flow of micropolar fluid. Sains Malays. **40**(11), 1293–1296 (2011)
50. M. Lakestani, Numerical solution for the Falkner-Skan equation using Chebyshev cardinal functions. Acta Univ. Apulensis **27**, 229–238 (2011)
51. B.Y. Yun, New approximate analytical solutions of the Falkner-Skan equation, J. Appl. Math., Article ID 170802 (2012)
52. F.A. Hendi, M. Hussian, Analytic solution for MHD Falkner-Skan flow over a porous surface, J. Appl. Math., Article ID 123185 (2012)
53. V. Marinca, R.D. Ene, B. Marinca, Analytic approximate solution for Falkner-Skan equation, Sci. World J., Art. ID 617453 (2014)
54. B.C. Sakiadis, Boundary layer behavior on continuous solid surfaces. I. Boundary layer equations for two dimensional and axisymmetric flow. AIChE J. **7**, 26–28 (1961)
55. B.C. Sakiadis, Boundary layer behavior on continuous solid surfaces. II. Boundary layer on a continuous flat surface. AIChE J. **7**, 221–225 (1961)
56. L.J. Crane, Flow past a stretching plate. Z. Angew Math. Phys. (ZAMP) **21**, 645–647 (1970)
57. E.M. Sparrow, G.S. Beavers, L.Y. Hung, Flow about a porous-surfaced rotating disk. Int. J. Heat Mass Trans. **14**, 993–996 (1971)
58. E.M. Sparrow, G.S. Beavers, L.Y. Hung, Channel and tube flows with surface mass transfer and velocity slip. Phys. Fluids **14**, 1312–1319 (1971)
59. J.A.D. Ackroyd, A series method for the solution of laminar boundary layers on moving surfaces. Z. Angew. Math. Phys. (ZAMP) **29**, 729–741 (1978)
60. K.R. Rajagopal, T.Y. Na, A.S. Gupta, Flow of a viscoelastic fluid over a stretching sheet. Rheol. Acta **23**(2), 213–215 (1984)

61. P.D. Ariel, MHD flow of a viscoelastic fluid past a stretching sheet with suction. Acta Mech. **105**, 49–56 (1994)
62. H.I. Anderson, Slip flow of a viscoelastic fluid past a stretching surface. Acta Mech. **95**, 227–230 (1992)
63. C.Y. Wang, Analysis of viscous flow due to a stretching sheet with surface slip and suction. Nonlinear Anal. Real World Appl. **10**(1), 375–380 (2009)
64. H. Mirogolbabei, D.D. Ganji, M.M. Etghani, A. Sobati, Adapted varational iteration method and axisymmetric flow over a stretching sheet. World J. Model. Simul. **5**(4), 307–314 (2009)
65. M. Sajid, I. Ahmed, T. Hayat, M. Ayub, Series solution for unsteady axisymmetric flow and heat transfer over a radially stretching sheet. Commun. Nonlinear Sci. Numer. Simul. **13**(10), 2193–2202 (2008)
66. M. Sajid, T. Hayat, S. Asghar, K. Vajravelu, Analytic solution for axisymmetric flow over a nonlinearity stretching sheet. Arch. Appl. Mech. **78**(2), 127–134 (2007)
67. B. Sahoo, Effect of slip, viscous dissipation and Joule heating on the MHD flow and heat transfer of a second grade fluid past a radially stretching sheet. Appl. Math. Mech. Engl. Ed. **31**(2), 159–173 (2010)
68. M. Miklavčič, C.Y. Wang, Viscous flow due to a shrinking sheet. Q. Appl. Math. **64** (283–290) (2006)
69. K.V. Prasad, K. Vajravelu, P.S. Dutt, The effects of variable fluid properties on the hydromagnetic flow and heat transfer over a non-linearity stretching sheet. Int. J. Therm. Sci. **40**, 603–610 (2010)
70. M. Turkyilmazoglu, Multiple solutions of heat and mass transfer of MHD slip flow for the viscoelastic fluid over a stretching sheet. Int. J. Therm. Sci. **50**, 2264–2276 (2011)
71. F. Labropulu, D. Li, I. Pop, Non-orthogonal stagnation-point flow over a stretching surface in a non-Newtonian fluid with heat transfer. Int. J. Therm. Sci. **49**, 1042–1050 (2010)
72. C.Y. Wang, C.-O. Ng, Slip flow due to a stretching cylinder. Int. J. Non-Linear Mech. **46**, 1191–1194 (2011)
73. M. Sajid, K. Mahmood, Z. Abbas, Axisymmetric stagnation-point flow with a general slip boundary condition over a lubricated surface. Chin. Phys. Lett. **29**(2), 307–310 (2012)
74. F.K. Tsou, E.M. Sparrow, R.J. Goldstein, Flow and heat transfer in the boundary layer on a continuous moving surface. Int. J. Heat Mass Transf. **10**, 219–235 (1967)
75. P.S. Gupta, A.S. Gupta, Heat and mass transfer on a stretching sheet with suction or blowing. Can. J. Chem. Eng. **55**, 744–746 (1977)
76. C.E. Maneschy, M. Massoudi, K.R. Rajagopal, Flow of a second-grade fluid over a porous elastic sheet due to stretching. J. Math. Phys. Sci. **27**, 353–367 (1993)
77. L.J. Grubka, K.M. Bobba, Heat transfer characteristics of a continuous stretching surface with variable temperature. J. Heat Mass Transf. **107**, 248–250 (1985)
78. C.Y. Wang, Liquid film on an unsteady stretching surface, Q. Appl. Math. XLVIII, 601–610 (1990)
79. R. Usha, S. Rukmani, The axisymmetric motion of a liquid film on an unsteady stretching surface, ASME. J. Fluids Eng. **117**, 81–85 (1995)
80. H.J. Anderson, J.B. Aaresth, D.S. Dandapat, Heat transfer in a liquid film on an unsteady stretching surface. Int. J. Heat Mass Transf. **43**, 69–74 (2000)
81. M.E. Ali, On thermal boundary layer on a power-law stretched surface with suction or injection. Int. J. Heat Fluid Flow **16**, 280–290 (1995)
82. E. Magyari, M.E. Ali, B. Keller, Heat and mass transfer characteristics of the self-similar boundary-layer flows induced by continuous surface stretched with rapidly decreasing velocities. Heat Mass Transf. **38**, 65–74 (2001)
83. K. Vajravelu, Viscous flow over a nonlinear stretching sheet. Appl. Math. Comput. **124**, 281–288 (2001)
84. E. Magyari, B. Keller, A direct method to calculate the heat transfer coefficient of steady similar boundary layer flows induced by continuous moving surfaces. Int. J. Therm. Sci. **44**, 245–254 (2005)

85. E.M. Elbashbeshy, M.A.A. Bazid, Heat transfer an unsteady stretching surface. Heat Mass Transf. **41**, 1–4 (2004)
86. B.S. Dandapat, A. Kitamura, B. Santra, Transient film profile on thin liquid film flow on a stretching surface. Z. Angew. Math. Phys. **57**, 623–635 (2006)
87. M.E. Ali, E. Magyari, Unsteady fluid and heat flow induced by a submerged stretching surface while its steady motion is slowed down gradually. Heat Mass Transf. **50**, 188–195 (2007)
88. I. Chung Liu, H.I. Anderson, Heat transfer in a liquid film on an unsteady stretching sheet. Int. J. Thermal Sciences **47**, 766–772 (2008)
89. C.H. Chen, Magneto-hydrodynamic mixed convection of a power-law fluid past a stretching surface in the presence of thermal radiation and internal heat generation/absorption. Int. J. Non-Linear Mech. **44**, 596–603 (2009)
90. B.S. Dandapat, B. Santra, S.K. Singh, Thin film flow over a nonlinear stretching sheet in presence of uniform transverse magnetic field. Z. Angew. Math. Phys. **61**, 685–695 (2010)
91. R. Cortell, Heat and fluid flow due to non-linearly stretching surfaces. Appl. Math. Comput. **217**, 7564–7572 (2011)
92. V. Marinca, N. Herisanu, *Nonlinear Dynamical Systems in Engineering* (Some Approximate Approaches, Springer, Berlin Heidelberg, 2011)
93. V. Marinca, N. Herisanu, The optimal homotopy asymptotic method for solving Blasius equation. Appl. Math. Comput. **231**, 134–139 (2014)
94. S.J. Liao, *Beyond Perturbation-Introduction to the Homotopy Analysis Method* (Chapman and Hall, CRC, Boca Raton, 2004)
95. S.J. Liao, An explicit, totally analytic approximate solution for Blasius viscous flow problems. Int. J. Non-Linear Mech. **34**, 759–778 (1999)
96. H. Schlichting, *Boundary Layer Theory* (McGraw, New York, NY, 1968)
97. H.K. Kuiken, A backward free-convective boundary layer. Q. J. Mech. Appl. Math. **34**, 397–413 (1981)
98. H. Blasius, Grenzschichten in Flussigkeiten mit kleiner Reibung. Z. Math. Phys. **56**, 1–37 (1908)
99. L. Howarth, On the solution of the laminar boundary layer equations. Proc. R. Soc. Lond. A **164**, 547–579 (1938)
100. J.H. He, A simple perturbation approach to Blasius equation. Appl. Math. Comput. **140**, 217–222 (2003)
101. A. Asaithambi, Solution of the Falkner-Skan equation by recursive evaluation of Taylor coefficients. J. Comput. Appl. Math. **176**, 203–214 (2005)
102. A.M. Wazwaz, The variational iteration method for solving two forms of Blasius equation on a half-infinite domain. Appl. Math. Comput. **188**, 485–491 (2007)
103. A.M. Wazwaz, A reliable algorithm for solving boundary value problems for higher-order integro-differential equations. Appl. Math. Comput. **118**, 327–342 (2001)
104. L. Wang, A new algorithm for solving classical Blasius equation. Appl. Math. Comput. **157**, 1–9 (2004)
105. I. Hashim, Comments on "A new algorithm for solving classical Blasius equation" by L. Wang. Appl. Math. Comput. **176**, 700–703 (2006)
106. R. Fazio, Numerical transformation methods: Blasius problem and its variants. Appl. Math. Comput. **215**, 1513–1521 (2009)
107. M. Parand, A. Dehghan, I. Pirkhedri, Sinc-collocation method for solving the Blasius equation. Phys. Lett. A **373**, 4060–4065 (2004)
108. B. Yao, J. Chen, A new analytical solution branch for the Blasius equation with a shrinking sheet. Appl. Math. Comput. **215**, 1146–1159 (2009)
109. B.I. Yun, Intuitive approach to the approximate analytical solutions for the Blasius problem. Appl. Math. Comput. **215**, 3489–3494 (2010)
110. E.M.A. Elbashbeshy, M.F. Dimian, Effect of radiation on the flow and heat transfer over a wedge with variable viscosity. Appl. Math. Comput. **132**, 445–454 (2002)

111. M.E.M. Ouaf, Exact solution of thermal radiation on MHD over a stretching porous sheet. Appl. Math. Comput. **170**, 1117–1125 (2003)
112. V. Marinca, N. Herişanu, Application of Optimal Homotopy Asymptotic Method for solving nonlinear equations arriving in heat transfer. Int. Commun. Heat Mass Transf. **35**, 710–715 (2008)
113. S. Abbasbandy, The application of homotopy analysis method to nonlinear equations arising in heat transfer. Phys. Lett. A **360**, 109–113 (2006)
114. S. Abbasbandy, Homotopy analysis method for heat radiation equations. Int. Commun. Heat Mass Transf. **34**(3), 380–387 (2007)
115. A. Rajabi, D.D. Ganji, N. Taherian, Application of homotopy perturbation method in nonlinear beat conduction and connection equations. Phys. Lett. A **360**, 570–573 (2007)
116. D.D. Ganji, The application of He's homotopy perturbation method to nonlinear equations arising in heat transfer. Phys. Lett. A **355**, 337–341 (2006)
117. C.W. Bert, Application of differential transform method in heat conduction in tapered fins. ASME J. Heat Transf. **124**, 208–209 (2002)
118. H.P. Chu, C.I. Chen, Hybrid differential transform and finite difference method to solve the nonlinear heat conduction problem. Commun. Nonlinear. Sci. Numer. Simul. **13**, 1605–1614 (2008)
119. D.D. Ganji, G.A.A. Afrouzi, R.A. Talarposhti, Application of variational iteration method and homotopy perturbation method for nonlinear heat diffusion and heat transfer equations. Phys. Lett. A **368**(450–457) (2007)
120. A. Aziz, T.Y. Na, *Perturbation method in Heat Transfer* (Hemisphere, New York, NY, 1984)
121. D.D. Ganji, A. Rajabi, Assessement of homotopy-perturbation method in heat radiation equations. Int. Comun. Heat Mass Transf. **33**, 397–400 (2006)
122. H. Yaghobi, M. Torabi, The application of differential transformation method to nonlinear equations arising in heat transfer. Int. Commun. Heat Mass Transf. **38**, 815–820 (2011)
123. V. Marinca, N. Herisanu, Optimal homotpy perturbation method for nonlinear differential equations. Nonlinear Sci. Lett. A **1**, 273–280 (2010)
124. L.M. Abia, O. Angulo, J.C. Lopez-Marcos, Age-structured population models and their numerical solution. Ecol. Modell. **188**, 112–136 (2005)
125. F.R. Sharpe, A.J. Lotka, A problem in age distribution. Philos. Mag. **21**, 435–438 (1911)
126. A.G. McKendick, Applications of the mathematics to medical problems. Proc. Edinburgh Math. Soc. **44**, 98–130 (1926)
127. A.J. Lotka, The structure of a growing population. Human Biol. **3**, 459–493 (1931)
128. T.R. Malthus, *An essay on the principle of population* (Cambridge University Press, Cambridge, 1992)
129. M.E. Gurtin, R.C. MacCamy, Nonlinear age-structured dependent population dynamics. Arch. Rat. Mech. Anal. **54**, 281–300 (1974)
130. E. Venturino, Nonlinearly interacting age-dependent population. Comput. Math. Appl. **13**, 901–911 (1987)
131. J. Douglas, F.A. Milner, Numerical methods for a model of population dynamics. Calcolo **24**, 247–254 (1987)
132. L.M. Abia, J.C. Lopez-Marcos, Runge-Kutta methods for age-structured population models. Appl. Numer. Math. **17**, 1–17 (1995)
133. M. Iannelli, F.A. Milner, On the approximate of the Lotka-Mc.Kendrick equations with finite life-span. J. Comput. Appl. Math. **136**, 245–254 (2001)
134. M.Y. Kim, Y. Kwon, A collocation method for the Gurtin Mc.Camy equation with finite life span, SIAM J Numer. Anal. **39**(6), 1914–1937 (2002)
135. M. Cai, Z. Chen, The exact solution of nonlinear age-structured population model. Nonlinear Anal. Real World Appl. **8**, 1096–1112 (2007)
136. X. Li, Variational iteration method for nonlinear age-structured population model. Comput. Math. Appl. **58**, 2177–2181 (2009)

137. O. Angulo, J.C. Lopez-Marcos, M.A. Lopez-Marcos, F.A. Milner, A numerical method for nonlinear age-structured population models with finite maximum age. J. Math. Anal. Appl. **361**, 150–160 (2010)
138. M. Goreishi, A.I.B. Md, A.K. Ismail, A.S.B. Alomaic, The comparison between homotopy analysis method and optimal homotopy asymptotic method for nonlinear age-structured population models. Commun. Nonlinear. Sci. Numer. Simul. **17**(1163–1177) (2012)
139. F.M. Scudo, Volterra and theoretical ecology. Theor. Populat. Biol. **2**, 1–23 (1971)
140. R.D. Small, *Population growth in a closed model Mathematical modelling Classroom notes in applied mathematics* (SIAM, Philadelphia, PA, 1989)
141. A.J. Lotka, *Elements of Physical Biology* (William and Wilkins, Baltimore, 1925)
142. V. Volterra, Variazioni e fluttuazioni del numero d'individui in specie animali conniventi. Mem. R. Acad. Naz. Lincei **2–3**, 30–111 (1926)
143. V. Marinca, N. Herisanu, Optimal parametric iteration method for solving multispecies Loka-Volterra equations. Discr. Dynam. Nat. Soc. Art. ID **842121** (2012). doi:10.1155/2012/842121
144. G. Bratu, Sur les equations integrales non lineaires. Bull. Soc. Math. France **43**, 113–142 (1914)
145. J. Liouville, Sur l'equation aux derivees partielles $\frac{d^2 \ln\lambda}{dudv} \pm 2\lambda a^2 = 0$. J. de Math. Pure Appl. **18**, 71–72 (1853)
146. J. Jacobsen, K. Schmitt, The Liouville-Bratu-Gelfand problem for radial operators. J. Diff. Equations **184**, 283–289 (2002)
147. S.G. Kenkatesh, S.K. Ayyaswany, S.R. Balachandar, The Legendre wavelet method for solving initial value problem of Bratu-type. Comput. Math. Appl. **63**, 1287–1295 (2012)
148. Y.Q. Wan, Q. Guo, H. Pan, Thermo-electro-hydrodynamic model for electrospinning process. Int. J. Nonlinear. Sci. Numer. Simul. **5**, 5–8 (2004)
149. H. Pan, R. Xing, Radial solutions for a prescribed mean curvature equation with exponential nonlinearity. Nonlinear Anal. **75**, 103–116 (2012)
150. J. Boyd, Chebyshev polynomial expansions for simultaneous approximation of two branches of a function with applications to the one-dimensional Bratu equation. Appl. Math. Comput. **143**, 189–200 (2003)
151. J. Boyd, One-point pseudospectral collocation for the one-dimensional Bratu equation. Appl. Math. Comput. **217**, 5553–5565 (2010)
152. J.H. He, H.M. Liu, Variational approach to nonlinear problems and a review on mathematical model of electrospinning. Nonlinear Anal. **63**, e919–e929 (2005)
153. M. Syam, The modified Broyden-variational method for solving nonlinear elliptic differential equations. Chaos Solitons Fractals **32**, 392–404 (2007)
154. R. Jalilian, Non-polynomial spline method for solving Bratu's problem. Comput. Phys. Commun. **181**, 1868–1872 (2010)
155. A.M. Wazwaz, Adomian decomposition method for a reliable treatment of the Bratu type equations. Appl. Math. Comput. **166**, 652–663 (2005)
156. S.A. Khuri, A new approach to Bratu's problem. App. Math. Comput. **147**, 131–136 (2004)
157. M.I. Syam, A. Hamdan, An efficient method for solving Bratu equations. Appl. Math. Comput. **176**, 704–713 (2006)
158. Y. Lin, J.A. Enszer, M.A. Stadherr, Enclosing all solutions of two-point boundary value problems for ODEs. Comput. Chem. Eng. **32**, 1714–1725 (2008)
159. S. Abbasbandy, E. Shivanian, prediction of multiplicity of solutions of nonlinear boundary value problems: Novel application of homotopy analysis method, Commun. Nonlinear. Sci. Numer. Simul. **15**, 3830-3846 (2010)
160. Y. Aksoy, M. Pakdemirli, New perturbation-iteration solutions for Bratu-type equations. Comput. Math. Appl. **59**, 2802–2808 (2010)
161. J.H. Lane, On the theoretical temperature of the sun, under the hypothesis of a gaseous mass maintaining its volume by its internal heat, and depending on the laws of gases as known to terrestrial experiment, Am. J. Sci. Arts, **L**, 57-74 (1870)

162. R. Emden, *Garkugeln Anwendungen der Mechan* (Warmtheorie, Druck and Verlag Von B. G., Leipzig and Berlin, 1907)
163. K. Parand, M. Shahimi, M. Dehghan, Rational Legendre pseudospectral approach for solving nonlinear differential equations of Lane–Emden type. J. Comput. Phys. **228**, 8830–8840 (2009)
164. J.I. Ramos, Piecewise quasilinearization techniques for singular boundary-value problems. Comput. Phys. Commun. **158**, 12–25 (2004)
165. H.T. Davis, *Introduction to Nonlinear Differential and Integral Equations* (Dover Publications, New York, NY, 1962)
166. S. Chandrasekhar, *Introduction to the Study of Stellar Structure* (Dover Publications, New York, 1967)
167. E. Momoniat, C. Harley, Approximate implicit solution of a Lane-Emden equation. New Astron. **11**(7), 520–526 (2006)
168. J.H. He, Variational approach to the Lane–Emden equation. Appl. Math. Comput. **143**, 539–541 (2003)
169. A.M. Wazwaz, A new algorithm for solving differential equations of Lane–Emden type. Appl. Math. Comput. **118**(2–3), 287–310 (2001)
170. S.A. Yousefi, Legendre wavelets method for solving differential equations of Lane–Emden type. Appl. Math. Comput. **181**, 1417–1422 (2006)
171. M.M. Hosseini, H. Nasabzadeh, On the convergence of Adomian decomposition method. Appl. Math. Comput. **182**, 536–543 (2006)
172. V. Marinca, N. Herisanu, An approximate solution for the nonlinear Lane-Emden type equation on a semi-infinite domain. AIP Conf. Proc. **1479**(2387–2390) (2012)
173. L. Cveticanin, I. Kovacic, Parametrically excited vibrations of an oscillator with strong cubic negative nonlinearity. J. Sound Vib. **304**, 201–212 (2007)
174. M.N. Hamdan, N.N. Shabaneh, On the large amplitude free vibrations of a restrained uniform beam carrying an intermediate lumped mass. J. Sound Vib. **199**, 711–726 (1997)
175. M.N. Hamdan, MHF Dado, Large amplitude free vibrations of a uniform cantilever beam carrying an intermediate lumped mass and rotary inertia. J. Sound Vib. **206**, 151–168 (1997)
176. N. Herisanu, V. Marinca, Explicit analytical approximation to large-amplitude non-linear oscillations of a uniform cantilever beam carrying an intermediate lumped mass and rotary inertia. Meccanica **45**, 847–855 (2010)
177. M.I. Qaisi, N.S.A. Huniti, Large amplitude free vibration of a conservative system with inertia and static non-linearity. J. Sound Vib. **242**(1), 1–7 (2001)
178. B.S. Wu, C.W. Lim, Y.F. Ma, Analytical approximation to large—amplitude oscillation of a nonlinear conservative system. Int. J. Non-Linear Mech. **38**, 1037–1043 (2003)
179. H.P. Gotlieb, Harmonic balance approach to limit cycles of nonlinear jerk equations. J. Sound Vib. **17**, 243–280 (2006)
180. B.S. Wu, C.W. Lim, W.P. Sun, Improved harmonic balance approach to periodic solutions of non-linear equations. Phys. Lett. A **354**, 95–100 (2006)
181. X. Ma, L. Wei, Z. Guo, He's homotopy perturbation method to periodic solutions of nonlinear jerk equations. J. Sound. Vib. **314**, 217–227 (2008)
182. H.P. Gotlieb, Harmonic balance approach to limit cycles of nonlinear jerk equations. J. Sound Vib. **271**, 671–685 (2004)
183. J. Awrejcewicz, I.V. Andrianov, Oscillations of non-linear system with restoring force close to sign(x). J. Sound Vib. **252**, 962–966 (2002)
184. A. Belendez, A. Hernandez, T. Belendez, C. Neipp, A. Marquez, Higher accuracy analytical approximations to a nonlinear oscillator with discontinuities by He's HPM. Phys. Lett. A. **372**, 2010–2006 (2008)
185. H.-M. Liu, Approximate period of nonlinear oscillators with discontinuities by modified Lindstedt-Poincare method. Chaos, Solitons Fractals **23**(577–579) (2005)
186. B.S. Wu, W.P. Sun, C.W. Lim, An analytical approximate technique for a class of strongly nonlinear oscillators. Int. J. Non-Linear Mech. **41**, 766–774 (2000)

187. M. Rafei, D.D. Ganji, N. Daniali, H. Pashaei, The variational iteration method for nonlinear oscillators with discontinuities. J. Sound Vib. **305**, 614–620 (2007)
188. V. Marinca, N. Herisanu, Optimal homotopy asymptotic approach to nonlinear oscillators with discontinuities. Sci. Res. Essays **8**, 161–167 (2013)
189. R.E. Mickens, K. Oyedeji, Construction of approximate analytical solutions to a new class of nonlinear oscillator equation. J. Sound Vib. **164**, 579–582 (1985)
190. R.E. Mickens, A generalized iteration procedure for calculating approximations to periodic solutions of "truly nonlinear oscillators". J. Sound Vib. **287**, 1045–1051 (2005)
191. R.E. Mickens, Iteration method solutions for conservative and limit-cycle $x^{1/3}$ force oscillators. J. Sound Vib. **292**, 964–968 (2006)
192. M. Senator, C.N. Bapat, A perturbation technique that works even when the non-linearity is not small. J. Sound. Vib. **164**, 1–27 (1993)
193. S.L. Das, A. Chatterjee, Multiple scales via Galerkin projections: approximate asymptotics for strongly nonlinear oscillations. Nonlinear. Dynam. **32**, 161–186 (2003)
194. A. Chatterjee, Harmonic balance based averaging: approximate realizations of an asymptotic technique. Nonlinear Dynam. **32**, 323–343 (2003)
195. R. Krivec, S. Mandelzweig, Numerical investigation of quasiliniarization method in quantum mechanics. Comput. Phys. Commun. **138**, 69–79 (2001)
196. C.W. Lim, B.S. Wu, A modified procedure for certain non-linear oscillators. J. Sound Vib. **257**, 202–206 (2002)
197. V. Marinca, N. Herisanu, Optimal variational method for truly nonlinear oscillators, J. Appl. Math. Art. ID 620267 (2013)
198. R.E. Mickens, *Oscillations in planar dynamic systems* (World Scientific, Singapore, 1966)
199. J. Beatty, R.E. Mickens, A qualitative study of the solutions to the differential equation $\ddot{x}+(1+\dot{x}^2)x=0$. J. Sound Vib. **283**, 471–477 (2005)
200. R.E. Mickens, Investigation of the properties of the period for the nonlinear oscillator $\ddot{x}+(1+\dot{x}^2)x=0$. J. Sound Vib. **292**, 1031–1035 (2006)
201. A. Belendez, T. Belendez, A. Hernandez, C. Neipp, A. Marquez, Asymptotic representations of the period for the nonlinear oscillator. J. Sound Vib. **299**, 403–408 (2007)
202. A. Belendez, T. Belendez, A. Hernandez, S. Gallego, M. Ortuno, C. Neipp, Comments on "Investigation of the properties of the period of nonlinear oscillator $\ddot{x}+(1+\dot{x}^2)x=0$. J. Sound Vib. **303**, 925–930 (2007)
203. A.H. Nayfeh, B. Balachandran, Modal interactions in dynamical and structural systems. Appl. Mech. Rev. **42**, 175–201 (1989)
204. A.N. Nayfeh, D. Mook, *Nonlinear Oscillations* (Wiley, New York, 1979)
205. H.P.W. Gottlieb, On the harmonic balance method for mixed-parity nonlinear oscillators. J. Sound Vib. **152**, 189–191 (1992)
206. M. Belhaq, F. Lakrad, On the elliptic harmonic balance method for mixed parity nonlinear oscillators. J. Sound Vib. **233**, 935–937 (2000)
207. H. Hu, Solution of a mixed parity nonlinear oscillator: harmonic balance. J. Sound Vib. **299**, 331–338 (2007)
208. V. Marinca, N. Herişanu, Periodic solutions for some strongly nonlinear oscillation by He's variational iteration method. Comput. Math. Appl. **54**(1188–1196) (2007)
209. V. Marinca, N. Herisanu, An optimal iteration method for strongly nonlinear oscillators, J. Appl. Math., Art. ID 906341 (2012)
210. N. Herişanu, V. Marinca, A modified variational iteration method for strongly nonlinear problems. Nonlinear Sci. Lett. A **1**(2), 183–192 (2010)
211. R.E. Mickens, Semi-classical quantization using the method of harmonic balance. Il Nuovo Cimento **101**, 359–366 (1988)
212. R.E. Mickens, Mathematical and numerical study of the Duffing-harmonic oscillator. J. Sound Vib. **244**, 563–567 (2001)
213. S.B. Tiwari, B. Nagesware Rao, N.S. Swamy, K.S. Sai, H.R. Nataraje, Analytical study on a Duffing-harmonic oscillator. J. Sound Vib. **285**, 1217–1222 (2005)

214. H. Hu, J.H. Tang, Solution of a Duffing-harmonic oscillator by the method of harmonic balance. J. Sound Vib. **294**, 637–639 (2006)
215. T. Ozis, A. Yildirim, Determination of the frequency-amplitude relation for a Duffing-harmonic oscillator by the energy balance method. Comput. Math. Appl. **54**, 1184–1187 (2007)
216. F. Dincă, C. Teodosiu, *Nonlinear and Random Vibrations* (Ed. Academiei R.S. România – Academic Press, New York, NY, 1973)
217. G. Duffing, *Erzwungene schwingungen bei veranderlicher eigenfrequenz und ihre technische bedentung, Series Sammlung Vieweg, 41/42* (Vieweg Sohen, Bramentweig, 1918)
218. M.J. Brennan, I. Kovacic, Examples of physical systems described by the Duffing equation, in The Duffing Equation: Nonlinear Oscillators and Their Behavior, ed. by I. Kovacic, M.J. Brennan (Wiley, New York, NY, 2011)
219. Y.K. Cheung, S.H. Chen, S.L. Lau, A modified Lindstedt-Poincare method for certain strongly non-linear oscillators. Int. J. Non-Linear Mech. **26**, 367–378 (1991)
220. S.E. Jones, Remarks on the perturbation process for certain conservative systems. Int. J. Non-Linear Mech. **13**, 121–132 (1978)
221. V. Marinca, N. Herişanu, A modified iteration perturbation method for some nonlinear oscillation problems. Acta Mech. **184**, 231–242 (2006)
222. S.J. Liao, A.T. Chwang, Application of homotopy analysis method in nonlinear oscillations. ASME J. Appl. Mech. **65**, 914–922 (1998)
223. J.H. He, Preliminary report on the energy balance for nonlinear oscillations. Mech. Res. Commun. **29**, 107–111 (2002)
224. V. Marinca, An approximate solution for one-dimensional weakly nonlinear oscillations. Int. J. Nonlinear Sci. Numer. Simul. **3**, 107–120 (2002)
225. J.H. He, Homotopy perturbation method: a new nonlinear analytical technique. Appl. Math. Comput. **133**, 73–79 (2003)
226. V. Marinca, N. Herisanu, Periodic solutions of Duffing equation with strong nonlinearity. Chaos Solitons Fractals **37**, 144–149 (2008)
227. H.S.Y. Chen, K.W. Chung, Z. Xu, A perturbation incremental method for strongly nonlinear oscillators. Int. J. Non-Linear Mech. **31**, 59–67 (1996)
228. B. Van der Pol, On relaxation-oscillations. Lond. Edinb. Dubl. Philos. Mag. J. Sci. **2**(7), 901–912 (1926)
229. L. Cveticanin, On the Van der Pol oscillator. An overview, Appl. Mech. Mat. **430**, 3-13 (2013)
230. B. Van der Pol, J. Van der Mark, The heartbeat considered as a relaxation oscillation and an electrical model of the hearth. Philos. Mag. Suppl. **6**, 763–775 (1928)
231. A. Liénard, Étude des oscillations entrenues, Rev. Génerale de l'Électricité, **23**, 946 – 554 (1928)
232. Q. Ding, A.Y.T. Leung, The number of limit cycle bifurcation diagrams for the generalized mixed Rayleigh-Lienard oscillator. J. Sound Vib. **322**, 393–400 (2004)
233. S. Lynch, C.J. Christopher, Limit cycles in highly non-linear differential equations. J. Sound Vib. **224**, 505–517 (1999)
234. L.D. Akulenko, L.I. Korovina, S.A. Kumakshev, S.V. Nesterov, Self-sustained oscillations of Rayligh and Van der Pol oscillators with moderately lerge feedback factors. J. Appl. Math. Mech. **68**, 241–248 (2004)
235. L. Cveticanin, Van der Pol oscillator with time-variable parameters. Acta Mech. **224**, 945–955 (2013)
236. M.P. Cartmell, S.W. Ziegler, R. Khanin, D.I.M. Forehand, Multiple scales analyses of the dynamics of weakly nonlinear systems, Appl. Mech. Rev. **56**, 455–492 (2003)
237. T.D. Burton, Non-linear oscillator limit cycle analysis using time transformation approach. Int. J. Non-Linear Mech. **17**, 7–19 (1982)
238. O.N. Minorsky, *Non-Linear Oscillations* (Princeton, van Nostrand, 1962)

239. S.J. Liao, An asymptotic approximate approach for free oscillations of self-excited system. Int. J. Non-Linear Mech. **39**, 271–280 (2004)
240. Y.M. Chen, J.K. Liu, A study of homotopy analysis method for limit cycle of Van der Pol equation. Commun. Nonlinear Sci. Numer. Simul. **14**, 1816–1821 (2009)
241. N. Herişanu, V. Marinca, An iteration procedure with application to Van der Pol oscillator. Int. J. Nonlinear Sci. Numer. Simul. **10**(3), 353–361 (2009)
242. V.D. Nguyen, Some properties of the generalized Van der Pol equation. J. Theor. Phys. **17**, 1183–1190 (1976)
243. C. Obi, Analytical theory of non-linear oscillations IV: The periodic oscillations of the equation $\ddot{x} - \varepsilon(1 - x^{2n+2})\dot{x} + x^{2n+1} = \varepsilon a \cos \omega t$, $a > 0$, $\omega > 0$ independent of ε. SIAM J. Appl. Math. **31**, 345–357 (1976)
244. G.M. Moremedi, D.P. Mason, V.M. Gorringe, On the limit cycle of a generalized Van der Pol equation. Int. J. Non-Linear Mech. **28**(2), 237–250 (1993)
245. I. Kovacic, R.E. Mickens, A generalized Van der Pol type oscillator: Investigation of the properties of its limit cycle. Math. Comput. Model. **55**, 645–653 (2012)
246. P. Fleming, J.V. Poplawski, Unbalance response prediction for rotors on ball bearings using speed and load-dependent nonlinear bearing stiffness. Int. J. Rotat. Mach. **11**, 53–59 (2005)
247. Y.P. Chang, S.C. Jen, S.H. Tu, S.S. Shyr, Y. Kang, Mode-locking quasi-period and chaos of rotors mounted on nonlinear bearings. Int. J. Rotat. Mach. **6**(191–200) (2000)
248. N. Herişanu, V. Marinca, T. Dordea, Gh. Madescu – A new analytical approach to nonlinear vibration of an electrical machine, Proceed. of the Romanian Academy, Series A, **9**, 229-236 (2008)
249. W.R. Finley, M.M. Hodowanec, W.G. Holter, An analytical approach to solving motor vibration problems. IEEE Trans. Ind. Appl. **36**(1467–1480) (2000)
250. N. Herisanu, V. Marinca, *Optimal homotopy perturbation method for a non-conservative dynamical system of a rotating electrical machine, Zeitschrift fur Naturforsch* ***67a****, 509-516,* 2012
251. V. Marinca, N. Herisanu, Nonlinear dynamic analysis of an electrical machine rotor-bearing system by optimal homotopy perturbation method. Comp. Math. Appl. **61**, 2019–2024 (2011)
252. C.W. Chang-Jian, C.K. Chen, Chaos and bifurcation of a flexible rotor supported by porous squeeze couple stress fluid film journal bearings with non-linear suspension. Chaos Solitons Fractals **35**, 358–375 (2008)

Chapter 5
The Third Alternative of the Optimal Homotopy Asymptotic Method

5.1 Overview

We apply an alternative to OHAM for the nonlinear differential equation

$$L[u(x)] + g(x) + N[u(x)] = 0, \quad x \in D \tag{5.1}$$

with the initial/boundary conditions

$$B\left[u(x), \frac{du(x)}{dx}\right] = 0. \tag{5.2}$$

Let $u_0(x)$ be an initial approximation of $u(x)$ such as

$$L[u_0(x)] + g(x) = 0 \tag{5.3}$$

and

$$B\left[u_0(x), \frac{du_0(x)}{dx}\right] = 0. \tag{5.4}$$

We remark that the linear operator L from Eqs. 5.1 and 5.3 is not unique.

Let us consider the functions $\varphi(x, p, C_i)$ (see Chap. 2) in the particular form

$$\varphi(x, p, C_i) = u_0(x) + pu_1(x, C_i) \tag{5.5}$$

where $p \in [0, 1]$ denotes an embedding parameter. It follows that the first-order approximate solution can be written as

V. Marinca, N. Herisanu, *The Optimal Homotopy Asymptotic Method*,
DOI 10.1007/978-3-319-15374-2_5

$$\overline{u}(x, C_2) = u_0(x) + u_1(x, C_i) \tag{5.6}$$

where $C_1, C_2, \ldots, C_s$ are arbitrary parameters, which will be determined later. The initial/boundary conditions are:

$$B\left[\overline{u}(x, C_i), \frac{d\overline{u}(x, C_i)}{dx}\right] = 0. \tag{5.7}$$

In what follows we construct a family of equations

$$\begin{aligned} &\mathbf{H}[L(\varphi(x,p,C_i) + g(x), H(x,C_i), N[\varphi(x,p,C_i)] = \\ &= L[u_0(x)] + g(x) + p[L(u_1(x,C_i) - H(x,C_i)N(u_0(x))] = 0 \end{aligned} \tag{5.8}$$

with the properties

$$\begin{aligned} &\mathbf{H}[L(\varphi(x,0,C_i) + g(x), H(x,C_i), N[\varphi(x,0,C_i)] = \\ &= L[\varphi(x,0)] + g(x) = L[u_0(x)] + g(x) = 0 \end{aligned} \tag{5.9}$$

$$\begin{aligned} &\mathbf{H}[L(\varphi(x,1,C_i) + g(x), H(x,C_i), N[\varphi(x,1,C_i)] = \\ &= H(x,C_i)[L[\overline{u}(x,C_i)] + g(x) + N(\overline{u}(x,C_i))] = 0 \end{aligned} \tag{5.10}$$

where $H(x, C_i) \neq 0$ is an arbitrary optimal auxiliary convergence-control function, and the terms in p^2 are neglected.

From Eqs. 5.5 and 5.6 one gets

$$\varphi(x, 0, C_i) = u_0(x) \tag{5.11}$$

$$\varphi(x, 1, C_i) = \overline{u}(x, C_i) \tag{5.12}$$

Now, equating only the coefficients of p^0 and p^1 into Eq. 5.8, we obtain the governing equation of $u_0(x)$ given by Eq. 5.3 and the governing equation of the first approximation $u_1(x, C_i)$, i.e.

$$L[u_1(x,C_i)] = H(x,C_i)N(u_0(x)), B\left[u_1(x,C_i), \frac{du_1(x,C_i)}{dx}\right] = 0, i = 1,2, \ldots, s \tag{5.13}$$

In general, the nonlinear operator may be written as

$$N(u_0(x)) = \sum_{i=1}^{n} h_i(x)g_i(x) \tag{5.14}$$

where the functions $h_i(x)$ and $g_i(x)$ are known and depend on the functions $u_0(x)$ and also on the nonlinear operator, n being a known integer. It is known that the general solution of nonhomogeneous linear Eq. 5.13 is equal to the sum of general solution of the corresponding homogeneous equation and some particular solutions of the nonhomogeneous equation. However, the particular solutions are readily select only in exceptional cases. In what follows, we do not solve Eq. 5.13, but from the theory of differential equations, taking into consideration the method of variation of parameters, Cauchy method, the method of influence function, the operator method [1] and so on, it is more convenient to consider the unknown function $u_1(x, C_j)$ in the form

$$u_1(x, C_j) = \sum_{i=1}^{m} H_i(x, h_j(x), C_j) g_i(x), \quad j = 1, 2, .., s \tag{5.15}$$

or

$$u_1(x, C_j) = \sum_{i=1}^{m} H_i\left(x, g_j, C_j\right) h_i(x), \quad j = 1, 2, .., s \tag{5.15'}$$

$$B\left(u_1(x, C_j), \frac{du_1(x, C_j)}{dx}\right) = 0 \tag{5.16}$$

where within the expression of $H_i(x, h_j, C_j)$ appear linear combinations of some functions h_i, some terms which are given by corresponding homogeneous equation and several unknown parameters C_j, j = 1,2,. . .,s, m being an arbitrary integer number.

For instance, if h_1 is a trigonometric function, then H_1 is a combination of trigonometric function. More precisely, if $h_1 = \sin \alpha x$, then $H_1(x, h_1, C_j) = C_1 \sin \alpha x + C_2 \cos \alpha x + C_3 \sin 2\alpha x + \ldots$, Similarly, if h_1 is a polynomial function such as $h_1 = x^3$, then $H_1(x, h_1, C_j)$ is a combination of polynomials, $H_1(x, h_1, C_j) = C_1 x + C_2 x^3 + C_3 x^7 + \ldots$. In the case when h_1 is a logarithmic function $h_1 = \ln x$ we can choose the optimal auxiliary function H_1 as a combination of logarithmic function $H_1(x, h_1, C_j) = C_1 \ln x + C_2 x \ln x + C_3 x^2 \ln 2x + \ldots$. We have a large freedom to choose the value of the integer positive m and the optimal auxiliary functions H_i. Note that some of the initial/boundary conditions could be fulfilled by Eq. 5.4 and the others by Eq. 5.16, so that finally, Eq. 5.7 responds to all initial/boundary conditions. We cannot demand that $u_1(x, C_j)$ given by Eq. 5.15 is a solution of Eq. 5.13, but $\bar{u}(x, C_i)$ given by Eq. 5.6 is the solution of Eqs. 5.1 and 5.2. The same considerations can be made for the Eq. 5.15′, where h_i and g_i are interchangeable.

Concluding, this is the underlying idea of our procedure—the third alternative of the Optimal Homotopy asymptotic Method.

5.1.1 The Convergence of the Approximate Solution 5.6

The convergence of the approximate solution $\bar{u}(x, C_i)$ given by Eq. 5.6 depends upon the optimal auxiliary functions (or optimal auxiliary convergence-control functions) $H_i(x, h_i(x), C_j)$, j = 1,2,. . .,s which appear in Eq. 5.15. There are many possibilities to choose these functions. We try to choose such functions H_i so that within Eq. 5.15 the terms under the sum from 1 to m be of the same shape with the terms found under the sum from Eq. 5.14 [2–7]. The first-order approximate solution $\bar{u}(x, C_i)$ also depends on the parameters C_j, j = 1,2. . .s. The values of these parameters can be optimally evaluated via the same various methods given in Sect. 2.4: the least-square method, Galerkin method, collocation method, Ritz method and so on. In this way, it is clear that the first-order approximate solutions given by Eq. 5.6 are well-determined. Since the optimal auxiliary functions are not unique, we have the possibility to determine multiple solutions for nonlinear differential equation 5.1. It should be emphasized that our procedure contains the optimal auxiliary functions $H_i(x, f_i(x), C_j)$, i = 1,2,. . .,m, j = 1,2,. . .,s which provide us with a simple way to adjust and control the convergence of the approximate solutions.

It is very important to properly choose these optimal functions H_i which appear in the construction of the first-order approximate solution.

In what follows we will present some applications of our procedure.

5.2 Thomas-Fermi Equation

From the earliest days of quantum mechanics, it has been clear that one could not hope to solve exactly most of the physically interesting systems, especially those with more particles. Thus, by 1930 (only 3 years after the first works of Thomas [8] and Fermi [9], and 5 years after the advent of the "new" quantum theory), a large variety of approximate methods had been developed to construct an approximate analytical solutions for nonlinear differential equations.

Thomas-Fermi equation describes the charge density in atoms of high atomic number. Thomas-Fermi atom model views the electrons in an atom as a gas and derives atomic structure in terms of the electrostatic potential. In the study of the Thomas-Fermi equation are used different methodologies as: modified Adomian decomposition method [10], optimal iteration method [11], optimal homotopy asymptotic method [12], series approach [13], δ -expansion method [14], homotopy analysis method [15, 16] and so on.

Bender et al. [14] consider that the Thomas-Fermi equation is quite unique in that the asymptotic methods that one would conventionally use to find an approximate solution to a nonlinear differential equation are dismal failures. In spite of this assertion, OHAM is very effective and accurate, demonstrating the general validity and the great potential for solving strongly and delicate nonlinear problems.

In what follows we suppose that all electrons of an atom are subject to the same conditions. Thus, each electron subject to the energy conservation law, has a potential energy $e\Phi(r)$ with $\Phi(r)$ the mean value of the potential energy owed to the nucleus and all other electrons. If $\rho(r)$ is the electronic charge density, then the Poisson equation can be written in the form:

$$\frac{1}{r}\frac{d^2}{dr^2}[r\Phi(r)] + 4\pi\rho(r) = 0 \tag{5.17}$$

if $\rho(r)$ and $\Phi(r)$ are spherically symmetric. From the energy conservation law one can get

$$\frac{p^2}{2m} - e\Phi(r) = E. \tag{5.18}$$

It follows that the maximum of the electron impulsion is obtained for $E = 0$ into Eq. 5.18:

$$p = \sqrt{2me\Phi(r)}. \tag{5.19}$$

The corresponding boundary condition are

$$\Phi(R) = 0, \quad \frac{d\Phi}{dr}(R) = \frac{d}{dr}\left(\frac{eZ}{r}\right)(R) = -\frac{eZ^2}{R^2} \tag{5.20}$$

where R is the radius of a sphere representing the atom. An another boundary condition it is obtain considering that the contribution of the electrons situated near the nucleus to the potential $\Phi(r)$ is null:

$$r\Phi(r) \to 0 \quad if \ \ r \to 0. \tag{5.21}$$

If p is the electron impulsion and $\hbar$ the Planck' constant, then the electronic charge density is defined as

$$\rho = -\frac{8\pi e}{3}\left(\frac{p}{\hbar}\right)^2. \tag{5.22}$$

From Eqs. 5.19 and 5.22 one retrieves

$$\rho = -\frac{8\pi e}{3}\frac{e}{\hbar^3}[2me\Phi(r)]^{3/2}. \tag{5.23}$$

If the first Bohr radius of the hydrogen atom is $a_B = \frac{\hbar^2}{4\pi^2 me^2}$ and making the transformations:

$$x = \frac{r}{a}, \quad y(x) = \frac{1}{Ze} r\Phi(r) \tag{5.24}$$

where $a = a_B \left(\frac{9\pi^2}{128Z}\right)^{1/3}$, then from Eqs. 5.17 and 5.24 we obtain so called the differential equation of Thomas-Fermi [13]:

$$\frac{d^2y}{dx^2} = \sqrt{\frac{y^3}{x}} \tag{5.25}$$

with boundary and subsidiary conditions, obtained from Eqs. 5.20 and 5.21:

$$y(0) = 1, \quad y(\infty) = 0, \quad y'(\infty) = 0 \tag{5.26}$$

where prime denotes the derivative with respect to x.

The Eq. 5.25 describes the charge density in atoms of high atomic number. The solution of Eqs. 5.25 and 5.26 can be found numerically with great difficulty [14] because the nonlinear differential equation has a singularity at $x = 0$.

5.2.1 *Approximate Solution of the Thomas-Fermi Equation by OHAM*

With the transformation

$$\sqrt{x} = t \tag{5.27}$$

one can get

$$y' = \frac{\dot{y}}{2t}, \quad y'' = \frac{t\ddot{y} - \dot{y}}{4t^3} \tag{5.28}$$

where prime and dot denote differentiation with respect to x and t, respectively. In this way, Eqs. 5.25 and 5.26 become

$$t\ddot{y} - \dot{y} - 4t^2y^{3/2} = 0, y(0) = 1, \quad y(\infty) = 0, \quad \dot{y}(\infty) = 0. \tag{5.29}$$

The linear operator is defined as

$$L(y(t)) = \ddot{y} - \lambda^2 y \tag{5.30}$$

where $\lambda > 0$ is an arbitrary parameter unknown at this moment. The nonlinear operator is $(g(t) = 0)$:

$$N(y(t)) = t\ddot{y} - \dot{y} - 4t^2y^{3/2} + \lambda^2 y - \ddot{y}. \tag{5.31}$$

Eqs. 5.3 and 5.4 can be written in the form

$$\ddot{y}_0 - \lambda^2 y_0 = 0, \quad y_0(0) = 1, y_0(\infty) = 0, \; \dot{y}_0(\infty) = 0 \tag{5.32}$$

whose solution is

$$y_0(t) = e^{-\lambda t} \tag{5.33}$$

The nonlinear operator given by Eq. 5.14 is obtained from Eqs. 5.31 and 5.33:

$$N(y_0(t)) = (\lambda^2 t + \lambda)e^{-\lambda t} - 4t^2 e^{-1.5\lambda t}. \tag{5.34}$$

Comparing Eqs. 5.34 and 5.14 it hold that

$$h_1(t) = \lambda^2 t + \lambda, \; g_1(t) = e^{-\lambda t}, \tag{5.35}$$

$$h_2(t) = -4t^2, \; g_2(t) = e^{-1.5\lambda t}, \tag{5.36}$$

It is natural to choose the expression of $y_1(t, C_i)$ from Eq. 5.15, in the form

$$\begin{aligned} &y_1(t, C_i) = H_1(t, h_1(t), C_i)g_1(t) + H_2(t, h_2(t), C_i)g_2(t), \\ &y_1(0) = 0, \; y_1(\infty) = 0, \; \dot{y}_1(\infty) = 0, \end{aligned} \tag{5.37}$$

with the optimal auxiliary functions H_1 and H_2 as follows:

$$H_1(t, h_1(t), C_i) = P_1(t) + P_2(t)e^{-\alpha t}, \quad H_2(t, h_2(t), C_i) = P_3(t), \tag{5.38}$$

$P_i(t), i = 1, \; 2, \; 3$ being arbitrary polynomials and α an unknown parameter. From Eq. 5.37 follows that

$$P_1(0) + P_2(0) + P_3(0) = 0 \tag{5.39}$$

The first-order approximate solution is obtained from Eqs. 5.33, 5.37, 5.38, 5.6 and 5.7 and therefore

$$\bar{y}(t, C_i) = [1 + P_1(t)]e^{-\lambda t} + P_2(t)e^{-(\lambda+\alpha)t} + P_3(t)e^{-1.5\lambda t}. \tag{5.40}$$

From Eq. 5.28, $\bar{y}(x)$ is derivable for $x = 0$ if the free term in $\bar{y}$ is null. But, from Eq. 5.40 we have this term in the form:

$$\bar{y}(0, C_i) = \dot{P}_1(0) + \dot{P}_2(0) + \dot{P}_3(0) - \lambda[1 + P_1(0) + P_2(0) + 1.5P_3(0)] \\ = 0. \tag{5.41}$$

There are many possibilities to choose the polynomials $P_i, i = 1, \ 2, \ 3$. Obviously, we should choose a simple case:

$$P_1(0) = P_2(0) = P_3(0) = \dot{P}_2(0) = \dot{P}_3(0) = 0. \tag{5.42}$$

From Eq. 5.41 one gets $\dot{P}_1(0) = \lambda$, and therefore the first-order approximate solution given by Eq. 5.40, becomes for the following two cases:

Case 5.1

$$\bar{y}(t, C_i) = (1 + \lambda t + C_1 t^2 + C_2 t^3 + C_3 t^{10})e^{-\lambda t} + (C_4 t^2 + C_5 t^3)e^{-(\lambda+\alpha)t} + \\ +(C_6 t^2 + C_7 t^3 + C_8 t^4)e^{-1.5\lambda t}. \tag{5.43}$$

Case 5.2

$$\bar{y}(t, C_i) = \left(1 + \lambda t + C_1 t^2 + C_2 t^3 + C_3 t^4 + C_4 t^5 + C_5 t^6 + C_6 t^7 + C_7 t^8 + \right. \\ \left. + C_8 t^9 + C_9 t^{10}\right)e^{-\lambda t} + \left(C_{10} t^2 + C_{11} t^3 + C_{12} t^4 + C_{13} t^5\right)e^{-(\lambda+\alpha)t} + \\ + \left(C_{14} t^2 + C_{15} t^3 + C_{16} t^4 + C_{17} t^5\right)e^{-1.5\lambda t}, \tag{5.44}$$

where C_i and α are unknown parameters.

5.2.2 *Numerical Examples*

In order to show the validity and accuracy of the OHAM, we compare previously obtained approximate solutions 5.43 and 5.44 with numerical integration results obtained by means of a fourth-order Runge–Kutta method in combination with shooting method, using Wolfram Mathematica 6.0 software. Using the transformation 5.27 and the least-square method for determination of the parameters C_i (more precisely optimal convergence-control parameters) we obtain for every case:

Case 5.2.2.A

$$\bar{y}(x) = \left(1 + 2.1169433085\sqrt{x} + 4.2890238832x + 11355.8788961130x\sqrt{x} + \right. \\ \left. +3.9370357382 \times 10^{-6}x^5\right) \times e^{-2.1169433085\sqrt{x}} + \left(2.6492880353x - \right. \\ \left. -11358.7791181328x\sqrt{x}\right)e^{-2.1169832169\sqrt{x}} + \left(- 6.2857387204x - \right. \\ \left. -4.1976563982x\sqrt{x} - 1.2938640524x^2\right)e^{-3.1754149627\sqrt{x}} \tag{5.45}$$

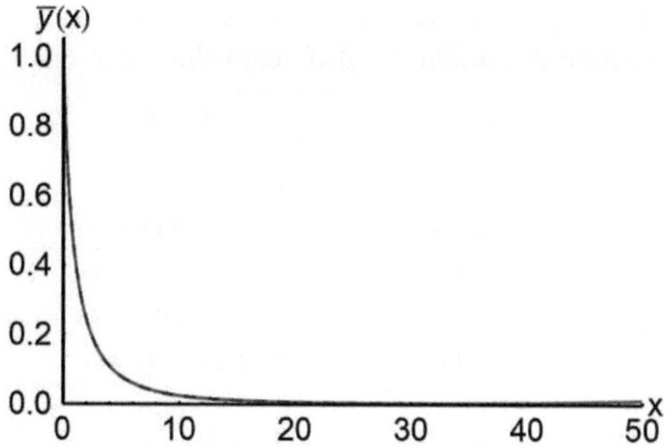

Fig. 5.1 Comparison between the results obtained by means of OHAM—Eq. (5.45)—and numerical results: (*red solid line*) numerical results and (*blue dashed line*) approximate solution (colour figure online)

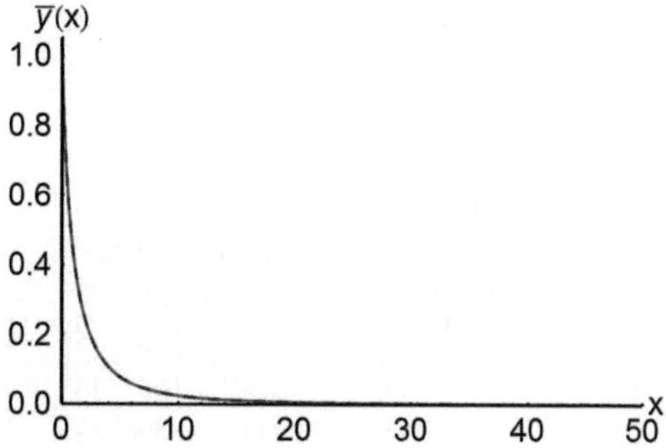

Fig. 5.2 Comparison between the results obtained by means OHAM—Eq. (5.46)—and numerical results: (*red solid line*) numerical results and (*blue dashed line*) approximate solution (colour figure online)

Case 5.2.2.B

$$\begin{aligned}
\overline{y}(x) = &\big(1 + 2.8440462072\sqrt{x} + 2.8385227981x + 408.2504889408x\sqrt{x} + \\
&+2684.0147299659x^2 - 2058.3106795493x^2\sqrt{x} + 694.4558078037x^3 - \\
&-136.4738073947x^3\sqrt{x} + 16.6503077715x^4 - 1.2030880433x^4\sqrt{x} + \\
&+0.0407110884x^5\big) \times e^{-2.8440462072\sqrt{x}} + \\
&+\big(-0.6666986241x - 186.2530445276x\sqrt{x} + 2044.6317622248x^2 - \\
&-2284.2358172213x^2\sqrt{x}\big)e^{-3.9157652153\sqrt{x}} + \big(0.2844032411x - \\
&-221.6578865121x\sqrt{x} - 5243.2544994816x^2 - \\
&-590.9937887039x^2\sqrt{x}\big)e^{-4.2660693108\sqrt{x}}
\end{aligned} \tag{5.46}$$

In Figs. 5.1 and 5.2 we present a comparison between the first-order approximate solutions given by Eqs. 5.45 and 5.46 respectively and numerical results.

Some values of the approximate solutions obtained by HAM [15], OHAM—Eq. 5.45, OHAM—Eq. 5.46 and numerical results for different values of x are given in Table 5.1.

In Table 5.2, we compare the relative error, $error = \left|y_{approx.}(x) - y_{numerical}(x)\right|$ with HAM [15], OHAM (Eq. 5.45), OHAM (Eq. 5.46) in 20 points in the range $0 \le x \le 100$.

On the other hand, from Eq. 5.45, the approximate initial slope $\overline{y}(0)$ becomes

Table 5.1 Approximate solutions obtained by HAM [15], OHAM Eq. 5.45, OHAM Eq. 5.46 and numerical solutions for different values of x

x	$y_{HAM}(x)$ [15]	$y_{OHAM}(x)$ 5.45	$y_{OHAM}(x)$ 5.46	$y_{numerical}(x)$
0	1	1	1	1
0.5	0.606987	0.606988435485	0.606986083475	0.606986374823
1	0.424008	0.424008595673	0.424007644164	0.424008039715
1.5	0.314778	0.314777002588	0.314776923695	0.314777447546
2	0.243009	0.243009307324	0.243008140827	0.243008484517
2.5	0.192795	0.192986464516	0.192983819450	0.192984090919
3	0.156633	0.156635746407	0.156632277419	0.156632627090
3.5	0.129370	0.129372540301	0.129369080972	0.129369533106
4	0.108404	0.108406587514	0.108403678244	0.108404170570
4.5	0.0919482	0.091949783668	0.091947558254	0.091948019541
5	0.0788078	0.078808972020	0.078807234369	0.078807630873
10	0.0243143	0.024322837681	0.024310868150	0.024313241504
15	0.0108054	0.010811989883	0.010798763188	0.010801392104
20	0.00578494	0.005782795820	0.005775234868	0.005773988601
25	0.00347375	0.003459567123	0.003449790557	0.003448745117
30	–	0.002225468475	0.002202137795	0.002205639738
35	–	0.001501637344	0.001458654549	0.001459192302
40	–	0.001045596788	0.000983936068	0.000957785351
45	–	0.000743310866	0.000012900540	0.000655659427
50	0.000632255	0.000535774498	0.000454954579	0.000441049796
100	0.000100243	0.000022731731	$8.98417326 \times 10^{-6}$	$7.36651319 \times 10^{-6}$

$$\bar{y}'(0) = -1.588151287649017 \tag{5.47}$$

and from Eq. 5.46 this is

$$\bar{y}'(0) = -1.5880719992608454 \tag{5.48}$$

Kobayashi [17] give the numerical result for the same problem as

$$y'_K(0) = -1.588071 \tag{5.49}$$

$(\varepsilon_{HAM} = |y'_{HAM}(0) - y'_{numerical}(0)|, \ \varepsilon_{OHAM} = |y'_{OHAM}(0) - y'_{numerical}(0)|)$

It is easy to verify the accuracy of the obtained solutions if we compare these analytical solutions 5.45 and 5.46 with numerical ones or with results obtained by HAM [15] procedure.

It can be seen from Tables 5.1 and 5.2 and Figs. 5.1 and 5.2 that the analytical solutions of the Thomas-Fermi equation obtained by OHAM are very accurate comparing to numerical solutions.

Table 5.2 Comparison between errors obtained by HAM [12], OHAM Eq. 5.45 and OHAM Eq. 5.46 for different values of x

x	*error* [15] $\lvert y_{HAM}(x) - y_{numerical}(x) \rvert$	*error* (5.45) $\lvert y_{OHAM}(x) - y_{numerical}(x) \rvert$	*error* (5.46) $\lvert y_{OHAM}(x) - y_{numerical}(x) \rvert$
0	0	0	0
0.5	6.25177×10^{-7}	2.060662×10^{-6}	2.913480×10^{-7}
1	3.9715×10^{-8}	5.559575×10^{-7}	3.955514×10^{-7}
1.5	5.52454×10^{-7}	4.449580×10^{-7}	5.238513×10^{-7}
2	5.15483×10^{-7}	8.228071×10^{-7}	3.436901×10^{-7}
2.5	1.890909×10^{-4}	2.373597×10^{-6}	2.714682×10^{-7}
3	3.7291×10^{-7}	3.119316×10^{-6}	3.496714×10^{-7}
3.5	4.66894×10^{-7}	3.007194×10^{-6}	4.521342×10^{-7}
4	1.7057×10^{-7}	2.416944×10^{-6}	4.923259×10^{-7}
4.5	1.80459×10^{-7}	1.764127×10^{-6}	4.612864×10^{-7}
5	1.69127×10^{-7}	1.341147×10^{-6}	3.965041×10^{-7}
10	1.058496×10^{-6}	9.596176×10^{-5}	2.373354×10^{-6}
15	–	1.059777×10^{-6}	2.628916×10^{-6}
20	1.095139×10^{-5}	8.807219×10^{-6}	1.246267×10^{-6}
25	2.500488×10^{-5}	1.082200×10^{-5}	1.045439×10^{-6}
30	–	1.982873×10^{-5}	3.501942×10^{-6}
35	–	4.244504×10^{-5}	5.377533×10^{-7}
40	–	8.781143×10^{-5}	2.615071×10^{-5}
45	–	1.6549741×10^{-4}	1.290054×10^{-5}
50	1.912002×10^{-4}	2.8392578×10^{-4}	1.390478×10^{-5}
100	9.287648×10^{-5}	2.273173×10^{-5}	1.617660×10^{-6}

The examples presented in this section lead to the very important conclusion that the accuracy of the obtained results growing along with increasing the number of parameters C_i in the optimal auxiliary functions H_1 and H_2. We mention that for HAM [15] one needed of 100 iterations while for OHAM was needed only one iteration.

A comparison between different approximations of the initial slope $y'(0)$ is presented in Table 5.3.

It is clear that the present results for the initial slope are better than those given in [15].

To the best of our knowledge, it is the first time when such an explicit analytic approximate solutions to the Thomas-Fermi equation are given. Some series solutions were reported, without providing an explicit analytical solution in the traditional meaning of the word "analytical".

Table 5.3 Comparison between different approximations of the initial slope $y'(0)$

Order of approximation	Liao [15]/ ε_{HAM}	Present Eq. 5.45/ ε_{OHAM}	Present Eq. 5.46/ ε_{OHAM}
10	$-1.50014/8.7 \times 10^{-2}$	$-1.5881512876/$ 8.02×10^{-5}	$-1.5880719992/$ 9.9×10^{-6}
20	$-1.54093/4.7 \times 10^{-2}$		
30	$-1.55595/3.2 \times 10^{-2}$		
40	$-1.56373/2.4 \times 10^{-2}$		
50	$-1.56848/1.9 \times 10^{-2}$		
60	$-1.57168/1.6 \times 10^{-2}$		
70	$-1.57399/1.4 \times 10^{-2}$		
80	$-1.57572/1.2 \times 10^{-2}$		
90	$-1.57708/1.09 \times 10^{-2}$		
100	$-1.57816/8.7 \times 10^{-3}$		

5.3 Multiple Solutions for the Upper-Convected Maxwell Fluid Over a Porous Stretching Plate

There are numerous constitutive equations that have been proposed to various type of non-Newtonian fluids. Majority of non-Newtonian fluids models are concerned with the fluids of grade two or three. One type of fluids in which the relaxation type phenomena can be considered is known as Maxwell model. Some investigations in this field are made by the mathematicians, engineers, physicians and computer scientists. Sakiadis [18], first studied various aspects of the stretching problem: the flow due to a semi-infinite horizontally moving wall in an ambient fluid. Sadeghy et al. [19] considered the problem of hydrodynamic Sakiadis flow of an upper-convected Maxwell fluid over a rigid plate moving steadily in an otherwise quiescent fluid. Homotopy analysis method is used by Hayat et al. [20] to solve nonlinear differential equation of the upper-convected Maxwell fluid over a porous stretching plate. Dual solutions in mixed convection flow near stagnation point on a vertical porous plate has been presented by Ishak et al. [21]. Sahoo [22] investigated the effects of partial slip in the MHD flow and mass transfer of an electrically conducting second grade fluid past an axisymmetric stretching sheet.

If we consider the steady incompressible two-dimensional flow, of an upper-convected Maxwell fluid over a porous stretching plate, then the constitutive equation for the Cauchy stress tensor $\mathbf{T}$ is given by

$$\mathbf{T} = -p\mathbf{I} + \mathbf{S} \tag{5.50}$$

such that the extra tensor $\mathbf{S}$ satisfies

$$\mathbf{S}+\lambda\left(\frac{d\mathbf{S}}{dt}-\mathbf{LS}-\mathbf{SL}^{T}\right)=\mu\mathbf{A}_{1} \tag{5.51}$$

where μ is the viscosity, λ is the relaxation time and the first Rivlin-Ericksen tensor $\mathbf{A}_1$ is defined as follows:

$$\mathbf{A}_{1}=(grad\mathbf{V})+(grad\mathbf{V})^{T}. \tag{5.52}$$

in which d/dt is the material time derivative and $\mathbf{V}$ is the velocity.

For the magnetohydrodynamic Sakiadis flow, by imposing an uniform magnetic field B_0 along the y-direction and neglecting the induced magnetic field, the equations which govern the steady flow can be written as

$$\frac{\partial u}{\partial x}+\frac{\partial v}{\partial y}=0 \tag{5.53}$$

$$\rho\left(u\frac{\partial u}{\partial x}+v\frac{\partial u}{\partial y}\right)=-\frac{\partial p}{\partial x}+\frac{\partial S_{xx}}{\partial x}+\frac{\partial S_{xy}}{\partial y}-\delta B_{0}^{2}u \tag{5.54}$$

$$\rho\left(u\frac{\partial v}{\partial x}+v\frac{\partial v}{\partial y}\right)=-\frac{\partial p}{\partial y}+\frac{\partial S_{yx}}{\partial x}+\frac{\partial S_{yy}}{\partial y} \tag{5.55}$$

where u, v are the velocity components, p is pressure, ρ is the density, δ is the electrical conductivity and S_{xx}, S_{xy}, S_{yx}, S_{yy} are the components of the extra tensor $\mathbf{S}$.

Using the boundary layer approximations [19, 23]

$$u=\mathrm{O}(1),\quad v=\mathrm{O}(\delta),\ x=\mathrm{O}(1),\quad y=\mathrm{O}(\delta) \tag{5.56}$$

$$\frac{T_{xx}}{\rho}=\mathrm{O}(1),\ \frac{T_{xy}}{\rho}=\mathrm{O}(\delta),\ \frac{T_{yy}}{\rho}=\mathrm{O}(\delta^{2}) \tag{5.57}$$

the flow in the absence of the pressure gradient is governed by Eq. 5.53 and

$$u\frac{\partial u}{\partial x}+v\frac{\partial v}{\partial x}+\lambda\left(u^{2}\frac{\partial^{2}u}{\partial x^{2}}+v^{2}\frac{\partial^{2}u}{\partial y^{2}}+2uv\frac{\partial^{2}u}{\partial x\partial y}\right)=\nu\frac{\partial^{2}u}{\partial y^{2}}-\frac{\delta B_{0}^{2}}{\rho}u \tag{5.58}$$

where ν is the kinematic viscosity of fluid.

The relevant initial/boundary conditions for the flow problem are

$$\begin{aligned}&u=Cx,\quad v=-V_{0},\quad at\ y=0\\&u\to 0,\quad as\ y\to\infty\end{aligned} \tag{5.59}$$

in which C is the stretching rate, and $V_0>0$ is the suction velocity and $V_0<0$ is the injection velocity.

Introducing the stream function Ψ such that

$$u = \frac{\partial \Psi}{\partial y}, \qquad v = -\frac{\partial \Psi}{\partial x} \tag{5.60}$$

then Eq. 5.53 is identically satisfied and Eq. 5.58 becomes

$$\begin{aligned} &\frac{\partial \Psi}{\partial y}\frac{\partial^2 \Psi}{\partial x \partial y} - \frac{\partial \Psi}{\partial x}\frac{\partial^2 \Psi}{\partial y^2} + \lambda\Big[\left(\frac{\partial \Psi}{\partial y}\right)^2 \frac{\partial^3 \Psi}{\partial x^2 \partial y} + \left(\frac{\partial \Psi}{\partial x}\right)^2 \frac{\partial^3 \Psi}{\partial y^3} - \\ &-2\frac{\partial \Psi}{\partial y}\frac{\partial \Psi}{\partial x}\frac{\partial^3 \Psi}{\partial x \partial y^2}\Big] = \nu\frac{\partial^3 \Psi}{\partial y^3} - \frac{\delta B_0^2}{\rho}\frac{\partial \Psi}{\partial y} \end{aligned} \tag{5.61}$$

Introducing the similarity transformations

$$\eta = \sqrt{\frac{C}{\nu}}y, \qquad \Psi = \sqrt{\nu C}xf(\eta) \tag{5.62}$$

the Eq. 5.60 becomes

$$u = Cxf'(\eta), \qquad v = -\sqrt{x}f(\eta). \tag{5.63}$$

Substituting Eq. 5.63 into Eq. 5.61, it is obtain the governing equation in the form

$$f''' - M^2 f' - f'^2 + ff'' + \beta\left(2ff'f'' - f^2 f'''\right) = 0 \tag{5.64}$$

where $M^2 = \frac{\delta B_0^2}{\rho C}$ and $\beta = \lambda C$.

Now, the initial/boundary conditions 5.59 are

$$f(0) = R, \qquad f'(0) = 1, \qquad f'(\infty) = 0 \tag{5.65}$$

with $R = \frac{V_0}{\sqrt{\nu C}}$, where $R > 0$ corresponds to suction velocity and $R < 0$ for injection velocity.

5.3.1 *Multiple Solutions with OHAM*

For the nonlinear diferential Eqs. 5.64 and 5.65 we choose the linear operator of the form

$$L[f(\eta)] = f'''(\eta) - K^2 f'(\eta). \tag{5.66}$$

We mention that the linear operator is not unique. Also, we have freedom to choose:

$$L[f(\eta)] = f'''(\eta) + Kf'(\eta). \tag{5.67}$$

$$L[f(\eta)] = f'''(\eta) + \frac{3K}{K\eta + 1}f''(\eta). \tag{5.68}$$

$$L[f(\eta)] = f'''(\eta) - \frac{6K^2}{(K\eta + 1)^2}f'(\eta). \tag{5.69}$$

where K is an unknown positive parameter and will be determined later. The initial approximation $f_0(\eta)$ can be obtained from Eq. 5.3 with initial/boundary conditions

$$f_0(0) = R, \quad f_0'(0) = 1, \quad f_0'(\infty) = 0. \tag{5.70}$$

The Eq. 5.3 with the linear operators 5.66 and 5.67 has the solution:

$$f_0(\eta) = R + \frac{1}{K}(1 - e^{-K\eta}) \tag{5.71}$$

while the Eq. 5.3 with the linear operators 5.68 and 5.69 has the solution:

$$f_0(\eta) = R + \frac{1}{K} - \frac{1}{K(K\eta + 1)} \tag{5.72}$$

In what follows we consider only the linear operator given by Eq. 5.66 and the initial approximation given by Eq. 5.71.

The nonlinear operator corresponding to nonlinear differential Eq. 5.64 is defined by

$$N[f(\eta)] = K^2 f' - M^2 f' - f'^2 + ff'' + \beta(2ff'f'' - f^2 f'''). \tag{5.73}$$

By substituting Eq. 5.71 into Eq. 5.73 it holds that

$$N[f_0(\eta)] = \left[K^2 - M^2 - 1 - KR - \beta(1 + KR)^2\right]e^{-K\eta} + \beta e^{-3K\eta} \tag{5.74}$$

Comparing Eqs. 5.14 and 5.74, one can get

$$\begin{aligned} &h_1(\eta) = K^2 - M^2 - 1 - KR - \beta(1 + KR)^2, \\ &g_1(\eta) = e^{-K\eta}, \quad h_2(\eta) = \beta, \quad g_2(\eta) = e^{-3K\eta}, \quad . \end{aligned} \tag{5.75}$$

The first approximation $f_1(\eta)$ given by Eq. 5.15 becomes

$$f_1(\eta, C_i) = H_1(\eta, C_i)e^{-K\eta} + H_2(\eta, C_i)e^{-3K\eta} \tag{5.76}$$

where we have freedom to choose a lot of possibilities for the unknown functions H_i, as follows:

$$H_1(\eta, C_i) = -\frac{C_1 + C_3 + C_5}{K} + C_1\eta + C_2\eta^2 + \left(\frac{C_1 + C_3 + C_5}{K} + \right.$$
$$\left. + C_3\eta + C_4\eta^2\right)e^{-K\eta} \qquad H_2(\eta, C_i) = C_5\eta + C_6\eta^2 \tag{5.77}$$

Substituting Eq. 5.77 into Eq. 5.76 we have

$$f_1(\eta, C_i) = -\left(\frac{C_1 + C_3 + C_5}{K} + C_1\eta + C_2\eta^2\right)e^{-K\eta} + \left(\frac{C_1 + C_3 + C_5}{K} + \right.$$
$$\left. + C_3\eta + C_4\eta^2\right)e^{-2K\eta} + (C_5\eta + C_6\eta^2)e^{-3K\eta} \tag{5.78}$$

The first-order approximate solution given by Eq. 5.6 is obtained from Eqs. 5.71 and 5.78

$$\begin{aligned}(A_1)\quad \overline{f}(\eta, C_i) = R + \frac{1}{K} + \left(-\frac{1 + C_1 + C_3 + C_5}{K} + C_1\eta + C_2\eta^2\right)e^{-K\eta} + \\ + \left(\frac{C_1 + C_3 + C_5}{K} + C_3\eta + C_4\eta^2\right)e^{-2K\eta} + (C_5\eta + C_6\eta^2)e^{-3K\eta}\end{aligned} \tag{5.79}$$

In this way, we can find and other solutions for other H_1 and H_2:

$$\begin{aligned}(A_2)\quad \overline{f}(\eta, C_i) = \frac{1}{K} + \left(2R - \frac{1 + C_1 + C_3 + C_5}{K} + C_1\eta + C_2\eta^2\right)e^{-K\eta} + \\ + \left(\frac{C_1 + C_3 + C_5}{K} - R + C_3\eta + C_4\eta^2\right)e^{-2K\eta} + (C_5\eta + C_6\eta^2)e^{-3K\eta}\end{aligned} \tag{5.80}$$

$$\begin{aligned}(A_3)\quad \overline{f}(\eta, C_i) = R + \frac{1}{K^2} + \left(-\frac{C_1 + C_3 + C_5}{2K} + \frac{K - 3}{2K^2} + C_1\eta + C_2\eta^2\right)e^{-K\eta} + \\ + (C_3\eta + C_4\eta^2)e^{-2K\eta} + \left(\frac{1 - K}{2K^2} + \frac{C_1 + C_3 + C_5}{2K} + C_5\eta + C_6\eta^2\right)e^{-3K\eta}\end{aligned} \tag{5.81}$$

$$\begin{aligned}(A_4)\quad \overline{f}(\eta, C_i) = R + \frac{1}{K} + \left(-\frac{1}{K} + C_1\eta^2 + C_2\eta^3 + C_3\eta^4\right)e^{-K\eta} + \\ + (C_4\eta^2 + C_5\eta^3 + C_6\eta^4)e^{-2K\eta}\end{aligned} \tag{5.82}$$

$$\begin{aligned}(A_5)\quad \overline{f}(\eta, C_i) = R + \frac{1}{K} + \left(-\frac{1}{K} + C_1\eta + C_2\eta^2 + C_3\eta^3 + C_4\eta^4\right)e^{-K\eta} + \\ + (-C_1\eta + C_5\eta^2 + C_6\eta^3)e^{-2K\eta}\end{aligned} \tag{5.83}$$

$$(A_6)\quad \overline{f}(\eta, C_i) = R + \frac{1}{K^2} + \left(-\frac{1}{K^2} + \left(1 - \frac{1}{K}\right)\eta + C_1\eta^2 + C_2\eta^3 + C_3\eta^4\right)e^{-K\eta} + $$
$$+ (C_4\eta^2 + C_5\eta^3 + C_6\eta^4)e^{-2K\eta} \tag{5.84}$$

$$(A_7)\quad \overline{f}(\eta, C_i) = R + \frac{1 - e^{-K\eta}}{K} + (C_1\eta^2 + C_2\eta^3 + C_3\eta^4)e^{-2K\eta} + $$
$$+ (C_4\eta^2 + C_5\eta^3 + C_6\eta^4)e^{-3K\eta} \tag{5.85}$$

$$(A_8)\quad \overline{f}(\eta, C_i) = R + \frac{1}{K}(1 - e^{-K\eta}) + (3C_1\eta + C_2\eta^2 + C_3\eta^3)e^{-2K\eta} + $$
$$+ (-2C_1\eta + C_4\eta^2 + C_5\eta^3 + C_6\eta^4)e^{-3K\eta} \tag{5.86}$$

$$(A_9)\quad \overline{f}(\eta, C_i) = R + \frac{1}{K^2}(1 - e^{-K\eta}) + \left[\left(1 - \frac{1}{K}\right)\eta + C_1\eta^2 + C_2\eta^3 + \right.$$
$$\left. + C_3\eta^4\right]e^{-2K\eta} + (C_4\eta^2 + C_5\eta^3 + C_6\eta^4)e^{-3K\eta} \tag{5.87}$$

$$(A_{10})\quad \overline{f}(\eta, C_i) = R + \frac{1}{K} + \left(-\frac{1}{K} + C_1\eta^2 + C_2\eta^3 + C_3\eta^4\right)e^{-K\eta} + $$
$$+ (C_4\eta^2 + C_5\eta^3 + C_6\eta^4)e^{-3K\eta} \tag{5.88}$$

$$(A_{11})\quad \overline{f}(\eta, C_i) = R + \frac{1}{K} + \left(-\frac{1}{K} + C_1\eta + C_2\eta^2\right)e^{-K\eta} + (C_3\eta + C_4\eta^2)e^{-2K\eta} + $$
$$+ (C_5\eta + C_6\eta^2)e^{-3K\eta} + [(-C_1 - C_3 - C_5)\eta + C_7\eta^2]e^{-4K\eta} \tag{5.89}$$

$$(A_{12})\quad \overline{f}(\eta, C_i) = R + \frac{1}{K} + \left(-\frac{1}{K} + C_1\eta^2\right)e^{-K\eta} + (C_2\eta^2 + C_3\eta^3)e^{-2K\eta} + $$
$$+ (C_4\eta^2 + C_5\eta^3 + C_6\eta^4)e^{-3K\eta} \tag{5.90}$$

It is clear that we can obtain many other solutions.

In the second case, we consider the linear operator given by Eq. 5.68 and the initial approximation given by Eq. 5.72.

The nonlinear operator corresponding to the nonlinear differential Eq. 5.64 is defined as

$$N[f(\eta)] = -\frac{3k}{kn + 1}f'' - M^2 f' - f'^2 + ff'' + \beta\left(2ff'f'' - f^2 f'''\right) \tag{5.91}$$

Substituting Eq. 5.72 into Eq. 5.91 one can put

$$N[f_0(\eta)] = -\frac{M^2}{(k\eta+1)^2} - \frac{2(1+kR)}{(k\eta+1)^3} + \frac{1+6k-6\beta(1+kR)^2}{(k\eta+1)^4} - \\ -\frac{16\beta(1+kR)}{(k\eta+1)^5} + \frac{10\beta}{(k\eta+1)^6} \tag{5.92}$$

Now, comparing Eqs. 5.14 and 5.92 one gets:

$$h_1(\eta) = -M^2, h_2(\eta) = -2(1+kR), h_3(\eta) = 1+6k^2-6\beta(1+kR)^2, \\ h_4(\eta) = -16\beta(1+kR), h_5(\eta) = 10\beta, g_1(\eta) = \frac{1}{(k\eta+1)^2}, g_2(\eta) = \frac{1}{(k\eta+1)^3} \\ g_3(\eta) = \frac{1}{(k\eta+1)^4}, g_4(\eta) = \frac{1}{(k\eta+1)^5}; g_5(\eta) = \frac{1}{(k\eta+1)^6} \tag{5.93}$$

The function $f_1(\eta)$ given by Eq. 5.13 can be written in the form

$$f_1(\eta, C_i) = H_1(\eta, C_i)\frac{1}{(k\eta+1)^2} + H_2(\eta, C_i)\frac{1}{(k\eta+1)^3} + \ldots + \\ +H_j(\eta, C_i)\frac{1}{(k\eta+1)^{j+1}} \\ f_1(0, C_i) = f_1'(0, C_i) = f_1'(\infty, C_i) = 0 \tag{5.94}$$

In what follows we consider only three cases for the optimal auxiliary functions H_i. If we choose $H_i = C_i\eta^2$ and j = 9, i = 1, 2,..., 9 one can get

$$f_1(\eta, C_i) = C_1\frac{\eta^2}{(k\eta+1)^2} + C_2\frac{\eta^2}{(k\eta+1)^3} + C_3\frac{\eta^2}{(k\eta+1)^4} + \ldots \\ + C_9\frac{\eta^2}{(k\eta+1)^{10}} \tag{5.95}$$

The first-order approximate solution given by Eq. 5.6 is obtained from Eqs. 5.72 and 5.95:

$$\overline{f}(\eta, C_i) = R + \frac{1}{k} - \frac{1}{k(k\eta+1)} + C_1\frac{\eta^2}{(k\eta+1)^2} + C_2\frac{\eta^2}{(k\eta+1)^3} + \\ (\mathrm{B1}): +C_3\frac{\eta^2}{(k\eta+1)^4} + C_4\frac{\eta^2}{(k\eta+1)^5} + C_5\frac{\eta^2}{(k\eta+1)^6} + C_6\frac{\eta^2}{(k\eta+1)^7} + \\ +C_7\frac{\eta^2}{(k\eta+1)^8} + C_8\frac{\eta^2}{(k\eta+1)^9} + C_9\frac{\eta^2}{(k\eta+1)^{10}} \tag{5.96}$$

If $H_i = C_i\eta^2$ and j = 10, i = 1,2,...,10 then the first-order approximate solution becomes

$$\bar{f}(\eta, C_i) = R + \frac{1}{k} - \frac{1}{k(k\eta+1)} + C_1\frac{\eta^2}{(k\eta+1)^2} + C_2\frac{\eta^2}{(k\eta+1)^3} +$$

$$\text{(B2)}: +C_3\frac{\eta^2}{(k\eta+1)^4} + C_4\frac{\eta^2}{(k\eta+1)^5} + C_5\frac{\eta^2}{(k\eta+1)^6} + C_6\frac{\eta^2}{(k\eta+1)^7} +$$

$$+C_7\frac{\eta^2}{(k\eta+1)^8} + C_8\frac{\eta^2}{(k\eta+1)^9} + C_9\frac{\eta^2}{(k\eta+1)^{10}} + C_{10}\frac{\eta^2}{(k\eta+1)^{11}} \quad (5.97)$$

In the last case we choose $H_i = C_i\eta^2$ and j = 11, i = 1,2,...,11 and therefore

$$\bar{f}(\eta, C_i) = R + \frac{1}{k} - \frac{1}{k(k\eta+1)} + C_1\frac{\eta^2}{(k\eta+1)^2} + C_2\frac{\eta^2}{(k\eta+1)^3} +$$

$$\text{(B3)}: +C_3\frac{\eta^2}{(k\eta+1)^4} + C_4\frac{\eta^2}{(k\eta+1)^5} + C_5\frac{\eta^2}{(k\eta+1)^6} + C_6\frac{\eta^2}{(k\eta+1)^7} +$$

$$+C_7\frac{\eta^2}{(k\eta+1)^8} + C_8\frac{\eta^2}{(k\eta+1)^9} + C_9\frac{\eta^2}{(k\eta+1)^{10}} +$$

$$+C_{10}\frac{\eta^2}{(k\eta+1)^{11}} + C_{11}\frac{\eta^2}{(k\eta+1)^{12}} \quad (5.98)$$

In this way we can obtain many other solutions.

5.3.2 Numerical Results

We illustrate the accuracy of our procedure for different values of the coefficients R, β and M. Also we represent graphically the behavior of the functions f and f' and we compare results obtained through our procedure with numerical results.

5.3.2.A First, we consider $\beta = 0.5, M = 0.75$, and $R = 0.25$ which corresponds to suction velocity. The optimal convergence-control parameters, are determined by means of the least-square method and are:

$$(A_1)\quad K = 1.7223999348,\ C_1 = 0.0203632630,\ C_2 = -0.0003551855,\\ C_3 = -0.0046442857,\ C_4 = 0.0021151819,\\ C_5 = -0.0040337933,\ C_6 = -0.0030441083 \tag{5.99}$$

$$(A_2)\quad K = 1.2039668671,\ C_1 = 0.0484595533,\ C_2 = -0.0035879577,\\ C_3 = -0.1933189832,\ C_4 = -0.0613273780,\\ C_5 = -0.0321447762,\ C_6 = -0.0048370121 \tag{5.100}$$

and so on.

The first-order approximate solutions obtained by means OHAM, for the above cases (A_1)–(A_{12}) become:

$$(A_1)\quad \overline{f}(\eta) = 0.8305852518 + \big(-0.5873694973 + 0.0203632630\eta-\\ -0.0003551855\eta^2\big)e^{-1.7223999348\eta} + \big(0.0067842454-\\ -0.0046442857\eta + 0.0021151819\eta^2\big)e^{-3.4447998696\eta}+\\ +(-0.0040337933\eta - 0.0030441083\eta^2)e^{-5.1671998044\eta} \tag{5.101}$$

$$(A_2)\quad \overline{f}(\eta) = 0.8305876410 + \big(-0.1835701350 + 0.0484595533\eta-\\ -0.0035879577\eta^2\big)e^{-1.2039668671\eta} + \big(-0.3970175060-\\ -0.1933189832\eta - 0.0613273780\eta^2\big)e^{-2.4079337342\eta}+\\ +(-0.0321447762\eta - 0.0048370121\eta^2)e^{-3.6119006013\eta} \tag{5.102}$$

$$(A_3)\quad \overline{f}(\eta) = 0.8305880207 + \big(-0.3420493544 + 0.0892570980\eta-\\ -0.0066950727\eta^2\big)e^{-1.3123992230\eta} + \big(-0.2899683613\eta-\\ -0.0166914996\eta^2\big)e^{-2.6247984460\eta} + \big(-0.2385386662-\\ -0.1873679245\eta - 0.0454910883\eta^2\big)e^{-3.9371976690\eta} \tag{5.103}$$

$$(A_4)\quad \overline{f}(\eta) = 0.8305846668 + \big(-0.5805846668 + 0.0113141332\eta^2-\\ -0.0026254854\eta^3 + 0.0002157623\eta^4\big)e^{-1.7224016705\eta} + \big(0.0194551449\eta^2+\\ +0.0055884706\eta^3 + 0.0021434013\eta^4\big)e^{-3.4448033410\eta} \tag{5.104}$$

$$(A_5)\quad \overline{f}(\eta) = 0.8305850786 + \big(-0.5805850786 + 0.0129138333\eta+\\ +0.0029826874\eta^2 - 0.0007022807\eta^3 + 0.0000575450\eta^4\big)e^{-1.7224004488\eta}+\\ +(-0.0129138333\eta + 0.0055128320\eta^2 - 0.0009540905\eta^3)e^{-3.4448008976\eta} \tag{5.105}$$

$$(A_6)\quad \overline{f}(\eta) = 0.8305896186 + \big(-0.5805896186 + 0.2380356841\eta-\\ -0.0458010823\eta^2 + 0.0050361161\eta^3 - 0.0002634636\eta^4\big)e^{-1.3123974169\eta}+\\ +(0.0276184306\eta^2 - 0.0042554588\eta^3 + 0.0040951218\eta^4)e^{-2.6247948339\eta} \tag{5.106}$$

$$(A_7)\quad \overline{f}(\eta) = 0.25 + 0.5805981621\left(1 - e^{-1.7223616352\eta}\right) + \\ +(3.7465003467\eta^2 - 2.5918747904\eta^3 + 0.5688873710\eta^4)e^{-3.4447232705\eta} + \\ +(-3.7216397470\eta^2 - 3.7194132780\eta^3 - 1.7743745762\eta^4)e^{-5.1670849057\eta} \tag{5.107}$$

$$(A_8)\quad \overline{f}(\eta) = 0.25 + 0.5806004947\left(1 - e^{-1.7223547156\eta}\right) + \\ +\left(28.6559933963\eta - 17.1574454736\eta^2 + \right. \\ \left. +3.0755892307\eta^3\right)e^{-3.4447094313\eta} + \left(- 28.6559933963\eta - \right. \\ \left. -32.1751588943\eta^2 - 15.8637665758\eta^3 - 4.4540759880\eta^4\right)e^{-5.1670641469\eta} \tag{5.108}$$

$$(A_9)\quad \overline{f}(\eta) = 0.25 + 0.5807600837\left(1 - e^{-1.3122047950\eta}\right) + \\ +\left(0.2379238333\eta + 14.3490983671\eta^2 - 7.4936309491\eta^3 + \right. \\ \left. +1.3346589985\eta^4\right)e^{-2.6244095900\eta} + \left(- 14.0803697555\eta^2 - \right. \\ \left. -10.6121821644\eta^3 - 3.7974679259\eta^4\right)e^{-3.9366143850\eta} \tag{5.109}$$

$$(A_{10})\quad \overline{f}(\eta) = 0.8305805525 + \left(- 0.5805805525 + 0.0163027398\eta^2 - \right. \\ \left. -0.0053622456\eta^3 + 0.0006113624\eta^4\right)e^{-1.7224138763\eta} + \left(0.0153758771\eta^2 + \right. \\ \left. +0.0170806180\eta^3 + 0.0369073209\eta^4\right)e^{-5.1672416289\eta} \tag{5.110}$$

$$(A_{11})\quad \overline{f}(\eta) = 0.8305850069 + \left(- 0.5805850069 + 0.0167016213\eta + \right. \\ \left. +0.0001742425\eta^2\right)e^{-1.7224006615\eta} + \left(- 0.0115884531\eta - \right. \\ \left. -0.0052672814\eta^2\right)e^{-3.4448013230\eta} + \left(- 0.0193966267\eta + \right. \\ \left. +0.0114385136\eta^2\right)e^{-5.1672019845\eta} + \left(0.0142834585\eta + \right. \\ \left. +0.0114385136\eta^2\right)e^{-6.8896026460\eta} \tag{5.111}$$

$$(A_{12})\quad \overline{f}(\eta) = 0.8305751642 + \left(- 0.5805751642 + \right. \\ \left. +0.0012101182\eta^3\right)e^{-1.7224298620\eta} + \left(- 0.4645923536\eta^2 + \right. \\ \left. +0.2718146543\eta^3\right)e^{-3.4448597240\eta} + \left(0.4983783090\eta^2 + \right. \\ \left. +0.5722116290\eta^3 + 0.4013983403\eta^4\right)e^{-5.1672895860\eta} \tag{5.112}$$

5.3.2.B In the second case, we consider $\beta = 0.5$, $M = 0.75$, and $R = -2$ for injection velocity. The optimal convergence-control parameters are:

$$(A_1)\quad K = 3.4039784470,\ C_1 = -0.0073489093, \\ C_2 = -0.0000300879, C_3 = 0.0019346837,\ C_4 = -0.0028434644, \\ C_5 = 0.0004870868,\ C_6 = 0.0008838343 \tag{5.113}$$

$$(A_2)\quad K = 0.3235777833,\ C_1 = 193.0767116481, \\ C_2 = -9.2315115218, C_3 = 269.5805006985,\ C_4 = 10.4423160474 \\ C_5 = -36.4073561981,\ C_6 = -8.0939438081 \tag{5.114}$$

and so on.

The first-order approximate solutions can be written for only three cases, in the form:

$$(A_1)\quad \overline{f}(\eta) = -1.7062261070 + \big(-0.2923264282 - 0.0073489093\eta - \\ -0.0000300879\eta^2\big)e^{-3.4039784470\eta} + \big(-0.0014474647 + 0.0019346837\eta - \\ -0.0028434644\eta^2\big)e^{-6.8079568940\eta} + \\ +(0.0004870868\eta + 0.0008838343\eta^2)e^{-10.2119353411\eta} \tag{5.115}$$

$$(A_4)\quad \overline{f}(\eta) = -1.7062260957 + \big(-0.2937739042 - \\ -0.0079483534\eta^2 + 0.0033559204\eta^3 - 0.0005109100\eta^4\big)e^{-3.4039783163\eta} + \\ +(-0.0123589233\eta^2 - 0.0091292547\eta^3 - 0.0057558384\eta^4)e^{-6.8079566326\eta} \tag{5.116}$$

$$(A_{11})\quad \overline{f}(\eta) = -1.7062260959 + \big(-0.2937739040 - 0.0058506315\eta - \\ -0.0004400135\eta^2\big)e^{-3.4039783189\eta} + \big(0.0043294258\eta + \\ +0.0042247893\eta^2\big)e^{-6.8079566378\eta} + \big(0.0072464953\eta - \\ -0.0092559326\eta^2\big)e^{-10.2119349567\eta} + \big(-0.0057252895\eta - \\ -0.0092559326\eta^2\big)e^{-13.6159132756\eta} \tag{5.117}$$

5.3.2.C In the third case, for $\beta = 0.5$, $M = 0.75$ and $R = 0.25$, for the above cases (B1–B3), the first-order approximate solutions are as follows:

(B1):

$$\overline{f}(\eta) = 1.7640282387 - \frac{1.5140282387}{1 + 0.6604896622\eta} - \frac{0.4038516600\eta^2}{(1 + 0.6604896622\eta)^2} - \\ - \frac{0.2651317383\eta^2}{(1 + 0.6604896622\eta)^3} + \frac{1.5875750484\eta^2}{(1 + 0.6604896622\eta)^4} - \frac{9.98303883215\eta^2}{(1 + 0.6604896622\eta)^5} + \\ + \frac{40.2722623404\eta^2}{(1 + 0.6604896622\eta)^6} - \frac{81.8453842441\eta^2}{(1 + 0.6604896622\eta)^7} + \frac{89.8542653196\eta^2}{(1 + 0.6604896622\eta)^8} - \\ \frac{51.6009371221\eta^2}{(1 + 0.6604896622\eta)^9} + \frac{12.2170946834\eta^2}{(1 + 0.6604896622\eta)^{10}} \tag{5.118}$$

(B2):

$$\overline{f}(\eta) = 1.5246445734 - \frac{1.2746445734}{1 + 0.7845324263\eta} - \frac{0.4232066131\eta^2}{(1 + 0.7845324263\eta)^2} - \\ - \frac{0.2248813142\eta^2}{(1 + 0.7845324263\eta)^3} + \frac{2.7575287469\eta^2}{(1 + 0.7845324263\eta)^4} - \frac{20.1233470298\eta^2}{(1 + 0.7845324263\eta)^5} + \\ + \frac{100.1896168227\eta^2}{(1 + 0.7845324263\eta)^6} - \frac{269.3616538738\eta^2}{(1 + 0.7845324263\eta)^7} + \frac{418.6971099614\eta^2}{(1 + 0.7845324263\eta)^8} - \\ - \frac{382.2371075125\eta^2}{(1 + 0.7845324263\eta)^9} + \frac{191.2755083010\eta^2}{(1 + 0.7845324263\eta)^{10}} - \frac{40.5986791066\eta^2}{(1 + 0.7845324263\eta)^{11}} \tag{5.119}$$

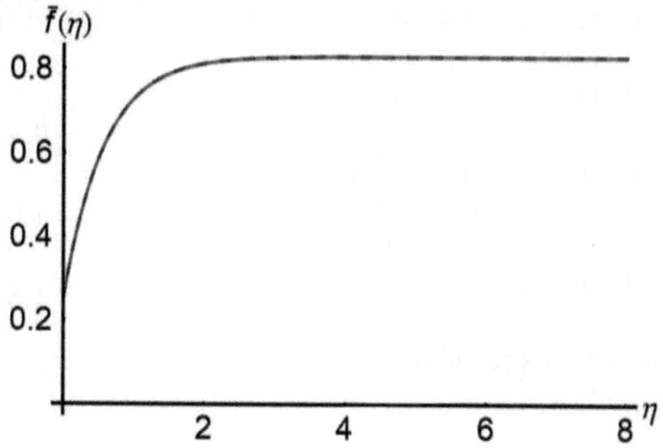

Fig. 5.3 Comparison between the approximate solution Eq. (5.101) and numerical solution in the case $\beta = 0.5$, $M = 0.75$, $R = 0.25$: (*red solid line*) numerical solution and (*blue dashed line*) approximate solution (colour figure online)

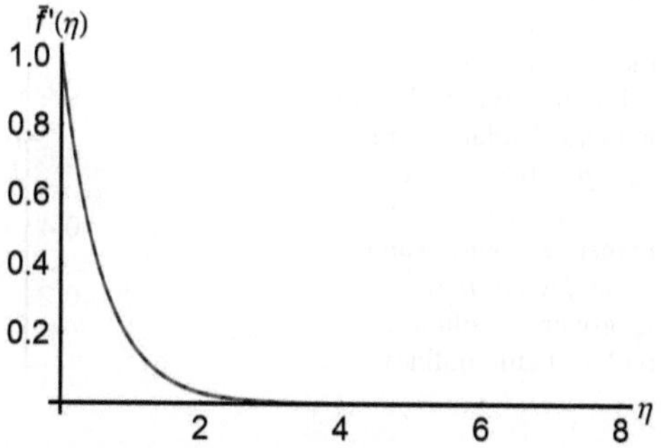

Fig. 5.4 Comparison between the derivative of the first-order approximate solution of Eq. (5.101) and numerical solution in the case $\beta = 0.5$, $M = 0.75$, $R = 0.25$: (*red solid line*) numerical solution and (*blue dashed line*) approximate solution (colour figure online)

(B3):

$$\begin{aligned}\overline{f}(\eta) &= 3.3822265399 - \frac{3.1322265399}{1+0.3192617096\eta} - \frac{0.2570668647\eta^2}{(1+0.3192617096\eta)^2} - \\ &- \frac{0.2781821246\eta^2}{(1+0.3192617096\eta)^3} + \frac{0.734477016\eta^2}{(1+0.3192617096\eta)^4} - \frac{5.8357903615\eta^2}{(1+0.3192617096\eta)^5} + \\ &+ \frac{23.9865472994\eta^2}{(1+0.3192617096\eta)^6} - \frac{65.8073457539\eta^2}{(1+0.3192617096\eta)^7} + \frac{118.7010136318\eta^2}{(1+0.3192617096\eta)^8} - \\ &- \frac{136.7897035346\eta^2}{(1+0.319261709\eta)^9} + \frac{97.2920818266\eta^2}{(1+0.319261709\eta)^{10}} - \frac{39.0230935985\eta^2}{(1+0.319261709\eta)^{11}} + \\ &+ \frac{6.765241722\eta^2}{(1+0.3192617096\eta)^{12}}\end{aligned} \tag{5.120}$$

In Figs. 5.3 and 5.4 are plotted a comparison between the first-order approximate solutions $\overline{f}(\eta)$ and $\overline{f}'(\eta)$ respectively for the suction velocity in the case (A_1) and numerical results, while in Figs 5.5 and 5.6 are plotted a comparison between the first-order approximate solutions $\overline{f}(\eta)$ and $\overline{f}'(\eta)$ respectively for the injection velocity in the case (A_1). In Tables 5.4, 5.5, 5.6, 5.7, 5.8 and 5.9 we present a comparison between the first-order approximate solutions 5.101–5.120 respectively, with numerical results for some values of variable η.

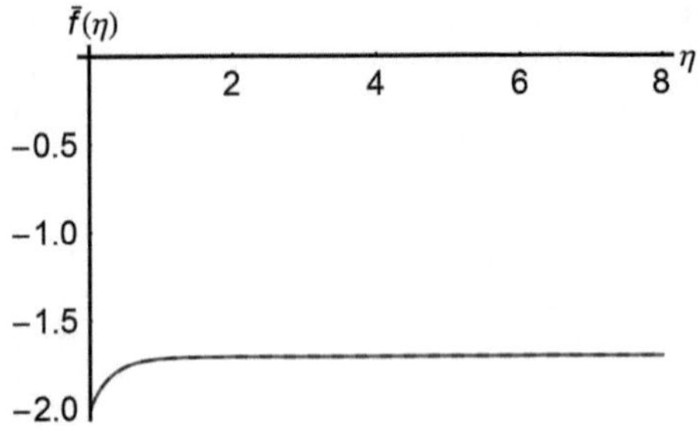

Fig. 5.5 Comparison between the approximate solution Eq. (5.115) and numerical solution in the case $\beta = 0.5$, $M = 0.75$, $R = -2$: (*red solid line*) numerical solution and (*blue dashed line*) approximate solution (colour figure online)

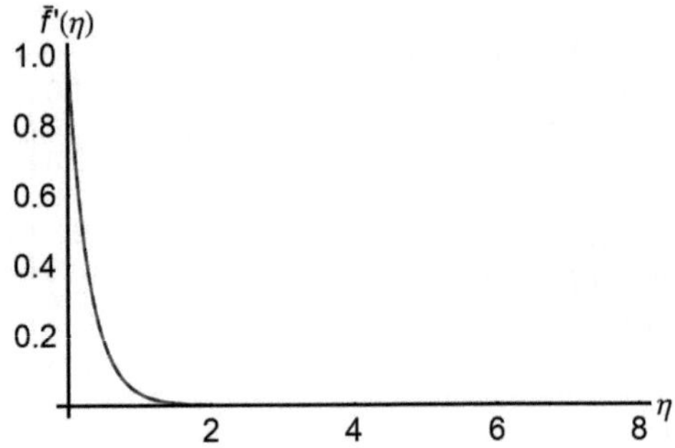

Fig. 5.6 Comparison between the derivative of the first-order approximate solution of Eq. 5.115 and numerical solution in the case $\beta = 0.5$, $M = 0.75$, $R = -2$: (*red solid line*) numerical solution and (*blue dashed line*) approximate solution (colour figure online)

Table 5.4 Comparison between the first-order approximate solution $\overline{f}$ obtained by OHAM with numerical results

η	$\overline{f}_{OHAM}$ Eq. 5.101	$\overline{f}_{OHAM}$ Eq. 5.102	$\overline{f}_{OHAM}$ Eq. 5.103	$f_{numerical}$
0	0.25	0.25	0.25	0.25
4/5	0.6867557456	0.6867558261	0.6867555458	0.6867557531
8/5	0.7952831540	0.7952832405	0.7952835769	0.7952831536
12/5	0.8219262765	0.8219262098	0.8219258666	0.8219262711
16/5	0.8284614196	0.8284613427	0.8284615570	0.8284614131
4	0.8300643020	0.8300645259	0.8300647355	0.8300642946
24/5	0.8304574558	0.8304574869	0.8304573214	0.8304574344
28/5	0.8305538969	0.8305537376	0.8305535912	0.8305538603
32/5	0.8305775573	0.8305776155	0.8305777588	0.8305775112
36/5	0.8305833631	0.8305839424	0.8305843430	0.8305833139
8	0.8305847881	0.8305859411	0.8305864617	0.8305847388

In Tables 5.10, 5.11, 5.12 and 5.13 we present the relative error $\varepsilon = \left|\overline{f}_{OHAM} - f_{numerical}\right|$ between the first-order approximate solutions 5.101–5.120 and numerical results, for some values of variable η.

Table 5.5 Comparison between the first-order approximate solution $\overline{f}$ obtained by OHAM with numerical results

η	$\overline{f}_{OHAM}$ Eq. 5.104	$\overline{f}_{OHAM}$ Eq. 5.105	$\overline{f}_{OHAM}$ Eq. 5.106	$f_{numerical}$
0	0.25	0.25	0.25	0.25
4/5	0.6867559559	0.6867558676	0.6867559421	0.6867557531
8/5	0.7952829566	0.7952830577	0.7952832693	0.7952831536
12/5	0.8219264714	0.8219263597	0.8219262460	0.8219262711
16/5	0.8284612774	0.8284613647	0.8284612078	0.8284614131
4	0.8300641916	0.8300642631	0.8300646204	0.8300642946
24/5	0.8304575489	0.8304574866	0.8304575594	0.8304574344
28/5	0.8305539626	0.8305539177	0.8305536600	0.8305538603
32/5	0.8305774335	0.8305775190	0.8305775904	0.8305775112
36/5	0.8305830455	0.8305832668	0.8305842233	0.8305833139
8	0.8305843381	0.8305846532	0.8305866484	0.8305847388

Table 5.6 Comparison between the first-order approximate solution $\overline{f}$ obtained by OHAM with numerical results

η	$\overline{f}_{OHAM}$ Eq. 5.107	$\overline{f}_{OHAM}$ Eq. 5.108	$\overline{f}_{OHAM}$ Eq. 5.109	$f_{numerical}$
0	0.25	0.25	0.25	0.25
4/5	0.6867604947	0.6867642071	0.6866887026	0.6867557531
8/5	0.7952711920	0.7952698391	0.7952830603	0.7952831536
12/5	0.8219400081	0.8219414386	0.8219258072	0.8219262711
16/5	0.8284434011	0.8284360789	0.8285164738	0.8284614131
4	0.8300472216	0.8300445299	0.8299931937	0.8300642946
24/5	0.8304557614	0.8304567567	0.8303854806	0.8304574344
28/5	0.8305615064	0.8305635876	0.8305702204	0.8305538603
32/5	0.8305888008	0.8305910952	0.8306729872	0.8305775112
36/5	0.8305957861	0.8305981136	0.8307238267	0.8305833139
8	0.8305975612	0.8305998931	0.8307459962	0.8305847388

It can be seen from the above tables that the solutions obtained by the proposed procedure are nearly identical with the numerical solutions obtained using a forth-order Runge–Kutta method in combination with the shooting method.

Also, we note that one can get a lot of other solutions by means of the initial approximation given by Eq. 5.71.

Table 5.7 Comparison between the first-order approximate solution $\overline{f}$ obtained by OHAM with numerical results

η	$\overline{f}_{OHAM}$ Eq. 5.110	$\overline{f}_{OHAM}$ Eq. 5.111	$\overline{f}_{OHAM}$ Eq. 5.112	$f_{numerical}$
0	0.25	0.25	0.25	0.25
4/5	0.6867577703	0.6867556858	0.6867565108	0.6867557531
8/5	0.7952835590	0.7952833690	0.7952904892	0.7952831536
12/5	0.8219267050	0.8219260684	0.8219182801	0.8219262711
16/5	0.8284593204	0.8284614530	0.8284624519	0.8284614131
4	0.8300649028	0.8300645204	0.8300733641	0.8300642946
24/5	0.8304590162	0.8304575585	0.8304617717	0.8304574344
28/5	0.8305540337	0.8305538374	0.8305514864	0.8305538603
32/5	0.8305757755	0.8305773944	0.8305708821	0.8305775112
36/5	0.8305801662	0.8305831505	0.8305746354	0.8305833139
8	0.8305807821	0.8305845549	0.8305752047	0.8305847388

Table 5.8 Comparison between the first-order approximate solution $\overline{f}$ obtained by OHAM with numerical results

η	$\overline{f}_{OHAM}$ Eq. 5.115	$\overline{f}_{OHAM}$ Eq. 5.116	$\overline{f}_{OHAM}$ Eq. 5.117	$f_{numerical}$
0	−2	−2	−2	0
4/5	−1.7258163099	−1.7258162625	−1.7258163629	−1.7258163121
8/5	−1.7075377409	−1.7075377235	−1.7075377264	−1.7075377364
12/5	−1.7063139221	−1.7063139346	−1.7063139675	−1.7063139251
16/5	−1.7062319851	−1.7062320229	−1.7062319899	−1.7062319847
4	−1.7062265004	−1.7062265071	−1.7062264918	−1.7062264984
24/5	−1.7062261333	−1.7062261259	−1.7062261225	−1.7062261309
28/5	−1.7062261087	−1.7062260981	−1.7062260977	−1.7062261063
32/5	−1.7062261071	−1.7062260959	−1.7062260960	−1.7062261046
36/5	−1.7062261070	−1.7062260957	−1.7062260959	−1.7062261044
8	−1.7062261070	−1.7062260957	−1.7062260959	−1.7062261044

5.4 Dual Solutions of the Unsteady Viscous Flow Over a Shrinking Cylinder

The flow of Newtonian and non-Newtonian fluids is important for engineers and applied mathematicians because of its several applications in engineering or industrial processes. In the last few decades, these fluids have attracted considerable attention from researchers in many branches of nonlinear dynamical systems in science and technology. The flow over a stretching/shrinking cylinder is an important problem in many engineering processes with applications in industries such as in plastic and metallurgy industries, glass-fiber production, wire drawing, and so

Table 5.9 Comparison between the first-order approximate solution $\overline{f}$ obtained by OHAM with numerical results

η	$\overline{f}_{OHAM}$ Eq. 5.118	$\overline{f}_{OHAM}$ Eq. 5.119	$\overline{f}_{OHAM}$ Eq. 5.120	$f_{numerical}$
0	0.25	0.25	0.25	0.25
4/5	0.6867508021	0.6867549888	0.6867557267	0.6867557531
8/5	0.7952864463	0.7952862406	0.7952832317	0.7952831536
12/5	0.8219220867	0.8219231492	0.8219262018	0.8219262711
16/5	0.8284650425	0.8284633218	0.8284614876	0.8284614131
4	0.8300669092	0.8300665681	0.8300643027	0.8300642946
24/5	0.8304557254	0.8304565209	0.8304574038	0.8304574344
28/5	0.8305528738	0.8305529772	0.8305539465	0.8305538603
32/5	0.8305813335	0.8505798428	0.8305776905	0.8305775112
36/5	0.8305918172	0.8305889777	0.8305834621	0.8305833139
8	0.8305950529	0.8305917010	0.8305847981	0.8305847388

Table 5.10 Comparison between the first-order approximate solution $\overline{f}$ obtained by OHAM with numerical results

η	ε Eq. 5.101	ε Eq. 5.102	ε Eq. 5.103	ε Eq. 5.104	ε Eq. 5.105
0	1.1×10^{-16}	1.1×10^{-16}	0	0	0
4/5	7.8×10^{-9}	7.2×10^{-8}	2.0×10^{-7}	2.0×10^{-7}	1.1×10^{-7}
8/5	9.0×10^{-10}	8.5×10^{-8}	4.2×10^{-7}	1.9×10^{-7}	9.7×10^{-8}
12/5	1.9×10^{-9}	6.4×10^{-8}	4.0×10^{-7}	1.9×10^{-7}	8.5×10^{-8}
16/5	4.0×10^{-10}	7.7×10^{-8}	1.4×10^{-7}	1.4×10^{-7}	5.5×10^{-8}
4	4.7×10^{-9}	2.1×10^{-7}	4.4×10^{-7}	1.1×10^{-7}	4.3×10^{-8}
24/5	1.2×10^{-9}	3.2×10^{-8}	1.1×10^{-7}	9.4×10^{-8}	3.1×10^{-8}
28/5	4.5×10^{-9}	1.5×10^{-7}	2.6×10^{-7}	7.0×10^{-8}	2.5×10^{-8}
32/5	3.5×10^{-9}	5.4×10^{-8}	2.4×10^{-7}	1.2×10^{-7}	4.1×10^{-8}
36/5	2.6×10^{-8}	5.5×10^{-7}	1.0×10^{-6}	3.4×10^{-7}	1.2×10^{-7}
8	6.4×10^{-8}	1.0×10^{-6}	1.7×10^{-6}	5.1×10^{-7}	1.9×10^{-7}

on. The pioneering works in the area of the flow inside a tube with time dependent diameter was [25, 26], where Uchidia and Aoki, and Skalak and Wang studied the internal flow velocity and pressure due to tube expansion or contraction. Miklavčič and Wang [27] investigated the flow over a shrinking sheet, obtaining an exact solution of the Navier–Stokes equations. Ishak et al. [28] reported that injection reduces the skin friction as well as the heat transfer rate at the surface while suction acts in the opposite manner. Fang et al. [29] obtained the exact solution of unsteady state Navier–Stokes equations.

Fang et al. [30] studied the viscous flow over a shrinking sheet by a newly proposed second order slip flow model. The exact solution of the full governing Navier–Stokes equation has two branches in a certain range of the parameters. The

Table 5.11 Comparison between the first-order approximate solution $\overline{f}$ obtained by OHAM with numerical results

η	ε Eq. 5.106	ε Eq. 5.107	ε Eq. 5.108	ε Eq. 5.109	ε Eq. 5.110
0	0	0	0	0	0
4/5	1.8×10^{-7}	4.7×10^{-6}	8.4×10^{-6}	6.7×10^{-5}	2.0×10^{-6}
8/5	1.1×10^{-7}	1.1×10^{-5}	1.3×10^{-5}	9.4×10^{-8}	4.0×10^{-7}
12/5	2.8×10^{-8}	1.3×10^{-5}	1.5×10^{-5}	4.6×10^{-7}	4.3×10^{-7}
16/5	2.1×10^{-7}	1.8×10^{-5}	2.5×10^{-5}	5.5×10^{-5}	2.0×10^{-6}
4	3.1×10^{-7}	1.7×10^{-5}	1.9×10^{-5}	7.1×10^{-5}	5.9×10^{-7}
24/5	1.0×10^{-7}	1.6×10^{-6}	6.9×10^{-7}	7.1×10^{-5}	1.5×10^{-6}
28/5	2.3×10^{-5}	7.6×10^{-6}	9.6×10^{-6}	1.6×10^{-5}	1.4×10^{-7}
32/5	2.9×10^{-8}	1.1×10^{-5}	1.3×10^{-5}	9.5×10^{-5}	1.7×10^{-6}
36/5	8.3×10^{-7}	1.2×10^{-5}	1.4×10^{-5}	1.4×10^{-5}	3.2×10^{-6}
8	1.7×10^{-6}	1.2×10^{-5}	1.5×10^{-5}	1.6×10^{-5}	4.0×10^{-6}

Table 5.12 Comparison between the first-order approximate solution $\overline{f}$ obtained by OHAM with numerical results

η	ε Eq. 5.111	ε Eq. 5.112	ε Eq. 5.115	ε Eq. 5.116	ε Eq. 5.117
0	0	0	0	0	0
4/5	6.7×10^{-8}	7.5×10^{-7}	2.4×10^{-9}	4.9×10^{-8}	5.0×10^{-8}
8/5	2.1×10^{-7}	7.3×10^{-6}	3.8×10^{-9}	1.3×10^{-8}	9.9×10^{-9}
12/5	2.0×10^{-7}	7.9×10^{-6}	4.0×10^{-9}	8.5×10^{-9}	4.2×10^{-8}
16/5	3.2×10^{-8}	1.0×10^{-6}	8.8×10^{-10}	3.6×10^{-8}	5.1×10^{-9}
4	2.1×10^{-7}	9.0×10^{-6}	4.9×10^{-10}	7.2×10^{-9}	6.6×10^{-9}
24/5	1.0×10^{-7}	4.3×10^{-6}	7.2×10^{-10}	6.6×10^{-9}	8.3×10^{-9}
28/5	5.4×10^{-8}	2.4×10^{-6}	6.8×10^{-10}	9.9×10^{-9}	8.5×10^{-9}
32/5	1.6×10^{-7}	6.6×10^{-6}	6.3×10^{-10}	1.0×10^{-8}	8.5×10^{-9}
36/5	2.3×10^{-7}	8.7×10^{-6}	6.0×10^{-10}	1.0×10^{-8}	8.5×10^{-9}
8	2.9×10^{-7}	9.6×10^{-6}	5.8×10^{-10}	1.0×10^{-8}	8.4×10^{-9}

problem of unsteady viscous flow over a permeable shrinking cylinder was solved by Zaimi et al. [31] numerically using the shooting method. The effect of suction and unsteadiness parameters on the flow velocity and the skin friction coefficient have been analyzed and presented graphically and the same authors in [32] studied the effects of the unsteadiness parameter and the Brownian motion parameter on the flow field and heat transfer characteristics. Dual solutions are found to exist in certain conditions.

In this section we assume an unsteady laminar boundary layer flow of a nanofluid over an infinite cylinder or a tube with a time dependent diameter in shrinking motion as show in Fig. 5.7.

Table 5.13 Comparison between the first-order approximate solution $\overline{f}$ obtained by OHAM with numerical results

η	ε Eq. 5.118	ε Eq. 5.119	ε Eq. 5.120
0	0	0	0
4/5	4.95×10^{-6}	7.64×10^{-7}	2.67×10^{-8}
8/5	3.29×10^{-6}	3.08×10^{-6}	7.68×10^{-8}
12/5	4.18×10^{-6}	3.12×10^{-6}	7.26×10^{-8}
16/5	3.62×10^{-6}	1.90×10^{-6}	6.76×10^{-8}
4	2.60×10^{-6}	2.26×10^{-6}	4.06×10^{-9}
24/5	1.72×10^{-6}	9.33×10^{-7}	5.07×10^{-8}
28/5	1.01×10^{-6}	9.15×10^{-7}	5.42×10^{-8}
32/5	3.77×10^{-6}	2.28×10^{-6}	1.29×10^{-7}
36/5	8.42×10^{-6}	5.5×10^{-6}	7.26×10^{-8}
8	1.02×10^{-5}	6.84×10^{-6}	5.47×10^{-8}

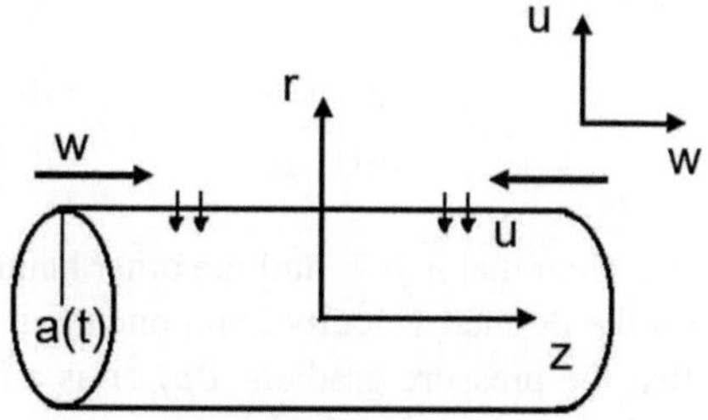

Fig. 5.7 A schematic model of flow an expanding cylinder with time dependent radius

Also we consider the three-dimensional unsteady Navier–Stokes equations for incompressible fluids without body force such that based on the axisymmetric flow assumption and there is no azimuthal velocity component we have:

$$\nabla \mathbf{v} = 0 \tag{5.121}$$

$$\frac{\partial \mathbf{v}}{\partial t} + \mathbf{v} \cdot \nabla \mathbf{v} = -\frac{1}{\rho} \nabla p + \nu \nabla^2 \mathbf{v} \tag{5.122}$$

where **v** is the velocity vector, $\rho(r)$ is the fluid density, p is the pressure and ν is the kinematic viscosity. The diameter of the cylinder is assumed as a function of time with unsteady radius $a(t) = a_0\sqrt{1-\beta t}$. For a positive value of β, the cylinder radius becomes smaller with time, e.g. contracting, while for a negative value of β, the diameter becomes layer with time, e.g. expanding. In cylindrical polar coordinates r and z are measured in the radial and axial directions respectively, Eqs. 5.121 and 5.122 can be written as [24], [29–33]:

$$\frac{1}{r}\frac{\partial(ru)}{\partial r}+\frac{\partial w}{\partial z}=0 \tag{5.123}$$

$$\frac{\partial u}{\partial r}+u\frac{\partial u}{\partial r}+w\frac{\partial u}{\partial z}=-\frac{1}{\rho}\frac{dp}{dr}+\nu\left(\frac{\partial^2 u}{\partial r^2}+\frac{1}{r}\frac{\partial u}{\partial r}+\frac{\partial^2 u}{\partial z^2}-\frac{u}{r^2}\right) \tag{5.124}$$

$$\frac{\partial w}{\partial z}+u\frac{\partial w}{\partial r}+w\frac{\partial w}{\partial z}=-\frac{1}{\rho}\frac{dp}{dz}+\nu\left(\frac{\partial^2 w}{\partial r^2}+\frac{1}{r}\frac{\partial w}{\partial r}+\frac{\partial^2 w}{\partial z^2}\right) \tag{5.125}$$

If we consider the constant mass transfer velocity U $(U<0)$ and a_0 a positive constant, then the boundary conditions are of the form:

$$u=\frac{U}{\sqrt{1-\beta t}},\qquad w=-\frac{4\nu z}{a_0^2(1-\beta t)}\ \ at\ \ r=a(t),\qquad w=0as\ \ r\to\infty. \tag{5.126}$$

By means of the similarity variables [32]

$$u=-\frac{2v}{a(t)}\frac{f(\eta)}{\sqrt{\eta}},\qquad w=\frac{4vz}{a^2(t)}f'(\eta),\quad \eta=\left(\frac{r}{a(t)}\right)^2. \tag{5.127}$$

it is clear that $\eta\geq 1$, and the other hand, Eq. 5.123 is satisfied automatically. Based on the defined velocity components, it is straightforward to derive from Eq. 5.124 that the pressure gradient $\partial p/\partial r$ is a function of t and r and is independent on z, such that, from Eq. 5.124 we obtain

$$\frac{p}{\rho}=G(t,z)+v\left(\frac{\partial u}{\partial r}+\frac{u}{r}\right)-\frac{1}{2}u^2+\int\frac{\partial u}{\partial t}\ dr \tag{5.128}$$

or using Eqs. 5.127 the pressure may be written as

$$\frac{p}{\rho}=G(t,z)-\frac{2v^2}{a^2(t)\left[f'(\eta)+\frac{1}{\eta}f^2(\eta)+2Sf(\eta)\right]} \tag{5.129}$$

where $G(t,z)$ is the constant of the integration on z and $S=\frac{a_0^2\beta}{4\nu}$ is the unsteady parameter for the expanding $(\beta<0)$ or contraction $(\beta>0)$ cylinder showing the strength of expansion or contraction. Substituting Eqs. 5.127 into Eq. 5.125 and rearranging terms, this becomes:

$$\eta f'''(\eta)+f''(\eta)+f(\eta)f''(\eta)-f'^2(\eta)-S[\eta f''(\eta)-f'(\eta)]=0 \tag{5.130}$$

with the boundary conditions transformed into the following

$$f(1) = \gamma, \quad f'(1) = -1, \quad f'(\infty) = 0 \tag{5.131}$$

where prime denotes differentiation with respect to η and $\gamma = -\frac{a_0 U}{2\nu} > 0$ is the dimensionless suction parameter.

5.4.1 Multiple-Dual Approximate Solutions of the Unsteady Viscous Flow by OHAM

The linear operator can be chose in the forms:

$$L(f(\eta)) = f'''(\eta) + Kf''(\eta) \tag{5.132}$$

$$L(f(\eta)) = f'''(\eta) - K^2 f'(\eta) \tag{5.133}$$

$$L(f(\eta)) = f'''(\eta) + \frac{2K}{K\eta + 1 - K} f''(\eta) \tag{5.134}$$

$$L(f(\eta)) = f'''(\eta) - \frac{4K^2}{(K\eta + 1 - K)^2} f'(\eta) \tag{5.135}$$

where $K > 0$ is an unknown positive parameter and will be determined later.

The initial approximation $f_0(\eta)$ can be obtained from Eq. 5.3, with boundary conditions

$$f_0(1) = \gamma, \quad f_0'(1) = -1, \quad f_0'(\infty) = 0 \tag{5.136}$$

Equation 5.3 with the linear operators 5.132 or 5.133 have the solutions:

$$f_0(\eta) = \gamma + \frac{e^{-K(\eta-1)} - 1}{K} \tag{5.137}$$

while the Eq. 5.3 with the linear operators 5.134 or 5.135 have the solutions:

$$f_0(\eta) = \gamma - \frac{1}{K} + \frac{1}{K(K\eta + 1 - K)} \tag{5.138}$$

The nonlinear operator corresponding to nonlinear differential Eq. 5.1 is defined as

$$N(f(\eta)) = (1 - K\eta)f''(\eta) + f(\eta)f''(\eta) - f'^2(\eta) - S\left[\eta f''(\eta) - f'(\eta)\right] \tag{5.139}$$

for linear operator defined by Eq. 5.132.

The same nonlinear operator for the linear operator defined by Eqs. 5.133, 5.134 and 5.135 are respectively:

$$N(f(\eta)) = f'''(\eta) + K^2 f'(\eta) + f(\eta) f''(\eta) - f'^2(\eta) - \\ -S\left[\eta f''(\eta) - f'(\eta)\right] \tag{5.140}$$

$$N(f(\eta)) = \left(1 - \frac{2K}{K\eta + 1 - K}\right) f'''(\eta) + f(\eta) f''(\eta) - f'^2(\eta) - \\ -S\left[\eta f''(\eta) - f'(\eta)\right] \tag{5.141}$$

$$N(f(\eta)) = f'''(\eta) + \\ + \frac{4K^2\eta}{(K\eta + 1 - K)^2} f'(\eta) + f(\eta) f''(\eta) - f'^2(\eta) - S\left[\eta f''(\eta) - f'(\eta)\right] \tag{5.142}$$

Substituting Eq. 5.137 into Eq. 5.139 it holds that

$$N(f_0(\eta)) = [-K(K+S)\eta + K(\gamma+1) + S - 1]e^{-K(\eta-1)} \tag{5.143}$$

Now, comparing Eqs. 5.14 and 5.143, one gets

$$h_1(\eta) = -K(K+S)\eta + K(\gamma+1) + S - 1, \;\; g_1(\eta) = e^{-K(\eta-1)}, \\ h_j(\eta) = g_j(\eta) = 0, \;\; for \quad j \geq 2. \tag{5.144}$$

The first approximation $f_1(\eta, C_i)$ can be written in the form

$$f_1(\eta, C_i) = H_1(\eta, C_i) e^{-K(\eta-1)}, \quad f_1(1) = f_1'(1) = f_1'(\infty) = 0 \tag{5.145}$$

where $H_1(\eta, C_i)$ are an arbitrary functions. Of course, we have freedom to choose such of functions with conditions obtained from Eq. 5.145:

$$H_1(0, C_i) = H_1'(0, C_i) = 0. \tag{5.146}$$

For example H_1 are given by:

$$H_1(\eta, C_i) = C_1(\eta-1)^2 + C_2(\eta-1)^3 + C_3(\eta-1)^4 + \\ +C_4(\eta-1)^2 e^{-(K+\alpha_1)(\eta-1)} + C_5(\eta-1)^2 e^{-(2K+\alpha_2)(\eta-1)} + \\ +C_6(\eta-1)^2 e^{-K\alpha_3(\eta-1)} + C_7(\eta-1)^2 e^{-K\alpha_4(\eta-1)} + C_8(\eta-1)^2 e^{-K\alpha_5(\eta-1)} \tag{5.147}$$

$$H_1(\eta, C_i) = C_1(\eta-1)^2 + C_2(\eta-1)^3 + C_3(\eta-1)^4 + C_4(\eta-1)^5 + \\ + \left[C_5(\eta-1)^2 + C_6(\eta-1)^3 + C_7(\eta-1)^4\right] e^{-K(\eta-1)} + \left[C_8(\eta-1)^2 + \right. \\ \left. +C_9(\eta-1)^3 + C_{10}(\eta-1)^4 + C_{11}(\eta-1)^5\right] e^{-2K(\eta-1)} \tag{5.148}$$

$$H_1(\eta, C_i) = C_1(\eta - 1) + C_2(\eta - 1)^2 + C_3(\eta - 1)^3 + C_4(\eta - 1)^4 +$$
$$+\left[-C_1(\eta - 1) + C_5(\eta - 1)^2 + C_6(\eta - 1)^3 + C_7(\eta - 1)^4 + C_8(\eta - 1)^5 +\right.$$
$$\left.+C_9(\eta - 1)^6\right]e^{-K(\eta-1)} + \left[C_{10}(\eta - 1)^2 + C_{11}(\eta - 1)^3\right]e^{-2K(\eta-1)} \tag{5.149}$$

Taking into consideration only the expression given by Eq. 5.147, from Eqs. 5.137, 5.145 and 5.6 we obtain the first-order approximate solution of Eqs. 5.130 and 5.131 in the form

$$\begin{aligned}\overline{f}(\eta, C_i) &= \gamma - \frac{1}{K} + \left[\frac{1}{K} + C_1(\eta - 1)^2 + C_2(\eta - 1)^3 +\right.\\ &\left.+C_3(\eta - 1)^4\right]e^{-K(\eta-1)} + C_4(\eta - 1)^2 e^{-(2K+\alpha_1)(\eta-1)} +\\ &+C_5(\eta - 1)^2 e^{-(3K+\alpha_2)(\eta-1)} + C_6(\eta - 1)^2 e^{-K(\alpha_3+1)(\eta-1)} +\\ &+C_7(\eta - 1)^2 e^{-K(\alpha_4+1)(\eta-1)} + C_8(\eta - 1)^2 e^{-K(\alpha_5+1)(\eta-1)}\end{aligned} \tag{5.150}$$

where $C_1, C_2, \ldots, \alpha_1, \alpha_2, \ldots$ are unknown parameters.

In the same way, many other approximate solutions can be yet obtained by means of the Eqs. 5.137, 5.138 and the nonlinear operators 5.140, 5.141 or 5.142.

5.4.2 Numerical Examples

We compare previously obtained approximate solutions 5.150 with numerical integration results obtained by means of a fourth-order Runge–Kutta method in combination with shooting method. Using the least-square method for determination of the parameters C_i and α_i, for the different values of the coefficients γ and S, we present the following four cases:

5.4.2.1. For $\gamma = 1$ and $S = -1$ we find dual solutions.

5.4.2.1.a)

$$\begin{aligned}&C_1 = 0.3333335814,\ C_2 = 0.0210985545,\ C_3 = 7.0349253510 \times 10^{-6},\\ &C_4 = -1.2078582888,\ C_5 = 1.1007711694,\ C_6 = 1.2527962825,\\ &C_7 = -1.4794094663,\ C_8 = 0.0003670989,\ K = 1,\\ &\alpha_1 = -1.0044596362,\ \alpha_2 = -1.9960995780,\ \alpha_3 = 0.0039364947,\\ &\alpha_4 = -0.0044067908,\ \alpha_5 = -0.0876855705\end{aligned}$$

The first expression of the first-order approximate solution given by Eq. 5.150 can be written in the form

$$\begin{aligned}\overline{f}(\eta) &= \left(1+0.3333335814(\eta-1)^2+0.0210985545(\eta-1)^3+\right.\\&\left.+7.034925351\times10^{-6}(\eta-1)^4\right)e^{1-\eta}+\\&+1.2527962825(\eta-1)^2e^{-1.0039364947(\eta-1)}+\\&+1.1007711694(\eta-1)^2e^{-1.0039004219(\eta-1)}-\\&-1.4794094663(\eta-1)^2e^{-0.9955932091(\eta-1)}-\\&-1.2078582888(\eta-1)^2e^{-0.9955403637(\eta-1)}+\\&+0.0003670989(\eta-1)^2e^{-0.9123144295(\eta-1)}\end{aligned}\tag{5.151}$$

5.4.2.1.b)

$$\begin{aligned}&C_1=-2.1103792246,\ C_2=-0.1209917376,\ C_3=0.0016464844,\\&C_4=-0.0401496995,\ C_5=5.1986534028,\ C_6=2.9102631733,\\&C_7=-3.1870328789,\ C_8=-3.0376107137,\ K=0.6170257079,\\&\alpha_1=1.5253390642,\ \alpha_2=-1.2342665630,\ \alpha_3=1.0337532632,\\&\alpha_4=0.0700064931,\ \alpha_5=1.0337537195.\end{aligned}$$

The second expression of the first-order approximate solution 5.150 is

$$\begin{aligned}\overline{f}(\eta) &= -0.6206780156+\left(1.6206780156-2.1103792246(\eta-1)^2-\right.\\&\left.-0.1209917376(\eta-1)^3+0.0016464844(\eta-1)^4\right)e^{-0.6170257079(\eta-1)}-\\&-0.0401496995(\eta-1)^2e^{-2.7593904801(\eta-1)}-\\&-3.0376107137(\eta-1)^2e^{-1.2548783286(\eta-1)}+\\&+2.9102631733(\eta-1)^2e^{-1.2548780470(\eta-1)}-\\&-3.1870328789(\eta-1)^2e^{-0.6602215139(\eta-1)}+\\&+5.1986534028(\eta-1)^2e^{-0.6168105608(\eta-1)}\end{aligned}\tag{5.152}$$

5.4.2.2. For $\gamma = 1$ and $S = -2$ the dual solutions are

5.4.2.2.a) The first expression is

$$\begin{aligned}\overline{f}(\eta) &= 0.5913897892+\left(0.4086102107-1.0365306560(\eta-1)^2-\right.\\&\left.-0.0340475469(\eta-1)^3-0.0208575441(\eta-1)^4\right)e^{-2.4473201442(\eta-1)}+\\&+0.9265758596(\eta-1)^2e^{-5.1900153052(\eta-1)}-\\&-0.9181338819(\eta-1)^2e^{-5.1900088912(\eta-1)}+\\&+0.3978132072(\eta-1)^2e^{-2.7650851899(\eta-1)}-\\&-0.2705600173(\eta-1)^2e^{-2.4471963928(\eta-1)}+\\&+0.9777857487(\eta-1)^2e^{-2.3078302052(\eta-1)}\end{aligned}\tag{5.153}$$

5.4.2.2.b) The second expression is

$$\begin{aligned}\overline{f}(\eta) &= -2.2653986945 + (3.2653986945 - 5.2607190849(\eta-1)^2- \\ &-0.0827319763(\eta-1)^3 - 0.0012312519(\eta-1)^4)e^{-0.3062413179(\eta-1)}- \\ &-0.0121539864(\eta-1)^2 e^{-8.2295440677\times10^6(\eta-1)}- \\ &-0.2925877070(\eta-1)^2 e^{-3.5914249798(\eta-1)}- \\ &-2.1949262210(\eta-1)^2 e^{-1.3550840661(\eta-1)}- \\ &-0.7516910873(\eta-1)^2 e^{-0.7893951308(\eta-1)}+ \\ &+5.1689397253(\eta-1)^2 e^{-0.2883024664(\eta-1)}\end{aligned} \quad (5.154)$$

5.4.2.3. For $\gamma = 2$ and $S = -1$, the corresponding dual solutions are

5.4.2.3.a) The first expression is

$$\begin{aligned}\overline{f}(\eta) &= 1.5235393146 + (0.4764606853 + 21.46184925584(\eta-1)^2- \\ &-4.43762024048(\eta-1)^3 + 0.29490275681(\eta-1)^4)e^{-2.0988090531(\eta-1)}- \\ &-2.53472639092(\eta-1)^2 e^{-3.6896808591(\eta-1)}+ \\ &+6.52289425339(\eta-1)^2 e^{-3.5588615870(\eta-1)}- \\ &-10.3740164628(\eta-1)^2 e^{-3.2031469638(\eta-1)}+ \\ &+16.2245606899(\eta-1)^2 e^{-2.7947822439(\eta-1)}- \\ &-31.0683634535(\eta-1)^2 e^{-2.4082455720(\eta-1)}\end{aligned} \quad (5.155)$$

5.4.2.3.b) The second expression is

$$\begin{aligned}\overline{f}(\eta) &= -0.9034898463 + (2.9034898463 - 10.4395166917(\eta-1)^2+ \\ &+0.0004971369(\eta-1)^3 - 5.0246397821\times10^{-6}(\eta-1)^4)e^{-0.3444131210(\eta-1)}- \\ &-0.1550351871(\eta-1)^2 e^{-3.6164143875(\eta-1)}- \\ &-0.3301513414(\eta-1)^2 e^{-1.7295917121(\eta-1)}- \\ &-0.5465749334(\eta-1)^2 e^{-0.9240575546(\eta-1)}- \\ &-0.1988704213(\eta-1)^2 e^{-0.5891517113(\eta-1)}+ \\ &+10.4244457170(\eta-1)^2 e^{-0.3444067343(\eta-1)}\end{aligned} \quad (5.156)$$

5.4.2.4. In the last case, we consider $\gamma = 2$ and $S = -2$. It holds that:

5.4.2.4.a) The first expression becomes

$$\begin{aligned}
\overline{f}(\eta) &= 1.7019613535 + \big(0.2980386464 - 0.8562787938(\eta-1)^2 + \\
&+0.0106882665(\eta-1)^3 + 0.0310882235(\eta-1)^4\big)e^{-3.3552695664(\eta-1)} - \\
&-0.8757867742(\eta-1)^2 e^{-6.3225573014(\eta-1)} + \\
&+0.9017666577(\eta-1)^2 e^{-6.3225571694(\eta-1)} + \\
&+1.1112669004(\eta-1)^2 e^{-3.3547932645(\eta-1)} - \\
&-0.9948349570(\eta-1)^2 e^{-2.7022990501(\eta-1)} + \\
&+0.8937777528(\eta-1)^2 e^{-2.6833589327(\eta-1)}
\end{aligned} \tag{5.157}$$

5.4.2.4.b) The second expression can be written as

$$\begin{aligned}
\overline{f}(\eta) &= -2.0992008391 + \big(4.0992008391 - 0.0204738147(\eta-1)^2 + \\
&+0.0008263633(\eta-1)^3 - 9.5217500205\times10^{-6}(\eta-1)^4\big)e^{-0.2439499890(\eta-1)} - \\
&-0.3773706801(\eta-1)^2 e^{-5.2745243636(\eta-1)} - \\
&-0.8947192974(\eta-1)^2 e^{-2.7755705539(\eta-1)} - \\
&-3.1860621314(\eta-1)^2 e^{-1.3987895675(\eta-1)} - \\
&-1.0872627477(\eta-1)^2 e^{-0.8222938433(\eta-1)} - \\
&-0.2399812958(\eta-1)^2 e^{-0.4723317954(\eta-1)}
\end{aligned} \tag{5.158}$$

In Tables 5.14, 5.15, 5.16, 5.17, 5.18, 5.19, 5.20, 5.21, 5.22, 5.23, 5.24, 5.25, 5.26, 5.27, 5.28 and 5.29 we present a comparison between every approximate solutions $\overline{f}(\eta)$ and $\overline{f'}(\eta)$ and numerical results obtained by Runge–Kutta method in combination with shooting method for different values of variable η and different values of coefficients γ and S. It can be observed that the solutions obtained by OHAM are in excellent agreement with numerical results.

Table 5.14 Comparison between the first expression of the first-order approximate solutions $\overline{f}(\eta)$ given by Eq. 5.151 obtained by OHAM with numerical results for $\gamma = 1$, $S = -1$

η	$\overline{f}_{OHAM}(\eta)$ Eq. 5.151	$f_{numerical}(\eta)$	error = $\lvert\overline{f}_{OHAM}(\eta) - f_{num}(\eta)\rvert$
1	1	1	0
9/5	0.4493290026	0.4493289954	7.1×10^{-9}
13/5	0.2018965295	0.2018965301	6.4×10^{-10}
21/5	0.0407622028	0.0407622034	6.02×10^{-10}
29/5	0.0082297496	0.0082297482	1.4×10^{-9}
37/5	0.0016615551	0.0016615578	2.6×10^{-9}
41/5	0.0007465838	0.0007465853	1.5×10^{-9}
9	0.0003354620	0.0003354620	1.4×10^{-11}
∞	6.661338×10^{-16}	8.764778×10^{-16}	2.1×10^{-11}

Table 5.15 Comparison between the derivative $\overline{f}'(\eta)$ obtained from Eq. 5.151 and numerical results for $\gamma = 1, S = -1$

η	$\overline{f}'_{OHAM}(\eta)$ from 5.151	$f'_{numerical}(\eta)$	error $= \lvert\overline{f}'(\eta) - f'_{num}(\eta)\rvert$
1	−1	−1	0
9/5	−0.4493289667	−0.4493289878	2.1×10^{-8}
13/5	−0.2018965563	−0.2018965347	2.1×10^{-8}
21/5	−0.0407621959	−0.0407621759	2.003×10^{-8}
29/5	−0.0082297502	−0.0082297476	2.5×10^{-9}
37/5	−0.0016615582	−0.0016615571	1.1×10^{-9}
41/5	−0.0007465846	−0.0007465884	3.8×10^{-9}
9	−0.0003354605	−0.0003354620	1.5×10^{-9}
∞	0	0	0

Table 5.16 Comparison between the second expression of the first-order approximate solutions $\overline{f}(\eta)$ given by Eq. 5.152 obtained by OHAM with numerical results for $\gamma = 1, S = -1$

η	$\overline{f}_{OHAM}(\eta)$ Eq. 5.152	$f_{numerical}(\eta)$	error $= \lvert\overline{f}(\eta) - f_{num}(\eta)\rvert$
1	1	1	0
9/5	0.3025681133	0.3025729017	4.7×10^{-6}
13/5	−0.1319503386	−0.1319549886	4.6×10^{-6}
21/5	−0.4960887571	−0.4961000384	1.1×10^{-5}
29/5	−0.5909291088	−0.5909208156	8.2×10^{-6}
37/5	−0.6138226907	−0.6138228346	1.4×10^{-7}
41/5	−0.6174089320	−0.6174180498	9.1×10^{-6}
9	−0.6191261345	−0.6191351201	8.9×10^{-6}
∞	−0.6206780156	−0.6206780156	4.5×10^{-14}

Table 5.17 Comparison between the derivative $\overline{f'}(\eta)$ obtained from Eq. 5.152 and numerical results for $\gamma = 1, S = -1$

η	$\overline{f}'_{OHAM}(\eta)$ from 5.152	$f'_{numerical}(\eta)$	error $= \lvert\overline{f}'(\eta) - f'_{num}(\eta)\rvert$
1	−1	−1	0
9/5	−0.7034979138	−0.7036237128	1.2×10^{-4}
13/5	−0.4016692734	−0.4016032225	6.6×10^{-5}
21/5	−0.1094954512	−0.1095146046	1.9×10^{-5}
29/5	−0.0270353065	−0.0270213000	1.4×10^{-5}
37/5	−0.0063314620	−0.0063462255	1.4×10^{-5}
41/5	−0.0030330955	−0.0030391496	6.05×10^{-6}
9	−0.0014527986	−0.0014468075	5.9×10^{-6}
∞	0	0	0

Table 5.18 Comparison between the first expression of the first-order approximate solutions $\overline{f}(\eta)$ given by Eq. 5.153 obtained by OHAM with numerical results for $\gamma = 1, S = -2$

η	$\overline{f}_{OHAM}(\eta)$ Eq. 5.153	$f_{numerical}(\eta)$	error $= \lvert\overline{f}(\eta) - f_{num}(\eta)\rvert$
1	1	1	0
9/5	0.6540357246	0.6540356881	3.6×10^{-8}
13/5	0.6019106890	0.6019108087	1.1×10^{-7}
21/5	0.5917231482	0.5917232409	9.2×10^{-8}
29/5	0.5914012763	0.5914011910	8.5×10^{-8}
37/5	0.5913902415	0.5913901904	5.1×10^{-8}
41/5	0.5913898860	0.5913898617	2.4×10^{-8}
9	0.5913898108	0.5913897995	1.1×10^{-8}
∞	0.5913897892	0.5913897892	6.9×10^{-14}

Table 5.19 Comparison between the derivative $\overline{f}'(\eta)$ obtained from Eq. 5.153 and numerical results for $\gamma = 1, S = -2$

η	$\overline{f}'_{OHAM}(\eta)$ from 5.153	$f'_{numerical}(\eta)$	error $= \lvert\overline{f}'(\eta) - f'_{num}(\eta)\rvert$
1	−1	−1	0
9/5	−0.1423061993	−0.1423072726	1.07×10^{-6}
13/5	−0.0231135261	−0.0231135957	6.9×10^{-8}
21/5	−0.0007095855	−0.0007094951	9.04×10^{-8}
29/5	−0.0000238338	−0.0000238925	5.8×10^{-8}
37/5	-8.861570×10^{-7}	-8.430734×10^{-7}	4.3×10^{-8}
41/5	-1.836854×10^{-7}	-1.588793×10^{-7}	2.4×10^{-8}
9	-4.011963×10^{-8}	-3.051385×10^{-8}	9.6×10^{-9}
∞	0	0	0

Table 5.20 Comparison between the second expression of the first-order approximate solutions $\overline{f}(\eta)$ given by Eq. 5.154 obtained by OHAM with numerical results for $\gamma = 1, S = -2$

η	$\overline{f}_{OHAM}(\eta)$ Eq. 5.154	$f_{numerical}(\eta)$	error $= \lvert\overline{f}(\eta) - f_{num}(\eta)\rvert$
1	1	1	0
9/5	−0.4931638904	−0.4931976737	3.3×10^{-5}
13/5	−1.5746868741	−1.5746416452	4.5×10^{-5}
21/5	−2.1944651392	−2.1944897035	2.4×10^{-5}
29/5	−2.2598561227	−2.2598257759	3.0×10^{-5}
37/5	−2.2649614997	−2.2650206599	5.9×10^{-5}
41/5	−2.2652663671	−2.2653038433	3.7×10^{-5}
9	−2.2653903391	−2.2653753555	1.4×10^{-5}
∞	−2.2653986945	−2.2653986945	5.1×10^{-14}

Table 5.21 Comparison between the derivative $\overline{f}'(\eta)$ obtained from Eq. 5.154 and numerical results for $\gamma = 1, S = -2$

η	$\overline{f}'_{OHAM}(\eta)$ from 5.154	$f'_{numerical}(\eta)$	error $= \lvert\overline{f}'(\eta) - f'_{num}(\eta)\rvert$
1	−1	−1	0
9/5	−1.8404107968	−1.8405104714	9.9×10^{-5}
13/5	−0.8900821972	−0.8900516447	3.05×10^{-5}
21/5	−0.1082080071	−0.1081247967	8.3×10^{-5}
29/5	−0.0091018331	−0.0091580377	5.6×10^{-5}
37/5	−0.0006415895	−0.0006482600	6.6×10^{-6}
41/5	−0.0002202957	−0.0001651469	5.5×10^{-5}
9	−0.0001070484	−0.0000411696	6.5×10^{-5}
∞	0	0	0

Table 5.22 Comparison between the first expression of the first-order approximate solutions $\overline{f}(\eta)$ given by Eq. 5.155 obtained by OHAM with numerical results for $\gamma = 2, S = -1$

η	$\overline{f}_{OHAM}(\eta)$ Eq. 5.155	$f_{numerical}(\eta)$	error $= \lvert\overline{f}(\eta) - f_{num}(\eta)\rvert$
1	2	2	0
9/5	1.6330858924	1.6330861523	2.5×10^{-7}
13/5	1.5550512928	1.5550531316	1.8×10^{-6}
21/5	1.5269871985	1.5269891353	1.9×10^{-6}
29/5	1.5239948037	1.5239925608	2.2×10^{-6}
37/5	1.5236020621	1.5236049410	2.8×10^{-6}
41/5	1.5235615146	1.5235649024	3.3×10^{-6}
9	1.5235467934	1.5235494172	2.6×10^{-6}
∞	1.5235393146	1.5235393145	1.07×10^{-11}

Table 5.23 Comparison between the derivative $\overline{f}'(\eta)$ obtained from Eq. 5.155 and numerical results for $\gamma = 2, S = -1$

η	$\overline{f}'_{OHAM}(\eta)$ from 5.155	$f'_{numerical}(\eta)$	error $= \lvert\overline{f}'(\eta) - f'_{num}(\eta)\rvert$
1	−1	−1	0
9/5	−0.1815921958	−0.1816077514	1.5×10^{-5}
13/5	−0.0465577528	−0.0465561725	1.5×10^{-6}
21/5	−0.0045275785	−0.0045260284	1.5×10^{-6}
29/5	−0.0005600263	−0.0005588247	1.2×10^{-6}
37/5	−0.0000797560	−0.0000778400	1.9×10^{-6}
41/5	−0.0000294916	−0.0000299133	4.2×10^{-7}
9	−0.0000104089	−0.0000116668	1.2×10^{-6}
∞	0	0	0

In Tables 5.30 and 5.31 we present a comparison between the skin friction coefficient $\overline{f}''(1)$ obtained by means of OHAM and numerical results. The

Table 5.24 Comparison between the second expression of the first-order approximate solutions $\overline{f}(\eta)$ given by Eq. 5.156 obtained by OHAM with numerical results for $\gamma = 2, S = -1$

η	$\overline{f}_{OHAM}(\eta)$ Eq. 5.156	$f_{numerical}(\eta)$	error $= \lvert\overline{f}(\eta) - f_{num}(\eta)\rvert$
1	2	2	0
9/5	0.9887179313	0.9886918458	2.6×10^{-5}
13/5	0.1773117221	0.1773449065	3.3×10^{-5}
21/5	−0.5977164367	−0.5977271839	1.07×10^{-5}
29/5	−0.8247609300	−0.8247529482	7.9×10^{-6}
37/5	−0.8842305620	−0.8842427763	1.2×10^{-5}
41/5	−0.8941040244	−0.8940998228	4.2×10^{-6}
9	−0.8989552832	−0.8989401737	1.5×10^{-5}
∞	−0.9034898463	−0.9034898463	2.5×10^{-13}

Table 5.25 Comparison between the derivative $\overline{f}'(\eta)$ obtained from Eq. 5.156 and numerical results for $\gamma = 2, S = -1$

η	$\overline{f}'_{OHAM}(\eta)$ from 5.156	$f'_{numerical}(\eta)$	error $= \lvert\overline{f}'(\eta) - f'_{num}(\eta)\rvert$
1	−1	−1	0
9/5	−1.2239052678	−1.2240905775	1.8×10^{-4}
13/5	−0.8011811985	−0.8011438870	3.7×10^{-5}
21/5	−0.2523667466	−0.2523093890	5.7×10^{-5}
29/5	−0.0682171457	−0.0682513159	3.4×10^{-5}
37/5	−0.0171900354	−0.0171741833	1.5×10^{-5}
41/5	−0.0084874339	−0.0084668931	2.05×10^{-5}
9	−0.0041430730	−0.0041382520	4.8×10^{-6}
∞	0	0	0

Table 5.26 Comparison between the first expression of the first-order approximate solutions $\overline{f}(\eta)$ given by Eq. 5.157 obtained by OHAM with numerical results for $\gamma = 2, S = -2$

η	$\overline{f}_{OHAM}(\eta)$ Eq. 5.157	$f_{numerical}(\eta)$	error $= \lvert\overline{f}(\eta) - f_{num}(\eta)\rvert$
1	2	2	0
9/5	1.7283791967	1.7283789517	2.4×10^{-7}
13/5	1.7050648140	1.7050648331	1.9×10^{-8}
21/5	1.7020220942	1.7020221347	4.04×10^{-8}
29/5	1.7019629186	1.7019628448	7.3×10^{-8}
37/5	1.7019614024	1.7019613948	7.5×10^{-9}
41/5	1.7019613622	1.7019613605	1.7×10^{-9}
9	1.7019613550	1.7019613548	2.1×10^{-10}
∞	1.7019613535	1.7019613535	4.4×10^{-13}

comparisons are found to be in very good agreement for the first and the second expressions of the solutions.

Table 5.27 Comparison between the derivative $\overline{f}'(\eta)$ obtained from Eq. 5.157 and numerical results for $\gamma = 2$, $S = -2$

η	$\overline{f}'_{OHAM}(\eta)$ from 5.157	$f'_{numerical}(\eta)$	error $= \lvert\overline{f}'(\eta) - f'_{num}(\eta)\rvert$
1	−1	−1	0
9/5	−0.0740545041	−0.0740547752	2.7×10^{-7}
13/5	−0.0079946060	−0.0079954465	8.4×10^{-7}
21/5	−0.0001438623	−0.0001440860	2.2×10^{-7}
29/5	-3.455048×10^{-6}	-3.391541×10^{-6}	6.3×10^{-8}
37/5	-1.05212×10^{-7}	-9.306626×10^{-8}	1.2×10^{-8}
41/5	-1.888261×10^{-8}	-1.520610×10^{-8}	3.6×10^{-9}
9	-3.332764×10^{-9}	-2.855009×10^{-9}	4.7×10^{-10}
∞	0	0	0

Table 5.28 Comparison between the second expression of the first-order approximate solutions $\overline{f}(\eta)$ given by Eq. 5.158 obtained by OHAM with numerical results for $\gamma = 2$, $S = -2$

η	$\overline{f}_{OHAM}(\eta)$ Eq. 5.158	$f_{numerical}(\eta)$	error $= \lvert\overline{f}(\eta) - f_{num}(\eta)\rvert$
1	2	2	0
9/5	0.0654321837	0.0654300951	2.08×10^{-6}
13/5	−1.2904284490	−1.2904456569	1.7×10^{-5}
21/5	−2.0213536720	−2.0213826614	2.8×10^{-5}
29/5	−2.0933993567	−2.0933669106	3.2×10^{-5}
37/5	−2.0987398049	−2.0988194297	7.9×10^{-5}
41/5	−2.0989655407	−2.0991066277	1.4×10^{-4}
9	−2.0990399214	−2.0991779752	1.3×10^{-4}
∞	−2.0992008391	−2.0992008391	1.08×10^{-13}

Table 5.29 Comparison between the derivative $\overline{f}'(\eta)$ obtained from Eq. 5.158 and numerical results for $\gamma = 2$, $S = -2$

η	$\overline{f}'_{OHAM}(\eta)$ from 5.158	$f'_{numerical}(\eta)$	error $= \lvert\overline{f}'(\eta) - f'_{num}(\eta)\rvert$
1	−1	−1	0
9/5	−2.3703131359	−2.3701583913	1.5×10^{-4}
13/5	−1.0812082251	−1.0812276578	1.9×10^{-5}
21/5	−0.1212490936	−0.1213008221	5.1×10^{-5}
29/5	−0.0097628901	−0.0097373799	2.5×10^{-5}
37/5	−0.0005551533	−0.0006619234	1.06×10^{-4}
41/5	−0.0001275175	−0.0001657752	3.8×10^{-5}
9	−0.0000825759	−0.0000406924	4.1×10^{-5}
∞	0	0	0

Table 5.30 Comparison between the skin friction coefficient $\overline{f}''_{OHAM}(1)$ and $f''_{numerical}(1)$

		The first expression of the first-order approximate solution		
γ	S	$\overline{f}''_{OHAM}(1)$	$f''_{numerical}(1)$	error = $\lvert\overline{f}''_{OHAM}(1) - f''_{numerical}(1)\rvert$
1	−1	1.0000007544	0.9999999999	7.5×10^{-7}
1	−2	2.6012206647	2.6012206637	9.9×10^{-10}
2	−1	2.5632048369	2.5632048269	9.9×10^{-9}
2	−2	3.7150911381	3.7150910381	9.9×10^{-8}

Table 5.31 Comparison between the skin friction coefficient $\overline{f}''_{OHAM}(1)$ and $f''_{numerical}(1)$

		The second expression of the first-order approximate solution		
γ	S	$\overline{f}''_{OHAM}(1)$	$f''_{numerical}(1)$	error = $\lvert\overline{f}''_{OHAM}(1) - f''_{numerical}(1)\rvert$
1	−1	0.0845138264	0.0845128264	1×10^{-6}
1	−2	–6.3800354048	–6.3800364048	1×10^{-6}
2	−1	−2.1469925948	−2.1469935948	1×10^{-9}
2	−2	−11.3677899455	−11.3677909455	9.9×10^{-7}

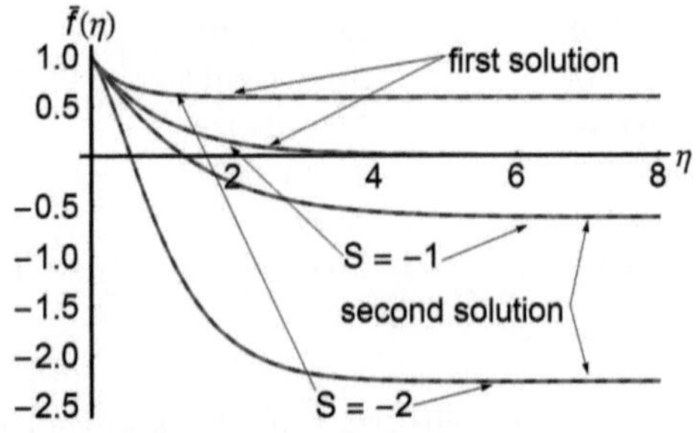

Fig. 5.8 Displacements for different values of S when $\gamma = 1$

From the Tables 5.30 and 5.31 it is seen that the magnitude of $\overline{f}''(1)$ increases as the parameters γ increases in the case of the first solutions given by the subcases 5.4.2.1.a), 5.4.2.2.a), 5.3.2.3.a) and 5.4.2.4.a). The opposite trend is observed for the variation of S, i.e. increasing S is to decrease the magnitude of the skin coefficient $\overline{f}''(1)$. In the case of the second expression of the solutions given by the subcases 5.4.2.1.b), 5.4.2.2.b), 5.4.2.3.b) and 5.4.2.4.b) the variation of the skin coefficient $\overline{f}''(1)$ is reverses.

Figures 5.8 and 5.9 present the displacement $\overline{f}(\eta)$ for different values of unsteadiness S, $\gamma = 1$ and $\gamma = 2$ respectively. It is seen that for fixed value of γ the displacement $\overline{f}(\eta)$ decreases as S increases for the first solutions. The opposite trend is observed for the second expression of the solutions.

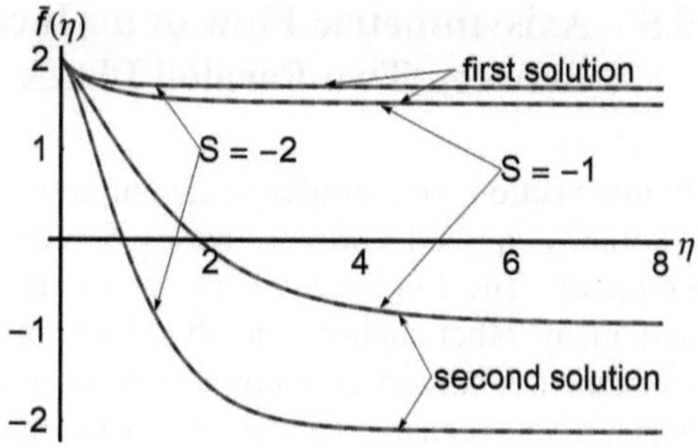

Fig. 5.9 Displacements for different values of S when $\gamma = 2$

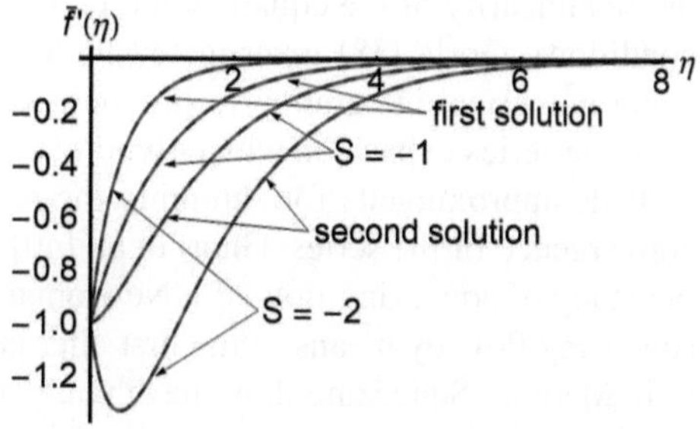

Fig. 5.10 Velocity profile for different values of S when $\gamma = 1$

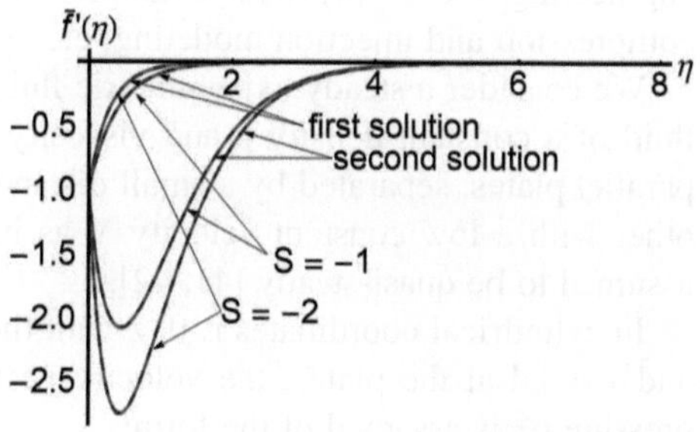

Fig. 5.11 Velocity profile for different values of S when $\gamma = 2$

Figures 5.10 and 5.11 depicts the velocity profiles $\overline{f}'(\eta)$ for fixed value of γ and some values of S. It is observed that in all cases, the velocity of fluid is damped faster as the magnitude of the unsteadiness parameter increases for the dual solutions. The velocity boundary layer thickness decreases as S decreases which implies the increase of the velocity gradient. For the first solution, the velocity gradient is positive, in contrast with the second solution. These conclusions are in concordance with results obtained in [29] and [32].

It is remarkable that this problem admits a lot of solutions depending on some optimal convergence-control parameters, and in a certain conditions ($S < 0$) every of these solutions admit dual solutions. The magnitude of the skin friction coefficient decreases with the increasing of the unsteadiness parameter. The flow velocity and the skin friction coefficient are influence by the parameters γ and S.

5.5 Axisymmetric Flow of an Incompressible Fluid Between Two Parallel Plates

In this Section we consider an incompressible Newtonian fluid which is forced by externally applied normal stresses or vertical velocities by means of a mobile boundary. The fundamental researches in this field were studied by Stefan [34] and many other authors, notably Proudman [35], Rasmussen [36]. Wang [37] carried out a numerical analysis of this problem by means of series expansion and asymptotic expansion techniques for moderately large Reynolds numbers. But the numerical methods are comparatively tedious and difficult to implement because of the nonlinearity of the equation and partly due to the need to satisfy the boundary conditions. Gorla [38] investigated the effect of mass inertia and lateral velocity using numerical integration techniques. Recently, Bujurke and Achar [39] developed the series expansion with polynomial coefficients of the solution function. Use of Padé approximants for summing the reverted series is found to accelerate the convergence of the series. Ghori et al. [40] used homotopy perturbation method in the study of squeezing flow of a Newtonian fluid, and Idrees et al. [41] studied the squeezing flow by means of the first alternative of the Optimal Homotopy Asymptotic Method. Squeezing flows have many applications including hydrostatic thrust bearing and air-cushioned vehicles, and in the food industry, especially in chemical engineering. Some practical examples of the model are: polymer processing, compression and injection modeling, etc.

We consider a steady axisymmetric fluid flow of an incompressible Newtonian fluid of a constant density ρ and viscosity μ, squeezed between two large planar parallel plates, separated by a small distance 2d, with the plates approaching each other with a low constant velocity V as in Fig. 5.12 and where the flow can be assumed to be quasi-steady [41, 42].

In cylindrical coordinates r, θ, z with the three-axis perpendicular in the plates and z = ±d at the plates, the velocity vector $u = (u_r, 0, u_z)$ and the generalized pressure p* is assumed of the form:

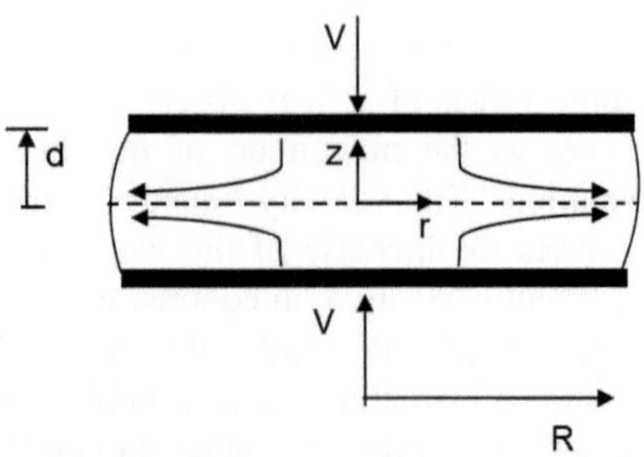

Fig. 5.12 A steady axisymmetric fluid flow between two parallel plates

$$p^* = p + \frac{1}{2}\rho\left(u^2\right) \tag{5.159}$$

$$\frac{1}{r}\frac{\partial(ru_r)}{\partial r} + \frac{\partial u_z}{\partial z} = 0 \tag{5.160}$$

The Navier–Stokes equations, if there is no tangential velocity $(u_\theta = 0)$, can be written as follows:

$$\rho\left(u_r\frac{\partial u_r}{\partial r} + u_z\frac{\partial u_r}{\partial z}\right) = -\frac{\partial \rho}{\partial r} + \mu\left(\frac{\partial^2 u_r}{\partial r^2} + \frac{1}{r}\frac{\partial u_r}{\partial r} - \frac{u_r}{r^2} + \frac{\partial^2 u_r}{\partial z^2}\right) \tag{5.161}$$

$$\rho\left(u_r\frac{\partial u_z}{\partial r} + u_z\frac{\partial u_z}{\partial z}\right) = -\frac{\partial \rho}{\partial z} + \mu\left(\frac{\partial^2 u_z}{\partial r^2} + \frac{1}{r}\frac{\partial u_z}{\partial r} + \frac{\partial^2 u_z}{\partial z^2}\right) \tag{5.162}$$

The boundary conditions of the problem are

$$u_r(0) = 0, \quad u_z(0) = 0 \tag{5.163}$$

$$u_r(d) = 0, \quad u_z(d) = -V \tag{5.164}$$

If $\psi(r,z)$ is the axisymmetric steady function, then

$$u_r = \frac{1}{r}\frac{\partial \psi}{\partial z}, \quad u_z = -\frac{1}{r}\frac{\partial \psi}{\partial r} \tag{5.165}$$

and therefore Eq. 5.160 is identically satisfied.

By means of Eq. 5.165 and if $E^2 = \frac{\partial^2}{\partial r^2} - \frac{1}{r}\frac{\partial}{\partial r} + \frac{\partial^2}{\partial z^2}$, then Eqs. 5.161 and 5.162 become

$$\frac{\partial \rho^2}{\partial r} + \frac{\rho}{r}\frac{\partial^2 \psi}{\partial t \partial z} - \rho\frac{\partial \psi}{\partial r}\frac{E^2\psi}{r^2} - \frac{\mu}{r}\frac{\partial E^2\psi}{\partial z} = 0 \tag{5.166}$$

$$\frac{\partial p^*}{\partial z} - \frac{\rho}{r}\frac{\partial^2 \psi}{\partial t \partial z} - \rho\frac{\partial \psi}{\partial z}\frac{E^2\psi}{r^2} + \frac{\mu}{r}\frac{\partial E^2\psi}{\partial z} = 0 \tag{5.167}$$

From the integrability conditions

$$\frac{\partial^2 p^*}{\partial r \partial z} = \frac{\partial^2 p^*}{\partial z \partial r} \tag{5.168}$$

we can obtain the compatibility equation in the form

$$\frac{1}{r}\frac{\partial E^2\psi}{\partial t} - \frac{\partial\left(\psi, \frac{E^2\psi}{r^2}\right)}{\partial(r,z)} = \frac{\mu}{\rho r}E^4\psi \tag{5.169}$$

or

$$\frac{\partial\left(\psi, \frac{E^2\psi}{r^2}\right)}{\partial(r,z)} + \frac{\mu}{\rho r}E^4\psi = 0 \tag{5.170}$$

If we choose the stream function ψ such as

$$\psi(r,z) = r^2 y(z) \tag{5.171}$$

and substituting Eq. 5.171 into Eq. 5.170 one obtains

$$y^{IV}(z) + \frac{\rho}{\mu}y(z)y'''(z) = 0 \tag{5.172}$$

where prime denotes differentiation with respect to z.

Introducing the non-dimensional parameters

$$y^* = \frac{y}{V}, \quad z^* = \frac{z}{d}, \quad R = \frac{\rho dV}{2\mu} \tag{5.173}$$

where R is the Reynolds number, and omitting the "*", Eqs. 5.172, 5.163 and 5.164 can be written as

$$y^{(IV)}(z) + Ry(z)y'''(z) = 0 \tag{5.174}$$

$$y(0) = 0, \quad y'(0) = 0, \quad y(1) = \frac{1}{2}, \quad y'(1) = 0 \tag{5.175}$$

5.5.1 Solutions of the Problem 5.174 and 5.175 with OHAM

The linear operator for the problem 5.174 and 5.175 is defined in the form:

$$L(y) = y^{(IV)} \tag{5.176}$$

with $g(x) = 0$. The nonlinear operator N becomes

$$N(y) = Ryy''' \tag{5.177}$$

The equations 5.3 and 5.4 are

$$y_0^{(IV)}(z) = 0, \quad y_0(0) = y_0'(0) = y_0'(1) = 0, \quad y(1) = \frac{1}{2} \tag{5.178}$$

From the last equation one an get

$$y_0(z) = -z^3 + \frac{3}{2}z^2 \tag{5.179}$$

Substituting Eq. 5.179 into Eq. 5.177 it follows that

$$N(y_0(z)) = -72Rz^2 + 48Rz^3 \tag{5.180}$$

Comparing Eqs. 5.180 and 5.14 it holds that

$$h_1(z) = -72R, \quad g_1(z) = z^2, \quad h_2(z) = 48R, \quad g_2(z) = z^3 \tag{5.181}$$

If we choose the expression of the first approximation $y_1(z, C_i)$ from Eq. 5.15 in the form

$$y_1(z, C_i) = P_1(z)z^2 + P_2(z)z^3, \quad y_1(0) = y_1'(0) = y_1(1) = y_1'(1) = 0 \tag{5.182}$$

where $P_1(z)$ and $P_2(z)$ are polynomials with the properties:

$$\begin{aligned} P_1(1, C_i) + P_2(1, C_i) = 0 \\ P_1'(1, C_i) + P_2'(1.C_i) + P_2(1, C_i) = 0 \end{aligned} \tag{5.183}$$

then, the first-order approximate solutions of Eqs. 5.174 and 5.175 are given by

$$\bar{y}(z, C_i) = \left[\frac{3}{2} + P_1(z, C_i)\right]z^2 + [-1 + P_2(z, C_i)]z^3 \tag{5.184}$$

where P_1 and P_2 verifies Eq. 5.183.

Of course, there exist many other possibilities to choose the polynomials P_1 and P_2. We present in the following only three cases.

Case 5.5.1.A

$$\begin{aligned} P_1(z) = C_1 + C_2 z + C_3 z^2 \\ P_2(z) = C_4 z^2 - (5C_1 + 4C_2 + 3C_3 + 2C_4)z^3 + (4C_1 + 3C_2 + 2C_3 + C_4)z^4 \end{aligned}$$

and therefore Eq. 5.184 becomes

$$\bar{y}(z, C_i) = \left(\frac{3}{2} + C_1\right)z^2 + (C_2 - 1)z^3 + C_3 z^4 + C_4 z^5 - (5C_1 + 4C_2 + \\ +3C_3 + 2C_4)z^6 + (4C_1 + 3C_2 + 2C_3 + C_4)z^7 \quad (5.185)$$

Case 5.5.1.B

$$\bar{y}(z, C_i) = \left(\frac{3}{2} + C_1\right)z^2 + (C_2 - 1)z^3 + C_3 z^4 + C_4 z^5 + C_5 z^6 - (6C_1 + \\ +5C_2 + 4C_3 + 3C_4 + 2C_5)z^7 + (5C_1 + 4C_2 + 3C_3 + 2C_4 + C_5)z^8 \quad (5.186)$$

Case 5.5.1.C

$$\bar{y}(z, C_i) = \left(\frac{3}{2} + C_1\right)z^2 + (C_2 - 1)z^3 + C_3 z^4 + C_4 z^6 - (6C_1 + 5C_2 + \\ +4C_3 + 2C_4)z^8 + (5C_1 + 4C_2 + 3C_3 + C_4)z^9 \quad (5.187)$$

5.5.2 *Numerical Examples*

Using the least square method, we determine the optimal values of the convergence-control parameters C_1, C_2,... for every case and $R = 2$.

For the Case 5.5.1.A, the values obtained for the optimal convergence-control parameters are:

$$C_1 = 0.09464039981813055, C_2 = -0.1337583177854468, \\ C_3 = -0.005347976047528237, C_4 = 0.01697059924199148$$

The first-order approximate solution of Eqs. 5.174 and 5.175 in this case gives

$$\bar{y}(z) = 1.5946403998181307\ z^2 - 1.133758317785447\ z^3 - \\ 0.005347976047528237\ z^4 + 0.01697059924199148\ z^5 + \\ 0.04393400170973623\ z^6 - 0.016438706936883242\ z^7 \quad (5.188)$$

For the Case 5.5.1.B we obtain

$$C_1 = 0.09498824140937175, C_2 = -0.13397704235502894, \\ C_3 = -0.014990183968127716, C_4 = 0.049828208868162915, C_5 = 0$$

and the solution

$$\bar{y}(z) = 1.5949882414093717\ z^2 - 1.133977042355029\ z^3 - \\ 0.014990183968127716\ z^4 + 0.049828208868162915\ z^5 + \\ 0.010431872586936314\ z^7 - 0.006281096541314368\ z^8 \quad (5.189)$$

In the last case we obtain

$$C_1 = 0.08162438140756319, C_2 = -0.1361179091796326 2,$$
$$C_3 = 0.04094928239882171, C_4 = 0.0005417698149299089$$

and the solution

$$\begin{aligned}\overline{y}(z) = 1.5816243814075632z^2 - 1.1361179091796325z^3 + \\ 0.04094928239882171z^4 + 0.0005417698149299089z^6 + \\ 0.02433727878284759z^8 - 0.01296011266931953z^9\end{aligned} \quad (5.190)$$

In Tables 5.32, 5.33 and 5.34 we present a comparison between every approximate solution given by Eqs. 5.188, 5.189 and 5.190 and numerical results obtained by a fourth-order Runge–Kutta method in combination with shooting method.

In Tables 5.35, 5.36 and 5.37 we present a comparison between the derivative of the approximate solution given by Eqs. 5.188, 5.189 and 5.190, respectively, and

Table 5.32 Comparison between the first-order OHAM solution 5.188 and numerical results

z	$y_{numerical}$	$y_{approx.}$	$\varepsilon = \|y_{num} - y_{app}\|$
0	-1.52×10^{-25}	0	1.52×10^{-25}
0.1	0.014812715309664	0.014812322878914	3.92×10^{-7}
0.2	0.054719179359675	0.054715024641183	4.15×10^{-6}
0.3	0.112945972474388	0.112932514095637	1.34×10^{-5}
0.4	0.182797275559084	0.182771822875611	2.54×10^{-5}
0.5	0.257728875590501	0.257694434333463	3.44×10^{-5}
0.6	0.331431720613536	0.331394889591429	3.68×10^{-5}
0.7	0.397912334036814	0.397877885640522	3.44×10^{-5}
0.8	0.451558997858817	0.451525580379190	3.34×10^{-5}
0.9	0.487185835715104	0.487146819483419	3.90×10^{-5}
1	0.500050996317850	0.499999999999999	5.09×10^{-5}

Table 5.33 Comparison between the first-order OHAM solution 5.189 and numerical results

z	$y_{numerical}$	$y_{approx.}$	$\varepsilon = \|y_{num} - y_{app}\|$
0	-1.52×10^{-25}	0	1.52×10^{-25}
0.1	0.014812715309664	0.014814905615806	2.19×10^{-6}
0.2	0.054719179359675	0.054719791498385	6.12×10^{-7}
0.3	0.112945972474388	0.112933092988456	1.28×10^{-5}
0.4	0.182797275559084	0.182763055264620	3.42×10^{-5}
0.5	0.257728875590501	0.257677138558307	5.17×10^{-5}
0.6	0.331431720613536	0.331375166803972	5.65×10^{-5}
0.7	0.397912334036814	0.397862613504903	4.97×10^{-5}
0.8	0.451558997858817	0.451517886056836	4.11×10^{-5}
0.9	0.487185835715104	0.487144937233494	4.08×10^{-5}
1	0.500050996317850	0.499999999999999	5.09×10^{-5}

Table 5.34 Comparison between the first-order OHAM solution 5.190 and numerical results

z	$y_{numerical}$	$y_{approx.}$	$\varepsilon = \lvert y_{num} - y_{app} \rvert$
0	-1.52×10^{-25}	0	1.52×10^{-25}
0.1	0.014812715309664	0.014684221605318	1.28×10^{-4}
0.2	0.054719179359675	0.054241641175827	4.77×10^{-4}
0.3	0.112945972474388	0.112004436591419	9.41×10^{-4}
0.4	0.182797275559084	0.181411427819532	1.38×10^{-3}
0.5	0.257728875590501	0.256028906782909	1.69×10^{-3}
0.6	0.331431720613536	0.329593777488089	1.83×10^{-3}
0.7	0.397912334036814	0.396083173819318	1.82×10^{-3}
0.8	0.451558997858817	0.449805723330611	1.75×10^{-3}
0.9	0.487185835715104	0.485495923331058	1.68×10^{-3}
1	0.500050996317850	0.498374690555210	1.67×10^{-3}

Table 5.35 Comparison between the derivatives of the first-order OHAM solution $\bar{y}'(z)$ obtained from Eq. 5.188 and numerical results

z	$y'_{numerical}$	$y'_{approx.}$	$\varepsilon = \lvert y'_{num} - y'_{app} \rvert$
0	2.15×10^{-20}	0	2.15×10^{-20}
0.1	0.284918838479259	0.284904944794647	1.38×10^{-5}
0.2	0.501912486671795	0.501846780095988	6.57×10^{-5}
0.3	0.651451033679262	0.651335892968404	1.15×10^{-4}
0.4	0.734653774735982	0.734539456612049	1.14×10^{-4}
0.5	0.753449455454865	0.753390627467757	5.88×10^{-5}
0.6	0.710607093441063	0.710614891238984	7.79×10^{-6}
0.7	0.609644877715945	0.609673557830788	2.86×10^{-5}
0.8	0.454642247229446	0.454624405205853	1.78×10^{-5}
0.9	0.249992760036392	0.249899472157543	9.32×10^{-5}
1	0.000139960890739	2.77×10^{-17}	1.39×10^{-4}

Table 5.36 Comparison between the derivatives of the first-order OHAM solution $\bar{y}'(z)$ obtained from Eq. 5.189 and numerical results

z	$y'_{numerical}$	$y'_{approx.}$	$\varepsilon = \lvert y'_{num} - y'_{app} \rvert$
0	2.15×10^{-20}	0	2.15×10^{-20}
0.1	0.284918838479259	0.284943358378015	2.45×10^{-5}
0.2	0.501912486671795	0.501841021559743	7.14×10^{-5}
0.3	0.651451033679262	0.651260490439670	1.90×10^{-4}
0.4	0.734653774735982	0.734438911498594	2.14×10^{-4}
0.5	0.753449455454865	0.753330100460701	1.19×10^{-4}
0.6	0.710607093441063	0.710628166444722	2.10×10^{-5}
0.7	0.609644877715945	0.609742411227910	9.75×10^{-5}
0.8	0.454642247229446	0.454698178241571	5.59×10^{-5}
0.9	0.249992760036392	0.249938325916923	5.44×10^{-5}
1	0.000139960890739	-8.60×10^{-16}	1.39×10^{-4}

Table 5.37 Comparison between the derivatives of the first-order OHAM solution $\bar{y}'(z)$ obtained from Eq. 5.190 and numerical results

z	$y'_{\text{numerical}}$	$y'_{\text{approx.}}$	$\varepsilon = \left\lvert y'_{num} - y'_{app} \right\rvert$
0	2.15×10^{-20}	0	2.15×10^{-20}
0.1	0.284918838479259	0.282405186945320	2.51×10^{-3}
0.2	0.501912486671795	0.497629214232627	4.28×10^{-3}
0.3	0.651451033679262	0.646688142555039	4.76×10^{-3}
0.4	0.734653774735982	0.730721763076866	3.93×10^{-3}
0.5	0.753449455454865	0.751177623525446	2.27×10^{-3}
0.6	0.710607093441063	0.710066047137085	5.41×10^{-4}
0.7	0.609644877715945	0.610219668566155	5.74×10^{-4}
0.8	0.454642247229446	0.455443981209560	8.01×10^{-4}
0.9	0.249992760036392	0.250398360741923	4.05×10^{-4}
1	0.000139960890739	1.52×10^{-16}	1.39×10^{-4}

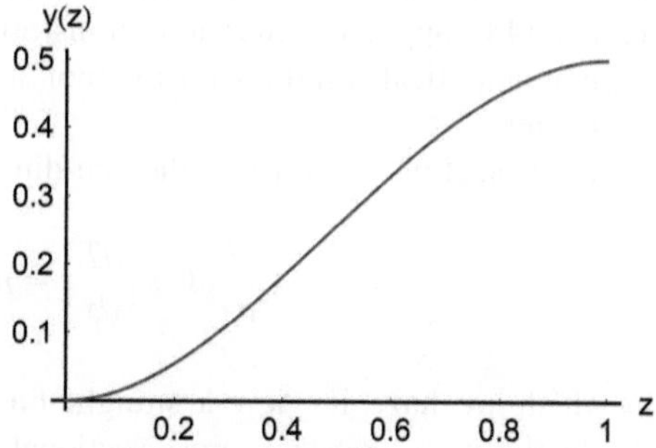

Fig. 5.13 Comparison between the approximate solution given by Eq. (5.188) and numerical results: (*blue dashed line*) approximate solution, (*red solid line*) numerical results (colour figure online)

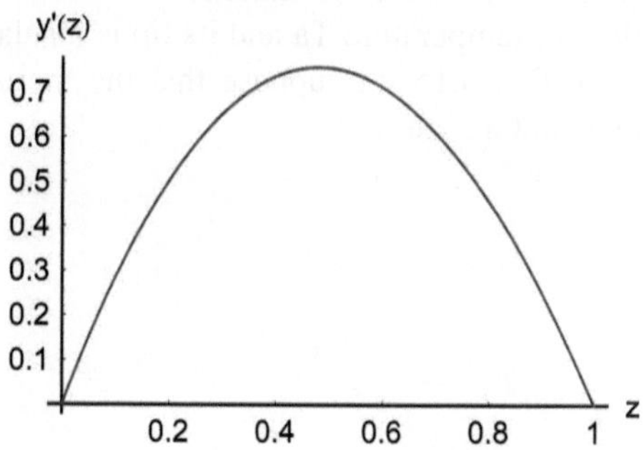

Fig. 5.14 Comparison between the derivative of the approximate solution obtained from (5.188) and numerical results: (*blue dashed line*) approximate solution and (*red solid line*) numerical results (colour figure online)

numerical results. In Figs. 5.13 and 5.14 we present a comparison between our approximate solution 5.188 and numerical solution. From these tables and figures it can be observed that the solutions obtained through OHAM are very accurate comparing to numerical results.

5.6 Heat Transfer with Variable Thermal Conductivity

In engineering, the thermal properties of most metallic materials depend on temperature. There exist cases of practice interest like nuclear reactors, cooling of gas turbine blades, and so on, where a combination of different kinds of heat transfer (conduction, convection and radiation) is present.

A substantial amount of research works has been invested in the study of this domain. Perturbation method and a numerical solution method to compute a closed form solution for a straight convecting fin with variable thermal conductivity was applied by Aziz and Hug [42]. Razelos and Imre [43] studied the variation of the convective heat transfer coefficient from the base of a convective fin to its tip. Homotopy perturbation method was applied by Ganji [44] and Rajabi et al. [45] in nonlinear heat conduction and convection equations, and homotopy analysis method was used by Abbasbandy [46] in the study of the cooling of a lumped system with variable specific heat and the temperature distribution equation.

Chu and Lo [47] presented a hybrid differential transformation-finite difference method to analytic nonlinear transient heat conduction problems and Yaghoobi and Torabi [48] applied differential transformation method to derive approximate explicit analytical solutions for two nonlinear heat transfer problems with a small parameter.

In this section we consider the one-dimensional energy balance equation

$$A_c \frac{d}{dx}\left[k(T)\frac{dT}{dx}\right] - ph(T_b - T_a) = 0 \tag{5.191}$$

in which we have in view a straight fin with a temperature-dependent thermal conductivity, a constant cross-sectional area Ac, perimeter p and length b (Fig. 5.15). The fin is attached to a base surface of temperature Tb, extends into a fluid of temperature Ta and its tip is insulated.

In Fig. 5.15 we suppose that the thermal conductivity is a linear function of temperature, i.e.

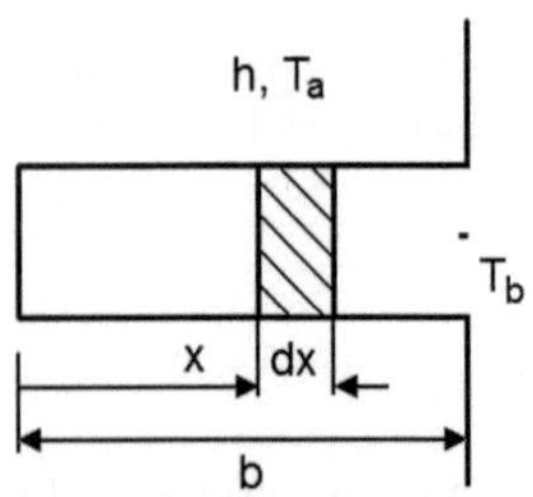

Fig. 5.15 Geometry of a straight fin

$$k(T) = K_a[1 + \lambda(T - T_a)] \tag{5.192}$$

where K_a is the thermal conductivity at the ambient fluid temperature and λ is the parameter describing the variation of the thermal conductivity.

Using the dimensionless parameters

$$\eta = \frac{x}{b}, \quad A = \lambda(T_b - T_a), \quad B = \frac{hPb^2}{K_a A_c}, \quad \theta = \frac{T - T_a}{T_b - T_a} \tag{5.193}$$

we find the following nonlinear differential equation

$$\theta''(\eta) + A\left[\theta(\eta)\theta''(\eta) + \theta'^2(\eta)\right] - B\theta(\eta) = 0 \tag{5.194}$$

with the boundary conditions

$$\theta'(0) = 0, \quad \theta(1) = 1 \tag{5.195}$$

where prime denotes differentiation with respect to η.

5.6.1 *Approximate Solutions of Eqs. 5.194 and 5.195 by OHAM*

For the problem 5.194 and 5.195, the linear operator is

$$L[\theta(\eta)] = \theta''(\eta) - 2 \tag{5.196}$$

while the nonlinear operator is given by

$$N[\theta(\eta)] = 2 + A\left[\theta(\eta)\theta''(\eta) + \theta'(\eta)\right] - B\theta(\eta) \tag{5.197}$$

From the equation

$$L[\theta_0(\eta)] = 0, \quad \theta_0'(0) = 0, \quad \theta_0(1) = 1 \tag{5.198}$$

yields

$$\theta_0(\eta) = \eta^2 \tag{5.199}$$

Substituting Eq. 5.199 into Eq. 5.197 we have

$$N[\theta_0(\eta)] = 2 + (6A - B)\eta^2 \quad (5.200)$$

From Eqs. 5.14 and 5.200 one gets

$$h_1(\eta) = 2, \quad g_1(\eta) = 1, \quad h_2(\eta) = 6A - B, \quad g_2(\eta) = \eta^2 \quad (5.201)$$

We recall that the choice of the linear operator is not unique. Also, this can be chosen as:

$$L[\theta(\eta)] = \theta'' \quad (5.202)$$

or

$$L[\theta(\eta)] = \theta'' - 1 \quad (5.203)$$

and so on.

The first approximation $\theta_1(\eta)$ is chosen in the form

$$\theta_1(\eta, C_i) = H_1(\eta, C_i) + H_2(\eta, C_i)\eta^2, \quad \theta_1'(0) = \theta_1(1) = 0 \quad (5.204)$$

where H_1 and H_2 are polynomials with the following properties:

$$H_1'(0, C_i) = 0, \quad H_1(1, C_i) + H_2(1, C_i) = 0 \quad (5.205)$$

We present only two possibilities to choose the optimal auxiliary functions H_1 and H_2:

$$\begin{aligned} H_1(\eta, C_i) &= C_1 + C_2(\eta^3 - 1) + C_3(\eta^5 - 1), \\ H_2(\eta, C_i) &= -C_1 + C_4(\eta^2 - 1) \end{aligned} \quad (5.206)$$

$$\begin{aligned} H_1(\eta, C_i) &= C_1 + C_2(\eta^5 - 1), \\ H_2(\eta, C_i) &= -C_1 + C_3(\eta - 1) + C_4(\eta^2 - 1) + C_5(\eta^4 - 1) \end{aligned} \quad (5.207)$$

Taking account of Eqs. 5.199, 5.204, 5.206 and 5.207 one can get the first-order approximate solutions, respectively

$$\overline{\theta}(\eta, C_i) = C_1 - C_2 - C_3 + (1 - C_1 - C_4)\eta^2 + C_2\eta^3 + C_4\eta^4 + C_3\eta^5 \quad (5.208)$$

$$\begin{aligned} \overline{\theta}(\eta, C_i) = C_1 - C_2 + (1 - C_1 - C_3 - C_4 - C_5)\eta^2 + \\ + C_3\eta^3 + C_4\eta^4 + C_2\eta^5 + C_5\eta^6 \end{aligned} \quad (5.209)$$

5.6.2 Numerical Examples

By means of the least square method, we obtain the values of the optimal convergence-control parameters C_1, C_2,... for each of the Eqs. 5.208 and 5.209 with $A = 0.25$, $B = 0.5$.

5.6.2.a) For the Eq. 5.208 we find the following values
$C_1 = 0.8257649215705$, $C_2 = -0.000081678843$, $C_3 = -0.000178607643$, $C_4 = 0.0030628491901$

The first-order approximate solution 5.208 of Eqs. 5.194 and 5.195 is given by

$$\overline{\theta}(\eta) = 0.8260252080572067 + 0.17117222923925016\eta^2 - 0.000081678843\eta^3 + 0.0030628491901\eta^4 - 0.000178607643\eta^5 \quad (5.210)$$

5.6.2.b) For the Eq. 5.209 the optimal convergence-control parameters are
$C_1 = 0.826011763824$, $C_2 = -0.000013446465$, $C_3 = -2.589872495116 \times 10^{-6}$, $C_4 = 0.002885471005$, $C_5 = -0.000054953567$

such that, in this case the first-order approximate solution becomes

$$\overline{\theta}(\eta) = 0.8260252102900947 + 0.17116030860945694\eta^2 - 2.5898724951169518 \times 10^{-6}\eta^3 + 0.0028854710059894 18\eta^4 - 0.000013446465\eta^5 - 0.000054953567\eta^6 \quad (5.211)$$

Figure 5.16 presents a comparison between the results obtained through the present procedure and the numerical integration results. In Fig. 5.17 the residual obtained from the initial equation 5.194 using the approximate solutions 5.210 and 5.211 are given in the considered cases. As it can be seen, a very good accuracy is achieved (Fig. 5.18).

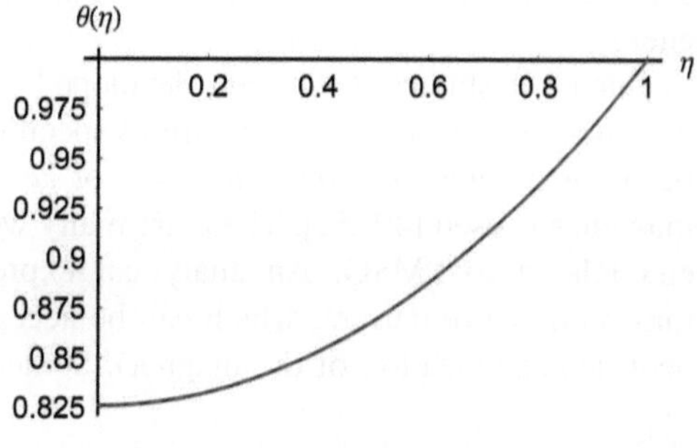

Fig. 5.16 Comparison between the approximate solution (5.210) and numerical results of Eq. 5.194 with $A = 0.25$, $B = 0.1$:(*blue dashed line*) approximate solution and (*red solid line*) numerical results (colour figure online)

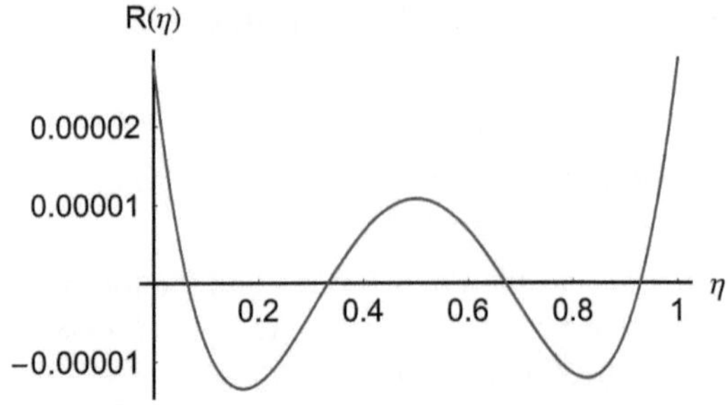

Fig. 5.17 The residual R(η) of the approximate solution (5.210)

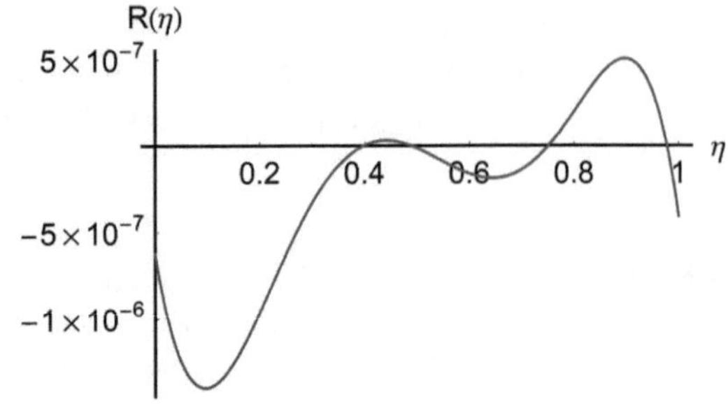

Fig. 5.18 The residual R(η) of the approximate solution (5.211)

5.7 Nonlinear Dynamical Model of a Permanent Magnet Synchronous Generator

The Permanent Magnet Synchronous Machines (PMSM) are rotating electrical machines where the stator is a classic three-phase stator like that of an induction motor and the rotor has surface mounted permanent magnets. Grace to their construction, they are characterized by a high reliability even at very high speeds. Basically, the permanent magnet synchronous machines, known for their simplicity and low maintenance, are highly efficient electrical machines which operate in either generator or motor mode, since the mode of operation is dictated by the sign of the mechanical torque, which is positive for motor mode and negative for generator mode. Permanent Magnet Synchronous Generators (PMSG) are widespread in wind energy field, which is the most prominent area of application for such devices. They are used to convert the mechanical wind energy into electrical energy.

Intensive studies have been developed in order to find optimal solutions for designing PMSG and to solve various specific aspects which their construction and behavior involve. To study this kind of generators, the basic theory of electrical machines is used [49–52]. There are many works that investigate various phenomena related to PMSG. An analytical expression is developed in [53] for the maximum stator current, which can be accepted without magnet damage (permanent demagnetization of the magnet). Some developments presented in [54] deal with the important case of the sudden short-circuit applied to large permanent magnet machines for assessing the risk of loss of magnetization in the permanent magnet and to avoid even partial demagnetization of the magnets. Some investigations of dynamic performances of PMSM are developed for minimizing braking motor torque in some applications or for maximizing electrodynamical braking torque in others.

In this section, an explicit analytical solution to a PMSG of 5 kW is developed and analyzed in order to investigate the transitory working regime of this generator that works in an actual wind power station. Generally, such an analytical approach which gives explicit analytical solutions is very useful for the study of stability or for determining some specific quantities dependent on the solution of the problem, as well as for investigating some phenomena modeled by nonlinear differential equations which cannot be easily solved numerically by traditional methods (e.g. due to multiple singularities).

5.7.1 Dynamic Model

The transient behavior of the PMSG can be predicted using the classical d–q equivalent circuit models. Starting from the equations of the smooth-air-gap synchronous machine, the d–q axis equations of permanent magnet synchronous machine, in the rotor reference frame (rotor d-q coordinates) takes the form [49–52, 58, 59]:

$$\begin{aligned} u_d &= R_s i_d + \frac{d\psi_d}{dt} - \omega\psi_q \\ u_q &= R_s i_q + \frac{d\psi_q}{dt} + \omega\psi_d \\ \psi_d &= L_s i_d + \psi_M \\ \psi_q &= L_s i_q \end{aligned} \tag{5.212}$$

where u_d and u_q are instantaneous values of d and q axis stator voltage components, i_d and i_q are instantaneous values of d and q axis stator current components, R_s and L_s are electrical resistance and synchronous inductance of the stator phase windings, ψ_d and ψ_q are instantaneous values of d and q axis stator flux components, ψ_M is the permanent magnet flux and ω is the angular stator frequency. The angular mechanical rotor speed is $\Omega_r = \omega/p_1$, where p_1 is the number of pole pairs of the electrical machine.

As the d-q voltage equations 5.212 were written for motoring, the PMSG output voltages are as follows:

$$\begin{aligned} u_d &= -R_L i_d \\ u_q &= -R_L i_q \end{aligned} \tag{5.213}$$

where R_L is the electrical resistance of the external load connected to the generator. For the sake of simplicity, we further denote: $R = R_s + R_L$, $L = L_s$, $\psi = \psi_M$.

Finally, the two-axis voltage equations 5.212 for the permanent magnet generator stator become:

$$\begin{aligned} \frac{di_d}{dt} &= \omega i_q - \frac{R}{L} i_d \\ \frac{di_q}{dt} &= -\omega i_d - \frac{R}{L} i_q - \frac{\psi}{L}\omega \end{aligned} \tag{5.214}$$

with the initial conditions:

$$i_d(0) = -\frac{\psi\omega_0^2 L}{R^2 + \omega_0^2 L^2}, \quad i_q(0) = -\frac{\psi\omega_0 R}{R^2 + \omega_0^2 L^2}, \quad \omega(0) = \omega_0 \tag{5.215}$$

In order to investigate the electromechanical transient operation, the generator motion equation has to be added [55]:

$$\frac{J_m}{p}\frac{d\omega}{dt} = T_m + T_e \tag{5.216}$$

where J_m is the axial moment of inertia (total inertia), T_m is the mechanical torque of the wind turbine, T_e is the electromagnetic torque of the generator, $T_e = \frac{3}{2} p_1 \psi i_q$.

The mechanical torque developed by the wind turbine may be expressed as [56, 57]:

$$T_m = \frac{1}{2}\rho\pi r^3 v^2 C_t(\lambda_r), \quad \lambda_r = \frac{r\Omega_r}{v} = \frac{r\omega}{p_1 v} \tag{5.217}$$

where ρ is the air density, r is the turbine radius, v is the wind speed, λ_r is the tip-speed ratio and $C_t(\lambda_r)$ is the torque coefficient. Considering $\rho = 1.225$ kg/m^3, $p_1 = 16$, $r = 3.5$ m and the torque coefficient provided by the turbine manufactures, $C_t(\lambda_r) = 0.0142 + 0.0621\lambda_r - 0.046\lambda_r^{2.5}$, the final form of the turbine torque becomes:

$$T_m = 1.17v^2 + 1.12v\omega - \frac{0.0085}{\sqrt{v}}\omega^{2.5} \tag{5.218}$$

5.7.2 Approximate Solution with OHAM

Now, to solve Eqs. 5.214, 5.215 and 5.216 we apply OHAM for two stages (time intervals) of the generator's dynamic behavior: the beginning of the transitory regime, which takes place immediately after the occurrence of the short-circuit, namely the first stage, and the ending of the transitory regime, before steady-state conditions (namely the second-stage). The first stage of the transitory regime shows the electromagnetic fast transitory regime and the second one generally emphasizes the mechanical slow transitory regime, cause by inertia. In the transitory regime the behavior is damped oscillatory such that we can choose the linear operators in the form:

$$L_1(i_d) = \frac{di_d}{dt} + \lambda i_d, \quad L_2(i_q) = \frac{di_d}{dt} + \lambda i_q, \quad L_3(\omega) = \frac{d\omega}{dt} + \lambda^* \omega \tag{5.219}$$

If we note the approximate solutions

$$\bar{i}_d = x_0 + x_1, \quad \bar{i}_q = y_0 + y_1, \quad \bar{\omega} = \omega_0 + \omega_1 \tag{5.220}$$

for the first stage, and

$$\bar{i}_d = \bar{x}, \quad \bar{i}_q = \bar{y}, \quad \bar{\omega} = \bar{\omega}_1 \tag{5.221}$$

for the second stage, taking into account Eq. 5.3 for each of the unknown functions i_d, i_q and ω, it is natural to choose

$$\begin{array}{lll} L_1(x_0) + g_1(t) = 0, & x_0(0) = X_0, & g_1(t) = -\lambda X_0 - ke^{-\lambda t}\cos kt \\ L_2(y_0) + g_2(t) = 0, & y_0(0) = Y_0, & g_2(t) = -\lambda Y_0 - ke^{-\lambda t}\cos kt \\ L_3(\omega_0) + g_3(t) = 0, & \omega_0(0) = \omega_0, & g_3(t) = -\lambda^* \omega_0 - k^* e^{-\lambda^* t}\cos k^* t \end{array} \tag{5.222}$$

where λ, λ^*, k, k^* are unknown and will be determined later.

From Eqs. 5.222 we find the following solutions for the initial approximations

$$\begin{array}{l} x_0(t) = X_0 + e^{-\lambda t}\sin kt, \quad y_0(t) = Y_0 + e^{-\lambda t}\sin kt, \\ \omega_0(t) = \omega_0 + e^{-\lambda^* t}\sin k^* t \end{array} \tag{5.223}$$

From Eqs. 5.214, 5.215 and 5.216, the nonlinear operators N_i become respectively:

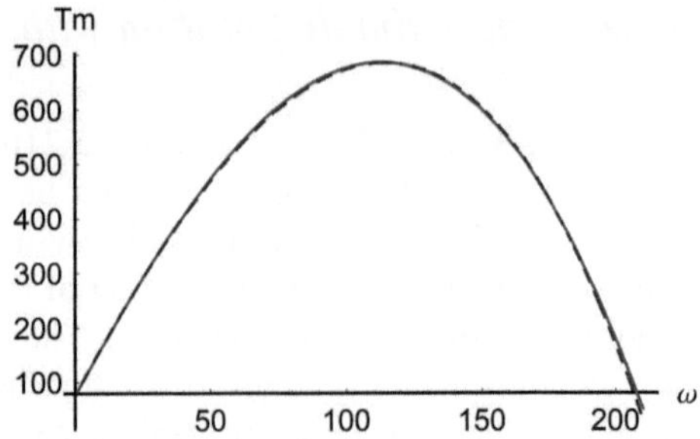

Fig. 5.19 The mechanical output torque of the wind turbine given by Eq. 5.218 (*red solid line*) and Eq. 5.225 (*blue dashed line*) (colour figure online)

$$\begin{aligned} N_1(i_d, i_q, \omega) &= \omega i_q + \left(\lambda - \frac{R}{L}\right) i_d - k e^{-kt} \cos kt - \lambda X_0, \\ N_2(i_d, i_q, \omega) &= -\omega i_d + \left(\lambda - \frac{R}{L}\right) i_q - k e^{-kt} \cos kt - \lambda Y_0 - \frac{\psi}{L}\omega, \\ N_3(i_d, i_q, \omega) &= \frac{p_1}{J_m}\left(1.17v^2 + 1.12v\omega - \frac{0.0085}{\sqrt{v}}\omega^{2.5}\right) + \\ &+ \frac{3\psi}{2J_m} i_q - \lambda^* \Omega_0 - k^* e^{-\lambda^* t} \cos k^* t \end{aligned} \tag{5.224}$$

In order to simplify Eq. 5.216, the mechanical output torque of the wind turbine, represented in Fig. 5.19 could be accurately approximated with the more adequate polynomial expression, corresponding to the wind speed v = 8.04 m/s:

$$T_m = 75.63 + 9.18781\omega - 0.0204959\omega^2 - 0.000115537\omega^3 \tag{5.225}$$

In this way, the last Eq. 5.224 can be written as follows:

$$\begin{aligned} N_3(i_d, i_q, \omega) &= \frac{p_1}{J_m}(75.63 + 9.18781\omega - 0.0204959\omega^2 - \\ &-0.000115537\omega^3) + \frac{3\psi}{2J_m} i_q \end{aligned} \tag{5.226}$$

The nonlinear operators are obtained from Eqs. 5.223 and 5.224:

$$
\begin{aligned}
&-N_1(x_0,y_0,\omega_0)=\omega_0 y_0+\left(\lambda-\frac{R}{L}\right)x_0+ke^{-\lambda t}\cos kt+\lambda X_0=\\
&=-\frac{R}{L}x_0+Y_0\Omega_0+e^{-\lambda t}\left[\left(\Omega_0+\lambda-\frac{R}{L}\right)\sin kt+k\cos kt\right]+\\
&+Y_0e^{-\lambda^* t}\sin k^*t+\frac{1}{2}e^{-(\lambda+\lambda^*)t}\left[\cos\left(k-k^*\right)t-\cos\left(k+k^*\right)t\right],\\
&-N_2(x_0,y_0,\omega_0)=\left(\lambda-\frac{R}{L}\right)y_0-\frac{\psi}{\mathrm{L}}\omega_0-\omega_0x_0+ke^{-\lambda t}\cos kt+\lambda Y_0=\\
&=-\frac{R}{L}Y_0+Y_0\Omega_0+e^{-\lambda t}\left[\left(\Omega_0+\lambda-\frac{R}{L}\right)\sin kt+k\cos kt\right]+\\
&+Y_0e^{-\lambda^* t}\sin k^*t+\frac{1}{2}e^{-(\lambda+\lambda^*)t}\left[\cos\left(k-k^*\right)t-\cos\left(k+k^*\right)t\right],\\
&-N_3(x_0,y_0,\omega_0)=\frac{p_1}{J_m}(75.63+9.18781\omega_0-0.0204959\omega_0^2-\\
&-0.00011553\omega_0^3)+\frac{3\psi}{2J_m}y_0=\frac{p_1}{J_m}(75.63+9.18781\Omega_0-0.0204959\Omega_0^2-\\
&-0.000115537\Omega_0^3)+\frac{3\psi}{2J_m}Y_0+e^{-\lambda^* t}(9.18781-0.0409918\Omega_0-\\
&-0.000346611\Omega_0^2)\sin k^*t+0.039709885e^{-2\lambda^* t}(1-\cos 2k^*t)-\\
&-3.8557\cdot 10^{-13}e^{-3\lambda^* t}\left(\sin 3k^*t-3\sin k^*t\right)+\frac{3\psi}{2J_m}e^{-\lambda t}\sin kt-\\
&-\lambda^*\Omega_0-k^*e^{-\lambda^* t}\cos k^*t
\end{aligned}
\tag{5.227}
$$

Comparing Eqs. 5.14 and 5.227, respectively, we have the freedom to choose Eqs. 5.15 in the forms:

$$
\begin{aligned}
&x_1(t,C_i)=e^{-\lambda t}\left[C_1(\cos kt-1)+C_2\sin kt+C_3(\cos 2kt-1)+\right.\\
&\left.+C_4\sin 2kt+C_5\right]-C_5,\quad x_1(0)=0,t\le t_0,\\
&y_1(t,C_i)=e^{-\lambda t}\left[C_6(\cos kt-1)+C_7\sin kt+C_8(\cos 2kt-1)+\right.\\
&\left.+C_9\sin 2kt+C_{10}\right]-C_{10},\quad y_1(0)=0,t\le t_0,\\
&\omega_1(t,C_i)=e^{-\lambda^* t}\left[C_{11}\left(\cos k^*t-1\right)+C_{12}\sin k^*t+C_{13}t^2+\right.\\
&\left.+C_{14}t+C_{15}\right]-C_{15},\quad \omega_1(0)=0,t\le t_0^*
\end{aligned}
\tag{5.228}
$$

and

$$
\begin{aligned}
&\overline{x}(t,K_i)=e^{-\lambda t}[K_1t^4+K_2t^3+K_3t^2+K_4t+K_5]+\overline{x}(t_0)-K_5,\quad t>t_0\\
&\overline{y}(t,K_i)=e^{-\lambda t}[K_6t^4+K_7t^3+K_8t^2+K_9t+K_{10}]+\overline{y}(t_0)-K_{10},\quad t>t_0\\
&\overline{\omega}(t,K_i)=e^{-\lambda^* t}[K_{11}t^4+K_{12}t^3+K_{13}t^2+K_{14}t+K_{15}]+\overline{\omega}\left(t_0^*\right)-K_{15},t>t_0^*
\end{aligned}
\tag{5.229}
$$

where $\overline{x}(t_0)=x_0(t_0)+x_1(t_0,C_i)$, $\overline{y}(t_0)=y_0(t_0)+y_1(t_0,C_i)$, $\overline{\omega}\left(t_0^*\right)=\omega_0\left(t_0^*\right)+\omega_1\left(t_0^*,C_i\right)$ are obtained from Eqs. 5.220, 5.223 and 5.228.

5.7.3 Results and Discussions

Based on the above analytical developments, four cases will be separately analyzed further for different moments of inertia $J_m = 160$ [kg m^2] and $J_m = 21.2$ [kg m^2], respectively for different electrical resistances R = 0.9 [ohm] and R = 6 [ohm], in the specific conditions of the actual generator given by $R_s = 0.9$ [ohm], L = 0.03 [H], ψ = 1.42 [Wb], $p_1 = 16$ pole pairs, with the initial conditions $i_d(0) = X_0 = -4.37$ [A], $i_q(0) = Y_0 = -13.7$ [A], $\omega(0) = \omega_0 = 170$ [rad/s]. For validation purposes, in each case, the analytical solutions are compared with numerical ones obtained using a forth-order Runge–Kutta method.

Case 5.7.3.a) We consider in the first case $J_m = 160$ [kgm^2], R = 0.9 [ohm], corresponding to a high inertia moment of the generator's rotor direct coupled with the wind turbine and small value of the electrical resistance ($R = R_s + R_L = 0.9 + 0.0$) due to the sudden short-circuit produced at the generator terminals. In this case the electrical resistance R_L is zero.

First stage: For the transitory regime, from Eq. 2.37 we obtain the following values of the parameters C_i:

$$\begin{aligned}
&C_1 = 41.58495128, C_2 = -6.346986159, C_3 = -0.178926996,\\
&C_4 = 0.09978393, C_5 = 41.6214537, \lambda = 30,\\
&k = 170.04561, C_6 = -5.291155935, C_7 = -43.17765268,\\
&C_8 = -0.69477905, C_9 = 0.022489318, C_{10} = -5.673964946,\\
&C_{11} = 0.852249547, C_{12} = -0.999921319,\\
&C_{13} = -1294.566669, C_{14} = -65.82033406,\\
&C_{15} = -2.360350666, \lambda^* = 30, k^* = 168.04591.
\end{aligned}$$

In this way, we can obtain the approximate solution of Eqs. 5.214, 5.215 and 5.216 for the transitory regime as follows:

$$\begin{aligned}
&i_d(t) = e^{-30t}\big(41.58495128\cos kt - 5.346986159\sin kt-\\
&-0.178926996\cos 2kt + 0.09978393\sin 2kt+\\
&+0.215429416\big) - 45.9914537, \quad t \le t_0
\end{aligned} \tag{5.230}$$

$$\begin{aligned}
&i_q(t) = e^{-30t}\big(-5.291155935\cos kt - 42.17765268\sin kt-\\
&-0.69477905\cos 2kt + 0.022489318\sin 2kt+\\
&+0.311970039\big) - 8.026035054, \quad t \le t_0
\end{aligned} \tag{5.231}$$

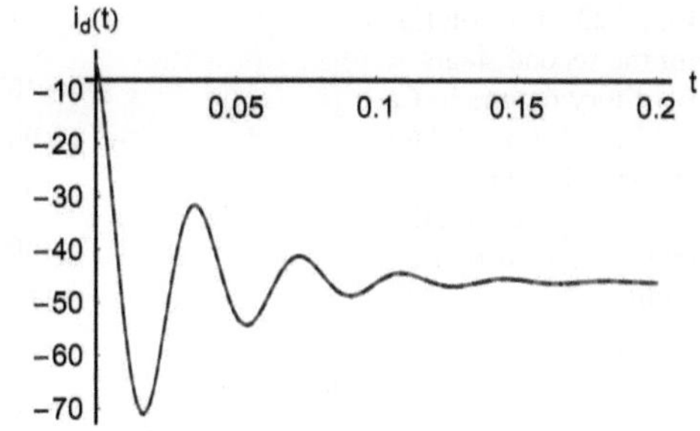

Fig. 5.20 The solution i_d for the first stage transitory regime in Case 5.7.3.a: (*red solid line*) numerical and (*blue dashed line*) analytical solution (5.230) (colour figure online)

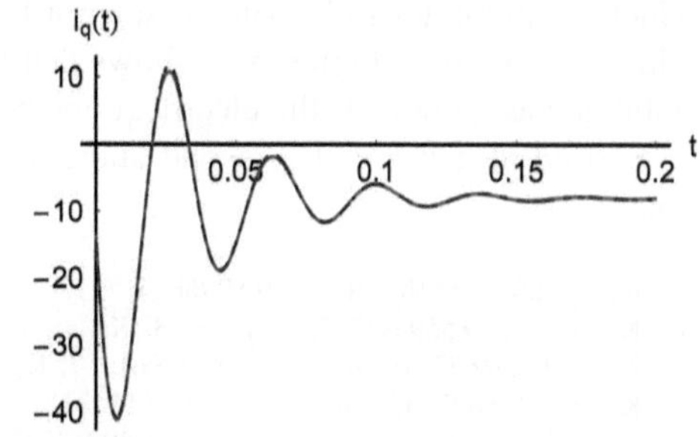

Fig. 5.21 The solution i_q for the first stage transitory regime in Case 5.7.3.a: (*red solid line*) numerical and (*blue dashed line*) analytical solution (5.231) (colour figure online)

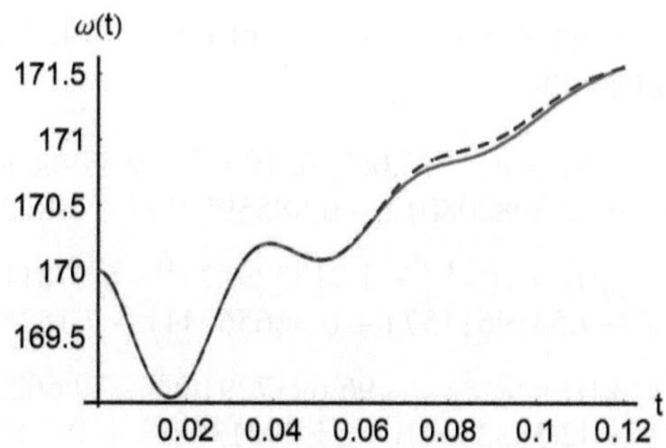

Fig. 5.22 The solution ω for the first stage mechanical transitory regime in Case 5.7.3.a: (*red solid line*) numerical and (*blue dashed line*) analytical solution (5.232) (colour figure online)

$$\begin{aligned}\omega(t) = e^{-30t}(0.852249547\cos k^*t + 0.0000786805072\sin k^*t-\\ -1294.566669t^2 - 65.82033406t - 3.212600513)+\\ +172.360350966, \quad t \le t_0^*\end{aligned} \tag{5.232}$$

Figures 5.20, 5.21 and 5.22 show the approximate solution of Eqs. 5.214–5.216 which is compared with the numerical solution obtained using a fourth-order Runge–Kutta method for the first stage of the transitory regime in the first case. In Figs. 5.20 and 5.21 we can see the rapidity of the electromagnetic transitory regime that becomes practically stable after around 0.2 s in this case. These kinds of solutions are very useful because they provide the maximum value of the current "spike" ($i_d \approx -70$ and $i_q \approx -40$) that could damage the permanent magnets. Moreover, they show the speed variation of the current (maximum current gradient),

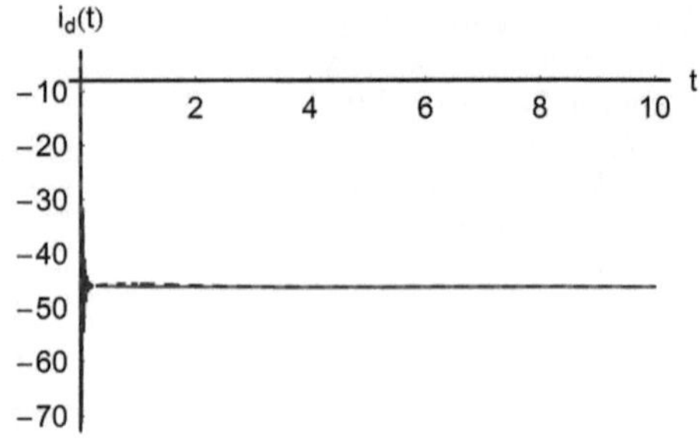

Fig. 5.23 The solution i_d for the second stage transitory regime in Case 5.7.3.a: (*red solid line*) numerical and (*blue dashed line*) analytical solution (5.233) (colour figure online)

which is important to be known since a large value can damage the generator's winding insulation. Figure 5.22 shows that the mechanical transitory regime is not stabilized as quickly as the electrical one because of high mass inertia.

Second stage: For the second stage of transitory regime, from Eqs. 2.37 we obtain:

$$\begin{aligned}
&K_1 = 2.082765106,\ K_2 = 9.879683617,\ K_3 = -4.94248071,\ K_4 = 5.839820801,\\
&K_5 = -0.885595917,\ \lambda = \lambda^* = 3,\ K_6 = -1.21832967,\ K_7 = -6.278116906,\\
&K_8 = 9.258677316,\ K_9 = -9.541961157,\ K_{10} = 0.36656044,\ K_{11} = -96.69172916,\\
&K_{12} = 79.66227371,\ K_{13} = -115.6174221,\\
&K_{14} = -42.38328341,\ K_{15} = -25.39313023
\end{aligned}$$

The approximate solutions of Eqs. 5.214, 5.215 and 5.216 for the second stage are given by:

$$\begin{aligned} i_d(t) = e^{-3\,t}\big(&2.082765106\, t^4 + 9.879683617\, t^3 - 4.94218071\, t^2 + \\ &+5.839820801\, t - 0.885595917\big) - 46.226685, \quad t > t_0 \end{aligned} \tag{5.233}$$

$$\begin{aligned} i_q(t) = e^{-3\,t}\big(&-1.21832967\, t^4 - 6.278116906\, t^3 + 9.258677316\, t^2 - \\ &-9.541961157\, t + 0.36656044\big) - 7.15259, \quad t > t_0 \end{aligned} \tag{5.234}$$

$$\begin{aligned} \omega(t) = e^{-3\,t}\big(&-96.69172916t^4 + 79.66227371t^3 - 115.6174221t^2 - \\ &-42.38328341t - 25.39313023\big) + 193.8975, \quad t > t_0^* \end{aligned} \tag{5.235}$$

Figures 5.23, 5.24 and 5.25 show the approximate solutions of Eqs. 5.214–5.216 which are compared with numerical ones in the second stage of the first case. Figures 5.23 and 5.24 prove that electromagnetic transitory regime is nearly finished after 0.2 s in this case, while the mechanical one is continued. Figure 5.25 shows how changes the speed of the generator (and also of the wind turbine) from the initial value $\omega_0 = 170$ [rad/s] to the final value corresponding to the steady-state after around 8 s.

Case 5.7.3.b) In the second case, we consider the moment of inertia $J = 160$ [kg m^2] and the resistance $R = 6$ [ohm]. This corresponds to a sudden change of

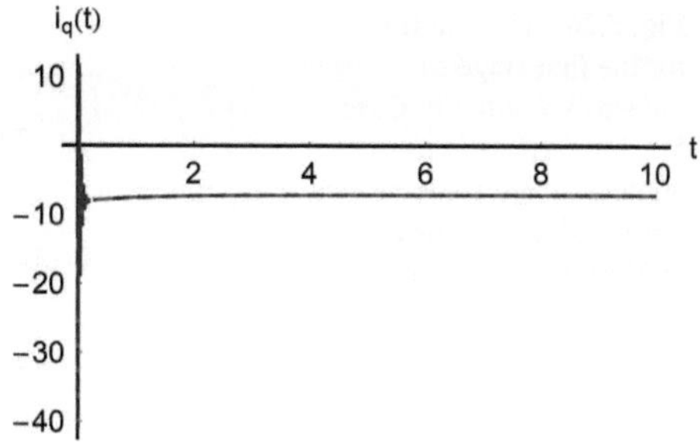

Fig. 5.24 The solution i_q for the second stage transitory regime in Case 5.7.3.a: (*red solid line*) numerical and (*blue dashed line*) analytical solution (5.234) (colour figure online)

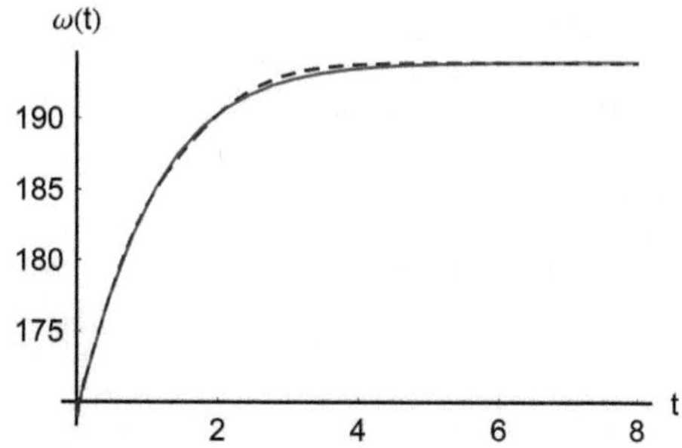

Fig. 5.25 The solution ω for the second stage mechanical transitory regime in Case 5.7.3.a: (*red solid line*) numerical and (*blue dashed line*) analytical solution (5.235) (colour figure online)

load from rated load to higher load ($R_L = 5.1$ [ohm]). Consequently, $R = R_S + R_L = 0.9 + 5.1 = 6$ [ohm].

First stage. For the incipient transitory regime, using the described procedure, one can get:

$$\begin{aligned} i_d &= e^{-200t}\big(13.43607007\cos kt + 11.49305076\sin kt + \\ &+0.849682982\cos 2kt - 0.303182819\sin 2kt + \\ &+1.104057368\big) - 19.75981042, \quad t \le t_0 \end{aligned} \tag{5.236}$$

$$\begin{aligned} i_q(t) &= e^{-200t}\big(12.38595273\cos kt - 11.20697234\sin kt - \\ &-0.096110793\cos 2kt - 0.753070987\sin 2kt - \\ &-2.670913452\big) - 23.31892849, \quad t \le t_0 \end{aligned} \tag{5.237}$$

$$\begin{aligned} \omega(t) &= e^{-2t}\big(-1549.939307\cos k^*t + 2885.868091\sin k^*t - \\ &-8407885.276t^2 - 8097985.418t - 4016050.973\big) + \\ &+4017770.913, \quad t \le t_0 \end{aligned} \tag{5.238}$$

where $k = 190.3995548$ and $k^* = 21.754322$.

Figures 5.26, 5.27 and 5.28 show a comparison between the present solution 5.236, 5.237 and 5.238, and numerical results for the transitory regime in the second case.

Second stage. For the ending of the transitory regime, the first-order approximate analytical solution is obtained as:

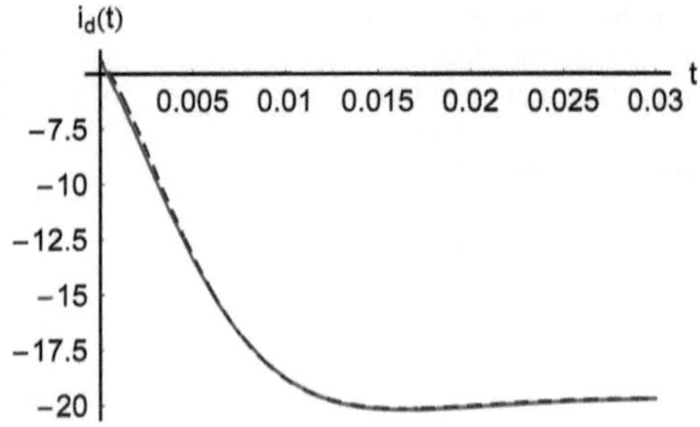

Fig. 5.26 The solution i_d for the first stage of transitory regime in Case 5.7.3.b: (*red solid line*) numerical and (blue dashed line) analytical solution (5.236) (colour figure online)

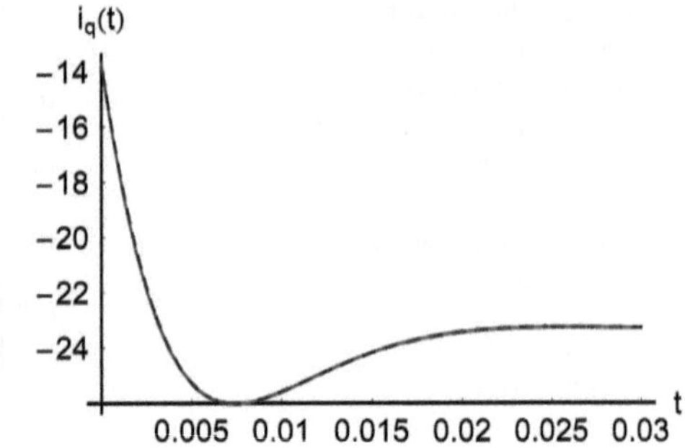

Fig. 5.27 The solution i_q for the first stage of transitory regime in Case 5.7.3.b: (red solid line) numerical and (blue dashed line) analytical solution (5.237) (colour figure online)

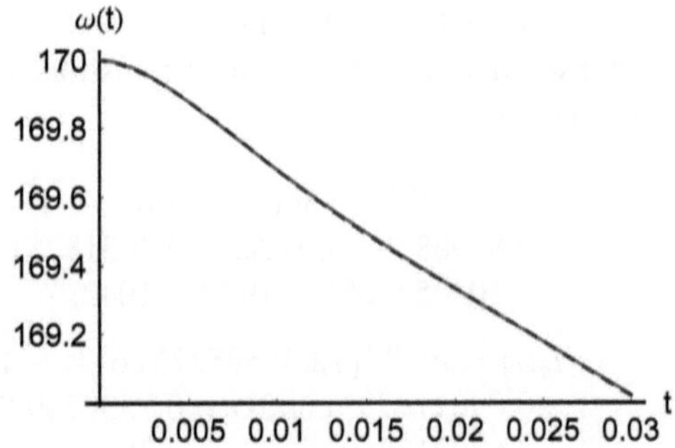

Fig. 5.28 The solution ω for the first stage of mechanical transitory regime in Case 5.7.3.b: (red solid line) numerical and (blue dashed line) analytical solution (5.238) (colour figure online)

$$i_d = e^{-t}(-0.221755861t^4 + 0.146949795t^3 - 2.022091269t^2 - \\ -4.275005236t - 8.735758817) - 11.139102, \quad t > t_0 \tag{5.239}$$

$$i_q(t) = e^{-t}(-0.041634569t^4 - 0.81185739t^3 + 1.441694975t^2 - \\ -4.296745884t - 3.173331249) - 20.0791, \quad t > t_0 \tag{5.240}$$

$$\omega(t) = e^{-t}(1.412628003t^4 - 0.755381542t^3 + 12.92658571t^2 + \\ +26.98459958t + 59.01826949) + 110.9501, \quad t > t_0 \tag{5.241}$$

Figures 5.29, 5.30 and 5.31 show the solutions 5.239, 5.240 and 5.241 for the second stage of the transitory regime in the second case, which are compared with corresponding numerical ones.

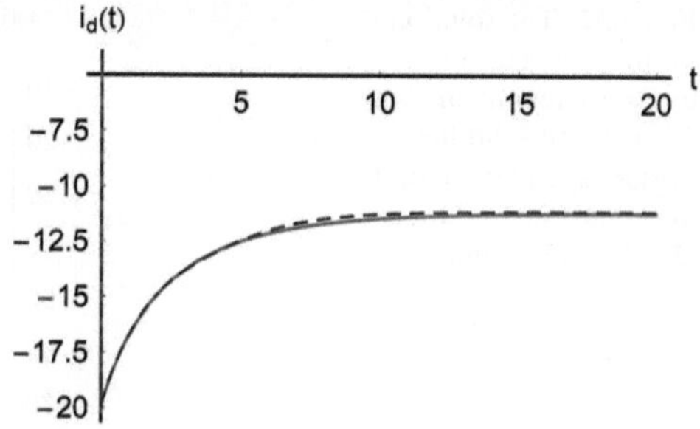

Fig. 5.29 The solution i_d for the second stage of transitory regime in Case 5.7.3.b: (red solid line) numerical and (blue dashed line) analytical solution (5.239) (colour figure online)

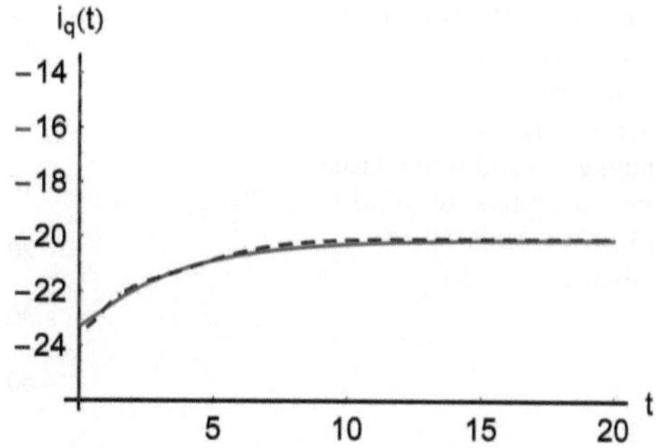

Fig. 5.30 The solution i_q for the second stage of transitory regime in Case 5.7.3.b: (red solid line) numerical and (blue dashed line) analytical solution (5.240) (colour figure online)

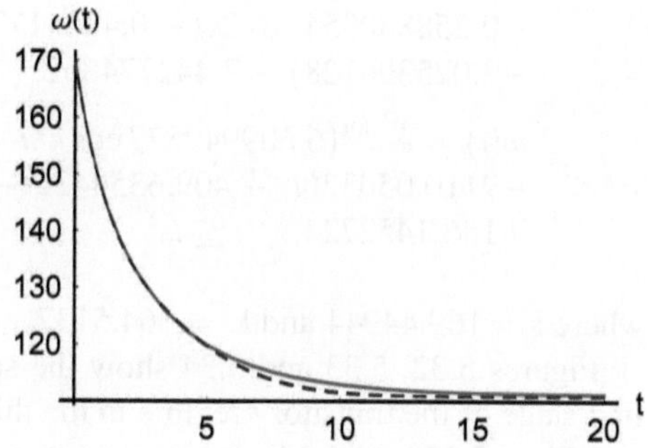

Fig. 5.31 The solution ω for the second stage of mechanical transitory regime in Case 5.7.3.b: (red solid line) numerical and (blue dashed line) analytical solution (5.241) (colour figure online)

Case 5.7.3.c) In the third case, we consider only the generator mass, without the turbine blades, with a moment of inertia J = 21.2 [kg m^2] and a full short-circuit at the generator terminals, that means R = 0.9 [ohm].

First stage. For the incipient transitory regime in this case we obtain:

$$\begin{aligned}i_d(t) = e^{-30}\big(&41.16412846\cos kt - 2.69415676\sin kt-\\&-0.638350767\cos 2kt + 0.832914459\sin 2kt+\\&+1.185070857\big) - 46.08084855, \quad t \le t_0\end{aligned} \tag{5.242}$$

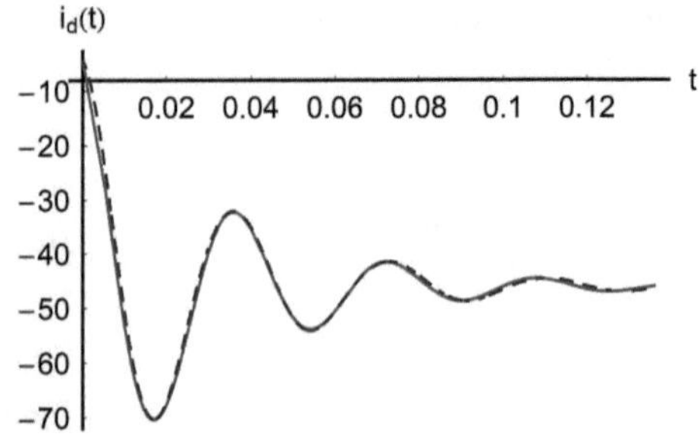

Fig. 5.32 The solution i_d for the first stage of transitory regime in Case 5.7.3.c: (red solid line) numerical and (blue dashed line) analytical solution (5.242) (colour figure online)

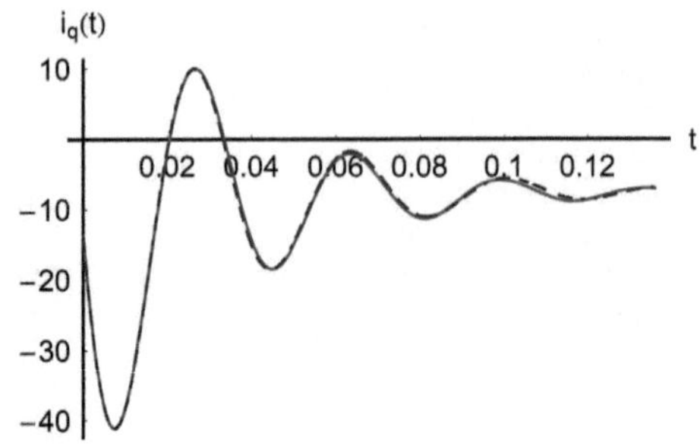

Fig. 5.33 The solution i_q for the first stage of transitory regime in Case 5.7.3.c: (red solid line) numerical and (blue dashed line) analytical solution (5.243) (colour figure online)

$$\begin{aligned}i_q(t) &= e^{-30t}\big(-3.972998056\cos kt - 40.71773812\sin kt-\\&-0.258830954\cos 2kt - 0.43841771\sin 2kt-\\&-2.025396128\big) - 7.442774862, \quad t \le t_0\end{aligned} \tag{5.243}$$

$$\begin{aligned}\omega(t) &= e^{-30t}\big(6.609949577\cos k^{*}t - 0.265468\sin k^{*}t-\\&-9110.050326t^2 - 409.6354422t - 22.75522194\big)+\\&+186.1452724, \quad t \le t_0\end{aligned} \tag{5.244}$$

where $k = 169.44944$ and $k^{*} = 164.5137$.

Figures 5.32, 5.33 and 5.34 show the solutions 5.242, 5.243 and 5.244 for the first stage of the transitory regime in the third case. Comparing Figs. 5.20 and 5.21 and Figs. 5.32 and 5.33, we can see that there is no significant influence of the moment of inertia on the electromagnetic transitory regime. As we expected, some influence of the inertia on the mechanical transitory regime can be emphasized comparing Figs. 5.22 and 5.34. Due to the low inertia, the angular speed changes more rapid in the same time interval.

Second stage. For the ending of the transitory regime we obtain:

$$\begin{aligned}i_d(t) &= e^{-3t}\big(0.7948211t^4 - 1.1179142t^3 + 0.540443488t^2-\\&-0.1077088t + 0.287524872\big) - 46.2279, \quad t > t_s\end{aligned} \tag{5.245}$$

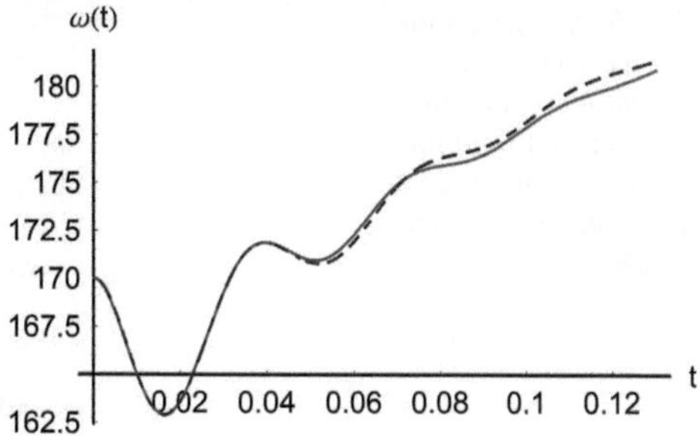

Fig. 5.34 The solution ω for the first stage of mechanical transitory regime in Case 5.7.3.c: (red solid line) numerical and (blue dashed line) analytical solution (5.244) (colour figure online)

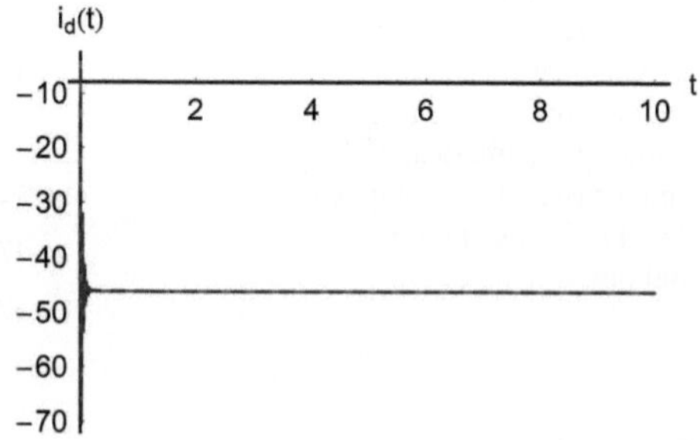

Fig. 5.35 The solution i_d for the second stage of transitory regime in Case 5.7.3.c: (red solid line) numerical,(blue dashed line) analytical (5.245) (colour figure online)

$$i_q(t) = e^{-3t}\big(1.820537509t^4 - 4.877639436t^3 + 4.431307636t^2 - \\ -1.515478371t - 0.25999871\big) - 7.15251, \quad t > t_s \tag{5.246}$$

$$\omega(t) = e^{-3t}\big(-49.65794988t^4 + 163.4024933t^3 - 211.1840176t^2 + \\ +130.196698t - 33.02074613\big) + 193.89012, \quad t > t_s \tag{5.247}$$

Figures 5.35, 5.36 and 5.37 show a comparison between the present solutions 5.245, 5.246 and 5.247 and the corresponding numerical ones for the second stage of the transitory regime in the third case. Figure 5.37 shows that the mechanical transitory regime is shorter that in the Case 5.7.3.a (Fig. 5.25), where the inertia was higher.

Case 5.7.3.d) In the last case, we consider J = 21.2 [kg m^2], R = 6 [ohm].

First stage. For the transitory regime it is obtained:

$$i_d(t) = e^{-200t}\big(18.52063597\cos kt + 13.64797046\sin kt + \\ +0.975607594\cos 2kt - 2.223811645\sin 2kt - \\ -4.471023617\big) - 19.36521988, \quad t \le t_s \tag{5.248}$$

$$i_q(t) = e^{-200t}\big(13.0242272\cos kt - 9.389570256\sin kt + \\ +0.80030287\cos 2kt - 1.115204977\sin 2kt - \\ -4.403169256\big) - 23.12136081, \quad t \le t_s \tag{5.249}$$

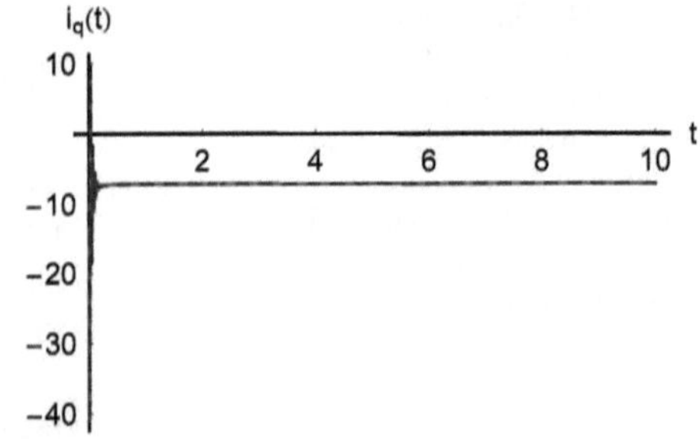

Fig. 5.36 The solution i_q for the second stage of transitory regime in Case 5.7.3.c: (red solid line) numerical and (blue dashed line) analytical solution (5.246)

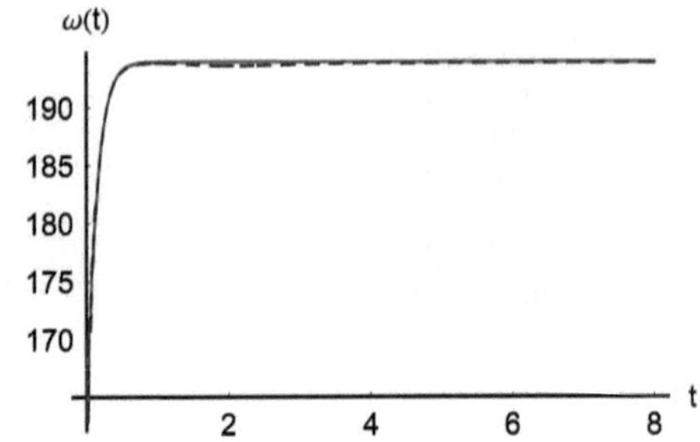

Fig. 5.37 The solution ω for the second stage of mechanical transitory regime in Case 5.7.3.c: (red solid line) numerical and (blue dashed line) analytical (5.247) (colour figure online)

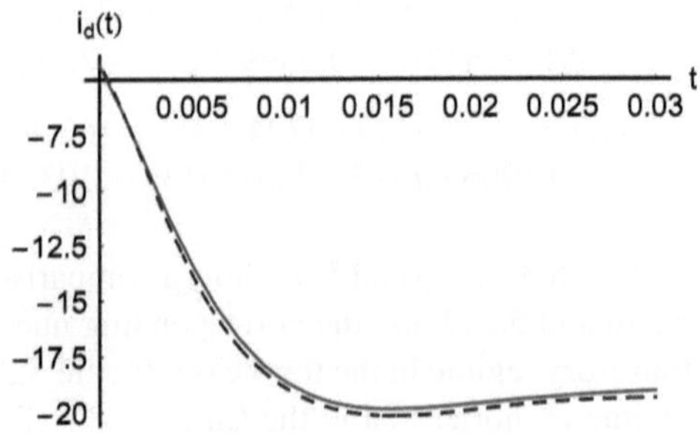

Fig. 5.38 The solution i_d for the first stage of transitory regime in Case 5.7.3.d: (red solid line) numerical and (blue dashed line) analytical solution (5.248) (colour figure online)

$$\begin{aligned}\omega(t) = e^{-200t}\big(&-10603.44583\cos k^*t + 14219.69362\sin k^*t-\\&-22313356.8t^2 - 8100874.06t - 1543748.305\big)+\\&+1554521.75083, \quad t \le t_s\end{aligned} \tag{5.250}$$

where k = 187.9514, k^* = 23.1431.

Figures 5.38, 5.39 and 5.40 show a comparison between the present solution 5.248, 5.249 and 5.250 and numerical results for the first stage of the transitory regime in the last case.

Second stage. For the stabilized regime, the corresponding analytical solutions will be:

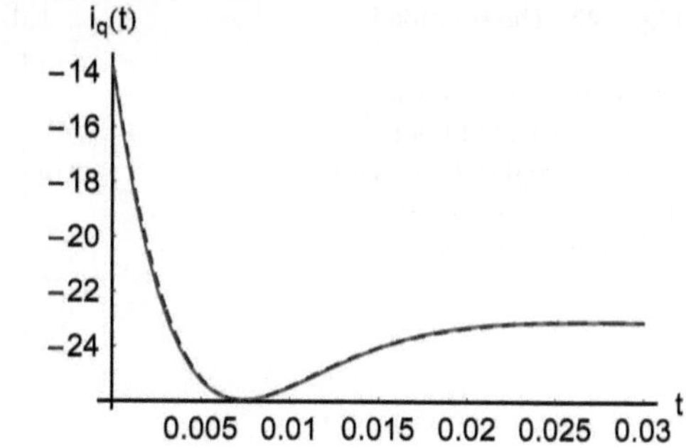

Fig. 5.39 The solution i_q for the first stage of transitory regime in Case 5.7.3.d: (red solid line) numerical and (blue dashed line) analytical solution (5.249) (colour figure online)

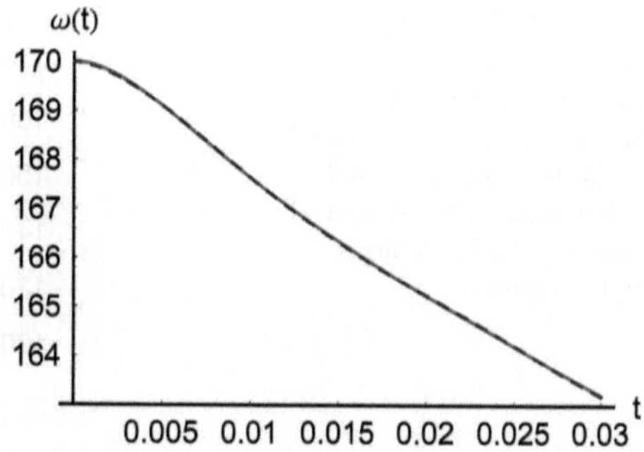

Fig. 5.40 The solution ω for the first stage of mechanical transitory regime in Case 5.7.3.d: (red solid line) numerical and (blue dashed line) analytical solution (5.250) (colour figure online)

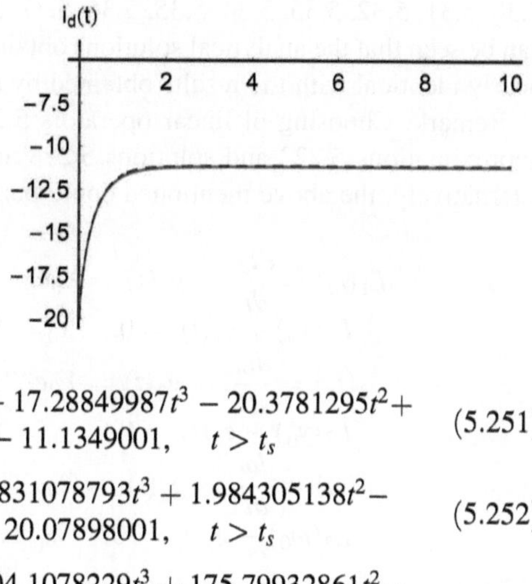

Fig. 5.41 The solution i_d for the second stage of transitory regime in Case 5.7.3.d: (red solid line) numerical and (blue dashed line) analytical solution (5.251) (colour figure online)

$$i_d(t) = e^{-200t}\big(-5.395866798t^4 + 17.28849987t^3 - 20.3781295t^2 + \\ +5.734121686t - 8.782366483\big) - 11.1349001, \quad t > t_s \tag{5.251}$$

$$i_q(t) = e^{-200t}\big(0.573876822t^4 - 2.831078793t^3 + 1.984305138t^2 - \\ -3.951015269t - 3.23826676\big) - 20.07898001, \quad t > t_s \tag{5.252}$$

$$\omega(t) = e^{-200t}\big(83.84126784t^4 - 104.1078229t^3 + 175.79932861t^2 - \\ -21.52561706t + 58.60847973\big) + 110.951101, \quad t > t_s \tag{5.253}$$

Figures 5.41, 5.42 and 5.43 show a comparison between the present solutions 5.251, 5.252 and 5.253 and numerical results for the second stage of the transitory regime

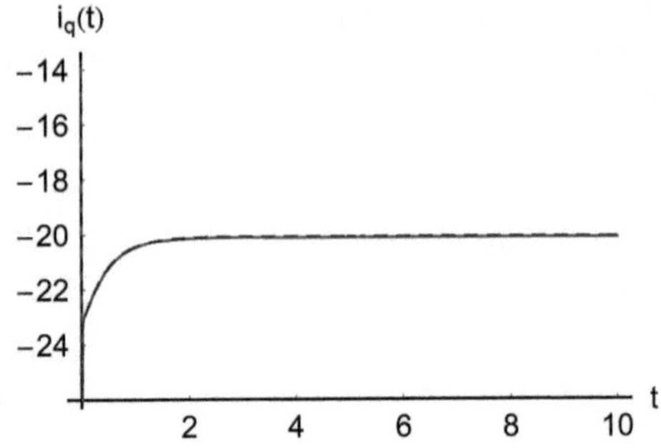

Fig. 5.42 The solution i_q for the second stage of transitory regime in Case 5.7.3.d: (red solid line) numerical and (blue dashed line) analytical solution (5.252) (colour figure online)

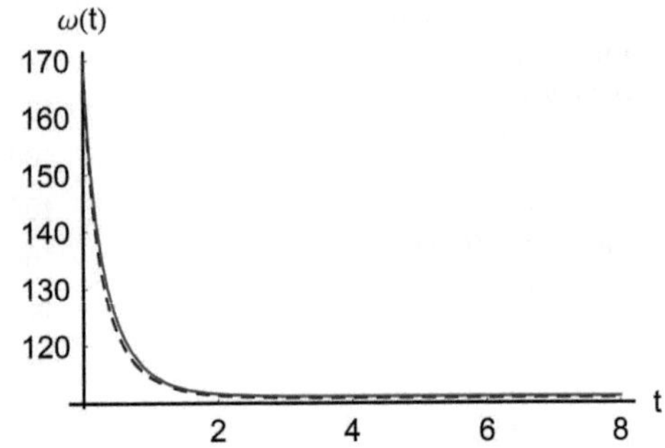

Fig. 5.43 The solution ω for the second stage of mechanical transitory regime in Case 5.7.3.d: (red solid line) numerical and (blue dashed line) analytical solution (5.253) (colour figure online)

in the last case. From Figs. 5.20, 5.21, 5.22, 5.23, 5.24, 5.25, 5.26, 5.27, 5.28, 5.29, 5.30, 5.31, 5.32, 5.33, 5.34, 5.35, 5.36, 5.37, 5.38, 5.39, 5.40, 5.41, 5.42 and 5.43 it can be seen that the analytical solutions obtained by OHAM are very accurate being nearly identical with the results obtained by a fourth-order Runge–Kutta method.

Remark: Choosing of linear operators 5.219, auxiliary functions 5.222, initial approximations 5.223 and solutions 5.228 and 5.229 is not unique. For example, alternatively, the above mentioned could be:

$$\begin{aligned}
L_1(i_d) &= \frac{di_d}{dt}, \quad g_1(t) = X_0 e^{-\lambda t}(\lambda \cos kt + k \sin kt),\\
&L_1(x_0) + g_1(t) = 0, \quad x_0 = X_0 e^{-\lambda t}\cos kt\\
L_2(i_q) &= \frac{di_q}{dt}, \quad g_2(t) = Y_0 e^{-\lambda t}(\lambda \cos kt + k \sin kt),\\
&L_2(y_0) + g_2(t) = 0, \quad y_0 = Y_0 e^{-\lambda t}\cos kt\\
L_3(\omega) &= \frac{d\omega}{dt}, \quad g_3(t) = \Omega_0 e^{-\lambda^* t}\left(\lambda^* \cos k^* t + k^* \sin k^* t\right),\\
&L_3(\omega_0) + g_3(t) = 0, \quad \omega_0 = \Omega_0 e^{-\lambda^* t}\cos k^* t
\end{aligned}$$

$$\begin{aligned}
&x_1(t, C_i) = e^{-\lambda t}\left[C_1(\cos kt - 1) + C_2 \sin kt + C_3 \sin 2kt + \right.\\
&\left. + C_4 \sin 3kt + C_5 \cos 2kt\right] - C_5, \quad x_1(0) = 0
\end{aligned}$$

$$y_1(t, C_i) = e^{-\lambda t}\big[C_6 \cos kt + C_7 \cos 2kt + C_8 \sin kt + C_9 \sin 2kt + \\ + C_{10} \sin 3kt + C_{11}\big] - C_6 - C_7 - C_{11}, y_1(0) = 0$$

$$\omega_1(t, C_i) = e^{-\lambda^* t}\big[C_{12} \cos k^* t + C_{13} \sin k^* t + C_{14} \sin 2kt + \\ + C_{15} t^3 + C_{16} t^2 + C_{17} t\big] - C_{12}, \quad \omega_1(0) = 0$$

The analytical approximate solutions obtained through OHAM are of considerable importance for a fast and practical analysis of electrical power systems dynamic behavior caused by possible perturbations generated by some short-circuits, sudden change of loads, disconnection of loads, other switching transients in power stations. Such analytical approach allows modeling and simulating turbine-generator systems for real-time computations, off-line applications or stability problems of electrical power systems.

References

1. L. Elsgolts, *Differential Equations and the Calculus of Variations* (Mir Publishers, Moscow, 1980)
2. V. Marinca, N. Herişanu, *Nonlinear Dynamical Systems in Engineering—Some Approximate Approaches* (Springer, Heidelberg, 2011)
3. V. Marinca, N. Herişanu, I. Nemeş, Optimal homotopy asymptotic method with application to thin film flow. Cent. Eur. J. Phys. **6**, 648–653 (2008)
4. V. Marinca, N. Herişanu, C. Bota, B. Marinca, An optimal homotopy asymptotic method applied to the steady flow of a fourth grade fluid past a porous plate. Appl. Math. Lett. **22**, 245–251 (2009)
5. V. Marinca, N. Herişanu, An optimal homotopy asymptotic approach applied to nonlinear MHD Jeffery-Hamel flow, Math. Probl. Eng. Article ID 169056 (2011)
6. V. Marinca, N. Herişanu, Determination of periodic solutions for the motion of a particle on a rotating parabola by means of the optimal homotopy asymptotic method. J. Sound Vibr. **329**, 1450–1459 (2010)
7. V. Marinca, N. Herişanu, The optimal homotopy asymptotic method for solving Blasius equation. Appl. Math. Comput. **231**, 134–139 (2014)
8. L.H. Thomas, The calculation of atomic fields. Proc. Camb. Philos. Soc. **23**, 542–548 (1927)
9. E. Fermi, Un metodo statistico par la determinizione di alcune Proprieta dell'atome, Rend. Accad. del Lincei, Cl. Sci. Fis. Mat. Nat. **6**, 602–607 (1927)
10. A.M. Wazwaz, The modified decomposition method and the Padé approximants for solving Thomas-Fermi equation. Math. Comput. **105**, 11–19 (1999)
11. V. Marinca, N. Herisanu, An optimal iteration method with application to the Thomas-Fermi equation. Cent. Eur. J. Phys. **9**(3), 891–895 (2011)
12. V. Marinca, R.D. Ene, Analitical approximate solutions to the Thomas-Fermi equation. Cent. Eur. J. Phys. **12**(7), 503–510 (2014)
13. M. Oulne, Variation and series approach to the Thomas–Fermi equation. Appl. Math. Comput. **218**, 303–307 (2011)
14. C.M. Bender, K.A. Milton, S.S. Pinsky, L.M. Simmons Jr., A new perturbative approach to nonlinear problems. J. Math. Phys. **30**(7), 1447–1455 (1989)
15. S.J. Liao, *Beyond Perturbation. Introduction to the Homotopy Analysis Method* (Chapman and Hall/CRC Press, Boca Raton, FL, 2003)

16. H.K. Khan, H. Xu, Series solution to the Thomas-Fermi equation. Phys. Lett. A **365**, 111–115 (2007)
17. S. Kobayashi et al., Some coefficients of the TFD function. J. Phys. Soc. Jpn. **10**, 759–765 (1955)
18. B.C. Sakiadis, Boundary layer behaviour an continuous solid surface I: Boundary layer equations for two dimensional and axisymmetric flow. AIChE J. **7**, 26–28 (1961)
19. K. Sadeghy, A.H. Najafi, M. Saffaripour, Sakiadis flow of an upper-convected Maxwell fluid. Int. J. Non-Linear Mech. **40**, 1220–1228 (2005)
20. T. Hayat, Z. Abbas, M. Sajid, Series solution for the upper-convected Maxwell fluid over a porous stretching plate. Phys. Lett. A **358**, 396–403 (2006)
21. A. Ishak, R. Nazar, N.M. Arifin, I. Pop, Dual solutions in mixed convection flow near a stagnation point on a vertical porous plate. Int. J. Therm. Sci. **47**, 417–422 (2008)
22. B. Sahoo, Effect of slip viscous dissipation and Joule heating on the MHD flow and heat transfer of a second grade fluid past a radially stretching sheet. Appl. Math. Mech. Engl. Ed. **31** (2), 159–173 (2010)
23. V. Marinca, R.D. Ene, B. Marinca, R. Negrea, Different approximations to the solution of upper-convected Maxwell fluid over a porous stretching plate, Abstract Appl. Anal. Article ID 139314 (2014)
24. H. Schlichting, *Boundary Layer Theory*, 6th edn. (McGraw-Hill, New York, 1964)
25. S. Uchida, H. Aoki, Unsteady flows in a semi-infinite contracting or expanding pipe. J. Fluid Mech. **82**, 371–387 (1977)
26. F.M. Skalak, C.Y. Wang, On the unsteady squeezing of viscous fluid from a tube, J. Aust. Math. Soc. B, 65–74 (1979)
27. M. Miclavčič, C.Y. Wang, Viscous flow due to a shrinking sheet. Q. Appl. Math. **64**, 283–290 (2006)
28. A. Ishak, R. Nazar, I. Pop, Uniform suction/blowing effect on flow and heat transfer due to a stretching cylinder. Appl. Math. Model. **32**, 2059–2066 (2008)
29. T.G. Fang, J. Zang, S.S. Yao, Viscous flow over an unsteady shrinking sheet with mass transfer. Chin. Phys. Lett. **26**(1), 1–4 (2009)
30. T.G. Fang, S.S. Yao, J. Zang, A. Aziz, Viscous flow over a shrinking sheet with a second order slip flow model. Commun. Nonlinear Sci. Numer. Simul. **15**, 1831–1842 (2010)
31. W.M.K.A. Wan Zaimi, A. Ishak, I. Pop, Unsteady viscous flow over a shrinking cylinder. J. Kind. Saud. Univ. Sci. **25**, 143–148 (2013)
32. K. Zaimi, A. Ishak, I. Pop, Unsteady flow due to a contracting cylinder in a nanofluid using Buongiorno's model. Int. J. Heat Mass Transf. **68**, 509–513 (2014)
33. V. Marinca, R.D. Ene, Dual solutions of the unsteady viscous flow over a shrinking cilynder with optimal homotopy asymptotic method, Adv. Math. Phys. Article ID 417643 (2014)
34. M.J. Stefan, Versuch Uber die scheinbare adhesion. Akad. Wiss. Math. Natur. **69**, 713–721 (1874)
35. L. Proudman, An example of steady laminar flow at large Reynolds number. ASME J. Appl. Mech. **9**, 593–601 (1960)
36. H. Rasmussen, Steady viscous flow between two porous slider. Z. Angew. Math. Phys. **21**, 187–195 (1970)
37. C.Y. Wang, Fluid dynamics of the circular porous slider. ASME J. Appl. Mech. **41**, 343–347 (1974)
38. R.S.R. Gorla, Flow and thermal characteristics of a circular porous slider bearing. Wear **94**, 157–174 (1984)
39. N.M. Bujurke, P.K. Achar, Computer extended series solution of the circular porous slider. Acta Mech. **101**, 81–92 (1993)
40. Q.K. Ghori, M. Ahmed, A.K. Siddiqui, Application of homotopy perturbation method to squeezing flow of a Newtonian fluid. Int. J. Nonlinear Sci. Numer. Simul. **8**(2), 179–184 (2007)

41. M. Idrees, S. Islam, S. Haq, S. Islam, Application of the Optimal Homotopy Asymptotic Method to squeezing flow. Comput. Math. Appl. **59**, 3858–3866 (2010)
42. A. Aziz, S.M.E. Hug, Perturbation solution for convecting fin with variable thermal conductivity. J. Heat Transf. **97**, 300–301 (1975)
43. P. Razelos, K. Imre, The optimum dimensions of circular fins with variable thermal parameters. J. Heat Transf. **102**, 420–425 (1980)
44. D.D. Ganji, The application of He's homotopy perturbation method to nonlinear equations arising in heat transfer. Phys. Lett. A **355**, 337–348 (2006)
45. A. Rajabi, D.D. Ganji, H. Taherian, Application of homotopy perturbation method to nonlinear heat conduction and convection equations. Phys. Lett. A **360**, 570–573 (2007)
46. S. Abbasbandy, The application of homotopy analysis method to nonlinear equations arising in heat transfer. Phys. Lett. A **360**, 109–113 (2006)
47. H.P. Chu, C.Y. Lo, Application of the hybrid differential transform-finite difference method to nonlinear transient heat conduction problems. Numer. Heat Transf. A **53**, 295–307 (2008)
48. H. Yaghoobi, M. Torabi, The application of differential transformation method to nonlinear equations arising in heat transfer. Int. Commun. Heat Mass Transf. **38**, 815–820 (2011)
49. P. Vas, *Electrical Machines and Drives. A Space-Vector Theory Approach* (Clarendon Press, Oxford, 1996)
50. I. Boldea, *Variable Speed Generators* (CRC Press, Boca Raton, FL, 2006)
51. C.-M. Ong, *Dynamics Simulation of Electric Machinery* (Prentice Hall, New Jersey, 1998)
52. A. Binder, *Elektrische Maschinen und Antriebe* (Springer, Berlin, Heidelberg, 2012)
53. T. Sebastian, G.R. Slemon, Transient Modelling and Performance of Variable-Speed Permanent-Magnet Motors. IEEE Trans. Ind. Appl. **25**(1), 101–106 (1989)
54. K.W. Klontz, T.J.E. Miller, M.I. McGilo, H. Karmaker, P. Zhong, Short-Circuit Analysis of Permanent-Magnet Generators. IEEE Trans. Ind. Appl. **47**(4), 1670–1680 (2011)
55. P.C. Krause, O. Wasynczuk, S.D. Sudhoff, *Analysis of Electrical Machines* (IEEE Press, New York, 1995)
56. S. Heier, *Grid integration of Wind Energy Conversion Systems* (Wiley, Chichester, 1998)
57. B.S. Borowy, Z.M. Salameh, Dynamic response of a stand-alone wind energy conversion system with battery energy storage to a wind gust. IEEE Trans. Energ. Convers. **12**(1), 73–78 (1997)
58. N. Herisanu, V. Marinca, T. Dordea, G. Madescu, A new analytical approach to nonlinear vibration of an electrical machine. Proc. Rom. Acad. Ser. A **9**, 229–236 (2008)
59. N. Herisanu, V. Marinca, Gh. Madescu, An analytical approach to non-linear dynamical model of a permanent magnet synchronous generator. (2014). Wind Energy, DOI:10.1002/we.1785

MIX
Papier aus verantwortungsvollen Quellen
Paper from responsible sources
FSC® C105338

If you have any concerns about our products,
you can contact us on
ProductSafety@springernature.com

In case Publisher is established outside the EU,
the EU authorized representative is:
Springer Nature Customer Service Center GmbH
Europaplatz 3, 69115 Heidelberg, Germany

Printed by Libri Plureos GmbH
in Hamburg, Germany